Springer Series in Operations Research and Financial Engineering

Series Editors:
Thomas V. Mikosch
Sidney I. Resnick
Stephen M. Robinson

Springer Series in Operations Research and Financial Engineering

Laurens de Haan
Ana Ferreira

Extreme Value Theory
An Introduction

 Springer

Laurens de Haan
Erasmus University
School of Economics
P.O. Box 1738
3000 DR Rotterdam
The Netherlands
ldehaan@few.eur.nl

Ana Ferreira
Instituto Superior de Agronomia
Departamento de Matemática
Tapada da Ajuda
1349-017 Lisboa
Portugal
anafh@isa.utl.pt

Series Editors:
Thomas V. Mikosch
University of Copenhagen
Laboratory of Actuarial Mathematics
DK-1017 Copenhagen
Denmark
mikosh@act.ku.dk

Stephen M. Robinson
University of Wisconsin-Madison
Department of Industrial
 Engineering
Madison, WI 53706
U.S.A.
smrobins@facstaff.wisc.edu

Sidney I. Resnick
Cornell University
School of Operations Research and Industrial Engineering
Ithaca, NY 14853
U.S.A.
sirl@cornell.edu

Mathematics Subject Classification (2000): 60G70, 60G99, 60A99

Library of Congress Control Number: 2006925909

ISBN-10: 0-387-23946-4   e-ISBN: 0-387-34471-3
ISBN-13: 978-0-387-23946-0

Printed on acid-free paper.

Printed in the United States of America.   (TXQ/MP)

9 8 7 6 5 4 3 2 1

springer.com

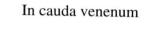
In cauda venenum

Preface

Approximately 40% of the Netherlands is below sea level. Much of it has to be protected against the sea by dikes. These dikes have to withstand storm surges that drive the seawater level up along the coast. The government, balancing considerations of cost and safety, has determined that the dikes should be so high that the probability of a flood (i.e., the seawater level exceeding the top of the dike) in a given year is 10^{-4}. The question is then how high the dikes should be built to meet this requirement. Storm data have been collected for more than 100 years. In this period, at the town of Delfzijl, in the northeast of the Netherlands, 1877 severe storm surges have been identified. The collection of high-tide water levels during those storms forms approximately a set of independent observations, taken under similar conditions (i.e., we may assume that they are independent and identically distributed). No flood has occurred during these 100 years.

At first it looks as if this is an impossible problem: in order to estimate the probability of a once-in-10 000 years event one needs more than observations over just 100 years. The empirical distribution function carries all the information acquired, and going beyond its range is impossible.

Yet it is easy to see that some information can be gained. For example, one can check whether the spacings (i.e., the difference between consecutive ordered observations) increase or decrease in size as one moves to the extreme observations. A decrease would point at a short tail and an increase at a long tail of the distribution.

Alternatively, one could try to estimate the first and second derivatives of the empirical distribution function near the boundary of the sample and extrapolate using these estimates.

The second option is where extreme value theory leads us. But instead of proceeding in a heuristic way, extreme value theory provides a solid theoretical basis and framework for extrapolation. It leads to natural estimators for the relevant quantities, e.g., those for extreme quantiles as in our example, and allows us to assess the accuracy of these estimators.

Extreme value theory restricts the behavior of the distribution function in the tail basically to resemble a limited class of functions that can be fitted to the tail of the distribution function. The two parameters that play a role, scale and shape, are based roughly on derivatives of the distribution function.

In order to be able to apply this theory some conditions have to be imposed. They are quite broad and natural and basically of a qualitative nature. It will become clear that the so-called extreme value condition is on the one hand quite general (it is not easy to find distribution functions that do not satisfy them) but on the other hand is sufficiently precise to serve as a basis for extrapolation.

Since we do not know the tail, the conditions cannot be checked (however, see Section 5.2). But this is a common feature in more traditional branches of statistics. For example, when estimating the median one has to assume that it is uniquely defined. And for assessing the accuracy one needs a positive density. Also, for estimating a mean one has to assume that it exists and for assessing the accuracy one usually assumes the existence of a second moment.

In these two cases it is easy to see what the natural conditions should be. This is not the case in our extrapolation problem. Nevertheless, some reflection shows that the "extreme value condition" is the natural one. For example (cf. Section 1.1.4), one way of expressing this condition is that it requires that a high quantile (beyond the scope of the available data) be asymptotically related in a linear way to an intermediate quantile (which can be estimated using the empirical distribution function).

The theory described in this book is quite recent: only in the 1980s did the contours of the statistical theory take shape. One-dimensional probabilistic extreme value theory was developed by M. Fréchet (1927), R. Fisher and L. Tippett (1928), and R. von Mises (1936), and culminated in the work of B. Gnedenko (1943). The statistical theory was initiated by J. Pickands III (1975).

The aim of this book is to give a thorough account of the basic theory of extreme values, probabilistic and statistical, theoretical and applied. It leads up to the current state of affairs. However, the account is by no means exhaustive for this field has become too vast. For these two reasons, the book is called an introduction.

The outline of the book is as follows. Chapters 1 and 2 discuss the extreme value condition. They are of a mathematical and probabilistic nature. Section 2.4 is important in itself and essential for understanding Sections 3.4, 3.6 and Chapter 5, but not for understanding the rest of the book. Chapter 3 discusses how to estimate the main (shape) parameter involved in the extrapolation and Chapter 4 explains the extrapolation itself. Examples are given.

In Chapter 5 some interesting but more advanced topics are discussed in a one-dimensional setting.

The higher-dimensional version of extreme value theory offers challenges of a new type. The model is explained in Chapter 6, the estimation of the main parameters (which are infinite-dimensional in this case) in Chapter 7, and the extrapolation in Chapter 8.

Chapter 9 (probabilistic) and Chapter 10 (statistical) treat the infinite-dimensional case.

Appendix B offers an introduction to the theory of regularly varying functions, which is basic for our approach. This text is partly based on the book *Regular Vari-*

ation, Extensions and Tauberian Theorems, by J.L. Geluk and L. de Haan, which is out of print. The authors wish to thank Jaap Geluk for his permission to use the text.

In a book of this extent it is possible that some errors may have escaped our attention. We are very grateful for feedback on any corrections, suggestions or comments (ldhaan@few.eur.nl, anafh@isa.utl.pt). We intend to publish possible corrections at Ana's webpage, http://www.isa.utl.pt/matemati/~anafh/anafh.html.

We wish to thank the statistical research unit of the University of Lisbon (CEAUL) for offering an environment conducive to writing this book. We acknowledge the support of FCT/POCTI/FEDER as well as the Gulbenkian foundation. We thank Holger Drees and the editors, Thomas Mikosch and Sidney Resnick, for their efforts to go through substantial parts of the book, which resulted in constructive criticism. We thank John Einmahl for sharing his notes on the material of Sections 7.3 and 10.4.2. The first author thanks the Université de Saint Louis (Sénegal) for the opportunity to present some of the material in a course. We are very grateful to Maria de Fátima Correia de Haan, who learned LATEX for the purpose of typing a substantial part of the text. Laurens de Haan also thanks Maria de Fátima for her unconditional support during these years. Ana Ferreira is greatly indebted to those who propitiated and encouraged her learning on the subject, especially to Laurens de Haan. Ana also thanks the long-enduring and unconditional support of her parents as well as her husband, Bernardo, and son, Pedro.

In a book of this extent it is possible that some errors may have escaped our attention. We are very grateful for feedback on any corrections, suggestions or comments (ldhaan@few.eur.nl, anafh@isa.utl.pt). We intend to publish possible corrections at Ana's webpage, http://www.isa.utl.pt/matemati/ anafh/anafh.html.

Lisbon,
2006

Laurens de Haan
Ana Ferreira

Contents

Part IV Appendix

List of Abbreviations and Symbols

Notation that is largely confined to sections or chapters is mostly excluded from the list below.

$=^d$	equality in distribution
$\to^d$	convergence in distribution
$\to^P$	convergence in probability
$a(t) \sim b(t)$	$\lim_t a(t)/b(t) = 1$
α	tail index
η	residual dependence index
γ	extreme value index
Γ	gamma function
μ	exponent measure
ϱ	metric $|1/x - 1/y|$
$1_{\{p\}}$	indicator function: equals 1 if p is true and 0 otherwise
$1 - F^{(n)}$	left-continuous empirical distribution function
2ERV	second-order extended regular variation
a_+	$\max(a, 0)$
a_-	$\min(a, 0)$
$a \vee b$	$\max(a, b)$
$a \wedge b$	$\min(a, b)$
$[a]$	largest integer less than or equal to a
$\lceil a \rceil$	smallest integer greater than or equal to a
a.s.	almost surely
$C[0, 1]$	space of continuous functions on $[0, 1]$ equipped with the supremum norm
$C^+[0, 1]$	$\{f \in C[0, 1] : f > 0\}$
$C_1^+[0, 1]$	$\{f \in C[0, 1] : f > 0, |f|_\infty = 1\}$
$\overline{C}_1^+[0, 1]$	$\{f \in C[0, 1] : f \geq 0, |f|_\infty = 1\}$

$\overline{C}_{\varrho}^{+}[0, 1]$	$(0, \infty] \times \overline{C}_1^{+}[0, 1]$ with the lower index ϱ meaning that the space $(0, \infty]$ is equipped with the metric ϱ
CSMS	complete separable metric space
D and D'	dependence conditions
$D[0, T]$	space of functions on $[0, T]$ that are right-continuous and have left-hand limits
$\mathcal{D}(G_\gamma)$	domain of attraction of G_γ
ERV	extended regular variation
f^-	left-continuous version of the function f
f^+	right-continuous version of the function f
$\bar{f}$	generalized inverse function of f
$f^{\leftarrow}$	(usually left-continuous) inverse function of f
$|f|_\infty$	$\sup_s |f(s)|$
F	distribution function
F_n	right-continuous empirical distribution function
G_γ	extreme value distribution function
GP	generalized Pareto
i.i.d.	independent and identically distributed
L	dependence function
$\mathbb{R}_+$	$[0, \infty)$
$\mathbb{R}_+^{2\star}$	$\mathbb{R}_+^2 \setminus \{(0, 0)\}$
$R(X_i)$	rank of X_i among $(X_1, X_2, \ldots, X_n)$
RV_α	regularly varying with index α
U	(usually left-continuous) inverse of $1/(1 - F)$
x^*	$\sup\{x : F(x) < 1\} = U(\infty)$
$_*x$	$\inf\{x : F(x) > 0\}$

Extreme Value Theory

Part I

One-Dimensional Observations

Limit Distributions and Domains of Attraction

1.1 Extreme Value Theory: Basics

1.1.1 Introduction

Partial Sums and Partial Maxima

The asymptotic theory of sample extremes has been developed in parallel with the central limit theory, and in fact the two theories bear some resemblance.

Let $X_1, X_2, X_3, \ldots$ be independent and identically distributed random variables. The central limit theory is concerned with the limit behavior of the partial sums $X_1 + X_2 + \cdots + X_n$ as $n \to \infty$, whereas the theory of sample extremes is concerned with the limit behavior of the sample extremes $\max(X_1, X_2, \ldots, X_n)$ or $\min(X_1, \ldots, X_n)$ as $n \to \infty$.

One may think of the two theories as concerned with failure. A tire of a car can fail in two ways. Every day of driving will wear out the tire a little, and after a long time the accumulated decay will result in failure (i.e., the partial sums exceed some threshold). But also when driving one may hit a pothole or one may accidentally hit the sidewalk. Such incidents have either no effect or the tire will be punctured. In the latter case it is just one big observation that causes failure, which means that partial maxima exceed some threshold.

In fact, in its early stages, the development of the theory of extremes was mainly motivated by intellectual curiosity.

Outline of This Chapter

Our interest is in finding possible limit distributions for (say) sample maxima of independent and identically distributed random variables. Let F be the underlying distribution function and x^* its right endpoint, i.e., $x^* := \sup\{x : F(x) < 1\}$, which may be infinite. Then

$$\max(X_1, X_2, \ldots, X_n) \overset{P}{\to} x^*, \quad n \to \infty,$$

where $\to^P$ means convergence in probability, since

$$P(\max (X_1, \ldots, X_n) \leq x) = P(X_1 \leq x, X_2 \leq x, \ldots, X_n \leq x) = F^n(x) ,$$

which converges to zero for $x < x^*$ and to 1 for $x \geq x^*$. Hence, in order to obtain a nondegenerate limit distribution, a normalization is necessary.

Suppose there exists a sequence of constants $a_n > 0$, and b_n real ($n = 1, 2, \ldots$), such that

$$\frac{\max (X_1, X_2, \ldots, X_n) - b_n}{a_n}$$

has a nondegenerate limit distribution as $n \to \infty$, i.e.,

$$\lim_{n \to \infty} F^n(a_n x + b_n) = G(x) , \tag{1.1.1}$$

say, for every continuity point x of G, and G a nondegenerate distribution function. In this chapter we shall find all distribution functions G that can occur as a limit in (1.1.1). These distributions are called *extreme value distributions*.

Next, for each of those limit distributions, we shall find necessary and sufficient conditions on the initial distribution F such that (1.1.1) holds. The class of distributions F satisfying (1.1.1) is called the *maximum domain of attraction* or simply *domain of attraction* of G. We are going to identify all extreme value distributions and their domains of attraction. But before doing so it is useful and illuminating to reformulate relation (1.1.1) in two other ways.

In (1.1.1) we have used a linear normalization. One could consider a wider class of normalizations. However, this linear normalization already leads to a sufficiently rich theory.

We shall always be concerned with sample maxima. Since

$$\min (X_1, X_2, \ldots, X_n) = - \max (-X_1, -X_2, \ldots, -X_n) ,$$

the results can easily be reformulated for sample minima.

1.1.2 Alternative Formulations of the Limit Relation

We are going to play a bit with condition (1.1.1). By taking logarithms left and right we get from (1.1.1) the equivalent relation that for each continuity point x for which $0 < G(x) < 1$,

$$\lim_{n \to \infty} n \log F(a_n x + b_n) = \log G(x) . \tag{1.1.2}$$

Clearly it follows that $F(a_n x + b_n) \to 1$, for each such x. Hence

$$\lim_{n \to \infty} \frac{- \log F(a_n x + b_n)}{1 - F(a_n x + b_n)} = 1,$$

and in fact (1.1.2) is equivalent to

$$\lim_{n \to \infty} n(1 - F(a_n x + b_n)) = - \log G(x) ,$$

or

$$\lim_{n\to\infty} \frac{1}{n\left(1 - F(a_n x + b_n)\right)} = \frac{1}{-\log G(x)} \ . \tag{1.1.3}$$

Next we are going to reformulate this condition in terms of the inverse functions. For any nondecreasing function f, let $f^{\leftarrow}$ be its *left-continuous inverse*, i.e.,

$$f^{\leftarrow}(x) := \inf\{y : f(y) \geq x\} \ .$$

For properties of the inverse function that will be used throughout, see Exercise 1.1.

Lemma 1.1.1 *Suppose f_n is a sequence of nondecreasing functions and g is a nondecreasing function. Suppose that for each x in some open interval (a, b) that is a continuity point of g,*

$$\lim_{n\to\infty} f_n(x) = g(x) \ . \tag{1.1.4}$$

Let $f_n^{\leftarrow}$, $g^{\leftarrow}$ be the left-continuous inverses of f_n and g. Then, for each x in the interval $(g(a), g(b))$ that is a continuity point of $g^{\leftarrow}$ we have

$$\lim_{n\to\infty} f_n^{\leftarrow}(x) = g^{\leftarrow}(x) \ . \tag{1.1.5}$$

Proof. Let x be a continuity point of $g^{\leftarrow}$. Fix $\varepsilon > 0$. We have to prove that for $n, n_0 \in \mathbb{N}, n \geq n_0$,

$$f_n^{\leftarrow}(x) - \varepsilon \leq g^{\leftarrow}(x) \leq f_n^{\leftarrow}(x) + \varepsilon \ .$$

We are going to prove the right inequality; the proof of the left-hand inequality is similar.

Choose $0 < \varepsilon_1 < \varepsilon$ such that $g^{\leftarrow}(x) - \varepsilon_1$ is a continuity point of g. This is possible since the continuity points of g form a dense set. Since $g^{\leftarrow}$ is continuous in x, $g^{\leftarrow}(x)$ is a point of increase for g; hence $g(g^{\leftarrow}(x) - \varepsilon_1) < x$. Choose $\delta < x - g(g^{\leftarrow}(x) - \varepsilon_1)$. Since $g^{\leftarrow}(x) - \varepsilon_1$ is a continuity point of g, there exists n_0 such that $f_n(g^{\leftarrow}(x) - \varepsilon_1) < g(g^{\leftarrow}(x) - \varepsilon_1) + \delta < x$ for $n \geq n_0$. The definition of the function $f_n^{\leftarrow}$ then implies $g^{\leftarrow}(x) - \varepsilon_1 \leq f_n^{\leftarrow}(x)$.   ∎

We are going to apply Lemma 1.1.1 to relation (1.1.3). Let the function U be the left-continuous inverse of $1/(1 - F)$. Note that $U(t)$ is defined for $t > 1$. It follows that (1.1.3) is equivalent to

$$\lim_{n\to\infty} \frac{U(nx) - b_n}{a_n} = G^{\leftarrow}\left(e^{-1/x}\right) =: D(x) \ , \tag{1.1.6}$$

for each positive x. This is encouraging since relation (1.1.6) looks simpler than (1.1.3). We are now going to make (1.1.6) more flexible in the following way:

Theorem 1.1.2 *Let $a_n > 0$ and b_n be real sequences of constants and G a nondegenerate distribution function. The following statements are equivalent:*

1.

$$\lim_{n\to\infty} F^n(a_n x + b_n) = G(x)$$

for each continuity point x of G.

2.

$$\lim_{t\to\infty} t\,(1 - F(a(t)x + b(t))) = -\log G(x)\,, \qquad (1.1.7)$$

for each continuity point x of G for which $0 < G(x) < 1$, $a(t) := a_{[t]}$, *and* $b(t) := b_{[t]}$ *(with* $[t]$ *the integer part of t).*

3.

$$\lim_{t\to\infty} \frac{U(tx) - b(t)}{a(t)} = D(x)\,, \qquad (1.1.8)$$

for each $x > 0$ *continuity point of* $D(x) = G^{\leftarrow}\left(e^{-1/x}\right)$, $a(t) := a_{[t]}$, *and* $b(t) := b_{[t]}$.

Proof. The equivalence of (2) and (3) follows from Lemma 1.1.1. We have already checked that (1) is equivalent to (1.1.6). So it is sufficient to prove that (1.1.6) implies (3). Let x be a continuity point of D. For $t \geq 1$,

$$\frac{U([t]x) - b_{[t]}}{a_{[t]}} \leq \frac{U(tx) - b_{[t]}}{a_{[t]}} \leq \frac{U\left([t]x\left(1 + 1/[t]\right)\right) - b_{[t]}}{a_{[t]}}\,.$$

The right-hand side is eventually less than $D(x')$ for any continuity point $x' > x$ with $D(x') > D(x)$. Since D is continuous at x, we obtain

$$\lim_{t\to\infty} \frac{U(tx) - b_{[t]}}{a_{[t]}} = D(x)\,.$$

This is (3). ∎

We shall see shortly (Section 1.1.4) the usefulness of these two alternative conditions for statistical applications.

1.1.3 Extreme Value Distributions

Now we are in a position to identify the class of nondegenerate distributions that can occur as a limit in the basic relation (1.1.1). This class of distributions was called the class of extreme value distributions.

Theorem 1.1.3 (Fisher and Tippet (1928), Gnedenko (1943)) *The class of extreme value distributions is* $G_\gamma(ax + b)$ *with* $a > 0$, b *real, where*

$$G_\gamma(x) = \exp\left(-(1 + \gamma x)^{-1/\gamma}\right), \qquad 1 + \gamma x > 0\,, \qquad (1.1.9)$$

with γ *real and where for* $\gamma = 0$ *the right-hand side is interpreted as* $\exp(-e^{-x})$.

Definition 1.1.4 The parameter γ in (1.1.9) is called the *extreme value index*.

Proof (of Theorem 1.1.3). Let us consider the class of limit functions D in (1.1.8). First suppose that 1 is a continuity point of D. Then note that for continuity points $x > 0$,

$$\lim_{t \to \infty} \frac{U(tx) - U(t)}{a(t)} = D(x) - D(1) =: E(x) . \tag{1.1.10}$$

Take $y > 0$ and write

$$\frac{U(txy) - U(t)}{a(t)} = \frac{U(txy) - U(ty)}{a(ty)} \frac{a(ty)}{a(t)} + \frac{U(ty) - U(t)}{a(t)} . \tag{1.1.11}$$

We claim that $\lim_{t \to \infty}(U(ty) - U(t))/a(t)$ and $\lim_{t \to \infty} a(ty)/a(t)$ exist. Suppose not. Then there are A_1, A_2, B_1, B_2 with $A_1 \neq A_2$ or $B_1 \neq B_2$, where B_i are limit points of $(U(ty) - U(t))/a(t)$ and A_i are limit points of $a(ty)/a(t)$, $i = 1, 2$, as $t \to \infty$. We find from (1.1.11) that

$$E(xy) = E(x) A_i + B_i, \tag{1.1.12}$$

$i = 1, 2$, for all x continuity points of $E(\cdot)$ and $E(\cdot \ y)$. For an arbitrary x take a sequence of continuity points x_n with $x_n \uparrow x$ ($n \to \infty$). Then $E(x_n y) \to E(xy)$ and $E(x_n) \to E(x)$ since E is left-continuous. Hence (1.1.12) holds for all x and y positive. Subtracting the expressions for $i = 1, 2$ from each other one obtains

$$E(x) (A_1 - A_2) = B_2 - B_1$$

for all $x > 0$. Since E cannot be constant (remember that G is nondegenerate) we must have $A_1 = A_2$ and hence also $B_1 = B_2$. Conclusion:

$$A(y) := \lim_{t \to \infty} \frac{a(ty)}{a(t)}$$

exists for $y > 0$, and for $x, y > 0$,

$$E(xy) = E(x) A(y) + E(y) .$$

Hence for $s := \log x, t := \log y$ ($x, y \neq 1$), and $H(x) := E(e^x)$, we have

$$H(t + s) = H(s)A(e^t) + H(t), \tag{1.1.13}$$

which we can write as (since $H(0) = 0$)

$$\frac{H(t + s) - H(t)}{s} = \frac{H(s) - H(0)}{s} A(e^t) . \tag{1.1.14}$$

There is certainly one t at which H is differentiable (since H is monotone); hence by (1.1.14) H is differentiable everywhere and

$$H'(t) = H'(0)A(e^t) . \tag{1.1.15}$$

Write $Q(t) := H(t)/H'(0)$. Note that $H'(0)$ cannot be zero: H cannot be constant since G is nondegenerate. Then $Q(0) = 0, Q'(0) = 1$. By (1.1.13),

$$Q(t + s) - Q(t) = Q(s)A(e^t),$$

and by (1.1.15),

$$Q(t + s) - Q(t) = Q(s)Q'(t) .$$ (1.1.16)

Subtracting the same expression with t and s interchanged we get

$$Q(t)\frac{Q'(s) - 1}{s} = \frac{Q(s)}{s}(Q'(t) - 1) ,$$

hence (let $s \to 0$)

$$Q(t) Q''(0) = Q'(t) - 1 .$$

It follows that Q is twice differentiable, and by differentiation,

$$Q''(0) Q'(t) = Q''(t) .$$

Hence

$$\left(\log Q'\right)'(t) = Q''(0) =: \gamma \in \mathbb{R} ,$$

for all t. It follows that (note that $Q'(0) = 1$)

$$Q'(t) = e^{\gamma t}$$

and (since $Q(0) = 0$)

$$Q(t) = \int_0^t e^{\gamma s} ds .$$

This means that

$$H(t) = H'(0)\frac{e^{\gamma t} - 1}{\gamma}$$

and

$$D(t) = D(1) + H'(0)\frac{t^\gamma - 1}{\gamma} .$$

Hence

$$D^\leftarrow(x) = \left(1 + \gamma\frac{x - D(1)}{H'(0)}\right)^{1/\gamma} .$$ (1.1.17)

Now $D(x) = G^\leftarrow\left(e^{-1/x}\right)$, and hence

$$D^\leftarrow(x) = \frac{1}{- \log G(x)} .$$ (1.1.18)

Combining (1.1.17) and (1.1.18) we obtain the statement of the theorem.

If 1 is not a continuity point of D, follow the proof with the function $U(tx_0)$ with x_0 a continuity point of D. ∎

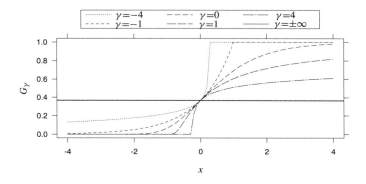

Fig. 1.1. Family of extreme value distributions G_γ.

Remark 1.1.5 Let $X_1, X_2, \ldots$ be independent and identically distributed random variables with distribution function F. The distribution function F is called *max-stable* if for some choice of constants $a_n > 0$ and b_n real,

$$P\left(\frac{\max(X_1, \ldots, X_n) - b_n}{a_n} \leq x\right) = P(X_1 \leq x)$$

for all x and $n = 1, 2, \ldots$. Note that the class of *max-stable distributions* is the same as the class of extreme value distributions (cf. Exercise 1.2).

The *parametrization* in Theorem 1.1.3 is due to von Mises (1936) and Jenkinson (1955). This theorem is an important result in many ways. It shows that the limit distribution functions form a simple explicit one-parameter family apart from the scale and location parameters. Figure 1.1 illustrates this family for some values of γ. Moreover, it shows that the class contains distributions with quite different features. Let us consider the subclasses $\gamma > 0$, $\gamma = 0$, and $\gamma < 0$ separately:

(a) For $\gamma > 0$ clearly $G_\gamma(x) < 1$ for all x, i.e., the right endpoint of the distribution is infinity. Moreover, as $x \to \infty$, $1 - G_\gamma(x) \sim \gamma^{-1/\gamma} x^{-1/\gamma}$, i.e., the distribution has a rather heavy right tail; for example, moments of order greater than or equal to $1/\gamma$ do not exist (cf. Exercise 1.16).
(b) For $\gamma = 0$ the right endpoint of the distribution equals infinity. The distribution, however, is rather light-tailed: $1 - G_0(x) \sim e^{-x}$ as $x \to \infty$, and all moments exist.
(c) For $\gamma < 0$ the right endpoint of the distribution is $-1/\gamma$ so it has a short tail, verifying $1 - G_\gamma(-\gamma^{-1} - x) \sim (-\gamma x)^{-1/\gamma}$, as $x \downarrow 0$.

An alternative parametrization is as follows:

(a) For $\gamma > 0$ use $G_\gamma((x-1)/\gamma)$ and get with $\alpha = 1/\gamma > 0$,

$$\Phi_\alpha(x) := \begin{cases} 0, & x \leq 0, \\ \exp(-x^{-\alpha}), & x > 0. \end{cases}$$

This class is often called the *Fréchet class of distributions* (Fréchet (1927)).
(b) The distribution function with $\gamma = 0$,

$$G_0(x) = \exp\left(-e^{-x}\right),$$

for all real x, is called the *double-exponential* or *Gumbel distribution*.
(c) For $\gamma < 0$ use $G_\gamma(-(1+x)/\gamma)$ and get with $\alpha = -1/\gamma > 0$,

$$\Psi_\alpha(x) := \begin{cases} \exp\left(-(-x)^\alpha\right) , & x < 0, \\ 1 , & x \geq 0 . \end{cases}$$

This class is sometimes called the *reverse-Weibull class of distributions*.

Recall that if relation (1.1.1) holds with $G = G_\gamma$ for some $\gamma \in \mathbb{R}$, we say that the distribution function F is in the domain of attraction of G_γ. Notation: $F \in \mathcal{D}(G_\gamma)$.

The result of Theorem 1.1.3 leads to the following reformulation of Theorem 1.1.2.

Theorem 1.1.6 *For $\gamma \in \mathbb{R}$ the following statements are equivalent:*

1. *There exist real constants $a_n > 0$ and b_n real such that*

$$\lim_{n \to \infty} F^n(a_n x + b_n) = G_\gamma(x) = \exp\left(-(1 + \gamma x)^{-1/\gamma}\right), \qquad (1.1.19)$$

for all x with $1 + \gamma x > 0$.
2. *There is a positive function a such that for $x > 0$,*

$$\lim_{t \to \infty} \frac{U(tx) - U(t)}{a(t)} = D_\gamma(x) = \frac{x^\gamma - 1}{\gamma}, \qquad (1.1.20)$$

where for $\gamma = 0$ the right-hand side is interpreted as $\log x$.
3. *There is a positive function a such that*

$$\lim_{t \to \infty} t \left(1 - F(a(t)x + U(t))\right) = (1 + \gamma x)^{-1/\gamma}, \qquad (1.1.21)$$

for all x with $1 + \gamma x > 0$.
4. *There exists a positive function f such that*

$$\lim_{t \uparrow x^*} \frac{1 - F(t + xf(t))}{1 - F(t)} = (1 + \gamma x)^{-1/\gamma} \qquad (1.1.22)$$

for all x for which $1 + \gamma x > 0$, where $x^* = \sup\{x : F(x) < 1\}$.

Moreover, (1.1.19) holds with $b_n := U(n)$ and $a_n := a(n)$. Also, (1.1.22) holds with $f(t) = a\left(1/(1 - F(t))\right)$.

Proof. The equivalence of (1), (2), and (3) has been established in Theorem 1.1.2. Next we prove that (2) implies (4).

It is easy to see that for $\varepsilon > 0$,

$$g\left(h^{\leftarrow}(t) - \varepsilon\right) \leq t \leq g\left(h^{\leftarrow}(t) + \varepsilon\right),$$

where g is a nondecreasing function and $h^{\leftarrow}$ its right-continuous inverse (cf. Exercise 1.1). It follows that

$$\frac{(1-\varepsilon)^{\gamma} - 1}{\gamma} \leftarrow \frac{U\left(\frac{1-\varepsilon}{1-F(t)}\right) - U\left(\frac{1}{1-F(t)}\right)}{a\left(\frac{1}{1-F(t)}\right)} < \frac{t - U\left(\frac{1}{1-F(t)}\right)}{a\left(\frac{1}{1-F(t)}\right)}$$

$$< \frac{U\left(\frac{1+\varepsilon}{1-F(t)}\right) - U\left(\frac{1}{1-F(t)}\right)}{a\left(\frac{1}{1-F(t)}\right)} \rightarrow \frac{(1+\varepsilon)^{\gamma} - 1}{\gamma},$$

as $t \uparrow x^*$, and consequently

$$\lim_{t \uparrow x^*} \frac{t - U\left(\frac{1}{1-F(t)}\right)}{a\left(\frac{1}{1-F(t)}\right)} = 0.$$

Hence by (2) for all $x > 0$,

$$\lim_{t \uparrow x^*} \frac{U\left(\frac{x}{1-F(t)}\right) - t}{a\left(\frac{1}{1-F(t)}\right)} = \frac{x^{\gamma} - 1}{\gamma},$$

and by Lemma 1.1.1,

$$\lim_{t \uparrow x^*} \frac{1 - F(t)}{1 - F\left(t + xa\left(\frac{1}{1-F(t)}\right)\right)} = (1 + \gamma x)^{1/\gamma},$$

i.e., (4) holds.

The converse (i.e., (4) implies (2)) is similar.   ∎

Example 1.1.7 Let F be the standard normal distribution. We are going to prove that (1.1.3) holds: for all $x > 0$,

$$\lim_{n \to \infty} n\left(1 - F(a_n x + b_n)\right) = e^{-x} \tag{1.1.23}$$

with

$$b_n := (2 \log n - \log \log n - \log(4\pi))^{1/2} \tag{1.1.24}$$

and

$$a_n := \frac{1}{b_n}. \tag{1.1.25}$$

Note first that $b_n/(2 \log n)^{1/2} \rightarrow 1$, $n \rightarrow \infty$; hence $\log b_n - 2^{-1} \log \log n - 2^{-1} \log 2 \rightarrow 0$ and hence

$$\frac{b_n^2}{2} + \log b_n - \log n + \frac{1}{2} \log(2\pi) \to 0 \, , \tag{1.1.26}$$

as $n \to \infty$. Now by (1.1.25),

$$-\frac{d}{dx} n \left(1 - F \left(a_n x + b_n \right) \right) = \frac{n}{b_n \sqrt{2\pi}} \exp \left(- \left(\frac{x}{b_n} + b_n \right)^2 / 2 \right)$$

$$= \exp \left\{ - \left(\frac{b_n^2}{2} + \log b_n - \log n + \frac{1}{2} \log(2\pi) \right) \right\} e^{-x^2/(2b_n^2)} e^{-x}$$

$$\to e^{-x}$$

for $x \in \mathbb{R}$. Hence

$$n \left(1 - F \left(a_n x + b_n \right) \right) = \frac{n}{b_n \sqrt{2\pi}} \int_x^\infty \exp \left(- \left(\frac{u}{b_n} + b_n \right)^2 / 2 \right) du$$

$$= \exp \left\{ - \left(\frac{b_n^2}{2} + \log b_n - \log n + \frac{1}{2} \log(2\pi) \right) \right\} \int_x^\infty e^{-u^2/(2b_n^2)} e^{-u} \, du$$

$$\to e^{-x}$$

by Lebesgue's theorem on dominated convergence. Hence (1.1.23) holds.

Since in the limit relation (1.1.23) we can replace a_n by a_n' and b_n by b_n' provided $a_n/a_n' \to 1$, $(b_n' - b_n)/a_n \to 0$, we can replace b_n, a_n from (1.1.24) and (1.1.25) by, e.g.,

$$b_n' = (2 \log n)^{1/2} - \frac{\log \log n + \log(4\pi)}{(2 \log n)^{1/2}}$$

and

$$a_n' = (2 \log n)^{-1/2} \, .$$

1.1.4 Interpretation of the Alternative Conditions; Case Studies

We want to introduce three problems in which extreme value theory can be fruitfully used, the first with γ near zero, the second with γ positive, and the third with γ negative. It shall become clear that the alternative conditions are the important ones. The present account is introductory. A more complete account will follow later on in the applications, Sections 3.7.3, 4.6.2, 7.2, 7.3, and 8.4.

Sea Level

As we related in the preface to this book, approximately 40% of the Netherlands is below the sea level. Much of it has to be protected against the sea by dikes. These dikes have to withstand storm surges that drive the seawater level up along the coast. The government, balancing considerations of cost and safety, has determined that the dikes should be so high that the probability of a flood (i.e., the seawater level exceeding the top of the dike) in a given year is 10^{-4}. The question is then how high the dikes should be built to meet this requirement. Storm data have been collected for more than 100 years. In this period, at the town of Delfzijl, in the northeast of the Netherlands, 1877 severe wind storms have been identified. The collection of high-tide water levels at Delfzijl during those storms forms approximately a set of independent observations, taken under similar conditions (i.e., we may assume that they are independent and identically distributed).

First we convert the 10^{-4} probability to a probability per storm (since that is what the data give us). Since there are 1877 storms in 111 years, we look for the level that is exceeded by one such storm with probability $(111/1877) \times 10^{-4}$. Let F be the distribution function of the high-tide water level during one such storm. Then we are looking for the $1 - (111/1877) \times 10^{-4}$ quantile, i.e., $F^{\leftarrow}\left(1 - (111/1877) \times 10^{-4}\right) = U\left((1877/111) \times 10^4\right) \approx U\left(17 \times 10^4\right)$.

Of course this is a simplification of what really goes on: if X is the maximum in a year and Z the maximum during a storm, we have $X = \max_{1 \leq i \leq N} Z_i$ with N the (random) number of storms in a year. We ignore the randomness of N.

Normally one would estimate a quantile by the empirical quantile, that is, one of the order statistics. But the highest order statistic corresponds in this case to $F^{\leftarrow}(1 - 1/1878) \approx U\left(19 \times 10^2\right)$. So we need to extrapolate beyond the range of the available data.

At this point Theorem 1.1.2(3) can help. In view of Theorem 1.1.3 the condition can be written as

$$\lim_{t \to \infty} \frac{U(tx) - U(t)}{a(t)} = \frac{x^{\gamma} - 1}{\gamma}$$

for $x > 0$, some real parameter γ, and an appropriate positive function a. Let us use the approximation with $t < 19 \times 10^2$ (so that we can estimate $U(t)$ using the empirical quantile function) and $tx := 17 \times 10^4$. Then the requested quantile is

$$U\left(17 \times 10^4\right) = U(tx) \approx U(t) + a(t)\frac{x^{\gamma} - 1}{\gamma}. \tag{1.1.27}$$

For the moment we just remark that Theorem 1.1.2(3) seems to provide a possibility to estimate an extreme quantile by fitting the function $(x^{\gamma} - 1)/\gamma$ (in the present case with γ close to zero) to the quantile type function U.

S&P 500

Daily price quotes of the S&P 500 total return index over the period from 01/01/1980 to 14/05/2002 are available, corresponding to 5835 observations. The daily price quotes, p_t say, are used to compute daily "continuously" compounded returns r_t by taking the logarithmic first differences of the price series, $r_t = \log(p_t/p_{t-1})$. Stock returns generally exhibit a positive mean due to positive growth of the economy. Therefore we shall focus only on the loss returns. We shall assume here that the observations are independent and identically distributed (cf. Jansen and de Vries (1991)).

Now consider the situation in which one has to decide on a big risky investment while one cannot afford to have a loss larger than a certain amount. Then it is of interest to know of the probability of the occurrence of such a loss.

If F is the distribution function of the log-loss returns and x is the critical (large) amount, the posed problem is the estimation of $1 - F(x)$. Then Theorem 1.1.6(3) suggests that for some positive function a and large x,

$$1 - F(x) \approx \frac{1}{t}\left(1 + \gamma \frac{x - U(t)}{a(t)}\right)^{-1/\gamma}$$

for some large t. This again motivates a tail probability estimator under the extreme value theory approach.

Life Span

There is some discussion among demographers and physicians about whether there is limited life span for humans; that is, if we consider the life span of a human as random, does its probability distribution have a finite endpoint? The problem can be considered from the point of view of extreme value theory.

A data set consists of the total lifespan (in days) of all people born in the Netherlands in the years 1877–1881, still alive on January 1, 1971, and who died as a resident of the Netherlands. This concerns about 10 000 people.

Now we want to decide whether the right endpoint of the distribution is finite. The endpoint, finite or not, is $\lim_{t \to \infty} U(t)$, which we denote by $U(\infty)$. It will be verified later on (Section 3.7) that for the given data set the hypothesis $\gamma < 0$ is not rejected. Moreover, it is known from Section 1.2 below that for γ negative we must have $U(\infty) < \infty$. Hence we shall believe that there is an age that cannot be exceeded by this cohort.

Next we estimate this maximal age $U(\infty)$. For this we use the limit relation (1.1.20) again:

$$\frac{U(tx) - U(t)}{a(t)} \to \frac{x^\gamma - 1}{\gamma} \qquad (t \to \infty),$$

but as we shall show later, if $\gamma < 0$, this relation is also valid for $x = \infty$, i.e.,

$$\frac{U(\infty) - U(t)}{a(t)} \to -\frac{1}{\gamma} \qquad (t \to \infty),$$

or

$$U(\infty) \approx U(t) - \frac{a(t)}{\gamma} \qquad (t \to \infty) .$$

This relation will be the basis for estimating $U(\infty)$.

1.1.5 Domains of Attraction: A First Approach

In this section we shall derive sufficient conditions on the distribution function F that ensure that there are sequences of constants $a_n > 0$ and b_n such that

$$\lim_{n \to \infty} F^n(a_n x + b_n) = G_\gamma(x) \tag{1.1.28}$$

for some given real γ and all x. These conditions, basically due to von Mises (1936), require the existence of one or two derivatives of F.

It is easy to see, using relation (1.1.6) from Section 1.1.2, that F cannot be in the domain of attraction of G_{γ_1} and G_{γ_2} with $\gamma_1 \neq \gamma_2$.

The following theorem states a sufficient condition for belonging to a domain of attraction. The condition is called *von Mises' condition*.

Theorem 1.1.8 *Let F be a distribution function and x^* its right endpoint. Suppose $F''(x)$ exists and $F'(x)$ is positive for all x in some left neighborhood of x^*. If*

$$\lim_{t \uparrow x^*} \left(\frac{1 - F}{F'} \right)' (t) = \gamma \tag{1.1.29}$$

or equivalently

$$\lim_{t \uparrow x^*} \frac{(1 - F(t)) F''(t)}{(F'(t))^2} = -\gamma - 1 , \tag{1.1.30}$$

then F is in the domain of attraction of G_γ.

Remark 1.1.9 Under (1.1.29) we have (1.1.28) with $b_n = U(n)$ and $a_n = nU'(n) = 1/(nF'(b_n))$.

Proof (of Theorem 1.1.8). Here, as elsewhere, the proof is much simplified by formulating everything in terms of the inverse function U rather than the distribution function F. By differentiating the relation

$$\frac{1}{1 - F(U(t))} = t$$

we obtain

$$U'(t) = \frac{[1 - F(U(t))]^2}{F'(U(t))} .$$

Differentiating once more, we find that

$$\frac{U''(t)}{U'(t)} = -2[1 - F(U(t))] - \frac{F''(U(t))[1 - F(U(t))]^2}{[F'(U(t))]^2} ,$$

so that

$$\frac{t\, U''(t)}{U'(t)} = -2 - \frac{F''(U(t))[1 - F(U(t))]}{[F'(U(t))]^2} .$$

By Theorems 1.1.2 and 1.1.6 the relation to be proved is equivalent to

$$\lim_{t \to \infty} \frac{U(tx) - U(t)}{a(t)} = \frac{x^\gamma - 1}{\gamma} \tag{1.1.31}$$

for all $x > 0$. So we need to prove that

$$\lim_{t\to\infty} \frac{t\,U''(t)}{U'(t)} = \gamma - 1$$

implies (1.1.31) for the same γ.

Since for $1 < x_0 < x$,

$$\log U'(x) - \log U'(x_0) = \int_{x_0}^{x} \frac{U''(s)}{U'(s)} ds,$$

we have for $x > 0, t, tx > 1$,

$$\log U'(tx) - \log U'(t) = \int_{1}^{x} A(ts)\frac{ds}{s}$$

with $A(t) := tU''(t)/U'(t)$. It follows that for $0 < a < b < \infty$,

$$\lim_{t\to\infty} \sup_{a\le x\le b} \left|\log \frac{U'(tx)}{U'(t)} - \log x^{\gamma-1}\right| = 0 .$$

Hence also, since $|e^s - e^t| < c\,|s - t|$ on a compact interval for some positive constant c,

$$\lim_{t\to\infty} \sup_{a\le x\le b} \left|\frac{U'(tx)}{U'(t)} - x^{\gamma-1}\right| = 0 .$$

This implies that

$$\frac{U(tx) - U(t)}{tU'(t)} - \frac{x^\gamma - 1}{\gamma} = \int_{1}^{x} \left(\frac{U'(ts)}{U'(t)} - s^{\gamma-1}\right) ds$$

converges to zero.   ■

For later use we next give von Mises' condition formulated in terms of U.

Corollary 1.1.10 *Condition* (1.1.29) *is equivalent to*

$$\lim_{t\to\infty} \frac{tU''(t)}{U'(t)} = \gamma - 1 , \tag{1.1.32}$$

which implies

$$\lim_{t\to\infty} \frac{U'(tx)}{U'(t)} = x^{\gamma-1} \tag{1.1.33}$$

locally uniformly in $(0, \infty)$ *and finally*

$$\lim_{t\to\infty} \frac{U(tx) - U(t)}{tU'(t)} = \frac{x^\gamma - 1}{\gamma},$$

so that by Theorem 1.1.6 we obtain $F \in \mathcal{D}(G_\gamma)$.

Simpler conditions are possible for $\gamma \ne 0$:

Theorem 1.1.11 $(\gamma > 0)$ *Suppose* $x^* = \infty$ *and* F' *exists. If*

$$\lim_{t \to \infty} \frac{t F'(t)}{1 - F(t)} = \frac{1}{\gamma} \tag{1.1.34}$$

for some positive γ, *then* F *is in the domain of attraction of* G_γ.

Proof. As in the proof of Theorem 1.1.8 we see that condition (1.1.34) is equivalent to

$$\lim_{t \to \infty} \frac{t U'(t)}{U(t)} = \gamma . \tag{1.1.35}$$

Further,

$$\log U(tx) - \log U(t) = \int_1^x \frac{ts U'(ts)}{U(ts)} \frac{ds}{s};$$

hence by (1.1.35) for $x > 0$,

$$\lim_{t \to \infty} \frac{U(tx)}{U(t)} = x^\gamma , \tag{1.1.36}$$

or

$$\lim_{t \to \infty} \frac{U(tx) - U(t)}{\gamma U(t)} = \frac{x^\gamma - 1}{\gamma} ,$$

which is condition (1.1.20) of Theorem 1.1.6. ∎

Corollary 1.1.12 $(\gamma > 0)$ *Condition* (1.1.34) *is equivalent to*

$$\lim_{t \to \infty} \frac{t U'(t)}{U(t)} = \gamma ,$$

which implies (1.1.33) *in view of* (1.1.36).

Theorem 1.1.13 $(\gamma < 0)$ *Suppose* $x^* < \infty$ *and* $F'(x)$ *exists for* $x < x^*$. *If*

$$\lim_{t \uparrow x^*} \frac{(x^* - t) F'(t)}{1 - F(t)} = -\frac{1}{\gamma} \tag{1.1.37}$$

for some negative γ, *then* F *is in the domain of attraction of* G_γ.

Proof. As before, we can see that condition (1.1.37) is equivalent to

$$\lim_{t \to \infty} \frac{t U'(t)}{U(\infty) - U(t)} = -\gamma . \tag{1.1.38}$$

Since

$$\log(U(\infty) - U(tx)) - \log(U(\infty) - U(t)) = -\int_1^x \frac{ts U'(ts)}{U(\infty) - U(ts)} \frac{ds}{s} ,$$

relation (1.1.38) implies

$$\lim_{t\to\infty} \frac{U(\infty) - U(tx)}{U(\infty) - U(t)} = x^\gamma \;,$$

i.e., for $x > 0$,

$$\lim_{t\to\infty} \frac{U(tx) - U(t)}{-\gamma(U(\infty) - U(t))} = \frac{x^\gamma - 1}{\gamma} \;.\qquad\blacksquare$$

Corollary 1.1.14 ($\gamma < 0$) *Condition* (1.1.37) *is equivalent to*

$$\lim_{t\to\infty} \frac{tU'(t)}{U(\infty) - U(t)} = -\gamma \;,$$

which implies (1.1.33).

To get an idea about the tail behavior of the distribution functions in the various domains of attraction note that for $x^* = \infty$ and $t > x_1$,

$$\log(1 - F(t)) = \log(1 - F(x_1)) - \int_{x_1}^t \frac{ds}{f(s)}$$

with $f(s) := (1 - F(s))/F'(s)$. Under condition (1.1.29) of Theorem 1.1.8,

$$\lim_{t\to\infty} \frac{f(t)}{t} = \lim_{t\to\infty} f'(t) = \gamma \;;$$

hence

$$\lim_{t\to\infty} \frac{\log(1 - F(t))}{\log t} = -\frac{1}{\gamma} \;,$$

i.e., for $\gamma > 0$ the function $1 - F(t)$ behaves roughly like $t^{-1/\gamma}$, which means a heavy tail. For $\gamma = 0$, however,

$$\lim_{t\to\infty} t^\alpha (1 - F(t)) = \lim_{t\to\infty} x_1^\alpha (1 - F(x_1)) \exp - \left\{ \int_{x_1}^t \left(\frac{s}{f(s)} - \alpha \right) \frac{ds}{s} \right\} = 0$$

for all $\alpha > 0$ and hence the tail is light. Similar reasoning reveals that for $\gamma < 0$ (in which case necessarily $x^* < \infty$, as we shall see later on in Theorem 1.2.1),

$$\lim_{t\uparrow x^*} \frac{\log(1 - F(t))}{\log(x^* - t)} = -\frac{1}{\gamma},$$

so that the function $1 - F(x^* - t)$ behaves roughly like $t^{-1/\gamma}$ as $t \downarrow 0$.

The reader may want to verify that the Cauchy distribution satisfies Theorem 1.1.11 with $\gamma = 1$ (Exercise 1.6); the normal, exponential, and any gamma distribution satisfy Theorem 1.1.8 with $\gamma = 0$ (Exercise 1.7), and a beta (μ, ν) distribution satisfies Theorem 1.1.13 with $\gamma = -\mu^{-1}$ (Exercise 1.8).

Remark 1.1.15 Conditions (1.1.29) for $\gamma = 0$ and (1.1.34) for $\gamma > 0$ are due to von Mises. Sometimes a condition, much less general in the case $\gamma = 0$, is referred to as von Mises' condition (cf., e.g., Falk, Hüsler, and Reiss (1994), Theorem 2.1.2).

1.2 Domains of Attraction

In this section we shall establish necessary and sufficient conditions for a distribution function F to belong to the domain of attraction of G_γ. Also we shall prove that the sufficient conditions from Section 1.1.5 are close in some sense to the necessary conditions.

The reader may have noticed that in the previous sections we were able to prove the results in a relatively elegant way, by reformulating the problem in terms of the function U, the inverse function of $1/(1 - F)$. In fact, as we shall see in this section, the function U plays a role in extreme value theory comparable to the role of the characteristic function in the theory of the stable distributions and their domains of attraction. We determine the domain of attraction conditions by starting from condition (1.1.8) of Theorem 1.1.2 with $D(x) = (x^\gamma - 1)/\gamma$ (cf. Theorems 1.1.3 and 1.1.6). That is,

$$\lim_{t \to \infty} \frac{U(tx) - U(t)}{a(t)} = \frac{x^\gamma - 1}{\gamma} \tag{1.2.1}$$

for all $x > 0$, where γ is a real parameter and a a suitable positive function.

We prove the following results.

Theorem 1.2.1 *The distribution function F is in the domain of attraction of the extreme value distribution G_γ if and only if*

1. for $\gamma > 0$: $x^* = \sup\{x : F(x) < 1\}$ is infinite and

$$\lim_{t \to \infty} \frac{1 - F(tx)}{1 - F(t)} = x^{-1/\gamma} \tag{1.2.2}$$

for all $x > 0$. This means that the function $1 - F$ is regularly varying at infinity with index $-1/\gamma$, see Appendix B;
2. for $\gamma < 0$: x^* is finite and

$$\lim_{t \downarrow 0} \frac{1 - F(x^* - tx)}{1 - F(x^* - t)} = x^{-1/\gamma} \tag{1.2.3}$$

for all $x > 0$;
3. for $\gamma = 0$: x^* can be finite or infinite and

$$\lim_{t \uparrow x^*} \frac{1 - F(t + xf(t))}{1 - F(t)} = e^{-x} \tag{1.2.4}$$

for all real x, where f is a suitable positive function. If (1.2.4) holds for some f, then $\int_t^{x^*} (1 - F(s))ds < \infty$ for $t < x^*$ and (1.2.4) holds with

$$f(t) := \frac{\int_t^{x^*} (1 - F(s))ds}{1 - F(t)}. \tag{1.2.5}$$

Theorem 1.2.2 *The distribution function F is in the domain of attraction of the extreme value distribution G_γ if and only if*

1. for $\gamma > 0$: $F(x) < 1$ for all x, $\int_1^\infty (1 - F(x))/x \, dx < \infty$, and

$$\lim_{t \to \infty} \frac{\int_t^\infty (1 - F(x)) \frac{dx}{x}}{1 - F(t)} = \gamma \; ; \tag{1.2.6}$$

2. for $\gamma < 0$: there is $x^* < \infty$ such that $\int_{x^*-t}^{x^*} (1 - F(x))/(x^* - x) \, dx < \infty$ and

$$\lim_{t \downarrow 0} \frac{\int_{x^*-t}^{x^*} (1 - F(x)) \frac{dx}{x^*-x}}{1 - F(x^* - t)} = -\gamma \; ; \tag{1.2.7}$$

3. for $\gamma = 0$ (here the right endpoint x^* may be finite or infinite): $\int_x^{x^*} \int_t^{x^*} (1 - F(s)) ds \, dt < \infty$ and the function h defined by

$$h(x) := \frac{(1 - F(x)) \int_x^{x^*} \int_t^{x^*} (1 - F(s)) \, ds \, dt}{\left(\int_x^{x^*} (1 - F(s)) \, ds \right)^2} \tag{1.2.8}$$

satisfies

$$\lim_{t \uparrow x^*} h(t) = 1 \; . \tag{1.2.9}$$

Remark 1.2.3 Limit (1.2.6) is equivalent to

$$\lim_{t \to \infty} E \left(\log X - \log t \,|\, X > t \right) = \gamma \; .$$

In fact,

$$\frac{\int_t^\infty 1 - F(x) \frac{dx}{x}}{1 - F(t)} = E \left(\log X - \log t \,|\, X > t \right),$$

since

$$\int_t^\infty \log x - \log t \, dF(x) = \int_t^\infty 1 - F(x) \frac{dx}{x} \; .$$

Relation (1.2.6) will be the basis for the construction of the Hill estimator of γ (cf. Section 3.2). Similarly, (1.2.7) can be interpreted as

$$\lim_{t \downarrow 0} E \left(\log(x^* - X) - \log t \,|\, X > x^* - t \right) = \gamma,$$

which will be the basis for the construction of the negative Hill estimator (Section 3.6.2), and (1.2.9) is equivalent to

$$\lim_{t \uparrow x^*} \frac{E \left((X - t)^2 | X > t \right)}{E^2 (X - t | X > t)} = \lim_{t \uparrow x^*} 2 \, h(t) = 2 \; ,$$

and this relation leads to the moment estimator of γ (Section 3.5).

Next we show how to find the normalizing constants $a_n > 0$ and b_n in the basic limit relation (1.1.1).

Corollary 1.2.4 *If F is in the domain of attraction of G_γ, then*

1. for $\gamma > 0$:
$$\lim_{n\to\infty} F^n(a_n x) = \exp\left(-x^{-1/\gamma}\right)$$

holds for $x > 0$ with
$$a_n := U(n) \; ;$$

2. for $\gamma < 0$:
$$\lim_{n\to\infty} F^n(a_n x + x^*) = \exp\left(-(-x)^{-1/\gamma}\right)$$

holds for $x < 0$ with
$$a_n := x^* - U(n) \; ;$$

3. for $\gamma = 0$:
$$\lim_{n\to\infty} F^n(a_n x + b_n) = \exp\left(-e^{-x}\right)$$

holds for all x with

$$a_n := f(U(n)) \, ,$$
$$b_n := U(n),$$

and f as in Theorem 1.2.1(3).

We reformulate Theorem 1.2.1 in a seemingly more uniform way.

Theorem 1.2.5 *The distribution function F is in the domain of attraction of the extreme value distribution G_γ if and only if for some positive function f,*

$$\lim_{t\uparrow x^*} \frac{1 - F(t + xf(t))}{1 - F(t)} = (1 + \gamma x)^{-1/\gamma} \tag{1.2.10}$$

for all x with $1 + \gamma x > 0$. If (1.2.10) holds for some $f > 0$, then it also holds with

$$f(t) = \begin{cases} \gamma t \, , & \gamma > 0, \\ -\gamma(x^* - t) \, , & \gamma < 0, \\ \int_t^{x^*} 1 - F(x)dx/(1 - F(t)) \, , & \gamma = 0 \, . \end{cases}$$

Furthermore, any f for which (1.2.10) holds satisfies

$$\begin{cases} \lim_{t\to\infty} f(t)/t = \gamma \, , & \gamma > 0, \\ \lim_{t\uparrow x^*} f(t)/(x^* - t) = -\gamma \, , & \gamma < 0, \\ f(t) \sim f_1(t) \text{ where } f_1(t) \text{ is some function} & \\ \text{for which } f_1'(t) \to 0 \, , \; t \uparrow x^* \, , & \gamma = 0 \, . \end{cases} \tag{1.2.11}$$

A useful representation is given next.

Theorem 1.2.6 *The distribution function F is in $\mathcal{D}(G_\gamma)$ if and only if there exist positive functions c and f, f continuous, such that for all $t \in (t_0, x^*)$, $t_0 < x^*$,*

$$1 - F(t) = c(t) \exp\left\{-\int_{t_0}^{t} \frac{ds}{f(s)}\right\}$$

with $\lim_{t\uparrow x^*} c(t) = c \in (0, \infty)$ and

$$\begin{cases} \lim_{t\to\infty} f(t)/t = \gamma , & \gamma > 0, \\ \lim_{t\uparrow x^*} f(t)/(x^* - t) = -\gamma , & \gamma < 0, \\ \lim_{t\uparrow x^*} f'(t) = 0 \text{ and } \lim_{t\uparrow x^*} f(t) = 0 \text{ if } x^* < \infty , \gamma = 0 . \end{cases}$$

Remark 1.2.7 The auxiliary functions f in Theorems 1.2.5 and 1.2.6 are asymptotically the same. If von Mises' condition is satisfied for $\gamma = 0$, then we can take $f(t) = (1 - F(t))/F'(t)$.

Remark 1.2.8 Note that $F_0(t) := \max(0, 1 - c \exp(-\int_{t_0}^{t} f^{-1}(s)\, ds))$ is a probability distribution function and that

$$1 - F(t) \sim 1 - F_0(t) , \qquad t \uparrow t^* . \tag{1.2.12}$$

It follows that for any $F \in \mathcal{D}(G_\gamma)$ there exists a distribution function F_0 with (1.2.12) such that F_0 satisfies von Mises' condition of Theorem 1.1.8 (for $\gamma = 0$), Theorem 1.1.11 (for $\gamma > 0$), or Theorem 1.1.13 (for $\gamma < 0$).

In order to prove the results of Theorems 1.2.1–1.2.6 we are going to study relation (1.2.1) for the inverse function U first.

Lemma 1.2.9 *Suppose* (1.2.1) *holds.*

1. If $\gamma > 0$, then $\lim_{t\to\infty} U(t) = \infty$ and

$$\lim_{t\to\infty} \frac{U(t)}{a(t)} = \frac{1}{\gamma} . \tag{1.2.13}$$

2. If $\gamma < 0$, then $\lim_{t\to\infty} U(t) < \infty$ and, with $U(\infty) := \lim_{t\to\infty} U(t)$,

$$\lim_{t\to\infty} \frac{U(\infty) - U(t)}{a(t)} = -\frac{1}{\gamma} . \tag{1.2.14}$$

In particular this implies that $\lim_{t\to\infty} a(t) = 0$.

3. If $\gamma = 0$, then

$$\lim_{t\to\infty} \frac{U(tx)}{U(t)} = 1 \tag{1.2.15}$$

for all $x > 0$ and $\lim_{t\to\infty} a(t)/U(t) = 0$. Moreover, if $U(\infty) < \infty$,

$$\lim_{t\to\infty} \frac{U(\infty) - U(tx)}{U(\infty) - U(t)} = 1 \tag{1.2.16}$$

for $x > 0$ and $\lim_{t\to\infty} a(t)/(U(\infty) - U(t)) = 0$. Further,

$$\lim_{t\to\infty} \frac{a(tx)}{a(t)} = 1 \tag{1.2.17}$$

for $x > 0$.

Corollary 1.2.10 1. For $\gamma > 0$ relation (1.2.1) is equivalent to

$$\lim_{t\to\infty} \frac{U(tx)}{U(t)} = x^\gamma \quad for \quad x > 0. \tag{1.2.18}$$

2. For $\gamma < 0$ relation (1.2.1) is equivalent to $U(\infty) < \infty$ and

$$\lim_{t\to\infty} \frac{U(\infty) - U(tx)}{U(\infty) - U(t)} = x^\gamma \quad for \quad x > 0. \tag{1.2.19}$$

Remark 1.2.11 Relation (1.2.18) means that the function U is regularly varying at infinity with index γ (see Appendix B); similarly for the relations (1.2.15)–(1.2.17) and (1.2.19). In (1.2.15)–(1.2.17) the index of regular variation is zero. In that case we say that the function is slowly varying (at infinity).

Lemma 1.2.12 Let F_1 and F_2 be two distribution functions with common upper endpoint x^*. Let F_1 be in the domain of attraction of G_γ, $\gamma \in \mathbb{R}$, that is,

$$\lim_{t\to\infty} \frac{U_1(tx) - U_1(t)}{a(t)} = \frac{x^\gamma - 1}{\gamma}, \quad x > 0, \tag{1.2.20}$$

where a is a suitable positive function and $U_i := (1/(1 - F_i))^\leftarrow$, $i = 1, 2$. The following statements are equivalent:

1.

$$\lim_{t\uparrow x^*} \frac{1 - F_2(t)}{1 - F_1(t)} = 1.$$

2.

$$\lim_{t\to\infty} \frac{U_2(t) - U_1(t)}{a(t)} = 0.$$

Moreover, each statement implies that F_2 is in the domain of attraction of G_γ.

Proof. Assume statement (1). Take $\varepsilon > 0$. The inequalities (valid for sufficiently large t):

$$(1 - \varepsilon)(1 - F_1(t)) \le 1 - F_2(t) \le (1 + \varepsilon)(1 - F_1(t))$$

are equivalent to the following inequalities for the inverses (for sufficiently large s):

$$U_1((1 - \varepsilon)s) \le U_2(s) \le U_1((1 + \varepsilon)s) .$$

Hence

$$\frac{U_1\left((1-\varepsilon)s\right)-U_1(s)}{a(s)} \le \frac{U_2(s)-U_1(s)}{a(s)} \le \frac{U_1\left((1+\varepsilon)s\right)-U_1(s)}{a(s)}.$$

The left- and right-hand sides converge respectively to $((1-\varepsilon)^\gamma-1)/\gamma$ and $((1+\varepsilon)^\gamma-1)/\gamma$. Hence statement (2) has been proved. The converse is similar. ∎

Proof (of Lemma 1.2.9 and Corollary 1.2.10). We prove the assertions for $\gamma > 0$. The proof of the other assertions is similar.

It is easy to see that if (1.2.1) and (1.2.13) hold, (1.2.18) is true and that (1.2.18) implies the other two.

Note that (1.2.1) implies for $x > 0$,

$$\lim_{t\to\infty}\frac{a(tx)}{a(t)}$$
$$= \lim_{t\to\infty}\left(\frac{U(txy)-U(t)}{a(t)}-\frac{U(tx)-U(t)}{a(t)}\right)\bigg/\left(\frac{U(txy)-U(tx)}{a(tx)}\right)$$
$$= x^\gamma.$$

Hence for $Z > 1$,

$$\lim_{k\to\infty}\frac{U(Z^{k+1})-U(Z^k)}{U(Z^k)-U(Z^{k-1})}$$
$$= \lim_{k\to\infty}\left(\frac{U(Z^{k+1})-U(Z^k)}{a(Z^k)}\right)\bigg/\left(\frac{U(Z^k)-U(Z^{k-1})}{Z^\gamma a(Z^{k-1})}\right)$$
$$= Z^\gamma,$$

that is, for $0 < \varepsilon < 1 - Z^{-\gamma}$, $k \ge n_0(\varepsilon)$,

$$\left(U(Z^k)-U(Z^{k-1})\right)Z^\gamma(1-\varepsilon) \le U(Z^{k+1})-U(Z^k)$$
$$\le \left(U(Z^k)-U(Z^{k-1})\right)Z^\gamma(1+\varepsilon). \tag{1.2.21}$$

From this, we have

$$\lim_{N\to\infty}U(Z^{N+1})-U(Z^{n_0})$$
$$= \lim_{N\to\infty}\sum_{n=n_0}^N\left(U(Z^{n_0})-U(Z^{n_0-1})\right)\prod_{k=n_0}^n\frac{U(Z^{k+1})-U(Z^k)}{U(Z^k)-U(Z^{k-1})}$$
$$\ge \lim_{N\to\infty}\sum_{n=n_0}^N\left(U(Z^{n_0})-U(Z^{n_0-1})\right)\prod_{k=n_0}^n Z^\gamma(1-\varepsilon)$$
$$= \left(U(Z^{n_0})-U(Z^{n_0-1})\right)\sum_{n=n_0}^\infty\left((1-\varepsilon)Z^\gamma\right)^{n-n_0+1} = \infty,$$

since $Z > 1$ and by assumption $\gamma > 0$. Hence $U(t) \to \infty, t \to \infty$.

In order to prove (1.2.13) add inequalities (1.2.21) for $k = n_0, \ldots, n$. Divide the result by $U(Z^n)$ and take the limit as $n \to \infty$. This gives

$$\lim_{n \to \infty} \frac{U(Z^{n+1})}{U(Z^n)} = Z^\gamma \tag{1.2.22}$$

for all $Z > 1$.

Next, for each $x > 1$, define $n(x) \in \mathbb{N}$ such that

$$Z^{n(x)} \leq x < Z^{n(x)+1} . \tag{1.2.23}$$

We then have for $t, x > 1$,

$$\frac{U(Z^{n(t)} Z^{n(x)})}{U(Z^{n(t)+1})} \leq \frac{U(tx)}{U(t)} \leq \frac{U(Z^{n(t)+1} Z^{n(x)+1})}{U(Z^{n(t)})} . \tag{1.2.24}$$

Now write the left-hand side as

$$\frac{U(Z^{n(t)} Z^{n(x)})}{U(Z^{n(t)})} \frac{U(Z^{n(t)})}{U(Z^{n(t)+1})} .$$

By (1.2.22) this converges to $Z^{\gamma n(x)} Z^{-\gamma}$ as $t \to \infty$, which by (1.2.23) is at least $(x/Z^2)^\gamma$. Similarly, the limsup of the right-hand side of (1.2.24) is at most $(xZ^2)^\gamma$. Now let $Z \downarrow 1$. Then we get (1.2.18) and hence (via (1.2.1)) relation (1.2.13). ∎

Proof (of Theorem 1.2.1 for $\gamma \neq 0$). We prove the theorem for $\gamma > 0$. The proof for $\gamma < 0$ is similar. From the definition of the inverse function $U(x)$ one sees that for any $\varepsilon > 0$,

$$U\left(\frac{1-\varepsilon}{1-F(t)}\right) \leq t \leq U\left(\frac{1+\varepsilon}{1-F(t)}\right) .$$

Hence

$$\frac{U\left(\frac{x}{1-F(t)}\right)}{U\left(\frac{1+\varepsilon}{1-F(t)}\right)} \leq t^{-1} U\left(\frac{x}{1-F(t)}\right) \leq \frac{U\left(\frac{x}{1-F(t)}\right)}{U\left(\frac{1-\varepsilon}{1-F(t)}\right)} . \tag{1.2.25}$$

Suppose (1.2.1) holds, i.e., we have (1.2.18). Then the right- and left-hand sides of (1.2.25) converge to $(x/(1+\varepsilon))^\gamma$ and $(x/(1-\varepsilon))^\gamma$ respectively. Hence, since the relation holds for all $\varepsilon > 0$, it implies

$$\lim_{t \to \infty} t^{-1} U\left(\frac{x}{1-F(t)}\right) = x^\gamma. \tag{1.2.26}$$

Next we apply Lemma 1.1.1 and get from (1.2.26),

$$\lim_{t \to \infty} \frac{1-F(t)}{1-F(tx)} = x^{1/\gamma},$$

i.e., (1.2.2). The proof of the converse implication is similar and is left to the reader. ∎

Proof (of Theorem 1.2.2 for $\gamma \neq 0$). We prove the theorem for $\gamma > 0$. The proof for $\gamma < 0$ is similar. First note that by (1.2.2) for any $\varepsilon > 0$ and sufficiently large t,

$$\frac{1 - F(te)}{1 - F(t)} \leq e^{\varepsilon - 1/\gamma} \ .$$

Hence

$$\frac{1 - F(te^n)}{1 - F(t)} = \prod_{k=1}^{n} \frac{1 - F(te^k)}{1 - F(te^{k-1})} \leq e^{(\varepsilon - 1/\gamma)n}$$

and for all $x > 1$,

$$\frac{1 - F(tx)}{1 - F(t)} \leq \frac{1 - F(te^{[\log x]})}{1 - F(t)} \leq e^{(\varepsilon - 1/\gamma)[\log x]}$$
$$\leq e^{(\varepsilon - 1/\gamma)(1 + \log x)} = e^{-1/\gamma + \varepsilon} x^{-1/\gamma + \varepsilon} \ .$$

The dominated convergence theorem now gives, in combination with (1.2.2),

$$\lim_{t \to \infty} \int_1^\infty \frac{1 - F(tx)}{1 - F(t)} \frac{dx}{x} = \int_1^\infty x^{-1/\gamma} \frac{dx}{x} = \gamma,$$

which is (1.2.6). In particular, we know that $\int_1^\infty (1 - F(x))/x \, dx < \infty$.
   For the converse statement assume that

$$\lim_{t \to \infty} a(t) = 1/\gamma$$

with

$$a(t) := \frac{1 - F(t)}{\int_t^\infty (1 - F(x))/x \, dx} \ .$$

Note that

$$-\log \int_t^\infty (1 - F(x)) \frac{dx}{x} + \log \int_1^\infty (1 - F(x)) \frac{dx}{x} = \int_1^t a(x) \frac{dx}{x} \ .$$

Hence, using the definition of the function a again, we have

$$1 - F(t) = a(t) \int_t^\infty (1 - F(x)) \frac{dx}{x}$$
$$= a(t) \int_1^\infty (1 - F(x)) \frac{dx}{x} \exp\left(-\int_1^t a(x) \frac{dx}{x} \right) \qquad (1.2.27)$$

and for $x > 0$, as $t \to \infty$,

$$\frac{1 - F(tx)}{1 - F(t)} = \frac{a(tx)}{a(t)} \exp\left(-\int_1^x a(ty) \frac{dy}{y} \right) \to \exp\left(-\frac{1}{\gamma} \int_1^x \frac{dy}{y} \right) = x^{-1/\gamma} \ .$$

∎

Remark 1.2.13 Representation (1.2.27) is sometimes useful. It expresses F in terms of the function a, which is of simpler structure.

Proof (of Corollary 1.2.4). For $\gamma > 0$ by Theorem 1.1.6 and Lemma 1.2.9, for $x > 0$,

$$\lim_{n \to \infty} \frac{U(nx)}{U(n)} = x^\gamma .$$

Then by Lemma 1.1.1,

$$\lim_{n \to \infty} n\{1 - F(xU(n))\} = x^{1/\gamma},$$

and hence by Theorem 1.1.2,

$$\lim_{n \to \infty} F^n(xU(n)) = \exp\left(-x^{1/\gamma}\right) .$$

The other cases are similar. ∎

Proof (of Theorem 1.2.5). Since U is the left-continuous inverse of $1/(1 - F)$, for $\varepsilon > 0$,

$$\frac{1 - F(U(t) + \varepsilon f(U(t)))}{1 - F(U(t))} \le \frac{1}{t\{1 - F(U(t))\}} \le \frac{1 - F(U(t) - \varepsilon f(U(t)))}{1 - F(U(t))} .$$

Since by (1.2.10) the left- and right-hand sides converge to $(1 \pm \gamma\varepsilon)^{-1/\gamma}$, it follows that

$$\lim_{t \to \infty} t\{1 - F(U(t))\} = 1. \tag{1.2.28}$$

With (1.2.10) this gives

$$\lim_{t \to \infty} t\{1 - F(U(t) + xf(U(t)))\} = (1 + \gamma x)^{-1/\gamma} .$$

Now, Theorem 1.1.6 tells us that $F \in \mathcal{D}(G_\gamma)$. This proves the first part of the theorem. The second statement just rephrases Theorem 1.2.1. Relation (1.2.11) for $\gamma \ne 0$ follows from the easily established fact that if (1.2.10) holds for $f = f_1$ and $f = f_2$, then $\lim_{t \uparrow x^*} f_1(t)/f_2(t) = 1$. For the case $\gamma = 0$ use Theorem 1.2.6. ∎

Proof (of Theorem 1.2.6 for $\gamma \ne 0$). For the "if" part just check directly that (1.2.2) or (1.2.3) of Theorem 1.2.1 is satisfied. Next suppose $F \in \mathcal{D}(G_\gamma)$ for $\gamma > 0$. Write

$$a(t) := \frac{1 - F(t)}{\int_t^\infty (1 - F(x))/x \, dx} .$$

Note that

$$\left(-\log \int_t^\infty (1 - F(x)) \frac{dx}{x}\right)' = \frac{a(t)}{t} ,$$

hence for $t > t_0$,

$$1 - F(t) = \frac{1 - F(t_0)}{a(t_0)} a(t) \exp\left(-\int_{t_0}^t a(s) \frac{ds}{s}\right) .$$

Theorem 1.2.2 states $\lim_{t \to \infty} a(t) = 1/\gamma$, hence the representation. The proof for $\gamma < 0$ is similar. ∎

For the proof of Theorems 1.2.1, 1.2.2, and 1.2.6 with $\gamma = 0$ we need some additional lemmas.

Lemma 1.2.14 *Suppose that for all $x > 0$, (1.2.1) holds with $\gamma = 0$, i.e.,*

$$\lim_{t \to \infty} \frac{U(tx) - U(t)}{a(t)} = \log x ,$$

where a is a suitable positive function. Then for all $\varepsilon > 0$ there exist $c > 0$, $t_0 > 1$ such that for $x \geq 1$, $t \geq t_0$,

$$\frac{U(tx) - U(t)}{a(t)} \leq cx^{\varepsilon} .$$

Proof. For all $Z > 1$, there exists $t_0 > 1$ such that for $t \geq t_0$,

$$\frac{U(te) - U(t)}{a(t)} \leq Z \quad \text{and} \quad \frac{a(te)}{a(t)} \leq Z$$

(use (1.2.17) for the last inequality). For $n = 1, 2, \ldots$ and $t \geq t_0$,

$$\frac{U(te^n) - U(t)}{a(t)} = \sum_{k=1}^{n} \frac{U(te^k) - U(te^{k-1})}{a(te^{k-1})} \prod_{r=1}^{k-1} \frac{a(te^r)}{a(te^{r-1})}$$

$$\leq \sum_{k=1}^{n} Z \prod_{r=1}^{k-1} Z \leq nZ^n .$$

Hence for $x > 1$ and $1 < Z < e$,

$$\frac{U(tx) - U(t)}{a(t)} \leq \frac{U\left(te^{[\log x]+1}\right) - U(t)}{a(t)}$$

$$\leq ([\log x] + 1) Z^{[\log x]+1} \leq (\log x + 2) Z^{\log x + 2}$$

$$\leq \frac{2Z^2}{\log Z} x^{2 \log Z}$$

(use for the last inequality $a + 2 \leq 2(\log Z)^{-1} Z^a$ for any $a > 0$ and $1 < Z < e$). ∎

Corollary 1.2.15 *If (1.2.1) holds for $\gamma = 0$, then $\int_1^{\infty} U(s)/s^2 \, ds < \infty$ and*

$$\lim_{t \to \infty} \frac{U_0(t) - U(t)}{a(t)} = 0 \tag{1.2.29}$$

with

$$U_0(t) := \frac{t}{e} \int_{t/e}^{\infty} U(s) \frac{ds}{s^2} = \int_1^{\infty} U(st/e) \frac{ds}{s^2} .$$

Note that U_0 is continuous and strictly increasing.

Proof.

$$\frac{U_0(t) - U(t)}{a(t)} = \int_1^\infty \frac{U(st/e) - U(t)}{a(t)} \frac{ds}{s^2} .$$

We can now apply Lebesgue's theorem on dominated convergence: (1.2.1) gives the pointwise convergence and Lemma 1.2.14 the uniform bound. In particular we have $\int_1^\infty U(s)/s^2 \, ds < \infty$. ∎

Corollary 1.2.16 *If $F \in \mathcal{D}(G_0)$, there exists a distribution function F_0, continuous and strictly increasing, such that*

$$\lim_{t \uparrow x^*} \frac{1 - F_0(t)}{1 - F(t)} = 1$$

and (1.2.29) holds.

Proof. Apply Corollary 1.2.15 with $U = (1/(1 - F))^{\leftarrow}$. Next apply Lemma 1.2.12. ∎

Proof (of Theorem 1.2.1 for $\gamma = 0$). We have proved already in Theorem 1.1.6 that (1.2.4) is necessary and sufficient for the domain of attraction. Since the function F is monotone and e^{-x} is continuous, relation (1.2.4) holds locally uniformly. Hence, in order to prove that (1.2.4) holds with the function f from (1.2.5), it is sufficient to prove that if (1.2.4) holds, then

$$f(t) \sim \frac{\int_t^{x^*} (1 - F(s)) \, ds}{1 - F(t)} , \quad t \uparrow x^*. \tag{1.2.30}$$

Take U_0 and F_0 from Corollaries 1.2.15 and 1.2.16. Note that (1.2.4) holds with F replaced by F_0, i.e.,

$$\lim_{t \uparrow x^*} \frac{1 - F_0(t + xf(t))}{1 - F_0(t)} = e^{-x} .$$

Since also (use l'Hôpital's rule)

$$\frac{\int_t^{x^*} (1 - F_0(s)) \, ds}{1 - F_0(t)} \sim \frac{\int_t^{x^*} (1 - F(s)) \, ds}{1 - F(t)}, \quad t \uparrow x^* ,$$

it is sufficient to prove the statement for F_0 rather than for F.

By dominated convergence, using the inequality from Lemma 1.2.14, we have

$$\lim_{z \to \infty} \frac{z \int_z^\infty U_0(s) \frac{ds}{s^2} - U_0(z)}{a(z)} = \lim_{z \to \infty} \int_1^\infty \frac{U_0(zx) - U_0(z)}{a(z)} \frac{dx}{x^2} = 1 .$$

If we substitute s by $1/(1 - F_0(u))$, i.e., $u = U_0(s)$ (since F_0 is continuous and strictly increasing!) and z by $1/(1 - F_0(t))$ in the left-hand side, we get

$$\lim_{t \uparrow x^*} \frac{\int_t^{x^*} (1 - F_0(u)) \, du}{a(1/(1 - F_0(t)))(1 - F_0(t))} = 1 .$$

Relation (1.2.30) follows by Theorem 1.1.6, last part, and the proof of Theorem 1.2.1 is complete. ∎

Proof (of Theorem 1.2.2 for $\gamma = 0$). Clearly relation (1.2.4) implies

$$\lim_{t \uparrow x^*} t + xf(t) = x^* \qquad \text{for all } x .$$

Now we replace the running variable t in (1.2.4) by $t' + yf(t')$ $(t' \uparrow x^*)$ and get

$$\lim_{t' \uparrow x^*} \frac{1 - F\left((t' + yf(t')) + xf(t' + yf(t'))\right)}{1 - F(t' + yf(t'))} \lim_{t' \uparrow x^*} \frac{1 - F\left(t' + yf(t')\right)}{1 - F(t')} = e^{-x}e^{-y},$$

that is,

$$\lim_{t' \uparrow x^*} \frac{1 - F\left(t' + f(t')\left\{y + x \frac{f(t' + yf(t'))}{f(t')}\right\}\right)}{1 - F(t')} = e^{-x}e^{-y} . \tag{1.2.31}$$

Now by (1.2.4), also

$$\lim_{t' \uparrow x^*} \frac{1 - F\left(t' + (x + y)f(t')\right)}{1 - F(t')} = e^{-x}e^{-y} . \tag{1.2.32}$$

Keep in mind that the convergence in (1.2.4) is locally uniform. It then follows from (1.2.31) and (1.2.32) that

$$\lim_{t' \uparrow x^*} \frac{f\left(t' + yf(t')\right)}{f(t')} = 1 \tag{1.2.33}$$

for all y (formally this is proved by contradiction: suppose that for some sequence $t'_n \uparrow x^*$ the limit in (1.2.33) equals $c \in [0, \infty]$, $c \neq 1$; then (1.2.31) cannot be true). This holds in particular for the function f from (1.2.5), i.e.,

$$\lim_{t \uparrow x^*} \frac{\int_{t + xf(t)}^{x^*}(1 - F(s))\, ds}{1 - F(t + xf(t))} \frac{1 - F(t)}{\int_t^{x^*}(1 - F(s))\, ds} = 1$$

for all x, which in combination with (1.2.4) gives

$$\lim_{t \uparrow x^*} \frac{\int_{t + xf(t)}^{x^*}(1 - F(s))\, ds}{\int_t^{x^*}(1 - F(s))\, ds} = e^{-x} \tag{1.2.34}$$

for all x. Now define the distribution function F_1 by

$$F_1(x) := \max\left(0, 1 - \int_x^{x^*}(1 - F(s))\, ds\right).$$

Relation (1.2.34), i.e.,

$$\lim_{t \uparrow x^*} \frac{1 - F_1(t + xf(t))}{1 - F_1(t)} = e^{-x} , \tag{1.2.35}$$

tells us by (1.2.4) that F_1 is in the domain of attraction of G_0. But then again by (1.2.4) and (1.2.5) we must have

$$\lim_{t \uparrow x^*} \frac{1 - F_1(t + x f_1(t))}{1 - F_1(t)} = e^{-x} \tag{1.2.36}$$

with

$$f_1(t) := \frac{\int_t^{x^*} (1 - F_1(s)) \, ds}{1 - F_1(t)}. \tag{1.2.37}$$

Since the convergence in (1.2.35) and (1.2.36) is locally uniform, the functions f and f_1 must be asymptotically equivalent:

$$f_1(t) \sim f(t) \qquad \text{as} \quad t \uparrow x^*. \tag{1.2.38}$$

This relation is the same as (1.2.9) of Theorem 1.2.2.

Conversely, suppose (1.2.9) holds. Note that for t large enough,

$$\frac{d}{dt} \{- \log(1 - F_2(t))\} = \frac{2h(t) - 1}{f_1(t)} > 0, \tag{1.2.39}$$

where

$$1 - F_2(t) := \frac{(1 - F_1(t))^2}{\int_t^{x^*} (1 - F_1(s)) \, ds}.$$

Moreover, by (1.2.9),

$$1 - F_2(t) \sim 1 - F(t) \qquad \text{as} \quad t \uparrow x^*. \tag{1.2.40}$$

Since F_2 is eventually monotone by (1.2.39), and $1 - F_2(t) \to 0$, $t \uparrow x^*$, by (1.2.40), there is a distribution function, $F_2^\star$ say, such that $F_2(t) = F_2^\star(t)$ for large t. So by Lemma 1.2.12 it is sufficient to prove that $F_2^\star$ is in the domain of attraction of G_0. Note that for $t > t_0$,

$$f_1(t) - f_1(t_0) = \int_{t_0}^t (h(s) - 1) \, ds \, ;$$

hence for all x,

$$\lim_{t \uparrow x^*} \frac{f_1(t + x f_1(t)) - f_1(t)}{f_1(t)} = \lim_{t \uparrow x^*} \int_0^x (h(t + u f_1(t)) - 1) \, du = 0$$

(by (1.2.9)) or

$$\lim_{t \uparrow x^*} \frac{f_1(t + x f_1(t))}{f_1(t)} = 1 \tag{1.2.41}$$

uniformly on bounded x intervals. Now by (1.2.39) for large t,

$$\frac{1 - F_2^\star(t + x f_1(t))}{1 - F_2^\star(t)} = \exp\left(-\int_0^x (2h(t + u f_1(t)) - 1) \frac{f_1(t)}{f_1(t + u f_1(t))} du\right).$$

Hence by (1.2.9) and (1.2.41),

$$\lim_{t \uparrow x^*} \frac{1 - F_2^\star(t + x f_1(t))}{1 - F_2^\star(t)} = e^{-x}. \tag{1.2.42}$$

∎

Proof (of Theorem 1.2.6 for $\gamma = 0$). Suppose $F \in \mathcal{D}(G_0)$. Define for $n = 1, 2, \ldots$ recursively

$$F_n(t) := \max \left(0, 1 - \int_t^{x^*} (1 - F_{n-1}(s)) \, ds \right) \tag{1.2.43}$$

and $F_0(t) := F(t)$. The integrals are finite: the arguments in the previous proof show that $F_n \in \mathcal{D}(G_0)$ for all n. Moreover, we have for all n, as $t \uparrow x^*$,

$$q_n(t) := 1 + \frac{d}{dt} \frac{1 - F_{n+1}(t)}{1 - F_n(t)} = \frac{(1 - F_{n-1}(t))(1 - F_{n+1}(t))}{(1 - F_n(t))^2} \to 1. \tag{1.2.44}$$

Write

$$Q_n(t) := \frac{1 - F_{n+1}(t)}{1 - F_n(t)}.$$

Now note that

$$1 - F(t) = q_1(t) q_2^2(t) q_3^3(t) \frac{(1 - F_3(t))^4}{(1 - F_4(t))^3}. \tag{1.2.45}$$

Define

$$1 - F_\star(t) := \frac{(1 - F_3(t))^4}{(1 - F_4(t))^3}.$$

Note that by (1.2.43), for sufficiently large t,

$$\frac{d}{dt} \{-\log(1 - F_\star(t))\} = 4 \frac{1 - F_2(t)}{1 - F_3(t)} - 3 \frac{1 - F_3(t)}{1 - F_4(t)} =: \frac{1}{f_\star(t)}. \tag{1.2.46}$$

Hence

$$\frac{1}{f_\star(t)} = \frac{1 - F_3(t)}{1 - F_4(t)} (4 \, q_3(t) - 3) > 0$$

and

$$\begin{aligned}
f_\star'(t) &= \frac{d}{dt} \left(\frac{4}{Q_2(t)} - \frac{3}{Q_3(t)} \right)^{-1} \\
&= - \left(\frac{4}{Q_2(t)} - \frac{3}{Q_3(t)} \right)^{-2} \left(\frac{-4 Q_2'(t)}{(Q_2(t))^2} + \frac{3 Q_3'(t)}{(Q_3(t))^2} \right) \\
&= - \left(4 - 3 \frac{Q_2(t)}{Q_3(t)} \right)^{-2} \left(-4 Q_2'(t) + 3 Q_3'(t) \left(\frac{Q_2(t)}{Q_3(t)} \right)^2 \right).
\end{aligned}$$

By (1.2.44) we have $Q_2(t)/Q_3(t) \to 1$, $Q_n'(t) \to 0$, $t \uparrow x^*$. Hence $f_\star'(t) \to 0$, $t \uparrow x^*$.

Note that $F_\star$ is monotone by (1.2.46) for large t and that $\lim_{t \uparrow x^*} F_\star(t) = 1$ by (1.2.45); hence $F_\star$ coincides with a distribution function for large t. Finally, $1 - F_\star(t) \sim 1 - F(t)$, $t \uparrow x^*$, by (1.2.44) and (1.2.45). Note that for F we derived the representation with

$$c(t) = \frac{1 - F(t)}{1 - F_\star(t)}$$

and

$$f(s) = f_\star(s) \ .$$

To prove the converse, first note that

$$\begin{cases} \lim_{t \to \infty} \frac{f(t)}{t} = 0 \ , \ x^* = \infty, \\ \lim_{t \uparrow x^*} \frac{f(t)}{x^* - t} = 0 \ , \ x^* < \infty, \end{cases}$$

since if $x^* = \infty$,

$$\frac{f(t) - f(t_0)}{t} = \frac{1}{t} \int_{t_0}^{t} f'(s) ds,$$

which converges to zero by hypothesis, and if $x^* < \infty$,

$$\frac{f(t)}{x^* - t} = \frac{-\int_t^{x^*} f'(s) ds}{x^* - t},$$

which again converges to zero by hypothesis. Hence we have that

$$t + x f(t) < x^* \tag{1.2.47}$$

for sufficiently large t and all real x. Obviously $t + x f(t) \to x^*$, $t \uparrow x^*$. Next note that

$$\frac{f(t + x f(t)) - f(t)}{f(t)} = \frac{1}{f(t)} \int_t^{t + x f(t)} f'(s) ds = \int_0^x f'(t + s f(t)) ds,$$

which converges to zero locally uniformly as $t \uparrow x^*$. Hence

$$\lim_{t \uparrow x^*} \frac{f(t + x f(t))}{f(t)} = 1 \tag{1.2.48}$$

locally uniformly. Combining (1.2.47) and (1.2.48) we have

$$\frac{1 - F(t + x f(t))}{1 - F(t)} = \frac{c(t + x f(t))}{c(t)} \exp\left(-\int_0^x \frac{f(t)}{f(t + x f(t))} ds\right) \to e^{-x}$$

as $t \uparrow x^*$. The result follows from Theorem 1.2.5. ∎

Exercises

1.1. Let f be any nondecreasing function and $f^{\leftarrow}$ its right- or left-continuous inverse, respectively $f^{\leftarrow}(y) := \inf\{s : f(s) > y\}$ or $f^{\leftarrow}(y) := \inf\{s : f(s) \geq y\}$. Check that:

(a) $(f^{\leftarrow})^{\leftarrow} = f^-$ if $f^{\leftarrow}$ is the left-continuous inverse, with f^- the left-continuous version of f.

(b) $(f^{\leftarrow})^{\leftarrow} = f^+$ if $f^{\leftarrow}$ is the right-continuous inverse, with f^+ the right-continuous version of f.

(c) $f^-(f^{\leftarrow}(t)) \leq t \leq f^+(f^{\leftarrow}(t))$ whether $f^{\leftarrow}$ is the right- or left-continuous inverse.

1.2. Verify that $G_{\gamma}^n(a_n x + b_n) = G_{\gamma}(x) = \exp\left(-(1 + \gamma x)^{-1/\gamma}\right)$, with $1 + \gamma x > 0$, for $a_n = n^{\gamma}$ and $b_n = (n^{\gamma} - 1)/\gamma$, for all n.

1.3. Consider the generalized Pareto distribution $H_{\gamma}(x) = 1 - (1 + \gamma x)^{-1/\gamma}$, $0 < x < (0 \vee (-\gamma))^{-1}$ (read $1 - e^{-x}$ if $\gamma = 0$). Determine $a(t)$ such that (1.1.20) holds for all t and x, i.e., $(U(tx) - U(t))/a(t) = (x^{\gamma} - 1)/\gamma$.

1.4. For $F(x) = 1 - 1/x$, $x > 1$, determine sequences a_n positive and b_n real such that $F^n(a_n x + b_n) \to \exp(-1/x)$, for $x > 0$.

1.5. Verify that the distribution function of X is in the domain of attraction of G_{γ} with $\gamma < 0$ if and only if $\tilde{X} = 1/(x^* - X)$ is in the domain of attraction of $G_{-\gamma}$.

1.6. Check that the Cauchy distribution $F(x) = 2^{-1} + \pi^{-1} \arctan x$, $x \in \mathbb{R}$, is in the Fréchet domain of attraction with $\gamma = 1$.

1.7. Check that the following distributions are in the Gumbel domain of attraction:

(a) Exponential distribution: $F(x) = 1 - e^{-x}$, $x > 0$.

(b) Any gamma (ν, α) distribution for which $F'(x) = (\Gamma(\nu))^{-1} \alpha^{\nu} x^{\nu-1} e^{-\alpha x}$, $\nu > 0$, $\alpha > 0$, $x > 0$. *Hint*: use l'Hôpital's rule to verify that $\lim_{t \to \infty}(1 - F(t))/F'(t) = \alpha^{-1}$.

1.8. Check that the beta (μ, ν) distribution, for which

$$F'(x) = \Gamma(\mu + \nu)(\Gamma(\mu))^{-1}(\Gamma(\nu))^{-1}(1 - x)^{\mu-1} x^{\nu-1},$$

$\mu > 0$, $\nu > 0$, $0 < x < 1$, is in the Weibull domain of attraction with $\gamma = -\mu^{-1}$.

1.9. Check domain of attraction conditions for $F(x) = e^x$, $x < 0$, and $F(x) = 1 - e^{1/x}$, $x < 0$.

1.10. Show that $F \in \mathcal{D}(G_{\gamma})$, for some real γ, is equivalent to $\lim_{t \to \infty}(V(tx) - V(t))/a(t) = (x^{\gamma} - 1)/\gamma$, $x > 0$, for some positive function a, where $V := (1/(-\log F))^{\leftarrow}$.

1.11. Suppose $F_i \in \mathcal{D}(G_{\gamma_i})$ for $i = 1, 2$ and $\gamma_1 < \gamma_2$. Suppose also that the two distributions have the same right endpoint x^*. Show that $1 - F_1(x) = o(1 - F_2(x))$, as $x \uparrow x^*$.

1.12. Let $F_i \in \mathcal{D}(G_{\gamma_i})$, $i = 1, 2$. Show that for $0 < p < 1$ the mixture $pF_1 + (1 - p)F_2 \in \mathcal{D}(G_{\max(\gamma_1, \gamma_2)})$ if:

(a) $\gamma_1 \neq \gamma_2$,
(b) $\gamma_1 = \gamma_2 \neq 0$.

Can you say something about the case $\gamma_1 = \gamma_2 = 0$?

1.13. Show that if $F \in \mathcal{D}(G_\gamma)$, then (for any γ)

$$\lim_{t \uparrow x^*} \frac{\lim_{s \uparrow t} 1 - F(s)}{\lim_{s \downarrow t} 1 - F(s)} = 1 \, .$$

Conclude that the geometric distribution $F(x) = 1 - e^{-[x]}$, $x > 0$, and also the Poisson distribution are in no domain of attraction.

1.14. Find a discrete distribution in the domain of attraction of an extreme value distribution.

1.15. Let $X_1, X_2, \ldots$ be an i.i.d. sample with distribution function F. Show that if F is in the domain of attraction of G_γ with γ negative and $c \in (0, \infty)$, there exist constants $a_n > 0$ such that $\left(\max(X_1, \ldots, X_n) - F^\leftarrow(1 - (cn)^{-1})\right)/a_n$ converges in distribution to $Y - c^\gamma/\gamma$, where Y has distribution function of the type $\exp\left(-(-y)^{-1/\gamma}\right)$, $y < 0$ (i.e., Weibull type).

1.16. Prove that if $F \in \mathcal{D}(G_\gamma)$ and X is a random variable with distribution function F, then for all $-\infty < x < U(\infty)$ ($U(\infty)$ is the right endpoint of F),

$$E|X|^\alpha 1_{\{X > x\}} < \infty$$

if $0 < \alpha < 1/\gamma_+$ with $\gamma_+ := \max(0, \gamma)$ and $E|X|^\alpha 1_{\{X > x\}} = \infty$ if $\alpha > 1/\gamma_+$. Recall that $U(\infty) < \infty$ if $\gamma < 0$ and $U(\infty) = \infty$ if $\gamma > 0$.

1.17. Let $F(x) := P(X \leq x)$ be in $\mathcal{D}(G_\gamma)$ with $\gamma > 0$. Let A be a positive random variable with $EA^{1/\gamma + \varepsilon} < \infty$ for some $\varepsilon > 0$. Let A and X be independent. Show that

$$\lim_{t \to \infty} \frac{P(AX > x)}{P(X > x)} = EA^{1/\gamma},$$

so that $P(AX \leq x)$ is also in $\mathcal{D}(G_\gamma)$.
Hint: By Appendix B, for $x \geq t_0$, $a \leq x/t_0$, we have $(1 - \varepsilon)a^{1/\gamma - \varepsilon} \leq P(aX > x)/P(X > x) \leq (1 + \varepsilon)a^{1/\gamma + \varepsilon}$. Hence

$$(1 - \varepsilon) \int_0^{x/t_0} a^{1/\gamma - \varepsilon} dP(A \leq a) \leq \frac{P(AX > x, A \leq x/t_0)}{P(X > x)}$$

$$\leq (1 + \varepsilon) \int_0^{x/t_0} a^{1/\gamma + \varepsilon} dP(A \leq a) \, .$$

Take the limits $x \to \infty$ and then $\varepsilon \downarrow 0$. Further use $P(AX > x, A \geq x/t_0) \leq P(A \geq x/t_0)$.

1.18. Show that $F(x) = 1 - e^{-x-\sin x}$, $x > 0$, is not in any domain of attraction (R. von Mises).

Hints: Show that $\lim_{k \to \infty} n_k (1 - F(x + \log n_k)) = e^{-x-\sin x}$ for all $x > 0$ with $n_k = [e^{2\pi k}]$ for $k = 1, 2, \ldots$, i.e., $\lim_{n \to \infty} U(n_k x) - \log n_k = U_1(x)$, where $U := (1/(1 - F))^{\leftarrow}$ and U_1 is the inverse of $e^{x+\sin x}$. Now proceed by contradiction.

2

Extreme and Intermediate Order Statistics

2.1 Extreme Order Statistics and Poisson Point Processes

The extreme value condition

$$\lim_{t \uparrow x^*} \frac{1 - F(t + xf(t))}{1 - F(t)} = (1 + \gamma x)^{-1/\gamma} \tag{2.1.1}$$

for each x, with $1 + \gamma x > 0$, or equivalently

$$\lim_{t \to \infty} \frac{U(tx) - U(t)}{a(t)} = \frac{x^\gamma - 1}{\gamma} \tag{2.1.2}$$

for each $x > 0$, where γ is a real constant called the *extreme value index*, is designed to allow convergence in distribution of normalized sample maxima, as in (1.1.1). But the conditions also imply convergence of other high-order statistics.

Let us start to derive the result for the exponential distribution. Suppose $E_1, E_2, \ldots$ are independent and identically distributed standard exponential and $E_{1,n} \leq E_{2,n} \leq \cdots \leq E_{n,n}$ are the nth order statistics. By Rényi's (1953) representation we have for fixed $k \leq n$,

$$\begin{aligned}
&\left(E_{1,n}, E_{2,n}, \ldots, E_{k,n} \right) \\
&\stackrel{d}{=} \left(\frac{E_1^\star}{n}, \frac{E_1^\star}{n} + \frac{E_2^\star}{n-1}, \ldots, \frac{E_1^\star}{n} + \frac{E_2^\star}{n-1} + \cdots + \frac{E_k^\star}{n-k+1} \right)
\end{aligned}$$

with $E_1^\star, \ldots, E_k^\star$ independent and identically distributed standard exponential. Hence

$$n \left(E_{1,n}, E_{2,n}, \ldots, E_{k,n} \right) \stackrel{d}{\to} \left(E_1^\star, E_1^\star + E_2^\star, \ldots, E_1^\star + \cdots + E_k^\star \right) . \tag{2.1.3}$$

This suggests, and we shall show this later on, that the point process of normalized lower *extreme-order statistics* converges to a homogeneous Poisson process.

Next we generalize the result (2.1.3) to the entire domain of attraction, and as usual, we formulate it for upper order statistics rather than lower ones.

Theorem 2.1.1 *Let $X_1, X_2, \ldots$ be i.i.d. with distribution function F. Suppose F is in the domain of attraction of G_γ for some $\gamma \in \mathbb{R}$. Let $X_{1,n} \le X_{2,n} \le \cdots \le X_{n,n}$ be the nth order statistics. Then with the normalizing constants $a_n > 0$ and b_n from (1.1.1) and fixed $k \in \mathbb{N}$,*

$$\left(\frac{X_{n,n} - b_n}{a_n}, \frac{X_{n-1,n} - b_n}{a_n}, \ldots, \frac{X_{n-k,n} - b_n}{a_n} \right)$$

converges in distribution to

$$\left(\frac{(E_1^\star)^{-\gamma} - 1}{\gamma}, \frac{(E_1^\star + E_2^\star)^{-\gamma} - 1}{\gamma}, \ldots, \frac{(E_1^\star + E_2^\star + \cdots + E_{k+1}^\star)^{-\gamma} - 1}{\gamma} \right),$$

where $E_1^\star, E_2^\star, \ldots$ are i.i.d. standard exponential.

From this representation the derivation of the (complicated) joint limit distributions is straightforward.

Proof. Note that if E is a random variable with standard exponential distribution, then

$$U\left(\frac{1}{1 - e^{-E}} \right)$$

has distribution function F. Hence

$$\left(X_{n,n}, X_{n-1,n}, \ldots, X_{n-k+1,n} \right)$$
$$\stackrel{d}{=} \left(U\left(\frac{1}{1 - e^{-E_{1,n}}} \right), U\left(\frac{1}{1 - e^{-E_{2,n}}} \right), \ldots, U\left(\frac{1}{1 - e^{-E_{k,n}}} \right) \right).$$

Next note that

$$\lim_{n \to \infty} \frac{U\left(\frac{1}{1 - e^{-x/n}} \right) - b_n}{a_n} = \lim_{n \to \infty} \frac{U\left(\frac{n}{n(1 - e^{-x/n})} \right) - U(n)}{a_n}$$
$$= \frac{\left(\lim_{n \to \infty} n \left(1 - e^{-x/n} \right) \right)^{-\gamma} - 1}{\gamma} = \frac{x^{-\gamma} - 1}{\gamma}.$$

Hence by (2.1.2), (2.1.3), and the fact that $n\left(1 - e^{-x/n} \right) \to x$, $n \to \infty$, for $x > 0$, we get the result. ∎

Under the conditions of Theorem 2.1.1 consider the random collection

$$\left\{ \left(\frac{i}{n}, \frac{X_i - b_n}{a_n} \right) \right\}_{i=1}^{\infty}$$

of points in $\mathbb{R}_+ \times \mathbb{R}$ and define a point process (random measure) N_n as follows: for each Borel set $B \subset \mathbb{R}_+ \times \mathbb{R}$,

$$N_n(B) = \sum_{i=1}^{\infty} 1_{\left\{ \left(\frac{i}{n}, \frac{X_i - b_n}{a_n} \right) \in B \right\}} \ .$$

Moreover, consider a Poisson point process N on $\mathbb{R}_+ \times (_*x, x^*]$, where $_*x$ and x^* are the lower and upper endpoints of the distribution function G_γ, with mean measure ν given by, with $0 < a < b$, $_*x < c < d \le x^*$,

$$\nu\left([a, b] \times [c, d] \right) = (b - a) \left[(1 + \gamma c)^{-1/\gamma} - (1 + \gamma d)^{-1/\gamma} \right] \ .$$

The following limit relation holds. For information about point processes see, e.g., Jagers (1974).

Theorem 2.1.2 *The sequence of point processes N_n converges in distribution to the Poisson point process N, i.e., for any Borel sets $B_1, \ldots, B_r \subset \mathbb{R}_+ \times (_*x, x^*]$ with $\nu(\partial B_i) = 0$ for $i = 1, 2, \ldots, r$,*

$$(N_n(B_1), \ldots, N_n(B_r)) \xrightarrow{d} (N(B_1), \ldots, N(B_r)) \ .$$

Proof. By Theorem 4.7 of Kallenberg (1983), see also Theorem A.1, p. 309, of Leadbetter, Lindgren, and Rootzén (1983), and Proposition 3.22, p. 156, of Resnick (1987), it is sufficient to check that for all half-open rectangles $I := (x_1, x_2] \times (y_1, y_2]$,

$$\lim_{n \to \infty} E N_n(I) = E N(I) \ , \tag{2.1.4}$$

and that for each $B = \bigcup_{i=1}^{k} I_i$, a finite union of half-open rectangles parallel to the axes,

$$\lim_{n \to \infty} P(N_n(B) = 0) = P(N(B) = 0) \ . \tag{2.1.5}$$

Now

$$E N_n(I) = \sum_{nx_1 < i \le nx_2} P\left(y_1 < \frac{X_i - b_n}{a_n} \le y_2 \right) \ .$$

For the proof of (2.1.4) it is sufficient to note that

$$nP\left(y_1 < \frac{X_i - b_n}{a_n} \le y_2 \right) \to (1 + \gamma y_1)^{-1/\gamma} - (1 + \gamma y_2)^{-1/\gamma}$$

by (1.1.7) of Theorem 1.1.2 and that

$$\frac{1}{n} \sum_{nx_1 < i \le nx_2} 1 \to x_2 - x_1, \qquad n \to \infty \ .$$

For relation (2.1.5) note that the rectangles I_i can be taken to be disjoint. In fact, by the independence of the X_i, $i = 1, 2 \ldots$, it is sufficient to consider a set B of disjoint half-open intervals with identical first coordinates, i.e., in a vertical strip. Then

$$P(N_n(B) = 0)$$

$$= \prod_{x_1 < i/n \le x_2} \left[1 - P\left(y_1^{(1)} < \frac{X_i - b_n}{a_n} \le y_2^{(1)} \text{ or} \cdots \right.\right.$$

$$\left.\left. \cdots \text{ or } y_1^{(k)} < \frac{X_i - b_n}{a_n} \le y_2^{(k)} \right) \right]$$

$$= \left(1 - \frac{\sum_{j=1}^{k} n P\left(y_1^{(j)} < \frac{X_i - b_n}{a_n} \le y_2^{(j)} \right)}{n} \right)^{\#\{i \,:\, nx_1 < i \le nx_2\}}$$

$$\to \exp\left(-(x_2 - x_1) \sum_{j=1}^{k} \left[\left(1 + \gamma y_1^{(j)} \right)^{-1/\gamma} - \left(1 + \gamma y_2^{(j)} \right)^{-1/\gamma} \right] \right)$$

$$= P(N(B) = 0) . \qquad \blacksquare$$

The result is quite helpful for developing intuition in extreme value theory: the larger order statistics can be thought of as points of a Poisson point process with mean measure determined by the extreme value distribution.

A clear and useful way to see what convergence in distribution of the point process means is the following (cf. Appendix A): there exists a sequence of point processes $\tilde{N}, \tilde{N}_1, \tilde{N}_2, \ldots$ defined on one sample space such that $\tilde{N} =^d N$ and $\tilde{N}_i =^d N_i$ for $i = 1, 2, \ldots$ and $\tilde{N}_i \to \tilde{N}$ a.s., as $i \to \infty$. That is, for every relatively compact set B whose boundary has zero mass under the limit measure, the number of points in B under $\tilde{N}_i$ converges to the number of points in B under $\tilde{N}$. Moreover (note that the numbers of points will be eventually equal) the position of all the points in B under $\tilde{N}_i$ will asymptotically coincide with the position of the points in B under $\tilde{N}$.

2.2 Intermediate Order Statistics

In the previous section we studied the asymptotic behavior of order statistics, that is, $X_{n-k,n}$, when $n \to \infty$ and k is fixed, along with an approximation by a Poisson point process. One can also consider $X_{n-k,n}$ with $k = k(n) \to \infty$ as $n \to \infty$. A commonly considered case is $k(n)/n \to p \in (0, 1)$ (the so-called *central order statistics*, see, e.g., Arnold, Balakrishnan, and Nagaraja (1992)). The normal distribution is then an appropriate limit distribution, and in fact, the stochastic process $X_{[ns],n}$, for some $0 < s \le 1$, properly normalized, can be approximated by a Brownian bridge (see, e.g., Proposition 2.4.9 below). But there is a case in between these two. Consider the order statistics $X_{n-k,n}$ with $n \to \infty$, $k = k(n) \to \infty$, and $k(n)/n \to 0$. Those are called *intermediate order statistics*. Their behavior can be connected with extreme value theory, and the stochastic process $X_{n-[ks],n}$, properly normalized, can be approximated by Brownian motions, as we shall see.

The following result shows that there is a connection between intermediate order statistics and extreme value theory.

Let $X_1, X_2, \ldots$ be independent and identically distributed random variables with distribution function F. Recall that $U = (1/(1-F))^{\leftarrow}$.

Theorem 2.2.1 *Suppose von Mises' condition for the domain of attraction of an extreme value distribution G_γ holds (cf. Section 1.1.5). Then, if $k = k(n) \to \infty$, $k/n \to 0$ as $n \to \infty$,*

$$\sqrt{k} \, \frac{X_{n-k,n} - U(\frac{n}{k})}{\frac{n}{k} U'(\frac{n}{k})}$$

is asymptotically standard normal.

In view of applications later on we state the following immediate corollary:

Corollary 2.2.2 *For $F_Y(y) = 1 - 1/y$, $y \geq 1$, as $n \to \infty$, $k \to \infty$, $k/n \to 0$,*

$$\sqrt{k} \left(\frac{k}{n} Y_{n-k,n} - 1 \right)$$

is asymptotically standard normal.

We first give the proof for the uniform distribution. In fact, we shall prove the following more general result due to Smirnov.

Lemma 2.2.3 (Smirnov (1949)) *Let $U_{1,n} \leq U_{2,n} \leq \cdots \leq U_{n,n}$ be the nth order statistics from a standard uniform distribution. Then, as $n \to \infty$, $k \to \infty$, $n - k \to \infty$,*

$$\frac{U_{k,n} - b_n}{a_n}$$

is asymptotically standard normal with

$$b_n := \frac{k-1}{n-1},$$

$$a_n := \sqrt{b_n(1-b_n)\frac{1}{n-1}}.$$

Proof. The density of $U_{k,n}$ is

$$\frac{n!}{(k-1)!(n-k)!} x^{k-1}(1-x)^{n-k};$$

hence the density of $(U_{k,n} - b_n)/a_n$ is

$$\left(\frac{n!}{(k-1)!(n-k)!} a_n b_n^{k-1} (1-b_n)^{n-k} \right) \left(1 + x\frac{a_n}{b_n} \right)^{k-1} \left(1 - x\frac{a_n}{1-b_n} \right)^{n-k}.$$

Using Stirling's formula for $n!$ one sees easily that the first factor tends to $(2\pi)^{-1/2}$. Next note that

$$(k-1)\log\left(1+x\frac{a_n}{b_n}\right) + (n-k)\log\left(1-x\frac{a_n}{1-b_n}\right)$$

$$= (k-1)\left(x\frac{a_n}{b_n} - \frac{x^2}{2}\left(\frac{a_n}{b_n}\right)^2 + \cdots\right)$$

$$+(n-k)\left(-x\frac{a_n}{1-b_n} - \frac{x^2}{2}\left(\frac{a_n}{1-b_n}\right)^2 + \cdots\right)$$

so the highest-order terms cancel. The coefficient of $-x^2/2$ is

$$(k-1)\left(\frac{a_n}{b_n}\right)^2 + (n-k)\left(\frac{a_n}{1-b_n}\right)^2 = 1\,.$$

The other terms are of smaller order. Since the sequence of densities converges pointwise, we have weak convergence of the probability distributions (Scheffé's theorem).   ∎

Proof (of Theorem 2.2.1). Smirnov's lemma implies that

$$\sqrt{k}\left(\frac{n}{kU_{k+1,n}} - 1\right)$$

converges to a standard normal distribution. Hence, since $(n/k)U_{k+1,n} \to^P 1$, also

$$\sqrt{k}\left(\frac{k}{nU_{k+1,n}} - 1\right)$$

converges to a standard normal distribution. Now note that

$$X_{n-k,n} \overset{d}{=} F^{\leftarrow}(1-U_{k+1,n}) = U\left(\frac{1}{U_{k+1,n}}\right)\,;$$

hence

$$\sqrt{k}\frac{X_{n-k,n} - U(\frac{n}{k})}{\frac{n}{k}U'(\frac{n}{k})} \overset{d}{=} \sqrt{k}\frac{U\left(\frac{n}{k}\frac{k}{nU_{k+1,n}}\right) - U(\frac{n}{k})}{\frac{n}{k}U'(\frac{n}{k})}$$

$$= \sqrt{k}\int_1^{k/(nU_{k+1,n})}\frac{U'\left(\frac{n}{k}s\right)}{U'(\frac{n}{k})}\,ds\,.$$

By (1.1.33) and Potter's inequalities (Proposition B.1.9), for $n \geq n_0$, $s \geq 1$,

$$(1-\varepsilon)s^{\gamma-1-\varepsilon'} < \frac{U'\left(\frac{n}{k}s\right)}{U'\left(\frac{n}{k}\right)} < (1+\varepsilon)s^{\gamma-1+\varepsilon'}\,.$$

Hence

$$(1 - \varepsilon) \sqrt{k} \; \frac{\left(\frac{k}{n \, U_{k+1,n}}\right)^{\gamma - \varepsilon'} - 1}{\gamma - \varepsilon'} \leq \sqrt{k} \; \frac{U\left(\frac{1}{U_{k+1,n}}\right) - U\left(\frac{n}{k}\right)}{\frac{n}{k} \, U'\left(\frac{n}{k}\right)}$$

$$\leq (1 + \varepsilon) \sqrt{k} \; \frac{\left(\frac{k}{n \, U_{k+1,n}}\right)^{\gamma + \varepsilon'} - 1}{\gamma + \varepsilon'}.$$

We already know that $k/(n \, U_{k+1,n}) \to^P 1$. Hence

$$\sqrt{k} \; \frac{\left(\frac{k}{n \, U_{k+1,n}}\right)^{\gamma \pm \varepsilon'} - 1}{\gamma \pm \varepsilon'}$$

has the same limit distribution as $\sqrt{k} \left(k/(nU_{k+1,n}) - 1\right)$. Since $\varepsilon > 0$ is arbitrary we find that

$$\sqrt{k} \; \frac{X_{n-k,n} - U\left(\frac{n}{k}\right)}{\frac{n}{k} \, U'\left(\frac{n}{k}\right)}$$

has the same limit distribution as $\sqrt{k} \left(k/(nU_{k+1,n}) - 1\right)$. ∎

So we see that the normal distribution is a natural limit distribution for intermediate order statistics. As in the case of extreme order statistics, where we made the connection with point processes, we want to put the present limit result in a wider framework, which in this case will be convergence toward a Brownian motion. However, for this result we need more than just the domain of attraction condition. One can consider the domain of attraction condition as a special kind of asymptotic expansion of U near infinity. For the approximation by Brownian motion, as well as for many statistical results as we shall see later on, it is very useful to have a higher-order expansion. We call this the second-order condition. This condition will be discussed in the next section. The extension (or rather analogue) of Theorem 2.2.1 in this framework will be discussed in Section 2.4.

2.3 Second-Order Condition

Once again we start with the extremal value condition (2.1.1), or equivalently

$$\lim_{t \to \infty} \frac{U(tx) - U(t)}{a(t)} = \frac{x^\gamma - 1}{\gamma} =: D_\gamma(x), \qquad (2.3.1)$$

for each $x > 0$, where $U = (1/(1 - F))^\leftarrow$.

We are going to develop a second-order condition related to (2.3.1). Suppose that there exists a function A not changing sign eventually and with $\lim_{t \to \infty} A(t) = 0$ such that for all $x > 0$,

$$\lim_{t \to \infty} \frac{\frac{U(tx) - U(t)}{a(t)} - D_\gamma(x)}{A(t)} \qquad (2.3.2)$$

exists. The function A could be either positive or negative. Write H for the limit function. Of course the case $H(x) = 0$ for all $x > 0$ is not very informative.

Let us rewrite relation (2.3.2) as follows: for all $x > 0$,

$$\lim_{t \to \infty} \frac{U(tx) - U(t) - a(t)D_\gamma(x)}{a_1(t)} = H(x) \tag{2.3.3}$$

with a as before and $a_1 = aA$. The first question is, which functions H are possible limit functions in (2.3.3)?

Note first that when we replace the function a by $a + ca_1$, for some constant c,

$$a(t) \sim a(t) + ca_1(t) ,$$

then we still have the limit relation (2.3.3) but with a new limit function $\tilde{H}$ satisfying

$$\tilde{H}(x) = H(x) - cD_\gamma(x) . \tag{2.3.4}$$

This means that we can always add a multiple of D_γ to the function H. It follows that if (2.3.3) holds with $H(x) = cD_\gamma(x)$, the relation is still not very informative. So we require that the function H in (2.3.3) not be a multiple of D_γ. In particular, H should not be identically zero.

Definition 2.3.1 The function U (or the probability distribution connected with it) is said to satisfy the *second-order condition* if for some positive function a and some positive or negative function A with $\lim_{t \to \infty} A(t) = 0$,

$$\lim_{t \to \infty} \frac{\frac{U(tx) - U(t)}{a(t)} - \frac{x^\gamma - 1}{\gamma}}{A(t)} =: H(x) , \quad x > 0 , \tag{2.3.5}$$

where H is some function that is not a multiple of the function $(x^\gamma - 1)/\gamma$. In particular, H should not be identically zero. Occasionally we shall refer to the functions a and A as (respectively) first-order and second-order *auxiliary functions*.

Remark 2.3.2 Note that the second-order condition implies the domain of attraction condition.

We have the following results. Proofs are given in the appendix on regular variation, Section B.3.

Theorem 2.3.3 *Suppose the second-order condition (2.3.5) holds. Then there exist constants $c_1, c_2 \in \mathbb{R}$ and some parameter $\rho \leq 0$ such that*

$$H(x) = c_1 \int_1^x s^{\gamma - 1} \int_1^s u^{\rho - 1} \, du \, ds + c_2 \int_1^x s^{\gamma + \rho - 1} \, ds . \tag{2.3.6}$$

Moreover, for $x > 0$,

$$\lim_{t \to \infty} \frac{\frac{a(tx)}{a(t)} - x^\gamma}{A(t)} = c_1 x^\gamma \frac{x^\rho - 1}{\rho} \tag{2.3.7}$$

and

$$\lim_{t \to \infty} \frac{A(tx)}{A(t)} = x^\rho . \tag{2.3.8}$$

For $\rho \neq 0$ we can write H as

$$H(x) = c_1 \frac{1}{\rho} \left(D_{\gamma+\rho}(x) - D_{\gamma}(x) \right) + c_2 D_{\gamma+\rho}(x), \tag{2.3.9}$$

and for $\rho = 0$ and $\gamma \neq 0$ we can write H as

$$H(x) = c_1 \frac{1}{\gamma} \left(x^{\gamma} \log x - D_{\gamma}(x) \right) + c_2 D_{\gamma}(x) . \tag{2.3.10}$$

For $\gamma = \rho = 0$ we get directly from (2.3.6)

$$H(x) = c_1 \frac{1}{2} (\log x)^2 + c_2 \log x . \tag{2.3.11}$$

Next we are going to simplify the limit function H by changing the functions a and a_1 a little as in (2.3.4). We work out one of the three cases. Suppose $\rho \neq 0$, so (2.3.9) holds. Replace a by $a + c_2 a_1$. Then the limit H changes to

$$(c_1 + \rho c_2) \frac{1}{\rho} \left(D_{\gamma+\rho}(x) - D_{\gamma}(x) \right) .$$

Next replace the function a_1 by $a_1(c_1 + \rho c_2)$. Then the limit H changes to

$$\frac{1}{\rho} \left(D_{\gamma+\rho}(x) - D_{\gamma}(x) \right) = \int_1^x s^{\gamma-1} \int_1^s u^{\rho-1} \, du \, ds ,$$

which is just the first term in (2.3.6) with $c_1 = 1$. Notice that in the process we may have changed the positive function a_1 into a negative one. One can simplify relations (2.3.10) and (2.3.11) as well. We formulate our result.

Corollary 2.3.4 *Suppose relation (2.3.3) holds for all $x > 0$ and the function H is not a multiple of D_{γ}. Then there exist (possibly different) functions a, positive, and a_1, positive or negative, such that*

$$\lim_{t \to \infty} \frac{U(tx) - U(t) - a(t) D_{\gamma}(x)}{a_1(t)} = \int_1^x s^{\gamma-1} \int_1^s u^{\rho-1} \, du \, ds, \tag{2.3.12}$$

or equivalently, there exist functions a, positive, and A, positive or negative, such that

$$\lim_{t \to \infty} \frac{\frac{U(tx)-U(t)}{a(t)} - D_{\gamma}(x)}{A(t)} = \int_1^x s^{\gamma-1} \int_1^s u^{\rho-1} \, du \, ds =: H_{\gamma,\rho}(x) . \tag{2.3.13}$$

Sometimes we shall write the limit function as

$$H_{\gamma,\rho}(x) := \frac{1}{\rho} \left(\frac{x^{\gamma+\rho} - 1}{\gamma + \rho} - \frac{x^{\gamma} - 1}{\gamma} \right) , \tag{2.3.14}$$

which for the cases $\gamma = 0$ and $\rho = 0$ is understood to be equal to the limit of (2.3.14) as $\gamma \to 0$ or $\rho \to 0$, respectively, that is,

$$\begin{cases} \frac{1}{\gamma}\left(x^{\gamma}\log x - \frac{x^{\gamma}-1}{\gamma}\right), & \rho = 0 \neq \gamma, \\ \frac{1}{\rho}\left(\frac{x^{\rho}-1}{\rho} - \log x\right), & \rho \neq 0 = \gamma, \\ \frac{1}{2}(\log x)^2, & \rho = 0 = \gamma. \end{cases}$$

Corollary 2.3.5 *Suppose relation* (2.3.3) *holds for all* $x > 0$ *and the function* H *is not a multiple of* D_{γ}. *Then there exist functions* $a_{\star}$, *positive, and* $A_{\star}$, *positive or negative, such that*

$$\lim_{t \to \infty} \frac{\frac{U(tx)-U(t)}{a_{\star}(t)} - D_{\gamma}(x)}{A_{\star}(t)} = \Psi_{\gamma,\rho}(x), \tag{2.3.15}$$

where

$$\Psi_{\gamma,\rho}(x) := \begin{cases} \frac{x^{\gamma+\rho}-1}{\gamma+\rho}, & \gamma+\rho \neq 0, \ \rho < 0, \\ \log x, & \gamma+\rho = 0, \ \rho < 0, \\ \frac{1}{\gamma}x^{\gamma}\log x, & \rho = 0 \neq \gamma, \\ \frac{1}{2}(\log x)^2, & \rho = 0 = \gamma, \end{cases} \tag{2.3.16}$$

$$a_{\star}(t) := \begin{cases} a(t)\left(1 - \frac{1}{\rho}A(t)\right), & \rho < 0, \\ a(t)\left(1 - \frac{1}{\gamma}A(t)\right), & \rho = 0 \neq \gamma, \\ a(t), & \rho = 0 = \gamma, \end{cases}$$

and

$$A_{\star}(t) := \begin{cases} \frac{1}{\rho}A(t), & \rho < 0, \\ A(t), & \rho = 0, \end{cases}$$

with a *and* A *from Corollary 2.3.4.*

Next we consider the limit relation for fixed $\Psi_{\gamma,\rho}$. The limit relation entails a very useful set of uniform inequalities.

Theorem 2.3.6 *Suppose* (2.3.15) *holds for some fixed* $\gamma \in \mathbb{R}$ *and* $\rho \leq 0$. *Then there are functions* a_0 *and* A_0 *satisfying, as* $t \to \infty$,

$$A_0(t) \sim A_{\star}(t),$$

$$\frac{a_0(t)}{a_{\star}(t)} - 1 = o(A_{\star}(t)),$$

with the following property: for any $\varepsilon, \delta > 0$ *there exists* $t_0 = t_0(\varepsilon, \delta)$ *such that for all* $t, tx \geq t_0$,

$$\left| \frac{\frac{U(tx)-U(t)}{a_0(t)} - \frac{x^{\gamma}-1}{\gamma}}{A_0(t)} - \Psi_{\gamma,\rho}(x) \right| \leq \varepsilon x^{\gamma+\rho} \max(x^{\delta}, x^{-\delta}), \tag{2.3.17}$$

and

$$\left| \frac{\frac{a_0(tx)}{a_0(t)} - x^\gamma}{A_0(t)} - x^\gamma \frac{x^\rho - 1}{\rho} \right| \le \varepsilon x^{\gamma+\rho} \max(x^\delta, x^{-\delta}) . \tag{2.3.18}$$

The functions a_0 and A_0 can be chosen as follows:

$$a_0(t) := \begin{cases} ct^\gamma , & \rho < 0, \\ -\gamma(U(\infty) - U(t)) , & \gamma < \rho = 0, \\ \gamma U(t) , & \gamma > \rho = 0, \\ \hat{U}(t) + \bar{\bar{U}}(t) , & \gamma = \rho = 0, \end{cases}$$

with $c := \lim_{t\to\infty} t^{-\gamma} a(t) > 0$,

$$\bar{U}(t) := \begin{cases} U(t) - c \frac{t^\gamma - 1}{\gamma} , & \rho < 0, \\ t^{-\gamma}(U(\infty) - U(t)) , & \gamma < \rho = 0, \\ t^{-\gamma} U(t) , & \gamma > \rho = 0, \\ \hat{U}(t) , & \gamma = \rho = 0 , \end{cases}$$

and for some integrable function g,

$$\hat{g}(t) := g(t) - \frac{1}{t} \int_0^t g(s) \, ds$$

and

$$A_0(t) := \begin{cases} -(\gamma + \rho)\frac{\bar{U}(\infty)-\bar{U}(t)}{a_0(t)} , & \gamma + \rho < 0 , \ \rho < 0, \\ (\gamma + \rho)\frac{\bar{U}(t)}{a_0(t)} , & \gamma + \rho > 0 , \ \rho < 0, \\ \frac{\hat{\bar{U}}(t)}{a_0(t)} , & \gamma + \rho = 0 , \ \rho < 0, \\ \frac{\hat{\bar{U}}(t)}{\bar{U}(t)} , & \gamma \ne \rho = 0, \\ \frac{\hat{\bar{U}}(t)}{a_0(t)} , & \gamma = \rho = 0 . \end{cases}$$

The next corollary is an alternative formulation of the result of the last theorem that is sometimes useful.

Corollary 2.3.7 *Under the conditions of Theorem 2.3.6, with the same functions a_0 and A_0 satisfying, as $t \to \infty$, $A_0(t) \sim A_\star(t)$ and $a_0(t)/a_\star(t) - 1 = o(A_\star(t))$, for any $\varepsilon, \delta > 0$ there exists $t_0 = t_0(\varepsilon, \delta)$ such that for all t, $tx \ge t_0$,*

$$\left| \frac{\frac{U(tx)-b_0(t)}{a_0(t)} - \frac{x^\gamma-1}{\gamma}}{A_0(t)} - \Psi_{\gamma,\rho}(x) \right| \le \varepsilon x^{\gamma+\rho} \max(x^\delta, x^{-\delta}) , \tag{2.3.19}$$

where

$$b_0(t) := \begin{cases} U(t) - \frac{1}{\gamma+\rho} a_0(t) A_0(t) \, , & \gamma + \rho \neq 0, \rho < 0, \\ U(t) \, , & \text{otherwise} \, , \end{cases}$$

and

$$\overline{\Psi}_{\gamma,\rho}(x) := \begin{cases} \frac{x^{\gamma+\rho}}{\gamma+\rho} \, , & \gamma + \rho \neq 0 \, , \ \rho < 0, \\ \log x \, , & \gamma + \rho = 0 \, , \ \rho < 0, \\ \frac{1}{\gamma} x^{\gamma} \log x \, , & \rho = 0 \neq \gamma, \\ \frac{1}{2}(\log x)^2 \, , & \rho = 0 = \gamma \, . \end{cases} \tag{2.3.20}$$

Next we formulate the second-order condition in terms of the distribution function, rather than in terms of the function U.

Theorem 2.3.8 *Suppose* (2.3.5) *holds. Then for all x with $1 + \gamma x > 0$,*

$$\lim_{t \uparrow x^*} \frac{\frac{1-F(t+xf(t))}{1-F(t)} - Q_\gamma(x)}{\alpha(t)} = (Q_\gamma(x))^{1+\gamma} H_{\gamma,\rho}\left(Q_\gamma^{-1}(x)\right), \tag{2.3.21}$$

where $f(t) := a(1/(1 - F(t)))$, $\alpha(t) := A(1/(1 - F(t)))$, and $Q_\gamma(x) := (1 + \gamma x)^{-1/\gamma}$. Conversely, (2.3.21) *implies* (2.3.5).

For convenience we state the simpler corresponding results for the case $\gamma > 0$ separately.

Theorem 2.3.9 *Suppose for some γ positive and positive or negative function A,*

$$\lim_{t \to \infty} \frac{\frac{U(tx)}{U(t)} - x^\gamma}{A(t)} := K(x)$$

exists for all $x > 0$ and K is not identically zero. Then for a possibly different function A, positive or negative,

$$\lim_{t \to \infty} \frac{\frac{U(tx)}{U(t)} - x^\gamma}{A(t)} = x^\gamma \frac{x^\rho - 1}{\rho} \tag{2.3.22}$$

for all $x > 0$ with $\rho \leq 0$. Moreover, for any $\varepsilon, \delta > 0$ there exists $t_0 = t_0(\varepsilon, \delta) > 1$ such that for all t, $tx \geq t_0$,

$$\left| \frac{\frac{U(tx)}{U(t)} - x^\gamma}{A_0(t)} - x^\gamma \frac{x^\rho - 1}{\rho} \right| \leq \varepsilon x^{\gamma+\rho} \max(x^\delta, x^{-\delta}), \tag{2.3.23}$$

with

$$A_0(t) := \begin{cases} \rho \left(1 - \frac{\lim_{s \to \infty} s^{-\gamma} U(s)}{t^{-\gamma} U(t)} \right) \, , & \rho < 0, \\ 1 - \frac{\int_0^t s^{-\gamma} U(s) \, ds}{t^{1-\gamma} U(t)} \, , & \rho = 0 \, . \end{cases}$$

Relation (2.3.22) *is equivalent to*

$$\lim_{t \to \infty} \frac{\frac{1-F(tx)}{1-F(t)} - x^{-1/\gamma}}{\alpha(t)} = x^{-1/\gamma} \frac{x^{\rho/\gamma} - 1}{\gamma\rho} \tag{2.3.24}$$

for all $x > 0$ with $\rho \leq 0$ and $\alpha(t) := A(1/(1 - F(t)))$.

Remark 2.3.10 For the equivalence of (2.3.22) and (2.3.24) see Exercise 2.11.

Example 2.3.11 *The function* $U(t) = c_0 t^\gamma + c_1$, *with c_0 and γ positive, and $c_1 \neq 0$, satisfies the second-order condition of Theorem 2.3.9 but not the second-order condition of Definition 2.3.1.*

It is interesting to observe that if (2.3.22) holds with $\rho < 0$, then for some positive constant c the function $|U(t) - ct^\gamma|$ is regularly varying with index $\gamma + \rho$. In particular; $U(t) \sim ct^\gamma$, $t \to \infty$. So the second-order condition with $\rho < 0$ makes the first-order relation particularly simple.

Finally, we provide the sufficient *second-order condition of von Mises* type.

Theorem 2.3.12 *Suppose the function $U = (1/(1-F))^\leftarrow$ is twice differentiable. Write*

$$A(t) := \frac{tU''(t)}{U'(t)} - \gamma + 1 .$$

If the function A has constant sign for large t, $\lim_{t\to\infty} A(t) = 0$, and the function $|A|$ is regularly varying with index $\rho \leq 0$, then for $x > 0$,

$$\lim_{t\to\infty} \frac{\frac{U(tx)-U(t)}{tU'(t)} - \frac{x^\gamma - 1}{\gamma}}{A(t)} = H_{\gamma,\rho}(x) .$$

Proof. First note that

$$\log U'(tx) - \log U'(t) = \int_1^x \frac{tu U''(tu)}{U'(tu)} \frac{du}{u} \to \int_1^x (\gamma - 1) \frac{du}{u} , \qquad (2.3.25)$$

since the function $A(t)$ vanishes at $t \to \infty$, i.e., $U' \in RV_{\gamma-1}$. Now

$$\frac{\frac{U(tx)-U(t)}{tU'(t)} - \frac{x^\gamma-1}{\gamma}}{A(t)} = \frac{\int_1^x \left(\frac{U'(ts)}{U'(t)} - s^{\gamma-1} \right) ds}{A(t)}$$

$$= \frac{\int_1^x s^{\gamma-1} t^{\gamma-1} \int_1^s \frac{U''(tu)(tu)^{1-\gamma} + (1-\gamma)(tu)^{-\gamma} U'(tu)}{U'(t)} t \, du \, ds}{A(t)}$$

$$= \frac{\int_1^x s^{\gamma-1} \int_1^s \left(\frac{tu U''(tu)}{U'(t)} - \gamma + 1 \right) \frac{U'(tu)}{U'(t)} u^{-\gamma} \, du \, ds}{A(t)}$$

$$= \int_1^x s^{\gamma-1} \int_1^s \frac{A(tu)}{A(t)} \frac{U'(tu)}{U'(t)} u^{-\gamma} \, du \, ds .$$

Since $A(tu)/A(t) \to u^\rho$ and $U'(tu)/U'(t) \to u^{\gamma-1}$ locally uniformly for $u > 0$ (Theorem B.1.4), the result follows. ∎

2.4 Intermediate Order Statistics and Brownian Motion

We continue the discussion on the behavior of intermediate order statistics under extreme value conditions. We have seen that a sequence of intermediate order statistics is asymptotically normal (when properly normalized) under von Mises' extreme value condition. However, we want to consider many intermediate order statistics at the same time; hence we want to consider the tail (empirical) quantile process.

It is instructive to start by proving the main result of Section 2.2, Theorem 2.2.1, i.e., the asymptotic normality of a sequence of intermediate order statistics, again, now not under von Mises' conditions but under the second-order condition.

Theorem 2.4.1 *Let* $X_{1,n} \leq X_{2,n} \leq \cdots \leq X_{n,n}$ *be the nth order statistics from an i.i.d. sample with distribution function F. Suppose that the second-order condition (2.3.21), or equivalently (2.3.5), holds for some* $\gamma \in \mathbb{R}$, $\rho \leq 0$. *Then*

$$\sqrt{k}\frac{X_{n-k,n} - U\left(\frac{n}{k}\right)}{a\left(\frac{n}{k}\right)}$$

is asymptotically standard normal provided the sequence $k = k(n)$ *is such that* $k(n) \to \infty$, $n \to \infty$, *and*

$$\lim_{n\to\infty} \sqrt{k}A\left(\frac{n}{k}\right)$$

exists and is finite.

Proof. Take independent and identically distributed random variables $Y_1, Y_2, \ldots$ with distribution function $1 - 1/y$, $y > 1$. Let $Y_{1,n} \leq Y_{2,n} \leq \cdots \leq Y_{n,n}$ be the nth order statistics. Then on the one hand, by Corollary 2.2.2

$$\sqrt{k}\left(\frac{k}{n}Y_{n-k,n} - 1\right) \xrightarrow{d} N \qquad (2.4.1)$$

with N standard normal. On the other hand,

$$X_{n-k,n} \overset{d}{=} U(Y_{n-k,n}) \ .$$

Hence by Theorem 2.3.6,

$$\sqrt{k}\frac{X_{n-k,n} - U(\frac{n}{k})}{a_0(\frac{n}{k})} \overset{d}{=} \sqrt{k}\frac{U(\frac{n}{k}\frac{k}{n}Y_{n-k,n}) - U(\frac{n}{k})}{a_0(\frac{n}{k})}$$

$$= \sqrt{k}\frac{\left(\frac{k}{n}Y_{n-k,n}\right)^{\gamma} - 1}{\gamma} + \sqrt{k}A_0\left(\frac{n}{k}\right)\Psi_{\gamma,\rho}\left(\frac{k}{n}Y_{n-k,n}\right)$$

$$+ o_P(1)\sqrt{k}A_0\left(\frac{n}{k}\right)\left(\frac{k}{n}Y_{n-k,n}\right)^{\gamma+\rho}$$

$$\max\left(\left(\frac{k}{n}Y_{n-k,n}\right)^{\delta}, \left(\frac{k}{n}Y_{n-k,n}\right)^{-\delta}\right) \ .$$

Now note that by Cramér's delta method,

$$\sqrt{k}\frac{\left(\frac{k}{n}Y_{n-k,n}\right)^{\gamma} - 1}{\gamma}$$

has the same limit distribution as $\sqrt{k}(kY_{n-k,n}/n-1)$, which is asymptotically standard normal; moreover, since $kY_{n-k,n}/n \to^P 1$ by (2.4.1), the other terms go to zero by assumption (note that $\Psi_{\gamma,\rho}(1) = 0$). ∎

The last result can be vastly generalized and yields the following, relating the tail quantile process to Brownian motion in a strong sense.

Theorem 2.4.2 (Drees (1998), Theorem 2.1) *Suppose* $X_1, X_2, \ldots$ *are i.i.d. random variables with distribution function* F. *Suppose that* F *satisfies the second-order extreme value condition* (2.3.21) *for some* $\gamma \in \mathbb{R}$ *and* $\rho \leq 0$. *Let* $X_{1,n} \leq X_{2,n} \leq \cdots \leq X_{n,n}$ *be the nth order statistics. We can define a sequence of Brownian motions* $\{W_n(s)\}_{s>0}$ *such that for suitably chosen functions* a_0 *and* A_0 *and each* $\varepsilon > 0$,

$$\sup_{k^{-1} \leq s \leq 1} s^{\gamma+1/2+\varepsilon} \left| \sqrt{k} \left(\frac{X_{n-[ks],n} - U(\frac{n}{k})}{a_0(\frac{n}{k})} - \frac{s^{-\gamma} - 1}{\gamma} \right) \right.$$
$$\left. - s^{-\gamma-1} W_n(s) - \sqrt{k} A_0 \left(\frac{n}{k} \right) \Psi_{\gamma,\rho}(s^{-1}) \right| \xrightarrow{P} 0, \quad (2.4.2)$$

$n \to \infty$, *provided* $k = k(n) \to \infty$, $k/n \to 0$, *and* $\sqrt{k} A_0(n/k) = O(1)$.

Definition 2.4.3 Let $X_{1,n} \leq X_{2,n} \leq \cdots \leq X_{n,n}$ be the nth order statistics and $k = k(n)$ a sequence satisfying $k \to \infty$, $k/n \to 0$, as $n \to \infty$. We define the *tail (empirical) quantile process* to be the stochastic process $\{X_{n-[ks],n}\}_{s\geq0}$.

Remark 2.4.4 It may happen that the convergence of $(U(tx) - U(t))/a(t)$ to $(x^\gamma - 1)/\gamma$ (cf. Section 2.3, relation (2.3.1)) is faster than any negative power of t, i.e.,

$$\lim_{t \to \infty} t^\alpha \left(\frac{U(tx) - U(t)}{a(t)} - \frac{x^\gamma - 1}{\gamma} \right) = 0$$

for all $x > 0$ and $\alpha > 0$. In that case the result of Theorem 2.4.2 holds with the bias part $\sqrt{k} A_0(n/k) \Psi_{\gamma,\rho}(s^{-1})$ replaced by zero provided $k(n) = o(n^{1-\varepsilon})$ for some $\varepsilon > 0$. A similar remark can be made in connection with the convergence results for the various estimators of Chapter 3.

One can extend the interval of definition of s as follows.

Corollary 2.4.5 *Define*

$$B_0 \left(\frac{n}{k} \right) := \begin{cases} U \left(\frac{n}{k} \right), & \gamma \geq -\frac{1}{2}, \\ X_{n,n} + \frac{a_0(\frac{n}{k})}{\gamma}, & \gamma < -\frac{1}{2}. \end{cases}$$

Then, under the conditions of Theorem 2.4.2,

$$\sup_{0<s\leq 1} s^{\gamma+1/2+\varepsilon} \left| \sqrt{k} \left(\frac{X_{n-[ks],n} - B_0(\frac{n}{k})}{a_0(\frac{n}{k})} - \frac{s^{-\gamma} - 1}{\gamma} \right) \right.$$
$$\left. - s^{-\gamma-1} W_n(s) - \sqrt{k} A_0 \left(\frac{n}{k} \right) \Psi_{\gamma,\rho}(s^{-1}) \right| \xrightarrow{P} 0.$$

We shall usually apply this result in the following form:

Corollary 2.4.6 *Under the conditions of Theorem 2.4.2, for each $\varepsilon > 0$,*

$$\sup_{0<s\leq 1} \min\left(1, s^{\gamma+1/2+\varepsilon}\right) \left| \sqrt{k}\left(\frac{X_{n-[ks],n} - X_{n-k,n}}{a_0(\frac{n}{k})} - \frac{s^{-\gamma}-1}{\gamma}\right)\right.$$
$$\left. - s^{-\gamma-1}W_n(s) + W_n(1) - \sqrt{k}A_0\left(\frac{n}{k}\right)\Psi_{\gamma,\rho}(s^{-1})\right| \xrightarrow{P} 0 .$$

Remark 2.4.7 A somewhat similar approximation to the tail empirical distribution function will be proved in Section 5.1.

A simpler version of Theorem 2.4.2 is valid when γ is positive.

Theorem 2.4.8 *Suppose $X_1, X_2, \ldots$ are i.i.d random variables with distribution function F. Suppose that F satisfies the second-order extreme value condition (2.3.24) for some $\gamma > 0$ and $\rho \leq 0$. Let $X_{1,n} \leq X_{2,n} \leq \cdots \leq X_{n,n}$ be the nth order statistics. We can define a sequence of Brownian motions $\{W_n(s)\}_{s\geq 0}$ such that with a suitable function A_0, and for $\varepsilon > 0$ sufficiently small,*

$$\sup_{0<s\leq 1} s^{\gamma+1/2+\varepsilon}\left| \sqrt{k}\left(\frac{X_{n-[ks],n}}{U\left(\frac{n}{k}\right)} - s^{-\gamma}\right)\right.$$
$$\left. - \gamma s^{-\gamma-1}W_n(s) - \sqrt{k}A_0\left(\frac{n}{k}\right)s^{-\gamma}\frac{s^{-\rho}-1}{\rho}\right| \xrightarrow{P} 0 ,$$

$n \to \infty$, *provided $k = k(n) \to \infty$, $k/n \to 0$, and $\sqrt{k}A_0\left(n/k\right) = O(1)$.*
Moreover,

$$\sup_{0<s\leq 1} s^{1/2+\varepsilon}\left| \sqrt{k}\left(\frac{\log X_{n-[ks],n} - \log U\left(\frac{n}{k}\right)}{\gamma} + \log s\right)\right.$$
$$\left. - s^{-1}W_n(s) - \sqrt{k}A_0\left(\frac{n}{k}\right)\frac{1}{\gamma}\frac{s^{-\rho}-1}{\rho}\right| \xrightarrow{P} 0 .$$

The remainder of this section is devoted to proving the above results. It is instructive to prove the result of Theorem 2.4.2 first for the special case $F(x) = 1 - (1 + \gamma x)^{-1/\gamma}$, $\gamma \in \mathbb{R}$, and all x with $1 + \gamma x > 0$. The next proposition will be used in its proof and in the proof of Theorem 2.4.2. Let $Q(t) := F^{\leftarrow}(t)$, $_*x$ and let x^* be the left and right endpoints of F respectively, and $\lceil x\rceil$ the smallest integer greater than or equal to x.

Proposition 2.4.9 (Csörgő and Horváth (1993), Theorem 6.2.1) *Let $X_1, X_2, \ldots$ be i.i.d. random variables with distribution function F and assume:*

1. *F is twice differentiable on $(_*x, x^*)$, $-\infty \leq _*x < x^* \leq \infty$,*
2. *$F'(x) = f(x) > 0$, $x \in (_*x, x^*)$,*

3. *for some* $C > 0$,

$$\sup_{0 < t < 1} t(1 - t)\frac{|f'(Q(t))|}{f^2(Q(t))} \le C .$$

Let $0 \le \varepsilon < \frac{1}{2}$. *Then we can define a sequence of Brownian bridges* $\{B_n(t)\}$ *such that*

$$\sup_{1/(n+1) \le t \le n/(n+1)} n^\varepsilon t^{\varepsilon-1/2}(1 - t)^{\varepsilon-1/2} \left| \sqrt{n} f(Q(t)) \left(Q(t) - X_{\lceil nt \rceil, n} \right) - B_n(t) \right|$$

$$= O_P(1) , \qquad n \to \infty . \tag{2.4.3}$$

Lemma 2.4.10 *Let* $Y_1, Y_2, \ldots$ *be i.i.d. random variables with distribution function* $1 - 1/y$, $y \ge 1$. *Consider the nth order statistics* $Y_{1,n} \le Y_{2,n} \le \cdots \le Y_{n,n}$. *For each* $\gamma \in \mathbb{R}$ *we can define a sequence of Brownian motions* $\{W_n(s)\}_{s \ge 0}$ *such that for each* $\varepsilon > 0$,

$$\sup_{k^{-1} \le s \le 1} s^{\gamma+1/2+\varepsilon} \left| \sqrt{k} \left(\frac{\left(\frac{k}{n} Y_{n-[ks],n}\right)^\gamma - 1}{\gamma} - \frac{s^{-\gamma} - 1}{\gamma} \right) - s^{-\gamma-1} W_n(s) \right|$$

$$= o_P(1) ,$$

$n \to \infty$, $k = k(n) \to \infty$, *and* $k/n \to 0$.

Proof. The sequence $(Y_1^\gamma - 1)/\gamma$, $(Y_2^\gamma - 1)/\gamma, \ldots$ has distribution function $1 - (1+\gamma x)^{-1/\gamma}$, for which the conditions of Proposition 2.4.9 hold. Hence for $0 \le \varepsilon < \frac{1}{2}$ and $\{B_n(t)\}$ a sequence of Brownian bridges,

$$\sup_{1/(n+1) \le t \le n/(n+1)} n^\varepsilon t^{\varepsilon-1/2}(1 - t)^{\varepsilon-1/2}$$

$$\left| n^{1/2}(1 - t)^{\gamma+1} \left(\frac{Y_{\lceil nt \rceil, n}^\gamma - 1}{\gamma} - \frac{(1 - t)^{-\gamma} - 1}{\gamma} \right) - B_n(t) \right| = O_P(1) .$$

Replace t by $1 - ks/n$ and note that $(1 - ks/n)^{\varepsilon-1/2} \to 1$ uniformly for $k^{-1} \le s \le 1$. After some rearrangements we get

$$\sup_{k^{-1} \le s \le 1} k^\varepsilon \left| \left(\frac{k}{n}\right)^\gamma s^{\gamma+1/2+\varepsilon} \sqrt{k} \left(\frac{Y_{n-[ks],n}^\gamma - 1}{\gamma} - \frac{\left(\frac{ks}{n}\right)^{-\gamma} - 1}{\gamma} \right) \right.$$

$$\left. - \left(\frac{k}{n}\right)^{-1/2} s^{\varepsilon-1/2} B_n\left(1 - \frac{ks}{n}\right) \right| = O_P(1) .$$

Next note that for all $\gamma \in \mathbb{R}$,

$$\left(\frac{k}{n}\right)^\gamma \left(\frac{Y_{n-[ks],n}^\gamma - 1}{\gamma} - \frac{\left(\frac{ks}{n}\right)^{-\gamma} - 1}{\gamma} \right) = \frac{\left(\frac{k}{n} Y_{n-[ks],n}\right)^\gamma - 1}{\gamma} - \frac{s^{-\gamma} - 1}{\gamma}$$

and that

$$B_n \left(1 - \frac{ks}{n}\right) \overset{d}{=} B_n \left(\frac{ks}{n}\right) \overset{d}{=} W_n \left(\frac{ks}{n}\right) - \frac{ks}{n} W_n(1) . \tag{2.4.4}$$

It follows that

$$k^\varepsilon \left| s^{\gamma+1/2+\varepsilon} \left\{ \sqrt{k} \left(\frac{\left(\frac{k}{n} Y_{n-[ks],n}\right)^\gamma - 1}{\gamma} - \frac{s^{-\gamma} - 1}{\gamma} \right) - s^{-\gamma-1} \sqrt{\frac{n}{k}} \, W_n \left(\frac{ks}{n}\right) \right\} \right.$$
$$\left. + \left(\frac{ks}{n}\right)^{1/2} s^\varepsilon W_n(1) \right| = O_P(1)$$

uniformly for $k^{-1} \le s \le 1$. Hence for $\varepsilon > 0$ the part within the absolute value is $o_P(1)$ uniformly. Moreover, notice that

$$\sup_{k^{-1} \le s \le 1} \left| \left(\frac{ks}{n}\right)^{1/2} s^\varepsilon \, W_n(1) \right| = o_P(1) .$$

The result follows.   ∎

The next step is to prove the result for sufficiently differentiable distribution functions. That is, we use von Mises' second-order condition.

Lemma 2.4.11 *Let $X_1, X_2, \ldots$ be i.i.d. random variables with distribution function F. Suppose $U = (1/(1-F))^{\leftarrow}$ satisfies von Mises' second-order condition of Theorem 2.3.12, Section 2.3. Then the result of Theorem 2.4.2 holds, provided $k = k(n) \to \infty$, $k/n \to 0$, and $\sqrt{k} A_0(n/k) = O(1)$.*

Proof. We are going to apply Proposition 2.4.9 but only for the right tail. The conditions of the lemma are that F'' exists, $F' > 0$, and

$$\sup_{0 < t < 1} t(1-t) \frac{\left|F''(Q(t))\right|}{(F'(Q(t)))^2} < \infty .$$

Since we are interested only in the right tail, we may without loss of generality change our distribution near the left endpoint in such a way that

$$\sup_{t \downarrow 0} t(1-t) \frac{\left|F''(Q(t))\right|}{(F'(Q(t)))^2} < \infty .$$

It remains to verify that

$$\sup_{t \uparrow 1} t(1-t) \frac{\left|F''(Q(t))\right|}{(F'(Q(t)))^2} < \infty,$$

or equivalently,

$$\sup_{t \uparrow 1} -(1-t) \frac{Q''(t)}{Q'(t)} < \infty .$$

Since $Q(t) = U(1/(1 - t))$, this is the same as

$$\sup_{s \to \infty} \left| 2 + \frac{sU''(s)}{U'(s)} \right| < \infty .$$

Now by assumption,

$$\lim_{s \to \infty} 2 + \frac{sU''(s)}{U'(s)} = 1 + \gamma ;$$

hence the conditions are fulfilled.

Next we apply Proposition 2.4.9. Since we can be sure of the behavior only near the right endpoint, in (2.4.3) we replace t by $1 - ks/n$ throughout:

$$\sup_{k^{-1} \le s \le 1} n^{\varepsilon} \left(\frac{ks}{n} \right)^{\varepsilon - 1/2} \left| n^{1/2} \left(\frac{n}{ks} \right)^{-1} \left(\frac{X_{n-[ks],n} - U(\frac{n}{ks})}{a(\frac{n}{ks})} \right) - B_n \left(1 - \frac{ks}{n} \right) \right|$$
$$= O_P(1) ,$$

with $a(t) = tU'(t)$, or after some rearrangement,

$$\sup_{k^{-1} \le s \le 1} k^{\varepsilon} \left| s^{1/2+\varepsilon} \left(\sqrt{k} \frac{X_{n-[ks],n} - U(\frac{n}{ks})}{a(\frac{n}{ks})} \right) \right.$$
$$\left. - s^{-1/2+\varepsilon} \left(\frac{n}{k} \right)^{1/2} B_n \left(1 - \frac{ks}{n} \right) \right| = O_P(1) . \quad (2.4.5)$$

Now take $\varepsilon > 0$. Then the expression within the absolute value must be $o_P(1)$ uniformly in s. Next we look at the Brownian bridge part. Recall (2.4.4) with $\{W_n\}$ Brownian motion. Further note that

$$\sup_{k^{-1} \le s \le 1} \frac{ks}{n} s^{-1/2+\varepsilon} \left(\frac{n}{k} \right)^{1/2} W_n(1) = o_P(1) . \quad (2.4.6)$$

Combining (2.4.4), (2.4.5), and (2.4.6), we get as $n \to \infty$,

$$\sup_{k^{-1} \le s \le 1} s^{1/2+\varepsilon} \left| \sqrt{k} \frac{X_{n-[ks],n} - U\left(\frac{n}{ks}\right)}{a\left(\frac{n}{ks}\right)} - \frac{W_n(s)}{s} \right| = o_P(1) . \quad (2.4.7)$$

It is not difficult to see that (2.4.7) still holds if we replace the function a with any function a_1 provided $a_1(t) \sim a(t), t \to \infty$. In fact, we shall use the function a_0 from Theorem 2.3.6.

Finally, we can handle the expansion in the statement of the theorem:

$$s^{\gamma+1/2+\varepsilon} \left\{ \sqrt{k} \left(\frac{X_{n-[ks],n} - U(\frac{n}{k})}{a_0(\frac{n}{k})} - \frac{s^{-\gamma} - 1}{\gamma} \right) \right.$$
$$\left. - s^{-\gamma-1} \sqrt{\frac{n}{k}} W_n \left(\frac{ks}{n} \right) - \sqrt{k} A_0 \left(\frac{n}{k} \right) \Psi_{\gamma,\rho}(s^{-1}) \right\}$$

$$\stackrel{d}{=} s^{\gamma+1/2+\varepsilon}\left\{\sqrt{k}\left(\frac{U(\frac{n}{ks})-U(\frac{n}{k})}{a_0(\frac{n}{k})}-\frac{s^{-\gamma}-1}{\gamma}\right)\right.$$

$$\left.-\sqrt{k}A_0\left(\frac{n}{k}\right)\Psi_{\gamma,\rho}(s^{-1})\right\} \tag{2.4.8}$$

$$+\frac{s^{\gamma}a_0(\frac{n}{ks})}{a_0(\frac{n}{k})}\left\{s^{1/2+\varepsilon}\left(\sqrt{k}\frac{X_{n-[ks],n}-U(\frac{n}{ks})}{a_0(\frac{n}{ks})}-\frac{W_n(s)}{s}\right)\right\} \tag{2.4.9}$$

$$+s^{\gamma+1/2+\varepsilon}\left(\frac{a_0(\frac{n}{ks})}{a_0(\frac{n}{k})}-s^{-\gamma}\right)\frac{W_n(s)}{s}, \tag{2.4.10}$$

where we used the identity

$$\frac{X_{n-[ks],n}-U(\frac{n}{k})}{a_0(\frac{n}{k})}=\frac{U(\frac{n}{ks})-U(\frac{n}{k})}{a_0(\frac{n}{k})}+\frac{a_0(\frac{n}{ks})}{a_0(\frac{n}{k})}\frac{X_{n-[ks],n}-U(\frac{n}{ks})}{a_0(\frac{n}{ks})}.$$

Expression (2.4.8) tends to zero uniformly in s by the inequalities of Theorem 2.3.6, Section 2.3.

Since $s^{\gamma}a_0(n/(ks))/a_0(n/k)\leq cs^{-\varepsilon'}$ for $n\geq n_0$ uniformly for $0<s\leq 1$ by Potter's inequalities (Proposition B.1.9), relation (2.4.7) implies that (2.4.9) tends to zero in probability uniformly in s. Note that (2.4.7) still holds with the function a replaced by a_0.

Again by (2.3.7) and (2.3.23) we have

$$\frac{s^{\gamma}a_0(\frac{n}{ks})}{a_0(\frac{n}{k})}-1=A_0\left(\frac{n}{k}\right)\frac{s^{-\rho}-1}{\rho}+o(1)A_0\left(\frac{n}{k}\right)s^{-\rho-\varepsilon}$$

uniformly for $0<s\leq 1$. Also recall that

$$\sup_{0<s\leq 1}\frac{W_n(s)}{s^{1/2-\varepsilon/2}}<\infty \qquad \text{a.s.}$$

Hence (2.4.10) tends to zero in probability uniformly in s. ∎

Proof (of Theorem 2.4.2). Since U satisfies the second-order condition, there exists a function U_1 satisfying von Mises' second-order condition such that

$$\lim_{t\to\infty}\frac{U(t)-U_1(t)}{tU_1'(t)A_1(t)}=0 \tag{2.4.11}$$

with $A_1(t)=tU_1''(t)/U_1'(t)-\gamma+1$ (Theorem B.3.13). With $q(t)=tU_1'(t)A_1(t)$ (note that $|q(t)|\in RV_{\gamma+\rho}$),

$$s^{\gamma+1/2+\varepsilon}\sqrt{k}\frac{U(Y_{n-[ks],n})-U_1(Y_{n-[ks],n})}{\frac{n}{k}U_1'(\frac{n}{k})}$$

$$=\left(\sqrt{k}A_1\left(\frac{n}{k}\right)\right)\frac{U(Y_{n-[ks],n})-U_1(Y_{n-[ks],n})}{q(Y_{n-[ks],n})}\left(s^{\gamma+1/2+\varepsilon}\frac{q(Y_{n-[ks],n})}{q(\frac{n}{k})}\right). \tag{2.4.12}$$

The first factor is bounded by assumption and the second one tends to zero by (2.4.11). For the last factor recall Potter's inequalities (Proposition B.1.9(5)): for $t \geq t_0, x \geq \frac{1}{2}$

$$(1 - \varepsilon) \, x^{\gamma + \rho} \min \left(x^{-\varepsilon'}, x^{\varepsilon'} \right) \leq \frac{q(tx)}{q(t)} \leq (1 + \varepsilon) \, x^{\gamma + \rho} \max \left(x^{-\varepsilon'}, x^{\varepsilon'} \right) . \quad (2.4.13)$$

We apply this with $t := n/k$ and $x := (k/n)Y_{n-[ks],n} \geq (k/n)Y_{n-k,n} \to^P 1, n \to \infty$. Now from Lemma 2.4.10 we have that

$$s^{\gamma + 1/2 + \varepsilon} \left(\frac{k}{n} Y_{n-[ks],n} \right)^{\gamma + \rho + \varepsilon'} = s^{-\rho - 1/2 + \varepsilon - \varepsilon'} \frac{W_n(s)}{\sqrt{k}} + \frac{o_P(1)}{\sqrt{k}} s^{-\rho + \varepsilon - \varepsilon' - \varepsilon''} .$$

$$(2.4.14)$$

Combining (2.4.13) and (2.4.14) and the fact that

$$\sup_{0 < s \leq 1} s^{-1/2 + \varepsilon} W_n(s)$$

is bounded a.s., we find that the third factor of (2.4.12) is bounded as well. Hence

$$\sup_{k^{-1} \leq s \leq 1} s^{\gamma + 1/2 + \varepsilon} \sqrt{k} \frac{U(Y_{n-[ks],n}) - U_1(Y_{n-[ks],n})}{\frac{n}{k} U_1'(\frac{n}{k})} = o_P(1) , \quad n \to \infty .$$

$$(2.4.15)$$

We already know from Lemma 2.4.11 that the result of Theorem 2.4.2 holds with $X_{n-[ks],n}$ replaced by $U_1(Y_{n-[ks],n})$. Relation (2.4.15) then implies that the result of Theorem 2.4.2 also holds with $X_{n-[ks],n}$ replaced by $U(Y_{n-[ks],n})$. This completes the proof. ∎

Proof (of Corollary 2.4.5). The range of t values in (2.4.3) is $(n + 1)^{-1} \leq t \leq n(n+1)^{-1}$. For the result of Theorem 2.4.2 we used only the range $n^{-1} \leq t \leq 1 - n^{-1}$. By taking $t = n/(n + 1)$ in (2.4.3) and following the lines of the proof of Theorem 2.4.2 with $s = n/(k(n + 1))$ we obtain

$$\left(\frac{n}{k(n + 1)} \right)^{\gamma + 1/2 + \varepsilon} \left\{ \sqrt{k} \left(\frac{X_{n,n} - U(\frac{n}{k})}{a_0(\frac{n}{k})} - \frac{\left(\frac{n}{k(n+1)} \right)^{-\gamma} - 1}{\gamma} \right) \right.$$

$$\left. - \left(\frac{n}{k(n + 1)} \right)^{-\gamma - 1} W_n \left(\frac{n}{k(n + 1)} \right) - \sqrt{k} A_0 \left(\frac{n}{k} \right) \Psi_{\gamma, \rho} \left(\frac{k(n + 1)}{n} \right) \right\} \xrightarrow{P} 0 .$$

$$(2.4.16)$$

Let us consider the case $\gamma \geq -\frac{1}{2}$ first. Since

$$\sup_{0 < s < k^{-1}} s^{-1/2 + \varepsilon} |W_n(s)| \xrightarrow{P} 0 , \quad (2.4.17)$$

$$\sup_{0 < s < k^{-1}} s^{\gamma + 1/2 + \varepsilon} \left| \Psi_{\gamma, \rho}(s^{-1}) \right| \xrightarrow{P} 0 , \quad (2.4.18)$$

and

$$\sup_{0<s<k^{-1}} s^{\gamma+1/2+\varepsilon} \left| \frac{s^{-\gamma}-1}{\gamma} \right| \xrightarrow{P} 0 , \tag{2.4.19}$$

as $n \to \infty$, (2.4.16) implies that, for $\gamma \neq 0$,

$$k^{-\gamma-\varepsilon} \left| \frac{X_{n,n} - U(\frac{n}{k})}{a_0(\frac{n}{k})} + \frac{1}{\gamma} \right| \xrightarrow{P} 0 . \tag{2.4.20}$$

Since

$$\sup_{0<s<k^{-1}} s^{\gamma+1/2+\varepsilon} \leq k^{-\gamma-1/2-\varepsilon}$$

and $X_{n-[ks],n} = X_{n,n}$ for $s < k^{-1}$, we get, using (2.4.17)–(2.4.20),

$$\sup_{0<s<k^{-1}} s^{\gamma+1/2+\varepsilon} \left| \sqrt{k} \left(\frac{X_{n-[ks],n} - U(\frac{n}{k})}{a_0(\frac{n}{k})} - \frac{s^{-\gamma}-1}{\gamma} \right) \right.$$
$$\left. - s^{-\gamma-1} W_n(s) - \sqrt{k} A_0 \left(\frac{n}{k} \right) \Psi_{\gamma,\rho}(s^{-1}) \right| \xrightarrow{P} 0 .$$

For $\gamma = 0$ a similar proof applies.

Next we consider the case $\gamma < -\frac{1}{2}$. We need to prove

$$\sup_{k^{-1}\leq s\leq 1} s^{\gamma+1/2+\varepsilon} \left| \sqrt{k} \left(\frac{X_{n-[ks],n} - X_{n,n}}{a_0(\frac{n}{k})} - \frac{s^{-\gamma}}{\gamma} \right) \right.$$
$$\left. - s^{-\gamma-1} W_n(s) - \sqrt{k} A_0 \left(\frac{n}{k} \right) \Psi_{\gamma,\rho}(s^{-1}) \right| \xrightarrow{P} 0 \tag{2.4.21}$$

and

$$\sup_{0<s<k^{-1}} s^{\gamma+1/2+\varepsilon} \left| \sqrt{k} \frac{s^{-\gamma}}{\gamma} - s^{-\gamma-1} W_n(s) - \sqrt{k} A_0 \left(\frac{n}{k} \right) \Psi_{\gamma,\rho}(s^{-1}) \right| \xrightarrow{P} 0 . \tag{2.4.22}$$

Now, (2.4.21) is dominated by the sum of two terms: the left-hand side of (2.4.2), which goes to zero, and

$$\sup_{k^{-1}\leq s\leq 1} s^{\gamma+1/2+\varepsilon} \sqrt{k} \left| \frac{X_{n,n} - U(\frac{n}{k})}{a_0(\frac{n}{k})} + \frac{1}{\gamma} \right|$$

$$\leq k^{-\gamma-1/2-\varepsilon} \sqrt{k} \left| \frac{X_{n,n} - U(\frac{n}{k})}{a_0(\frac{n}{k})} + \frac{1}{\gamma} \right|$$

$$\leq k^{-\gamma-1/2-\varepsilon} \left| \sqrt{k} \left(\frac{X_{n,n} - U(\frac{n}{k})}{a_0(\frac{n}{k})} - \frac{\left(\frac{n}{k(n+1)} \right)^{-\gamma} - 1}{\gamma} \right) \right.$$

$$\left. - \left(\frac{n}{k(n+1)} \right)^{-\gamma-1} W_n \left(\frac{n}{k(n+1)} \right) - \sqrt{k} A_0 \left(\frac{n}{k} \right) \Psi_{\gamma,\rho} \left(\frac{k(n+1)}{n} \right) \right|$$

$$+ k^{-\gamma - \varepsilon} \frac{\left(\frac{n}{k(n+1)} \right)^{-\gamma}}{\gamma}$$

$$+ k^{-\gamma - 1/2 - \varepsilon} \left(\frac{n}{k(n+1)} \right)^{-\gamma - 1} \left| W_n \left(\frac{n}{k(n+1)} \right) \right|$$

$$+ k^{-\gamma - 1/2 - \varepsilon} \left| \sqrt{k} A_0 \left(\frac{n}{k} \right) \Psi_{\gamma, \rho} \left(\frac{k(n+1)}{n} \right) \right| .$$

The first term of this expression tends to zero by (2.4.16). One easily checks that the other three terms tend to zero too. Similarly one checks that (2.4.22) tends to zero by considering the three terms separately. ∎

Proof (of Theorem 2.4.8). As before, we prove the result with $X_{n-[ks],n}$ replaced with $U\left(Y_{n-[ks],n} \right)$, where $\{ Y_{i,n} \}$ are the nth order statistics from the distribution function $1 - 1/x$, $x \geq 1$. Theorem 2.3.9 tells us that for $\varepsilon > 0$,

$$\frac{U\left(Y_{n-[ks],n} \right)}{U\left(\frac{n}{k} \right)} = \left(\frac{k}{n} Y_{n-[ks],n} \right)^{\gamma}$$

$$+ A_0 \left(\frac{n}{k} \right) \left\{ \left(\frac{k}{n} Y_{n-[ks],n} \right)^{\gamma} \frac{\left(\frac{k}{n} Y_{n-[ks],n} \right)^{\rho} - 1}{\rho} + o_p(1) \left(\frac{k}{n} Y_{n-[ks],n} \right)^{\gamma + \rho + \varepsilon} \right\} .$$

$$(2.4.23)$$

Now Lemma 2.4.10 tells us that, uniformly for $k^{-1} \leq s \leq 1$,

$$\left(\frac{k}{n} Y_{n-[ks],n} \right)^{\gamma} = s^{-\gamma} + \frac{\gamma}{\sqrt{k}} \left(s^{-\gamma - 1} W_n(s) + o_p(1) s^{-\gamma - 1/2 - \varepsilon} \right) . \quad (2.4.24)$$

Similarly, we get from Lemma 2.4.10,

$$\frac{\left(\frac{k}{n} Y_{n-[ks],n} \right)^{\rho} - 1}{\rho} = \frac{s^{-\rho} - 1}{\rho} + \frac{1}{\sqrt{k}} \left(s^{-\rho - 1} W_n(s) + o_p(1) s^{-\rho - 1/2 - \varepsilon} \right) . \quad (2.4.25)$$

Now note that in fact the product of the right-hand sides of (2.4.24) and (2.4.25) can be written as

$$s^{-\gamma} \frac{s^{-\rho} - 1}{\rho} + o_p(1) s^{-\gamma - 1/2 - \varepsilon} .$$

Hence

$$\left(\frac{k}{n} Y_{n-[ks],n} \right)^{\gamma} \frac{\left(\frac{k}{n} Y_{n-[ks],n} \right)^{\rho} - 1}{\rho} = s^{-\gamma} \frac{s^{-\rho} - 1}{\rho} + o_p(1) s^{-\gamma - 1/2 - \varepsilon} . \quad (2.4.26)$$

Similarly one checks that, for $\varepsilon < -\rho + \frac{1}{2}$,

$$\left(\frac{k}{n} Y_{n-[ks],n} \right)^{\gamma + \rho + \varepsilon} = O_p \left(s^{-\gamma - \frac{1}{2} - \varepsilon} \right) . \quad (2.4.27)$$

It follows by combining (2.4.23), (2.4.24), (2.4.26), and (2.4.27) that the supremum over $k^{-1} \leq s \leq 1$ of the expression in the first statement of the theorem is $o_p(1)$. The rest of the proof is like that of Corollary 2.4.5.

For the second statement note that the second-order condition (2.3.22) is equivalent to

$$\lim_{t \to \infty} \frac{\log U(tx) - \log U(t) - \gamma \log x}{A(t)} = \frac{x^\rho - 1}{\rho} .$$

Moreover, we have the uniform inequalities (Theorem B.2.18)

$$\left| \frac{\log U(tx) - \log U(t) - \gamma \log x}{A_0(t)} - \frac{x^\rho - 1}{\rho} \right| \leq \varepsilon x^\rho \max \left(x^\delta, x^{-\delta} \right) .$$

The rest of the proof is similar to that of the first statement.   ∎

Exercises

2.1. Let $X_1, X_2, \ldots$ be i.i.d. random variables with distribution function F and $X_{1,n} \leq X_{2,n} \cdots \leq X_{n,n}$ the nth order statistics. Let F be in the domain of attraction of G_γ with $\gamma > 0$. Prove that $X_{n,n}/X_{n-1,n} \to^d Y^\gamma$ as $n \to \infty$, where Y has distribution function $1 - 1/x$, $x \geq 1$.

Hint: Rényi's representation (Section 2.1) implies that for exponential order statistics, $(E_{n-1,n}, E_{n,n} - E_{n-1,n})$ are independent and $E_{n,n} - E_{n-1,n}$ has a standard exponential distribution. Hence $(X_{n,n}/X_{n-1,n}) =^d (U(Y^\star Y_{n-1,n})/U(Y_{n-1,n})$, where $Y^\star$ and $Y_{n-1,n}$ are independent, $Y^\star$ has distribution function $1 - 1/x$, $x \geq 1$, and $Y_{n-1,n}$ is the second maximum of a sample of size n with distribution function $1 - 1/x$, $x \geq 1$. Finally, use Corollary 1.2.10.

Remark: The converse statement is also true (see Smid and Stam (1975)).

2.2. (Beirlant and Teugels (1986)) Let $X_1, X_2, \ldots$ be i.i.d. random variables with distribution function F with $x^* > 0$. Define $M_{n,k}^{(1)} := k^{-1} \sum_{i=0}^{k-1} \log X_{n-i,n} - \log X_{n-k,n}$. If F is in the domain of attraction of some extreme value distribution G_γ with auxiliary function $a(n)$ and $k \leq n$ is a fixed integer, then

$$\frac{M_{n,k}^{(1)}}{a(n)/U(n)} \xrightarrow{d} \begin{cases} \frac{1}{k} \sum_{i=0}^{k-1} Z_i , & \gamma \geq 0, \\ Q_k^{-\gamma} \frac{1}{k} \sum_{i=0}^{k-1} \frac{\exp\left(\gamma \sum_{j=i}^{k-1} Z_i/j\right) - 1}{\gamma} , & \gamma < 0 , \end{cases}$$

as $n \to \infty$, with $U := (1/(1-F))^\leftarrow$, $Q_k, Z_0, Z_1, \ldots, Z_{k-1}$ independent, Q_k gamma distributed with k degrees of freedom, and $Z_i, i = 0, 1, \ldots, k-1$, i.i.d. exponential.

2.3. Derive the limit distribution of $X_{n,n}$ from the point process convergence of Theorem 2.1.2. Do the same for the joint distribution of $(X_{n-1,n}, X_{n,n})$.

2.4. What are the possible limit distributions of $(X_{n-1,n}, X_{n,n})$?

2.5. Let $Y_1, Y_2, \ldots$ be independent and identically distributed with distribution function $1 - 1/x$, $x \geq 1$. Using the point process convergence of Theorem 2.1.2 find the limit distribution under a trend, i.e., the limit distribution of $\max_{1 \leq i \leq n}(X_i - i)$.
Hint: Recall what convergence of point process means (cf. last paragraph of Section 2.1).

2.6. Let $U_{1,n} \leq U_{2,n} \leq \cdots \leq U_{n,n}$ be the order statistics from a standard uniform distribution. Let $k = k(n)$ be a sequence of integers such that for some $p \in (0, 1)$, $\lim_{n \to \infty} \sqrt{n}(k/n - p) = 0$. Prove that $\sqrt{n}(U_{k,n} - p)/\sqrt{p(1 - p)}$ has a standard normal limit distribution as $n \to \infty$.

2.7. Show that the distribution function F defined by $1 - F(x) = x^{-1}(1 + x^{-1} \exp(\sin \log x))$ satisfies the domain of attraction condition but not the second-order relation (2.3.24).

2.8. Find the second-order relation for the Cauchy distribution.

2.9. Check the second-order condition for the normal distribution: note that with Φ the standard normal distribution function,

$$1 - \Phi(t) = (2\pi)^{-1/2} e^{-t^2/2}(1/t - 1/t^3 + o(1/t^3)) .$$

Write $\psi(t) := 1/(1 - \Phi(t))$. Prove that for $x \in \mathbb{R}$,

$$\lim_{t \to \infty} t^2 \left(\psi(t + x/t)/\psi(t) - e^x \right) = \left(x^2/2 + x \right) e^x$$

locally uniformly and conclude that for $x > 0$,

$$\lim_{t \to \infty} (\Psi(t))^3 \{ \Psi(tx) - \Psi(t) - (\log x)/\Psi(t) \} = -(\log x)^2/2 - \log x , \quad x > 0 ,$$

with Ψ the inverse function of $\psi(t)$. Finally, for $x > 0$,

$$\lim_{t \to \infty} (2 \log t)^{3/2} \left(\Psi(tx) - \Psi(t) - \frac{\log x}{(2 \log t - \log \log t - \log 4\pi)^{1/2}} \right)$$
$$= -(\log x)^2/2 - \log x .$$

Hint: For the last step use (and prove)

$$\Psi(t) = (2 \log t - \log \log t - \log 4\pi)^{1/2} + o\left((2 \log t)^{-1/2} \right) , \quad t \to \infty .$$

2.10. Check that the gamma distribution satisfies the second-order regular variation condition with $\gamma = \rho = 0$ and determine possible auxiliary functions a and A.

2.11. Prove the equivalence of (2.3.22) and (2.3.24) by noting that (2.3.22) is equivalent to

$$\lim_{t \to \infty} \frac{\frac{U(U^{\leftarrow}(t) x^{1/\gamma})}{t} - x}{A(U^{\leftarrow}(t))} = x \frac{x^{\rho/\gamma} - 1}{\gamma}$$

(with $U^{\leftarrow}$ the left-continuous inverse of U) and then applying Vervaat's lemma (Appendix A).

2.12. Let $U(t) = t^\gamma - k/\gamma + t^{\gamma+\tau}/\tau + o(t^{\gamma+\tau})$ for $\gamma > 0$ and $\tau < 0, t \to \infty$. Check that $U(t)$ satisfies the second-order condition for γ positive (2.3.22) with $A(t) = t^\tau$ if $k = 0$ or $\gamma + \tau > 0$ and $A(t) = kt^{-\gamma}$ if $\gamma + \tau < 0$. Discuss the case $\gamma + \tau = 0$.

2.13. Let $\gamma > 0$. Check that for $\gamma + \rho > 0$, or $\gamma + \rho < 0$ and $\lim_{t\to\infty} U(t) - a(t)/\gamma = 0$, if $A^\star(t)$ is the auxiliary function in (2.3.22) then possible first- and second-order auxiliary functions for (2.3.5) are $a(t) = \gamma U(t)(1 + A^\star(t)/\gamma)$ and $A(t) = (\gamma + \rho)A^\star(t)/\gamma$ respectively (and vice versa).

2.14. Verify that if $U(t) = c_0 + c_1 t^\gamma (1 + c_2 t^\rho + o(t^\rho))$ for $\gamma < 0, \rho < 0, \gamma + \rho \neq 0$, $c_0, c_2 \neq 0$, and $c_1 < 0$, as $t \to \infty$, then the second-order condition (2.3.5) holds with $A(t) = \rho\gamma^{-1}(\gamma + \rho)c_2 t^\rho$ and $a(t) = \gamma c_1 t^\gamma (1 + \rho^{-1} A(t))$.

2.15. The Student t-distribution with ν degrees of freedom satisfies

$$1 - F(t) = c_\nu \frac{\nu^{\nu/2}}{t^\nu} + d_\nu \frac{\nu^{\nu/2+1}}{t^{\nu+2}} + O(t^{-\nu-4}),$$

$t \to \infty$, where

ν	1	2	3	4	5
c_ν	$1/\pi$	$1/4$	$2/(3\pi)$	$3/16$	$8/(15\pi)$
d_ν	$-1/(3\pi)$	$-3/16$	$-4/(5\pi)$	$-10/32$	$-8/(7\pi)$

(Martins (2000)). Hence this model satisfies the second-order condition (2.3.5) with $\gamma = 1/\nu$ and $\rho = -2/\nu$. Obtain the auxiliary functions.

2.16. Lemma 2.4.10 implies that

$$\sqrt{k}\left(\frac{k}{n}Y_{n-[ks],n} - s^{-1}\right) \xrightarrow{d} s^{-1}W(s)$$

in $D(0, 1]$, with W denoting Brownian motion. Let F_n be the empirical distribution function. Note that $(n/k)\{1 - F_n(n/(kx))\}$ is the inverse function of $\left((k/n)Y_{n-[ks],n}\right)^{-1}$. Use Vervaat's lemma (Lemma A.0.2) to conclude that

$$\sqrt{k}\left\{\frac{n}{k}\left(1 - F_n\left(\frac{n}{k}x\right)\right) - \frac{1}{x}\right\} \xrightarrow{d} -W\left(\frac{1}{x}\right)$$

in $D[1, \infty)$.

2.17. Let s be some fixed positive constant. Deduce under the conditions of Theorem 2.4.2 that

$$\sqrt{k}\,\frac{X_{n-[ks],n} - U\left(\frac{n}{k}s^{-1}\right)}{a\left(\frac{n}{k}\right)}$$

converges to a normal random variable with mean zero and variance $s^{-2\gamma-1}$ as $n \to \infty$.

2.18. Formulate an analogous weak convergence result for the empirical distribution function in the situation of Theorem 2.4.2. Assume $\sqrt{k}A_0(n/k) \to 0$. Prove the result.

2.19. Prove that under the conditions of Theorem 2.4.8 and $\sqrt{k}A_0(n/k) \to \lambda$,

$$\sqrt{k}\left(\left(\log X_{n-2k,n} - \log X_{n-k,n}\right)/\log 2 - \gamma\right)$$

has asymptotically a normal distribution. This is a first example of an estimator of γ. Notice that $\left(\log X_{n-2k,n} - \log X_{n-k,n}\right)/\log 2$ is a consistent and asymptotically normally distributed estimator of γ.

3

Estimation of the Extreme Value Index and Testing

3.1 Introduction

The alternative conditions of Theorem 1.1.6 (Section 1.1.3) serve as a basis for statistical applications of extreme value theory.

Consider relation (1.1.22) (Section 1.1.3): there exists a positive nondecreasing function f such that

$$\lim_{t \uparrow x^*} \frac{1 - F(t + xf(t))}{1 - F(t)} = (1 + \gamma x)^{-1/\gamma} \qquad (3.1.1)$$

for all x for which $1 + \gamma x > 0$, where $x^* = \sup\{x : F(x) < 1\}$.

Let X be a random variable with distribution function F and let $F \in \mathcal{D}(G_\gamma)$ for some real γ. Then (3.1.1) tells us that for $x > 0$, $x < (0 \vee (-\gamma))^{-1}$,

$$\lim_{t \uparrow x^*} P\left(\frac{X - t}{f(t)} > x \mid X > t\right) = (1 + \gamma x)^{-1/\gamma}. \qquad (3.1.2)$$

That is, the conditional distribution of $(X - t)/f(t)$ given $X > t$ has the limit distribution, as $t \uparrow x^*$,

$$H_\gamma(x) := 1 - (1 + \gamma x)^{-1/\gamma}, \quad 0 < x < (0 \vee (-\gamma))^{-1}, \qquad (3.1.3)$$

where for $\gamma = 0$ the right-hand side is interpreted as $1 - e^{-x}$. This class of distribution functions is called the class of the *generalized Pareto distributions* (GP). Figure 3.1 illustrates this class for some values of γ.

Relation (3.1.1) means loosely speaking that from some high threshold t onward (i.e., $X > t$) the distribution function can be written approximately as

$$1 - F(x) \approx (1 - F(t))\left\{1 - H_\gamma\left(\frac{x - t}{f(t)}\right)\right\}, \quad x > t,$$

which is a parametric family of distribution tails. One can expect this approximation to hold for intermediate and extreme order statistics. Let $X_1, X_2, \ldots$ be independent

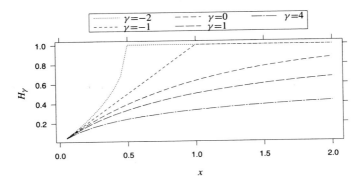

Fig. 3.1. Family of GP distributions: for $\gamma = -2$ and $\gamma = -1$ the right endpoints are 1 and 0.5 respectively, for $\gamma \geq 0$ the right endpoint equals infinity.

and identically distributed random variables with distribution function F, and F_n the corresponding *empirical distribution function*, i.e., $F_n(x) := n^{-1} \sum_{i=1}^{n} 1_{\{X_i \leq x\}}$. Let us apply the last approximation with $t := X_{n-k,n}$, where we choose $k = k(n) \to \infty$, $k/n \to 0$, $n \to \infty$. Then

$$1 - F(x) \approx \left(1 - F(X_{n-k,n})\right) \left\{1 - H_\gamma \left(\frac{x - X_{n-k,n}}{f(\frac{n}{k})}\right)\right\}$$

and, since $1 - F(X_{n-k,n}) \approx 1 - F_n(X_{n-k,n}) = k/n$,

$$1 - F(x) \approx \frac{k}{n} \left\{1 - H_\gamma \left(\frac{x - X_{n-k,n}}{f(\frac{n}{k})}\right)\right\} . \tag{3.1.4}$$

In order to make this approximation applicable, we need to estimate γ and the function f at the point n/k.

This approximation is valid for any x larger than $X_{n-k,n}$ and can be used even for $x > X_{n,n}$, i.e., outside the range of the observations, and this is in fact the basis for applications of extreme values.

Next we consider a similar application of relation (1.1.20) (Section 1.1.3), that there exists a positive function a such that for all $x > 0$,

$$\lim_{t \to \infty} \frac{U(tx) - U(t)}{a(t)} = \frac{x^\gamma - 1}{\gamma} . \tag{3.1.5}$$

Relation (3.1.5) leads to the following approximation:

$$U(x) \approx U(t) + a(t) \frac{\left(\frac{x}{t}\right)^\gamma - 1}{\gamma} , \quad x > t . \tag{3.1.6}$$

This approximation is useful when one wants to estimate a quantile $F^{\leftarrow}(1 - p) = U(1/p)$ with p very small, since this quantile is then related to a much lower quantile

$U(t) = F^{\leftarrow}(1 - 1/t)$, which can be estimated by an intermediate order statistic. Hence we choose $t := n/k$ with $k = k(n) \to \infty$, $k/n \to 0$, $n \to \infty$. Then for large y, for example $y = 1/p$ with p small,

$$U(y) \approx U\left(\frac{n}{k}\right) + a\left(\frac{n}{k}\right)\frac{\left(\frac{yk}{n}\right)^{\gamma} - 1}{\gamma}. \tag{3.1.7}$$

In order to make this approximation applicable, we need to estimate γ, the function a at the point n/k, and $U(n/k)$. The latter quantity can be estimated by an intermediate order statistic. Again the approximation will be used not only for $y < n$ but also for extrapolation outside the sample.

Let us go back to the examples of applications of extreme value theory discussed in Section 1.1.4.

Sea Level

As mentioned previously, the Dutch government requires the sea dikes to be so high that in any year a flood occurs with probability $1/10\,000$. In order to estimate the height of the dike that corresponds to that probability, 1877 high-tide water levels are available, monitored at the coast, one for each severe storm of a certain type, over a period of 111 years. The observations can be considered as realizations of independent and identically distributed random variables. So, if F is the distribution of these random variables (the ones taken during storms), we need to estimate $U((1877/111) \times 10^4) \approx U(17 \times 10^4)$. The largest observation roughly corresponds to $U(1878) \approx U(19 \times 10^2)$. This means that we have to estimate the quantile function outside the range of the available data, and for this the extreme value conditions can be used. So we suppose $F \in \mathcal{D}(G_{\gamma})$ for some $\gamma \in \mathbb{R}$. Then by (3.1.7),

$$U(17 \times 10^4) \approx U\left(\frac{n}{k}\right) + a\left(\frac{n}{k}\right)\frac{\left(\frac{17k \times 10^4}{n}\right)^{\gamma} - 1}{\gamma}. \tag{3.1.8}$$

We propose the estimator

$$\hat{U}(17 \times 10^4) = \hat{U}\left(\frac{n}{k}\right) + \hat{a}\left(\frac{n}{k}\right)\frac{\left(\frac{17k \times 10^4}{n}\right)^{\hat{\gamma}} - 1}{\hat{\gamma}} \tag{3.1.9}$$

based on suitable estimators $\hat{\gamma}$, $\hat{U}(n/k)$, and $\hat{a}(n/k)$. In the rest of this and in the next chapter we shall meet various estimators of these quantities for which the vector

$$\sqrt{k}\left(\hat{\gamma} - \gamma, \frac{\hat{U}(\frac{n}{k}) - U(\frac{n}{k})}{a(\frac{n}{k})}, \frac{\hat{a}(\frac{n}{k})}{a(\frac{n}{k})} - 1\right) \tag{3.1.10}$$

is asymptotically normal under suitable conditions. Using this relation we will prove in Chapter 4 that

$$\hat{U}(17 \times 10^4) - U(17 \times 10^4) \, ,$$

suitably normalized, is asymptotically normal. This leads to an asymptotic confidence interval for $U(17 \times 10^4)$.

S&P 500

On the basis of observations of loss returns of the S&P 500 index, we want to estimate the probability that a certain given large loss is exceeded.

If F is the distribution of the loss log-returns and supposing $F \in \mathcal{D}(G_\gamma)$, relation (3.1.4) suggests

$$(1 - F(x))\widehat{} = \frac{k}{n} \left\{ 1 + \hat{\gamma} \left(\frac{x - X_{n-k,n}}{\hat{a}(\frac{n}{k})} \right) \right\}^{-1/\hat{\gamma}}$$

with $\hat{\gamma}$ and $\hat{a}(n/k)$ suitable estimators. Recall the relation $f(t) = a(1/(1 - F(t)))$ (cf. Theorem 1.1.6) with a the positive function in (3.1.5). Then for $t := X_{n-k,n}$ we get $f(X_{n-k,n}) = a(1/(1 - F(X_{n-k,n}))) \approx a(n/k)$.

Again we see that the estimation of γ is a crucial step, which is the main subject in the present chapter. Next, in Chapter 4 we shall prove asymptotic normality of the tail estimator suitably normalized.

Life Span

The life span of people born in the Netherlands in the years 1877–1881 is assumed to be random. Based on life spans of about $10\,000$ people, we want to decide whether the underlying distribution has a finite upper limit $U(\infty)$, and if so, we want to estimate $U(\infty)$. The asymptotic normality of (3.1.10) provides a confidence interval for γ that enables us to test the hypothesis $H_0 : \gamma \geq 0$ versus $H_1 : \gamma < 0$. It will later turn out that the null hypothesis is rejected. Then we want to estimate the finite value $U(\infty)$ and for that we use the limit relation (1.2.14) of Lemma 1.2.9:

$$\lim_{t \to \infty} \frac{U(\infty) - U(t)}{a(t)} = -\frac{1}{\gamma} \, ,$$

i.e.,

$$U(\infty) \approx U(t) - \frac{a(t)}{\gamma} \, .$$

We propose the estimator

$$\hat{U}(\infty) = \hat{U}\left(\frac{n}{k}\right) - \frac{\hat{a}\left(\frac{n}{k}\right)}{\hat{\gamma}},$$

and we will prove, using the joint asymptotic normality in (3.1.10), that $\hat{U}(\infty) - U(\infty)$, suitably normalized, is asymptotically normal. An asymptotic confidence interval for $U(\infty)$ ensues.

The above examples show that it is useful to develop estimators of γ, $U(n/k)$, $a(n/k)$, $U(1/p)$ with p small, $U(\infty)$ and $1 - F(x)$ with x large. In this chapter we discuss estimators for γ (and occasionally for $a(n/k)$), and in the next chapter we shall discuss estimators for the other quantities.

3.2 A Simple Estimator for the Tail Index ($\gamma > 0$): The Hill Estimator

In order to introduce the Hill estimator, a simple and widely used estimator, let us start from Theorem 1.2.1: $F \in \mathcal{D}(G_\gamma)$ for $\gamma > 0$ if and only if

$$\lim_{t \to \infty} \frac{1 - F(tx)}{1 - F(t)} = x^{-1/\gamma} , \quad \gamma > 0 .$$

In this case the parameter $\alpha := 1/\gamma > 0$ is called the *tail index* of F. Theorem 1.2.2 gives an equivalent form of this condition:

$$\lim_{t \to \infty} \frac{\int_t^\infty (1 - F(x)) \frac{dx}{x}}{1 - F(t)} = \gamma .$$

Now partial integration yields

$$\int_t^\infty (1 - F(s)) \frac{ds}{s} = \int_t^\infty (\log u - \log t) \, dF(u) .$$

Hence we have

$$\lim_{t \to \infty} \frac{\int_t^\infty (\log u - \log t) \, dF(u)}{1 - F(t)} = \gamma . \tag{3.2.1}$$

In order to develop an estimator based on this asymptotic result, replace in (3.2.1) the parameter t by the intermediate order statistic $X_{n-k,n}$ and F by the empirical distribution function F_n. We then get *Hill's* (1975) *estimator* $\hat{\gamma}_H$, defined by

$$\hat{\gamma}_H := \frac{\int_{X_{n-k,n}}^\infty \log u - \log X_{n-k,n} \, dF_n(u)}{1 - F_n(X_{n-k,n})} ,$$

or

$$\hat{\gamma}_H := \frac{1}{k} \sum_{i=0}^{k-1} \log X_{n-i,n} - \log X_{n-k,n} . \tag{3.2.2}$$

For the proof of the following theorems we need this auxiliary result.

Lemma 3.2.1 *Let* $Y_1, Y_2, \ldots$ *be i.i.d. random variables with distribution function* $1 - 1/y$, $y \geq 1$, *and let* $Y_{1,n} \leq Y_{2,n} \leq \cdots \leq Y_{n,n}$ *be the nth order statistics. Then with* $k = k(n)$,

$$\lim_{n \to \infty} Y_{n-k,n} = \infty \quad a.s. , \quad n \to \infty ,$$

provided $k(n) = o(n)$.

Proof. The strong law of large numbers implies

$$\frac{1}{n} \sum_{i=1}^{n} 1_{\{Y_i > r\}} \to \frac{1}{r} \qquad \text{a.s.,} \quad n \to \infty,$$

for $r = 1, 2, \ldots$. We proceed by contradiction. Suppose that for some r we have $Y_{n-k,n} < r$ infinitely often. This implies

$$\frac{k}{n} = \frac{1}{n} \sum_{i=1}^{n} 1_{\{Y_i > Y_{n-k,n}\}} > \frac{1}{n} \sum_{i=1}^{n} 1_{\{Y_i > r\}} > \frac{1}{2r}$$

infinitely often, which is the desired contradiction. ∎

Theorem 3.2.2 *Let $X_1, X_2, \ldots$ be i.i.d. random variables with distribution function F. Suppose $F \in \mathcal{D}(G_\gamma)$ with $\gamma > 0$. Then as $n \to \infty$, $k = k(n) \to \infty$, $k/n \to 0$,*

$$\hat{\gamma}_H \xrightarrow{P} \gamma .$$

Proof. By Corollary 1.2.10, $F \in \mathcal{D}(G_\gamma)$ with $\gamma > 0$ implies

$$\lim_{t \to \infty} \frac{U(tx)}{U(t)} = x^\gamma$$

for $x > 0$, i.e. (Proposition B.1.9), for $x \geq 1$ and $t \geq t_0$,

$$(1 - \varepsilon) x^{\gamma - \varepsilon'} < \frac{U(tx)}{U(t)} < (1 + \varepsilon) x^{\gamma + \varepsilon'},$$

or equivalently,

$$\log(1 - \varepsilon) + (\gamma - \varepsilon') \log x < \log U(tx) - \log U(t)$$
$$< \log(1 + \varepsilon) + (\gamma + \varepsilon') \log x . \qquad (3.2.3)$$

Let $Y_1, Y_2, \ldots$ be independent and identically distributed, with common distribution $1 - 1/y$, $y \geq 1$. Note that $U(Y_i) =^d X_i$, $i = 1, 2, \ldots$. So it is sufficient to prove the result for $\hat{\gamma}_H := k^{-1} \sum_{i=0}^{k-1} \log U(Y_{n-i,n}) - \log U(Y_{n-k,n})$. Apply (3.2.3) with $t = Y_{n-k,n}$, $x = Y_{n-i,n}/Y_{n-k,n}$. Since by Lemma 3.2.1, $Y_{n-k,n} \to \infty$ a.s., $n \to \infty$, we have eventually,

$$\log(1 - \varepsilon) + (\gamma - \varepsilon') \log \left(\frac{Y_{n-i,n}}{Y_{n-k,n}} \right) < \log U(Y_{n-i,n}) - \log U(Y_{n-k,n})$$
$$< \log(1 + \varepsilon) + (\gamma + \varepsilon') \log \left(\frac{Y_{n-i,n}}{Y_{n-k,n}} \right)$$

for $i = 0, 1, \ldots, k - 1$; hence

$$\log(1-\varepsilon) + (\gamma - \varepsilon')\frac{1}{k}\sum_{i=0}^{k-1}\log\left(\frac{Y_{n-i,n}}{Y_{n-k,n}}\right) < \hat{\gamma}_H$$

$$< \log(1+\varepsilon) + (\gamma + \varepsilon')\frac{1}{k}\sum_{i=0}^{k-1}\log\left(\frac{Y_{n-i,n}}{Y_{n-k,n}}\right).$$

It now suffices to prove that as $n \to \infty$,

$$\frac{1}{k}\sum_{i=0}^{k-1}\log\left(\frac{Y_{n-i,n}}{Y_{n-k,n}}\right) \xrightarrow{P} 1 .$$

This is part of a separate lemma (note that $\log Y_i$ has a standard exponential distribution), which we give next. ∎

Lemma 3.2.3 *Let $E, E_1, E_2, \ldots$ be i.i.d. standard exponential and let $E_{1,n} \leq E_{2,n} \leq \cdots \leq E_{n,n}$ be the nth order statistics. Let f be such that $\mathrm{Var}\, f(E) < \infty$. Then*

$$\sqrt{k}\left(\frac{1}{k}\sum_{i=0}^{k-1}f(E_{n-i,n} - E_{n-k,n}) - Ef(E)\right)$$

is independent of $E_{n-k,n}$ and asymptotically normal with mean zero and variance $\mathrm{Var}\, f(E)$ as $n \to \infty$, provided $k = k(n) \to \infty$ and $k/n \to 0$.

Proof. Rényi's (1953) representation implies the independence statement and it gives for each n,

$$\left\{E_{n-i,n} - E_{n-k,n}\right\}_{i=0}^{k-1} \stackrel{d}{=} \left\{\sum_{j=i+1}^{k}\frac{E^\star_{n-j+1}}{j}\right\}_{i=0}^{k-1}$$

with $E_1^\star, E_2^\star, \ldots$ independent and identically distributed standard exponential. It follows that the distribution of the left-hand side does not depend on n and that

$$\left\{E_{n-i,n} - E_{n-k,n}\right\}_{i=0}^{k-1} \stackrel{d}{=} \left\{E_{k-i,k}\right\}_{i=0}^{k-1} .$$

It follows that

$$\sqrt{k}\left(\frac{1}{k}\sum_{i=0}^{k-1}f(E_{n-i,n} - E_{n-k,n}) - Ef(E)\right)$$

$$\stackrel{d}{=} \sqrt{k}\left(\frac{1}{k}\sum_{i=0}^{k-1}f(E_{k-i,k}) - Ef(E)\right)$$

$$= \sqrt{k}\left(\frac{1}{k}\sum_{j=1}^{k}f(E_j) - Ef(E)\right)$$

since we take the average of all order statistics.

The result follows from the central limit theorem. ∎

A somewhat surprising converse of Theorem 3.2.2 was proved by Mason (1982). The following is a somewhat stronger result, with different proof.

Theorem 3.2.4 *Let $X_1, X_2, \ldots$ be i.i.d. random variables with distribution function F. Suppose that for some sequence of integers $k = k(n) \to \infty$, $k(n)/n \to 0$, and $k(n+1)/k(n) \to 1$, as $n \to \infty$,*

$$\hat{\gamma}_H \xrightarrow{P} \gamma > 0 .$$

Then $F \in \mathcal{D}(G_\gamma)$.

Proof. Let F_n be the empirical distribution function of $X_1, X_2, \ldots, X_n$ and G_n the empirical distribution function of $Y_1, Y_2, \ldots, Y_n$, which are independent and identically distributed $1 - 1/x$, $x \geq 1$. Then for each n,

$$1 - F_n(x) \stackrel{d}{=} 1 - G_n \left(\frac{1}{1 - F(x)} \right) .$$

We write

$$
\begin{aligned}
\hat{\gamma}_H &= \frac{n}{k} \int_{X_{n-k,n}}^{\infty} (1 - F_n(u)) \, \frac{du}{u} \\
&= \frac{n}{k} \int_{X_{n-k,n}}^{\infty} \left\{ 1 - G_n \left(\frac{1}{1 - F(u)} \right) \right\} \frac{du}{u} \\
&= \frac{n}{k} \int_{Y_{n-k,n}}^{\infty} (1 - G_n(s)) \, d \log U(s)
\end{aligned}
$$

with $Y_{1,n} \leq Y_{2,n} \leq \cdots \leq Y_{n,n}$ the order statistics of $Y_1, Y_2, \ldots, Y_n$.

We are going to use the following results:

1.

$$P \left(\sup_{s \geq 1} s \, (1 - G_n(s)) > b \right) = \frac{1}{b} , \qquad \text{for } b > 1,$$

$$P \left(\inf_{1 \leq s \leq Y_{n,n}} s \, (1 - G_n(s)) < a \right) \leq \frac{e}{a} \, e^{-1/a} , \qquad \text{for } 0 < a < 1$$

(Shorack and Wellner (1986), pp. 345 and 415).

2. Note that

$$
\begin{aligned}
\inf_{Y_{n-k,n} \leq s \leq Y_{n,n}} s \, (1 - G_n(s)) &= \inf_{0 \leq s \leq 1} \frac{ks}{n} Y_{n-[ks],n} \\
&= \left(\inf_{0 \leq s \leq 1} s \frac{Y_{n-[ks],n}}{Y_{n-k,n}} \right) \left(\frac{k}{n} Y_{n-k,n} \right) ,
\end{aligned}
$$

where the two factors are independent. Hence this is the same in distribution to

$$\left(\inf_{0 \leq s \leq 1} s Y^\star_{k-[ks],n} \right) \frac{k}{n} Y_{n-k,n} ,$$

with the $Y^\star$'s independent of and equal in distribution to the Y's, that is,

$$\left(\inf_{1 \le s \le Y^*_{k,k}} s\,(1 - G_n(s)) \right) \frac{k}{n} Y_{n-k,n} \; .$$

3. From Corollary 2.2.2,

$$\lim_{n \to \infty} P\left(1 - \varepsilon \le \frac{k}{n} Y_{n-k,n} \le 1 + \varepsilon \right) = 1 \; .$$

For sufficiently large t let $n = n(t)$ be the integer satisfying

$$\frac{n(t)}{k(n(t))} \le t \le \frac{n(t) + 1}{k(n(t) + 1)} \; .$$

Consider

$$P\left(\frac{1 - \varepsilon}{(1 + \varepsilon)^2} \, t \, (1 + \varepsilon) \int_{t(1+\varepsilon)}^{\infty} \frac{d \log U(s)}{s} \le \frac{n}{k}(1 - \varepsilon) \int_{\frac{n}{k}(1+\varepsilon)}^{\infty} \frac{d \log U(s)}{s} \right.$$

$$\le Y_{n-k,n} \int_{Y_{n-k,n}}^{\infty} \frac{d \log U(s)}{s} \le \frac{n}{k} \int_{Y_{n-k,n}}^{\infty} \frac{s\,(1 - G_n(s))}{1 - \varepsilon} \, \frac{d \log U(s)}{s}$$

$$\left. \le \frac{\gamma + \varepsilon}{1 - \varepsilon} \right) \; .$$

The first inequality is true by definition. The second and fourth inequalities are true with probabilities tending to 1 (for the second we use result (3); the fourth is true by assumption). By results (1) and (2) the third inequality is true with probability at least $1 - e(1 - \varepsilon)^{-1} \exp(-1/(1 - \varepsilon)) > 0$.

Hence we reach the conclusion that for each $\varepsilon > 0$,

$$P\left(t \int_{t}^{\infty} \frac{d \log U(s)}{s} \le \gamma + \varepsilon \right) > 0$$

for t sufficiently large. Hence for sufficiently large t, $t \int_{t}^{\infty} s^{-1}\, d \log U(s) \le \gamma + \varepsilon$. We get the other inequality in a similar fashion.

Hence

$$\lim_{t \to \infty} t \int_{t}^{\infty} \frac{d \log U(s)}{s} = \gamma \; .$$

Now by partial integration

$$t \int_{t}^{\infty} \frac{1}{s}\, d \log U(s) = t \int_{t}^{\infty} \log U(s)\, \frac{ds}{s^2} - \log U(t) \; .$$

Hence by Remark B.2.14(2) we find that for $x > 0$,

$$\lim_{t \to \infty} (\log U(tx) - \log U(t)) = \gamma \log x \; .$$

That is, U is regularly varying with index γ, which implies (Proposition B.1.9) that the function $1 - F$ is regularly varying with index $-1/\gamma$. ∎

Next we formulate conditions that lead to asymptotic normality for $\hat{\gamma}_H$.

Theorem 3.2.5 *Suppose that the distribution function F satisfies the second-order condition of Section 2.3, i.e., for $x > 0$,*

$$\lim_{t\to\infty} \frac{\frac{U(tx)}{U(t)} - x^\gamma}{A(t)} = x^\gamma \frac{x^\rho - 1}{\rho}, \tag{3.2.4}$$

or equivalently,

$$\lim_{t\to\infty} \frac{\frac{1-F(tx)}{1-F(t)} - x^{-1/\gamma}}{A\left(\frac{1}{1-F(t)}\right)} = x^{-1/\gamma} \frac{x^{\rho/\gamma} - 1}{\gamma\rho}, \tag{3.2.5}$$

where $\gamma > 0$, $\rho \leq 0$, and A is a positive or negative function with $\lim_{t\to\infty} A(t) = 0$. Then

$$\sqrt{k}(\hat{\gamma}_H - \gamma) \xrightarrow{d} N\left(\frac{\lambda}{1-\rho}, \gamma^2\right)$$

with N standard normal, provided $k = k(n) \to \infty$, $k/n \to 0$, $n \to \infty$, and

$$\lim_{n\to\infty} \sqrt{k} A\left(\frac{n}{k}\right) = \lambda \tag{3.2.6}$$

with λ finite.

Remark 3.2.6 It may happen that the convergence of $U(tx)/U(t)$ to x^γ is faster than any negative power of t, i.e.,

$$\lim_{t\to\infty} t^\alpha \left(\frac{U(tx)}{U(t)} - x^\gamma\right) = 0$$

for all $x > 0$ and $\alpha > 0$. In that case, the result of Theorem 3.2.5 holds with the bias part $\lambda/(1 - \rho)$ replaced by zero. A similar remark can be made for all the other estimators.

Proof (of Theorem 3.2.5). We write the second-order condition as

$$\lim_{t\to\infty} \frac{\frac{x^{-\gamma} U(tx)}{U(t)} - 1}{A(t)} = \frac{x^\rho - 1}{\rho}.$$

Since $\lim_{t\to\infty} A(t) = 0$, this is equivalent to

$$\lim_{t\to\infty} \frac{\log U(tx) - \log U(t) - \gamma \log x}{A(t)} = \frac{x^\rho - 1}{\rho}.$$

We apply the inequality given in Theorem B.2.18: for a possibly different function A_0, with $A_0(t) \sim A(t)$, $t \to \infty$, and for each $\varepsilon > 0$ there exists a t_0 such that for $t \geq t_0$, $x \geq 1$,

$$\left| \frac{\log U(tx) - \log U(t) - \gamma \log x}{A_0(t)} - \frac{x^\rho - 1}{\rho} \right| \leq \varepsilon x^{\rho+\varepsilon}. \tag{3.2.7}$$

As in the proof of Theorem 3.2.2, we note that

$$\hat{\gamma}_H \stackrel{d}{=} \frac{1}{k} \sum_{i=0}^{k-1} \log U(Y_{n-i,n}) - \log U(Y_{n-k,n}),$$

where the Y_i's are independent and have common distribution function $1 - 1/x$, $x \geq 1$. Hence we work with this representation for $\hat{\gamma}_H$.

Apply (3.2.7) with $t := Y_{n-k,n} \to \infty$ a.s. $n \to \infty$ (Lemma 3.2.1), and $x := Y_{n-i,n}/Y_{n-k,n}$. Then we get eventually, as in the proof of Theorem 3.2.2,

$$\hat{\gamma}_H \stackrel{d}{=} \frac{1}{k} \sum_{i=0}^{k-1} \log U(Y_{n-i,n}) - \log U(Y_{n-k,n})$$

$$= \frac{\gamma}{k} \sum_{i=0}^{k-1} \log \left(\frac{Y_{n-i,n}}{Y_{n-k,n}} \right) + A_0(Y_{n-k,n}) \frac{1}{k} \sum_{i=0}^{k-1} \frac{\left(\frac{Y_{n-i,n}}{Y_{n-k,n}} \right)^{\rho} - 1}{\rho}$$

$$+ o_P(1) |A_0(Y_{n-k,n})| \frac{1}{k} \sum_{i=0}^{k-1} \left(\frac{Y_{n-i,n}}{Y_{n-k,n}} \right)^{\rho+\varepsilon},$$

hence

$$\sqrt{k}(\hat{\gamma}_H - \gamma) = \gamma \sqrt{k} \left(\frac{1}{k} \sum_{i=0}^{k-1} \log \frac{Y_{n-i,n}}{Y_{n-k,n}} - 1 \right)$$

$$+ \sqrt{k} A_0(Y_{n-k,n}) \frac{1}{k} \sum_{i=0}^{k-1} \frac{\left(\frac{Y_{n-i,n}}{Y_{n-k,n}} \right)^{\rho} - 1}{\rho}$$

$$+ o_P(1) \sqrt{k} |A_0(Y_{n-k,n})| \frac{1}{k} \sum_{i=0}^{k-1} \left(\frac{Y_{n-i,n}}{Y_{n-k,n}} \right)^{\rho+\varepsilon}.$$

The first term, suitably normalized, is asymptotically normal by Lemma 3.2.3. As in the proof of Lemma 3.2.3, we have

$$\frac{1}{k} \sum_{i=0}^{k-1} \frac{\left(\frac{Y_{n-i,n}}{Y_{n-k,n}} \right)^{\rho} - 1}{\rho} \stackrel{d}{=} \frac{1}{k} \sum_{i=0}^{k-1} \frac{Y_i^{\rho} - 1}{\rho}. \tag{3.2.8}$$

Hence (3.2.8) tends to $E(Y_1^{\rho} - 1)/\rho = (1 - \rho)^{-1}$ by the law of the large numbers. Similarly,

$$\frac{1}{k} \sum_{i=0}^{k-1} \left(\frac{Y_{n-i,n}}{Y_{n-k,n}} \right)^{\rho+\varepsilon} \stackrel{P}{\to} EY_1^{\rho+\varepsilon} = \frac{1}{1 - \rho - \varepsilon}.$$

It remains to prove that

$$\frac{A_0(Y_{n-k,n})}{A_0 \left(\frac{n}{k} \right)} \stackrel{P}{\to} 1.$$

This follows from Lemma 2.2.3, the fact that the function $|A_0|$ is regular varying, and Potter's inequalities (Proposition B.1.9). ∎

Proof (Second proof of Theorem 3.2.5 via the tail quantile process). By the second statement of Theorem 2.4.8, with $\{W_n(s)\}_{s\geq 0}$ a sequence of Brownian motions and for each $\varepsilon > 0$,

$$\log X_{n-[ks],n} - \log X_{n-k,n} = -\gamma \log s + \frac{\gamma}{\sqrt{k}}\left(s^{-1}W_n(s) - W_n(1)\right)$$

$$+ A_0\left(\frac{n}{k}\right)\left(\frac{s^{-\rho} - 1}{\rho} + o_P(1)s^{-1/2-\varepsilon}\right) ,$$

where the o_P-term tends to zero uniformly for $0 < s \leq 1$. Hence

$$\hat{\gamma}_H = \int_0^1 \left(\log X_{n-[ks],n} - \log X_{n-k,n}\right) ds$$

$$= \gamma \int_0^1 (-\log s) \, ds + \frac{\gamma}{\sqrt{k}} \int_0^1 \left(s^{-1}W_n(s) - W_n(1)\right) ds$$

$$+ A_0\left(\frac{n}{k}\right) \int_0^1 \frac{s^{-\rho} - 1}{\rho} \, ds + o_P(1) A_0\left(\frac{n}{k}\right) \int_0^1 s^{-1/2-\varepsilon} \, ds .$$

It follows that

$$\sqrt{k}\left(\hat{\gamma}_H - \gamma\right) = \gamma \int_0^1 \left(s^{-1}W_n(s) - W_n(1)\right) ds + \sqrt{k}A_0\left(\frac{n}{k}\right)\frac{1}{1-\rho} + o_P(1) .$$

The result follows, since

$$\mathrm{Var}\left(\int_0^1 \left(s^{-1}W_n(s) - W_n(1)\right) ds\right) = 1 .$$ ∎

Remark 3.2.7 A third proof of Theorem 3.2.5, using an expansion for the tail empirical distribution function, will be given in Section 5.1.

Examples of distributions satisfying the second-order condition are abundant. For example, the Cauchy distribution satisfies

$$1 - F(x) = (x\pi)^{-1} - (3\pi)^{-1}x^{-3} + o(x^{-3}) , \qquad x \to \infty .$$

Hence it satisfies the second-order condition (3.2.5) with $\gamma = 1$ and $\rho = -2$. In fact, it is easy to see that if

$$1 - F(x) = c_1 x^{-1/\gamma} + c_2 x^{-1/\gamma + \rho/\gamma}(1 + o(1)) , \qquad x \to \infty , \qquad (3.2.9)$$

for constants $c_1 > 0$, $c_2 \neq 0$, $\gamma > 0$, and $\rho < 0$, then the second-order condition (3.2.5) holds with the indicated γ and ρ.

The second-order framework used in Theorem 3.2.5 provides the most natural approach to the asymptotic normality of estimators like Hill's estimator. However, next we discuss some of the problems related to second-order conditions of this type.

The parameter ρ controls the speed of convergence to asymptotic normality of $\hat{\gamma}_H$. For instance, a distribution function F satisfying (3.2.9) satisfies the second-order condition (3.2.5) with $A(1/(1 - F(t))) = \rho\gamma^{-1}c_2 c_1^{-1}t^{\rho/\gamma}$ and hence (3.2.4) with $A(t) = \rho\gamma^{-1}c_2 c_1^{\rho-1}t^\rho$. Moreover, if (3.2.6) holds with $\lambda \neq 0$, then a simple calculation shows that this is true if and only if

$$k(n) \sim \left(\frac{\lambda\gamma}{\rho c_2}c_1^{1-\rho}\right)^{2/(1-2\rho)} n^{-2\rho/(1-2\rho)} . \tag{3.2.10}$$

Then the convergence rate in Theorem 3.2.5 is of order $n^{\rho/(1-2\rho)}$.

Now consider three types of sequences:

1. Suppose $\sqrt{k}\,|A\,(n/k)| \to \infty$. Then it is not difficult to see using the inequalities in the proof of Theorem 3.2.5 that

$$\frac{\hat{\gamma}_H - \gamma}{A\left(\frac{n}{k}\right)} \overset{P}{\to} \frac{1}{1-\rho} .$$

Since for large n, k must be much larger than $n^{-2\rho/(1-2\rho)}$, we have for large n, n/k much smaller than $n^{1+2\rho/(1-2\rho)} = n^{1/(1-2\rho)}$. Hence the convergence rate $|A\,(n/k)|$ is slower than the rate $n^{\rho/(1-2\rho)}$ found after (3.2.10).

2. Suppose $\sqrt{k}A\,(n/k) \to 0$. Then

$$k(n) = o\left(n^{-2\rho/(1-2\rho)}\right),$$

and the convergence rate $1/\sqrt{k}$ is again slower than $n^{\rho/(1-2\rho)}$.

3. Suppose $\sqrt{k}A\,(n/k) \to \lambda \neq 0, \infty$. Then by (3.2.10) the convergence rate $1/\sqrt{k}$ is of order $n^{\rho/(1-2\rho)}$. This is the optimal situation.

The above discussion leads to the question, what is the best choice for λ? Theorem 3.2.5 tells us that if $\sqrt{k}A\,(n/k) \to \lambda$, then

$$\sqrt{k}(\hat{\gamma}_H - \gamma) \overset{d}{\to} \gamma N + \frac{\lambda}{1-\rho} \tag{3.2.11}$$

with N standard normal and hence

$$\hat{\gamma}_H - \gamma \overset{d}{\approx} \frac{\gamma N}{\sqrt{k}} + \frac{\lambda}{(1-\rho)\sqrt{k}} \overset{d}{\approx} \frac{\gamma N}{\sqrt{k}} + \frac{A\left(\frac{n}{k}\right)}{(1-\rho)} . \tag{3.2.12}$$

We want to know for which choices of $k = k(n)$ this approximation is best, i.e., for which k its mean square error

$$\frac{\gamma^2}{k} + \frac{A^2\left(\frac{n}{k}\right)}{(1-\rho)^2} \tag{3.2.13}$$

is minimal. For the time being we continue to consider the special case $A(t) = ct^\rho$. Write $r := n/k$. This leads us to

$$\text{argmin}_{(n/r)=1,2,...} \left(\frac{r\gamma^2}{n} + \frac{c^2 r^{2\rho}}{(1-\rho)^2} \right),$$

and for simplicity we consider

$$\text{argmin}_{t>0} \left(\frac{t\gamma^2}{n} + \frac{c^2 t^{2\rho}}{(1-\rho)^2} \right).$$

The infimum is attained by setting the derivative equal to zero, i.e.,

$$\frac{\gamma^2}{n} = \frac{-2\rho \, c^2 t^{2\rho-1}}{(1-\rho)^2},$$

or

$$t = \left(\frac{\gamma^2(1-\rho)^2}{-2\rho \, c^2} \right)^{1/(2\rho-1)} n^{1/(1-2\rho)}.$$

Equivalently, by setting $t = r = n/k$,

$$k = \left[\left(\frac{\gamma^2(1-\rho)^2}{-2\rho \, c^2} \right)^{1/(1-2\rho)} n^{-2\rho/(1-2\rho)} \right],$$

where $[x]$ means the integer part of x. Hence

$$k_0(n) = \left[\left(\frac{\gamma^2(1-\rho)^2}{-2\rho \, c^2} \right)^{1/(1-2\rho)} n^{-2\rho/(1-2\rho)} \right], \qquad (3.2.14)$$

and we call $k_0(n)$ the optimal choice for the sequence $k(n)$ under the given conditions.

Now we go back to (3.2.12). We would like to consider $\min_k E(\hat\gamma_H - \gamma)^2$, but since the expectation may not exist, we consider the minimum of the substitute expression

$$E \left(\frac{\gamma N}{\sqrt{k}} + \frac{A\left(\frac{n}{k}\right)}{(1-\rho)} \right)^2, \qquad (3.2.15)$$

and the sequence k_0 that optimizes (3.2.15) will serve as the optimal choice for the estimator $\hat\gamma_H$ too. Note that

$$\lim_{n\to\infty} \sqrt{k_0} A \left(\frac{n}{k_0} \right) = \lim_{n\to\infty} c \, n^\rho \, k^{1/2-\rho} = \frac{\text{sign}(c) \, \gamma \, (1-\rho)}{\sqrt{-2\rho}}, \qquad (3.2.16)$$

where $\text{sign}(c) = 1$ if $c > 0$ and $\text{sign}(c) = -1$ if $c < 0$. Hence for this choice of k we have

$$\sqrt{k_0} \left(\hat\gamma_H - \gamma \right) \overset{d}{\to} N \left(\frac{\text{sign}(c) \, \gamma}{\sqrt{-2\rho}}, \gamma^2 \right)$$

and

$$\lim_{n \to \infty} k_0 \min_k E \left(\frac{\gamma N}{\sqrt{k}} + \frac{A\left(\frac{n}{k}\right)}{(1-\rho)} \right)^2 = \lim_{n \to \infty} k_0 \min_k \left(\frac{\gamma^2}{k} + \frac{A^2\left(\frac{n}{k}\right)}{(1-\rho)^2} \right)$$

by (3.2.13), which equals

$$\lim_{n \to \infty} k_0 \left(\frac{\gamma^2}{k_0} + \frac{A^2\left(\frac{n}{k_0}\right)}{(1-\rho)^2} \right) = \gamma^2 + \frac{1}{(1-\rho)^2} \left(\frac{\text{sign}(c)\gamma(1-\rho)}{\sqrt{-2\rho}} \right)^2$$

$$= \gamma^2 \left(1 - \frac{1}{2\rho} \right)$$

by (3.2.16).

So far we have considered only the special case $A(t) = ct^\rho$ with $\rho < 0$. This is often assumed in applications of extreme value theory. However, such simplification is not feasible in the case $\rho = 0$. So next we shall consider the optimality problem in the more general case of the second-order condition, not just the special case $A(t) = ct^\rho$.

As it will turn out in the end, if a sequence $k_0(n)$ is optimal then any sequence $k(n) \sim k_0(n)$, as $n \to \infty$, is also optimal. This implies that we can replace the function A by any function A^* with $A^*(t) \sim A(t)$, as $t \to \infty$, without loss of generality.

Similarly as before, we are faced with finding

$$\text{argmin}_{t>0} \left(\frac{t\gamma^2}{n} + \frac{A^2(t)}{(1-\rho)^2} \right), \tag{3.2.17}$$

where the function $|A|$ is regularly varying with index $\rho \le 0$. For $\rho = 0$ it is reasonable to assume that there exists a function A^* with $|A^*(t)| \sim |A(t)|$, as $t \to \infty$, and $|A^*|$ is monotone decreasing (see Theorem C.1 in Dekkers and de Haan (1993)). In that case we can assume without loss of generality that the function A^2 satisfies

$$\lim_{t \to \infty} \frac{A^2(tx) - A^2(t)}{q(t)} = -\log x , \quad x > 0 ,$$

with q a suitable positive function.

Then for each $\rho \le 0$ (cf. Proposition B.2.15) there exists a positive decreasing function $s \in RV_{2\rho-1}$ such that as $t \to \infty$,

$$A^2(t) \sim \int_t^\infty s(u) \, du . \tag{3.2.18}$$

We have for $c > 1$ and sufficiently large t,

$$\frac{t\gamma^2}{n} + \frac{c^{-1}}{(1-\rho)^2} \int_t^\infty s(u) \, du < \frac{t\gamma^2}{n} + \frac{A^2(t)}{(1-\rho)^2}$$

$$< \frac{t\gamma^2}{n} + \frac{c}{(1-\rho)^2} \int_t^\infty s(u) \, du . \tag{3.2.19}$$

The infimum over $t > 0$ for the right- and left-hand sides can be calculated by just setting the derivative equal to zero. For the right-hand side we get

$$\frac{\gamma^2(1-\rho)^2}{c\,n} = s(t),$$

which is equivalent to

$$t = s^{\leftarrow}\left(\frac{\gamma^2(1-\rho)^2}{c\,n}\right),$$

and the infimum is

$$\frac{\gamma^2}{n}\, s^{\leftarrow}\left(\frac{\gamma^2(1-\rho)^2}{c\,n}\right) + \frac{c}{(1-\rho)^2}\int_{s^{\leftarrow}(\gamma^2(1-\rho)^2/(c\,n))}^{\infty} s(u)\,du$$

$$= \frac{c}{(1-\rho)^2}\left\{\frac{\gamma^2(1-\rho)^2}{c\,n}\, s^{\leftarrow}\left(\frac{\gamma^2(1-\rho)^2}{c\,n}\right) + \int_{s^{\leftarrow}(\gamma^2(1-\rho)^2/(c\,n))}^{\infty} s(u)\,du\right\}$$

$$= \frac{c}{(1-\rho)^2}\int_0^{\gamma^2(1-\rho)^2/(c\,n)} s^{\leftarrow}(u)\,du,$$

where for the last step we used

$$v\,s^{\leftarrow}(v) + \int_{s^{\leftarrow}(v)}^{\infty} s(u)\,du = \int_0^v s^{\leftarrow}(u)\,du . \qquad (3.2.20)$$

For the left-hand side of (3.2.19) we have the same result but with c replaced by c^{-1}. It follows that the infimum (3.2.17) is

$$\frac{1}{(1-\rho)^2}\int_0^{\gamma^2(1-\rho)^2/n} s^{\leftarrow}(u)\,du,$$

and it is attained at

$$t \sim s^{\leftarrow}\left(\frac{\gamma^2(1-\rho)^2}{n}\right),$$

i.e. (since t replaced n/k),

$$k(n) \sim \frac{n}{s^{\leftarrow}\left(\frac{\gamma^2(1-\rho)^2}{n}\right)} .$$

Hence an optimal sequence $k_0 = k_0(n)$ in the sense of minimizing $\gamma^2/k + A^2(n/k)/(1-\rho)^2$ is given by

$$k_0 = \left[\frac{n}{s^{\leftarrow}\left(\frac{\gamma^2(1-\rho)^2}{n}\right)}\right].$$

What can we say about the asymptotic distribution of $\sqrt{k_0}(\hat{\gamma}_H - \gamma)$? As before,

$$\sqrt{k_0}(\hat{\gamma}_H - \gamma) \overset{d}{\approx} \gamma N + \frac{\sqrt{k_0}A\left(\frac{n}{k_0}\right)}{1 - \rho},$$

so we have to evaluate $\sqrt{k_0}A\,(n/k_0)$ for n large. By (3.2.18) and (3.2.20),

$$k_0 A^2\left(\frac{n}{k_0}\right) \sim \frac{n}{s^{\leftarrow}\left(\frac{\gamma^2(1-\rho)^2}{n}\right)} \int_{s^{\leftarrow}(\gamma^2(1-\rho)^2/n)}^{\infty} s(u)\,du$$

$$= \frac{n}{s^{\leftarrow}\left(\frac{\gamma^2(1-\rho)^2}{n}\right)} \left\{ \int_0^{\gamma^2(1-\rho)^2/n} s^{\leftarrow}(u)\,du - \frac{\gamma^2(1-\rho)^2}{n}\, s^{\leftarrow}\left(\frac{\gamma^2(1-\rho)^2}{n}\right) \right\}$$

$$= \gamma^2(1-\rho)^2 \left\{ \frac{\int_0^{\gamma^2(1-\rho)^2/n} s^{\leftarrow}(u)\,du}{\frac{\gamma^2(1-\rho)^2}{n}\, s^{\leftarrow}\left(\frac{\gamma^2(1-\rho)^2}{n}\right)} - 1 \right\}.$$

Now by Theorem B.1.5, since $(1/s)^{\leftarrow} \in RV_{1/(1-2\rho)}$,

$$\lim_{x\to\infty} \frac{\int_0^{1/x} s^{\leftarrow}(u)\,du}{\frac{1}{x}\, s^{\leftarrow}\left(\frac{1}{x}\right)} = \lim_{x\to\infty} \frac{\int_x^{\infty} \left(\frac{1}{s}\right)^{\leftarrow}(u)\,\frac{du}{u^2}}{\frac{1}{x}\left(\frac{1}{s}\right)^{\leftarrow}(x)} = \frac{1-2\rho}{-2\rho} \tag{3.2.21}$$

(note that $s^{\leftarrow}(1/x) = (1/s)^{\leftarrow}(x)$). Hence for $\rho < 0$,

$$\lim_{n\to\infty} k_0 A^2\left(\frac{n}{k_0}\right) = \gamma^2(1-\rho)^2\left(\frac{1-2\rho}{-2\rho} - 1\right) = \frac{\gamma^2(1-\rho)^2}{-2\rho}$$

and

$$\sqrt{k_0}(\hat{\gamma}_H - \gamma) \to N\left(\frac{\mathrm{sign}(A)\gamma}{\sqrt{-2\rho}}, \gamma^2\right).$$

For $\rho = 0$ the limit in (3.2.21) must be interpreted as infinity. This means that for $\rho = 0$, by minimizing the mean square error we get an optimal sequence k_0 for which

$$\sqrt{k_0}(\hat{\gamma}_H - \gamma) + b_n \overset{d}{\to} N(0, \gamma^2),$$

where b_n is a slowly varying sequence tending to plus or minus infinity. This statement is not useful for obtaining an asymptotic confidence interval for γ. All we can say is that

$$\frac{\sqrt{k_0}}{b_n}(\hat{\gamma}_H - \gamma) \overset{P}{\to} 1.$$

If for $\rho = 0$ we take the sequence $k(n)$ a bit smaller, we do get asymptotic normality. Take $k_\lambda := k_\lambda(n)$ such that

$$\lim_{n\to\infty} k_\lambda A^2\left(\frac{n}{k_\lambda}\right) = \lambda^2 > 0. \tag{3.2.22}$$

Write $f(t) := \lambda^2 t \,/ \int_t^{\infty} s(u)\,du$ with s as in (3.2.18). Then $f(t) = \lambda^2 / \int_1^{\infty} s(tu)\,du$ is increasing and is RV_1. Moreover, we have

$$n \sim f\left(\frac{n}{k_\lambda}\right) .$$

In contrast to the case $\rho < 0$ we have for the optimal choice

$$n \sim \frac{\gamma^2(1-\rho)^2}{s(n/k_0)} .$$

Now, the functions $f(t) = \lambda^2 t / \int_t^\infty s(u)\, du$ and $1/s(t)$ are both RV_1, but by Theorem B.1.5,

$$\lim_{t\to\infty} s(t)\, f(t) = 0 .$$

Consider as an example the distribution function for which $U(t) = t^\gamma \log t + 1$. Then

$$U(tx) - x^\gamma U(t) = t^\gamma x^\gamma \log x ,$$

hence for $x > 0$,

$$\lim_{t\to\infty} \frac{\frac{U(tx)}{U(t)} - x^\gamma}{\frac{1}{\log t}} = x^\gamma \log x .$$

Consequently,

$$A(t) = \frac{1}{\log t}$$

and

$$s(t) = \frac{2}{t(\log t)^3} .$$

It follows for the optimal sequence $k_0(n)$ that

$$k_0(n) \sim \frac{\gamma^2(1-\rho)^2}{2}(\log n)^3 ,$$

but the limit relation

$$kA^2\left(\frac{n}{k}\right) \to \lambda > 0$$

holds for the sequence

$$k(n) \sim \lambda \log n .$$

Remark 3.2.8 An adaptive choice of $k_0(n)$ is possible. That is, one can obtain an estimator $\hat{k}_0(n)$ such that $\hat{k}_0(n)/k_0(n) \to^P 1$. We refer to Drees and Kaufmann (1998), Danielsson, de Haan, Peng and de Vries (2001), and Beirlant, Vynckier, and Teugels (1996). Similarly, adaptive choices of $k_0(n)$ are possible for the other estimators of the extreme value index discussed in the next sections, as well as for estimating high quantiles and tail probabilities (cf. Chapter 4).

Some final words about the Hill estimator (but this is true for most other estimators too). As we have seen, in the case $\rho = 0$ of the second-order condition we have only asymptotic normality of $\sqrt{k}(\hat{\gamma}_H - \gamma)$ if $k(n)$ grows with n very slowly. This means that for even moderate sample sizes the estimator may give the wrong impression since the bias takes over very rapidly.

Another disadvantage of the Hill estimator is the fact that the estimator is not shift invariant. A shift of the observations does not affect the first-order parameter γ but it may affect the second-order parameter. Consider the special case

$$U(t) = c_0 + c_1 t^\gamma + c_2 t^{\gamma+\tau} + o\left(t^{\gamma+\tau}\right) \tag{3.2.23}$$

with c_1 positive, c_0 and c_2 not zero, $\gamma > 0$, and $\tau < 0$. Then

$$\frac{U(tx)}{U(t)} - x^\gamma \sim \frac{c_0}{c_1} t^{-\gamma} x^\gamma (x^{-\gamma} - 1) + \frac{c_2}{c_1} t^\tau x^\gamma (x^\tau - 1) .$$

For $\tau > -\gamma$ the second term dominates and for $\tau < -\gamma$ the first term dominates. Hence in the second case $(\tau < -\gamma)$ one can improve the rate of convergence by applying a shift $-c_0$ to the observations so that the first term of (3.2.23) disappears and the second-order parameter changes from $-\gamma$ to $\tau < -\gamma$. This simple trick (due to Holger Drees) works in surprisingly many cases and results in much less disturbing bias. Of course the trick also works when t^τ is replaced by any τ-varying function.

3.3 General Case $\gamma \in \mathbb{R}$: The Pickands Estimator

The simplest and oldest estimator for γ is the *Pickands estimator* (1975):

$$\hat{\gamma}_P := (\log 2)^{-1} \log \frac{X_{n-k,n} - X_{n-2k,n}}{X_{n-2k,n} - X_{n-4k,n}} . \tag{3.3.1}$$

We shall prove weak consistency and asymptotic normality of $\hat{\gamma}_P$.

Theorem 3.3.1 *Let $X_1, X_2, \ldots$ be i.i.d. random variables with distribution function F. Suppose $F \in \mathcal{D}(G_\gamma)$ with $\gamma \in \mathbb{R}$. Then as $n \to \infty$, $k = k(n) \to \infty$, $k/n \to 0$,*

$$\hat{\gamma}_P \xrightarrow{P} \gamma .$$

For the proof we need the following auxiliary result.

Lemma 3.3.2 *Let $Y_1, Y_2, \ldots$ be i.i.d. random variables with distribution function $1 - 1/x$, $x \geq 1$. Then as $n \to \infty$, $k = k(n) \to \infty$, $k/n \to 0$, the random vector*

$$\sqrt{2k} \left(\frac{1}{2} \frac{Y_{n-k,n}}{Y_{n-2k,n}} - 1, \frac{\sqrt{2}}{2} \frac{Y_{n-2k,n}}{Y_{n-4k,n}} - \sqrt{2} \right) \tag{3.3.2}$$

is asymptotically bivariate standard normal.

Proof. Since by Rényi's representation (with $Y_{0,n} := 1$)

$$\left\{\frac{Y_{n-i+1,n}}{Y_{n-i,n}}\right\}_{i=1}^n \stackrel{d}{=} \left\{\left(Y_{n-i+1}^\star\right)^{1/i}\right\}_{i=1}^n \tag{3.3.3}$$

with $Y_1^\star, Y_2^\star, \ldots, Y_n^\star$ independent and with common distribution function $1 - 1/x$, the two components of the random vector in (3.3.2) are independent. By restricting attention to $0 \le i \le k$ in (3.3.3) one also sees that the distribution of

$$\left\{\frac{Y_{n-i+1,n}}{Y_{n-k,n}}\right\}_{i=1}^k$$

does not depend on n. So the first component in (3.3.2) is equal in distribution to

$$\sqrt{2k}\left(\frac{1}{2}Y_{k,2k} - 1\right) \tag{3.3.4}$$

and the second component to

$$2\sqrt{k}\left(\frac{1}{2}Y_{2k,4k} - 1\right) . \tag{3.3.5}$$

Now use the fact that by Lemma 2.2.3,

$$\sqrt{2k}\left(\frac{1}{2}Y_{k,2k} - 1\right)$$

is asymptotically standard normal (note that $Y_{k,2k} =^d U_{k+1,2k}^{-1}$ with $U_{k,n}$ as in Lemma 2.2.3). Similarly for the second component in (3.3.2). ∎

Corollary 3.3.3 *Denote the limit vector of* (3.3.2) *by* (Q, R). *Then*

$$\sqrt{2k}\left(\frac{1}{4}\frac{Y_{n-k,n}}{Y_{n-4k,n}} - 1\right) \stackrel{d}{\to} Q + \frac{R}{\sqrt{2}}.$$

Proof.

$$\frac{1}{4}\frac{Y_{n-k,n}}{Y_{n-4k,n}} - 1 = \left(\frac{1}{2}\frac{Y_{n-k,n}}{Y_{n-2k,n}} - 1\right) + \left(\frac{1}{2}\frac{Y_{n-2k,n}}{Y_{n-4k,n}} - 1\right)$$
$$+ \left(\frac{1}{2}\frac{Y_{n-k,n}}{Y_{n-2k,n}} - 1\right)\left(\frac{1}{2}\frac{Y_{n-2k,n}}{Y_{n-4k,n}} - 1\right) . \qquad ∎$$

Proof (of Theorem 3.3.1). We use the domain of attraction condition (see Theorem 1.1.6, Section 1.1.3)

$$\lim_{t\to\infty} \frac{U(tx) - U(t)}{a(t)} = \frac{x^\gamma - 1}{\gamma}, \qquad x > 0 .$$

Since U is monotone, the relation holds locally uniformly. It follows that locally uniformly for $0 < x, y < \infty$,

$$\lim_{t\to\infty} \frac{U(tx) - U(t)}{U(ty) - U(t)} = \frac{x^\gamma - 1}{y^\gamma - 1} . \tag{3.3.6}$$

As in the proof of Theorem 3.2.2 we can write $U(Y_{n-i,n})$ for $X_{n-i,n}$, $i = 1, 2, \ldots, n$, with the Y_i's from Lemma 3.3.2. Now observe that

$$\frac{X_{n-k,n} - X_{n-2k,n}}{X_{n-2k,n} - X_{n-4k,n}} = \frac{X_{n-k,n} - X_{n-4k,n}}{X_{n-2k,n} - X_{n-4k,n}} - 1 \tag{3.3.7}$$

$$= \frac{U\left(\frac{Y_{n-k,n}}{Y_{n-4k,n}} Y_{n-4k,n}\right) - U(Y_{n-4k,n})}{U\left(\frac{Y_{n-2k,n}}{Y_{n-4k,n}} Y_{n-4k,n}\right) - U(Y_{n-4k,n})} - 1 ,$$

$Y_{n-4k,n} \to \infty$ a.s., $n \to \infty$, and that by Lemma 3.3.2,

$$\frac{Y_{n-k,n}}{Y_{n-4k,n}} \overset{P}{\to} 4 \quad \text{and} \quad \frac{Y_{n-2k,n}}{Y_{n-4k,n}} \overset{P}{\to} 2 . \tag{3.3.8}$$

Combining (3.3.6)–(3.3.8), we find that

$$\frac{X_{n-k,n} - X_{n-2k,n}}{X_{n-2k,n} - X_{n-4k,n}} \overset{P}{\to} \frac{4^\gamma - 1}{2^\gamma - 1} - 1 = 2^\gamma .$$

The result follows. ∎

Remark 3.3.4 Note that $(2^\gamma - 1)/\gamma$ is the median of the limiting generalized Pareto distribution (3.1.3) and $(4^\gamma - 1)/\gamma$ is its 0.75 quantile. Hence the Pickands estimator estimates γ via the *quantiles* of the limiting distribution, in contrast to the Hill estimator, which estimates γ via a *moment* of the limiting generalized Pareto distribution.

Theorem 3.3.5 *Let $X_1, X_2, \ldots$ be i.i.d. random variables with distribution function F. Suppose F satisfies the second-order condition of Theorem 2.3.8, i.e.,*

$$\lim_{t\uparrow x^*} \frac{\frac{1-F(t+xf(t))}{1-F(t)} - Q_\gamma(x)}{\alpha(t)} = \left(Q_\gamma(x)\right)^{1+\gamma} H_{\gamma,\rho}(Q_\gamma^{-1}(x))$$

with $Q_\gamma(x) := (1 + \gamma x)^{-1/\gamma}$, f some positive function, and α some positive or negative function with $\lim_{t\uparrow x^*} \alpha(t) = 0$. Recall the equivalent relation in terms of $U := (1/(1 - F))^{\leftarrow}$:

$$\lim_{t\to\infty} \frac{\frac{U(tx)-U(t)}{a(t)} - D_\gamma(x)}{A(t)} = H_{\gamma,\rho}(x) := \int_1^x s^{\gamma-1} \int_1^s u^{\rho-1} \, du \, ds$$

for all $x > 0$ with $D_\gamma(x) = (x^\gamma - 1)/\gamma$, $a(t) = f(U(t))$, and $A(t) = \alpha(U(t))$. Then, for $k = k(n) \to \infty$, $k/n \to 0$, and

$$\lim_{n\to\infty} \sqrt{k} A\left(\frac{n}{k}\right) = \lambda \tag{3.3.9}$$

with λ finite,

$$\sqrt{k}(\hat{\gamma}_P - \gamma) \xrightarrow{d} N\left(\lambda b_{\gamma,\rho}, var_\gamma\right)$$

with N standard normal, where

$$b_{\gamma,\rho} := \begin{cases} \frac{4^{-\rho}\gamma((4^{\gamma+\rho}-1)-(2^\gamma+1)(2^{\gamma+\rho}-1))}{\rho 2^\gamma(\gamma+\rho)(2^\gamma-1)\log 2} , & \rho < 0 \neq \gamma, \\[2mm] \frac{1-2^{-\rho+1}+4^{-\rho}}{\rho^2(\log 2)^2} , & \rho < 0 = \gamma, \\[2mm] 1 , & \rho = 0, \end{cases}$$

and

$$var_\gamma := \begin{cases} \frac{\gamma^2(2^{2\gamma+1}+1)}{4(\log 2)^2(2^\gamma-1)^2} , & \gamma \neq 0, \\[2mm] \frac{3}{4(\log 2)^4} , & \gamma = 0 . \end{cases}$$

Proof. We repeat the inequalities of Theorem 2.3.6: there exist a_0 and A_0 such that for any $\varepsilon, \delta > 0$ there exists t_0 such that for $t, tx \geq t_0$,

$$\left| \frac{\frac{U(tx)-U(t)}{a_0(t)} - \frac{x^\gamma-1}{\gamma}}{A_0(t)} - \Psi_{\gamma,\rho}(x) \right| \leq \varepsilon x^{\gamma+\rho} \max(x^\delta, x^{-\delta}) =: q_{\gamma,\rho,\varepsilon,\delta}(x)$$

with

$$\Psi_{\gamma,\rho}(x) = \begin{cases} \frac{x^{\gamma+\rho}-1}{\gamma+\rho} , & \gamma+\rho \neq 0 , \ \rho < 0, \\[2mm] \log x , & \gamma+\rho = 0 , \ \rho < 0, \\[2mm] \frac{1}{\gamma}x^\gamma \log x , & \rho = 0 \neq \gamma, \\[2mm] \frac{1}{2}(\log x)^2 , & \rho = 0 = \gamma . \end{cases}$$

It follows that

$$\sqrt{k}\frac{U(Y_{n-k,n}) - U(Y_{n-4k,n})}{a_0(Y_{n-4k,n})}$$

$$= \sqrt{k}\frac{U\left(\frac{Y_{n-k,n}}{Y_{n-4k,n}}Y_{n-4k,n}\right) - U(Y_{n-4k,n})}{a_0(Y_{n-4k,n})}$$

$$= \sqrt{k}\frac{\left(\frac{Y_{n-k,n}}{Y_{n-4k,n}}\right)^\gamma - 1}{\gamma} + \sqrt{k}A_0(Y_{n-4k,n})\Psi_{\gamma,\rho}\left(\frac{Y_{n-k,n}}{Y_{n-4k,n}}\right) \qquad (3.3.10)$$

$$+ o_P(1)\sqrt{k}A_0(Y_{n-4k,n})q_{\gamma,\rho,\varepsilon,\delta}\left(\frac{Y_{n-k,n}}{Y_{n-4k,n}}\right) . \qquad (3.3.11)$$

By Cramér's delta method and Corollary 3.3.3, we have for the first term in (3.3.10),

$$\sqrt{2k}\left(\frac{\left(\frac{Y_{n-k,n}}{Y_{n-4k,n}}\right)^\gamma - 1}{\gamma} - \frac{4^\gamma - 1}{\gamma}\right) \xrightarrow{d} 4^\gamma\left(Q + \frac{R}{\sqrt{2}}\right) .$$

Recall from Corollary 2.3.5, Theorem 2.3.6, and assumption (3.3.9) that if $\rho < 0$,

$$\lim_{t \to \infty} \sqrt{k} A_0 \left(\frac{n}{k} \right) = \lim_{t \to \infty} \sqrt{k} \frac{1}{\rho} A \left(\frac{n}{k} \right) = \frac{\lambda}{\rho} ,$$

and if $\rho = 0$,

$$\lim_{t \to \infty} \sqrt{k} A_0 \left(\frac{n}{k} \right) = \lim_{t \to \infty} \sqrt{k} A \left(\frac{n}{k} \right) = \lambda .$$

Hence the second term in (3.3.10) converges to $(1_{\{\rho<0\}}\rho^{-1} + 1_{\{\rho=0\}}) 4^{-\rho} \lambda \Psi_{\gamma,\rho}(4)$. Finally, it is easy to see that (3.3.11) is asymptotically negligible. Therefore

$$\sqrt{k} \left(\frac{U(Y_{n-k,n}) - U(Y_{n-4k,n})}{a_0(Y_{n-4k,n})} - \frac{4^\gamma - 1}{\gamma} \right)$$

$$\xrightarrow{d} \frac{4^\gamma}{\sqrt{2}} \left(Q + \frac{R}{\sqrt{2}} \right) + \left(1_{\{\rho<0\}} \frac{1}{\rho} + 1_{\{\rho=0\}} \right) 4^{-\rho} \lambda \Psi_{\gamma,\rho}(4) . \quad (3.3.12)$$

Similarly,

$$\sqrt{k} \left(\frac{U(Y_{n-2k,n}) - U(Y_{n-4k,n})}{a_0(Y_{n-4k,n})} - \frac{2^\gamma - 1}{\gamma} \right)$$

$$\xrightarrow{d} 2^{\gamma-1} R + \left(1_{\{\rho<0\}} \frac{1}{\rho} + 1_{\{\rho=0\}} \right) 4^{-\rho} \lambda \Psi_{\gamma,\rho}(2) . \quad (3.3.13)$$

Combining (3.3.12) and (3.3.13) we get

$$\sqrt{k} \left(2^{\hat{\gamma}_P} - 2^\gamma \right) = \sqrt{k} \left(\frac{U(Y_{n-k,n}) - U(Y_{n-4k,n})}{U(Y_{n-2k,n}) - U(Y_{n-4k,n})} - \frac{4^\gamma - 1}{2^\gamma - 1} \right)$$

$$= \sqrt{k} \left\{ \frac{U(Y_{n-k,n}) - U(Y_{n-4k,n})}{a_0(Y_{n-4k,n})} \frac{2^\gamma - 1}{\gamma} \right.$$
$$\left. - \frac{U(Y_{n-2k,n}) - U(Y_{n-4k,n})}{a_0(Y_{n-4k,n})} \frac{4^\gamma - 1}{\gamma} \right\}$$

$$\times \left\{ \frac{U(Y_{n-2k,n}) - U(Y_{n-4k,n})}{a_0(Y_{n-4k,n})} \frac{2^\gamma - 1}{\gamma} \right\}^{-1}$$

$$= \sqrt{k} \left\{ \left(\frac{U(Y_{n-k,n}) - U(Y_{n-4k,n})}{a_0(Y_{n-4k,n})} - \frac{4^\gamma - 1}{\gamma} \right) \frac{2^\gamma - 1}{\gamma} \right.$$
$$\left. - \left(\frac{U(Y_{n-2k,n}) - U(Y_{n-4k,n})}{a_0(Y_{n-4k,n})} - \frac{2^\gamma - 1}{\gamma} \right) \frac{4^\gamma - 1}{\gamma} \right\}$$

$$\times \left\{ \frac{U(Y_{n-2k,n}) - U(Y_{n-4k,n})}{a_0(Y_{n-4k,n})} \frac{2^\gamma - 1}{\gamma} \right\}^{-1}$$

$$\xrightarrow{d} \frac{\frac{2^\gamma - 1}{\gamma} \left(2^{-1/2} 4^\gamma \left(Q + \frac{R}{\sqrt{2}} \right) + \left(1_{\{\rho<0\}} \frac{1}{\rho} + 1_{\{\rho=0\}} \right) 4^{-\rho} \lambda \Psi_{\gamma,\rho}(4) \right)}{\left(\frac{2^\gamma - 1}{\gamma} \right)^2}$$

$$-\frac{\frac{4^\gamma-1}{\gamma}\left(2^{\gamma-1}R+\left(1_{\{\rho<0\}}\frac{1}{\rho}+1_{\{\rho=0\}}\right)4^{-\rho}\lambda\Psi_{\gamma,\rho}(2)\right)}{\left(\frac{2^\gamma-1}{\gamma}\right)^2}$$

$$=2^\gamma\frac{\gamma}{2^\gamma-1}(2^{\gamma-1/2}Q-2^{-1}R)$$

$$+\left(1_{\{\rho<0\}}\frac{1}{\rho}+1_{\{\rho=0\}}\right)4^{-\rho}\lambda\frac{\frac{2^\gamma-1}{\gamma}\Psi_{\gamma,\rho}(4)-\frac{4^\gamma-1}{\gamma}\Psi_{\gamma,\rho}(2)}{\left(\frac{2^\gamma-1}{\gamma}\right)^2}.$$

Now apply Cramér's delta method. It follows that

$$\sqrt{k}(\hat\gamma_P-\gamma)\overset{d}{\to}\frac{1}{\log 2}\left(\frac{\gamma}{2^\gamma-1}\right)\left(2^{\gamma-1/2}Q-2^{-1}R\right)$$

$$+\left(1_{\{\rho<0\}}\frac{1}{\rho}+1_{\{\rho=0\}}\right)4^{-\rho}\lambda\frac{\frac{2^\gamma-1}{\gamma}\Psi_{\gamma,\rho}(4)-\frac{4^\gamma-1}{\gamma}\Psi_{\gamma,\rho}(2)}{2^\gamma\log 2\left(\frac{2^\gamma-1}{\gamma}\right)^2}.$$

The result follows. The particular case $\gamma=0$ and $\rho<0$ is left to the reader (cf. Exercise 3.6). ∎

Proof (Second proof of Theorem 3.3.5 via the tail quantile process). Rewrite (3.3.1) as

$$\hat\gamma_P=(\log 2)^{-1}\log\left(\frac{X_{n-[k'/4],n}-X_{n-[k'/2],n}}{X_{n-[k'/2],n}-X_{n-k',n}}\right)$$

for some sequence of integers $k'=k'(n)$, where $k'=4k$. Using Theorem 2.4.2 with $\{W_n(s)\}_{s>0}$ a sequence of Brownian motions, $s=\frac{1}{4}$ and $s=\frac{1}{2}$,

$$\frac{X_{n-[k'/4],n}-X_{n-[k'/2],n}}{X_{n-[k'/2],n}-X_{n-k',n}}$$

$$=\left\{\frac{4^\gamma-2^\gamma}{\gamma}+\frac{1}{\sqrt{k'}}\left(4^{\gamma+1}W_n\left(\frac{1}{4}\right)\right.\right.$$

$$\left.-2^{\gamma+1}W_n\left(\frac{1}{2}\right)+\sqrt{k'}A_0\left(\frac{n}{k'}\right)(\Psi_{\gamma,\rho}(4)-\Psi_{\gamma,\rho}(2))+o_P(1)\right)\bigg\}$$

$$\times\left\{\frac{2^\gamma-1}{\gamma}+\frac{1}{\sqrt{k'}}\left(2^{\gamma+1}W_n\left(\frac{1}{2}\right)\right.\right.$$

$$\left.\left.-W_n(1)+\sqrt{k'}A_0\left(\frac{n}{k'}\right)\Psi_{\gamma,\rho}(2)+o_P(1)\right)\right\}^{-1}$$

$$=2^\gamma\left\{1+\frac{1}{\sqrt{k'}}\frac{\gamma}{2^\gamma-1}\left(2^{\gamma+2}W_n\left(\frac{1}{4}\right)+W_n(1)-2(2^\gamma+1)W_n\left(\frac{1}{2}\right)\right)\right.$$

$$\left.+A_0\left(\frac{n}{k'}\right)\frac{\gamma}{2^\gamma-1}\left(2^{-\gamma}(\Psi_{\gamma,\rho}(4)-\Psi_{\gamma,\rho}(2))-\Psi_{\gamma,\rho}(2)\right)+o_P(1)\right\}.$$

Hence,

$$\log \left(\frac{X_{n-[k'/4],n} - X_{n-[k'/2],n}}{X_{n-[k'/2],n} - X_{n-k',n}} \right)$$

$$= \gamma \log 2 + \frac{1}{\sqrt{k'}} \frac{\gamma}{2^\gamma - 1} \left(2^{\gamma+2} W_n \left(\frac{1}{4} \right) + W_n(1) - 2(2^\gamma + 1) W_n \left(\frac{1}{2} \right) \right)$$

$$+ A_0 \left(\frac{n}{k'} \right) \frac{\gamma}{2^\gamma - 1} \left(2^{-\gamma} \left(\Psi_{\gamma,\rho}(4) - \Psi_{\gamma,\rho}(2) \right) - \Psi_{\gamma,\rho}(2) \right) + o_P(1)$$

so that as $n \to \infty$,

$$\sqrt{k'} \left(\hat{\gamma}_P - \gamma \right)$$

$$- \frac{\gamma}{(2^\gamma - 1) \log 2} \left(2^{\gamma+2} W_n \left(\frac{1}{4} \right) + W_n(1) - 2(2^\gamma + 1) W_n \left(\frac{1}{2} \right) \right)$$

$$- \sqrt{k'} A_0 \left(\frac{n}{k'} \right) \frac{\gamma}{(2^\gamma - 1) \log 2} \left(2^{-\gamma} \left(\Psi_{\gamma,\rho}(4) - \Psi_{\gamma,\rho}(2) \right) - \Psi_{\gamma,\rho}(2) \right)$$

$$= o_P(1) .$$

The result follows. ∎

3.4 The Maximum Likelihood Estimator ($\gamma > -\frac{1}{2}$)

The class of distribution functions satisfying $F \in \mathcal{D}(G_\gamma)$, for some $\gamma \in \mathbb{R}$, cannot be parametrized by a finite number of parameters; hence a straightforward maximum likelihood estimator does not exist. However, let us look at the limit relation (3.1.2) given in the introduction of this chapter: for $0 < x < (0 \vee (-\gamma))^{-1}$ and f a positive nondecreasing function,

$$\lim_{t \uparrow x^*} P \left(\frac{X - t}{f(t)} > x \,\middle|\, X > t \right) = 1 - H_\gamma(x) := (1 + \gamma x)^{-1/\gamma} . \tag{3.4.1}$$

This relation suggests that the larger observations (reflected in the condition $X > t$) approximately follow a generalized Pareto (GP) distribution. Since the class of GP distributions is parametrized by just one parameter γ, this suggests that if we apply the maximum likelihood procedure to the largest observations using the GP distribution as a model, we could obtain a useful estimator for γ.

This idea that we are now going to explain in detail will lead to what is generally called the maximum likelihood estimator of γ in extreme value theory (although sometimes a slightly different definition is used). After determining the estimator, we shall (and have to, since the general asymptotic theory of maximum likelihood estimators does not apply for this approximate model) prove asymptotic normality. In order to use the condition "$X > t$" in (3.4.1) properly we need the following lemma.

Lemma 3.4.1 *Let* $X, X_1, X_2, \ldots, X_n$ *be i.i.d. random variables with common distribution function* F, *and let* $X_{1,n} \le X_{2,n} \le \cdots \le X_{n,n}$ *be the nth order statistics.*

The joint distribution of $\{X_{i,n}\}_{i=n-k+1}^{n}$ given $X_{n-k,n} = t$, for some $k = 1, \ldots, n-1$, equals the joint distribution of the set of order statistics $\{X_{i,k}^{\star}\}_{i=1}^{k}$ of i.i.d. random variables $\{X_i^{\star}\}_{i=1}^{k}$ with distribution function

$$F_t(x) = P(X \leq x | X > t) = \frac{F(x) - F(t)}{1 - F(t)}, \qquad x > t .$$

Proof. Let $E_{1,n} \leq \cdots \leq E_{n,n}$ be the order statistics from an independent and identically distributed sample, with standard exponential distribution, $P(E > x) = e^{-x}$, $x > 0$. Then it is easy to see that the conditional distribution of $(E_{n-k+1,n}, \ldots, E_{n,n})$ given $\{E_{n-k,n} = t\}$ equals the distribution of $(E_{1,k}^{\star}, \ldots, E_{k,k}^{\star})$ with

$$P(E^{\star} > x) = e^{-(x-t)}, \qquad x > t . \tag{3.4.2}$$

Hence with $V := (-\log(1 - F))^{\leftarrow}$, the conditional distribution of $(V(E_{n-k+1,n}), \ldots, V(E_{n,n}))$ given $(V(E_{n-k,n}) = V(t))$ equals the distribution of $(V(E_{1,k}^{\star}), \ldots, V(E_{k,k}^{\star}))$.

Now for $x > V(t)$,

$$P\left(V(E^{\star}) > x\right) = P\left(E^{\star} > -\log(1 - F(x))\right),$$

which, by (3.4.2), equals

$$e^{-\{-\log(1-F(x))-t\}} = (1 - F(x)) \, e^t ,$$

and with $t = V^{\leftarrow}(V(t)) = -\log(1 - F(V(t)))$ equals

$$\frac{1 - F(x)}{1 - F(V(t))} = P(X > x | X > V(t)) .$$

Hence we have proved the lemma for any distribution function F that is continuous and strictly increasing. We leave the more general case to the reader. ∎

Let $X_1, \ldots, X_n$ be an independent and identically distributed sample with common distribution function F. As in the previous sections, to estimate γ we shall concentrate on some set of upper order statistics $(X_{n-k,n}, X_{n-k+1,n}, \ldots, X_{n,n})$ or, equivalently, on $(Z_0, Z_1, \ldots, Z_k) := (X_{n-k,n}, X_{n-k+1,n} - X_{n-k,n}, \ldots, X_{n,n} - X_{n-k,n})$. The likelihood function is obtained from the conditional distribution of $(Z_1, \ldots, Z_k)$ given $Z_0 = t$, which, according to Lemma 3.4.1, equals the distribution of the kth order statistics from a sample $(Z_1^{\star}, \ldots, Z_k^{\star})$ with common distribution function $F_t(t+x) = (F(t+x) - F(t))/(1 - F(t))$, $x > 0$. That is, we disregard the marginal information contained in $Z_0 = X_{n-k,n}$ (this approach is commonly referred to as the conditional likelihood approach). Then recall that the order is irrelevant to the likelihood and since the X_i's are assumed to be independent, the $Z_i^{\star}$'s are independent as well. Consequently we consider the resulting k independent and identically distributed random variables with distribution function $F_t(x+t) = (F(x+t) - F(t))/(1 - F(t))$, $x > 0$.

Now consider the usual asymptotic setting, where $k = k(n) \to \infty$ and $n/k \to \infty$, as $n \to \infty$, and hence $X_{n-k,n} \to x^*$ a.s. Then in view of the generalized Pareto approximation we apply the maximum likelihood procedure to the limiting Pareto distribution, which is explicit. Hence, the *maximum likelihood estimator* of γ (and consequently of the scale) is obtained by maximizing with respect to γ (and σ) the approximative likelihood $\prod_{i=1}^{k} h_{\gamma,\sigma}(z_i)$ with $z_i = x_{n-i+1,n} - x_{n-k,n}$ and $h_{\gamma,\sigma}(x) = \partial H_\gamma(x/\sigma)/\partial x$.

Note that this approximative conditional likelihood function tends to ∞ if $\gamma < -1$ and $\gamma/\sigma \downarrow -(X_{n,n} - X_{n-k,n})^{-1}$, so that a maximum over the full range of possible values for (γ, σ) does not exist. We shall concentrate on the region $(\gamma, \sigma) \in (-1/2, \infty) \times (0, \infty)$, since the maximum likelihood estimator behaves irregularly if $\gamma \le -\frac{1}{2}$.

The likelihood equations are given in terms of the partial derivatives

$$\begin{cases} \dfrac{\partial \log h_{\gamma,\sigma}(z)}{\partial \gamma} = \dfrac{1}{\gamma^2} \log\left(1 + \dfrac{\gamma}{\sigma}z\right) - \left(\dfrac{1}{\gamma} + 1\right) \dfrac{\frac{z}{\sigma}}{1 + \frac{\gamma}{\sigma}z}, \\[2ex] \dfrac{\partial \log h_{\gamma,\sigma}(z)}{\partial \sigma} = -\dfrac{1}{\sigma} - \left(\dfrac{1}{\gamma} + 1\right) \dfrac{-\frac{\gamma}{\sigma^2}z}{1 + \frac{\gamma}{\sigma}z}, \end{cases}$$

where for $\gamma = 0$ these should be interpreted as

$$\begin{cases} \frac{1}{2}\left(\frac{z}{\sigma}\right)^2 - \frac{z}{\sigma}, \\[2ex] -\dfrac{1}{\sigma} + \dfrac{z}{\sigma^2}. \end{cases}$$

The resulting likelihood equations in terms of the excesses $X_{n-i+1,n} - X_{n-k,n}$ are as follows:

$$\begin{cases} \displaystyle\sum_{i=1}^{k} \dfrac{1}{\gamma^2} \log\left(1 + \dfrac{\gamma}{\sigma}(X_{n-i+1,n} - X_{n-k,n})\right) \\[2ex] \quad - \left(\dfrac{1}{\gamma} + 1\right) \dfrac{\frac{1}{\sigma}(X_{n-i+1,n} - X_{n-k,n})}{1 + \frac{\gamma}{\sigma}(X_{n-i+1,n} - X_{n-k,n})} = 0, \\[3ex] \displaystyle\sum_{i=1}^{k} \left(\dfrac{1}{\gamma} + 1\right) \dfrac{\frac{\gamma}{\sigma}(X_{n-i+1,n} - X_{n-k,n})}{1 + \frac{\gamma}{\sigma}(X_{n-i+1,n} - X_{n-k,n})} = k \end{cases} \qquad (3.4.3)$$

(with a similar interpretation when $\gamma = 0$), which for $\gamma \ne 0$ can be simplified to

$$\begin{cases} \dfrac{1}{k} \displaystyle\sum_{i=1}^{k} \log\left(1 + \dfrac{\gamma}{\sigma}(X_{n-i+1,n} - X_{n-k,n})\right) = \gamma, \\[2ex] \dfrac{1}{k} \displaystyle\sum_{i=1}^{k} \dfrac{1}{1 + \frac{\gamma}{\sigma}(X_{n-i+1,n} - X_{n-k,n})} = \dfrac{1}{\gamma + 1}. \end{cases} \qquad (3.4.4)$$

Note that the maximum likelihood estimator of γ is shift and scale invariant, and the maximum likelihood estimator of σ is shift invariant.

Theorem 3.4.2 *Let $X_1, X_2, \ldots$ be i.i.d. random variables with distribution function F. Suppose F satisfies the second-order condition of Theorem 2.3.8 with $\gamma > -\frac{1}{2}$, or equivalently,*

$$\lim_{t \to \infty} \frac{\frac{U(tx)-U(t)}{a(t)} - \frac{x^\gamma - 1}{\gamma}}{A(t)} = \int_1^x s^{\gamma-1} \int_1^s u^{\rho-1} \, du \, ds \qquad (3.4.5)$$

for all $x > 0$ and with $\gamma > -\frac{1}{2}$. Then, for $k = k(n) \to \infty$, $k/n \to 0$ $(n \to \infty)$, and

$$\lim_{n \to \infty} \sqrt{k} A\left(\frac{n}{k}\right) = \lambda \qquad (3.4.6)$$

with λ finite, the system of likelihood equations (3.4.3) has a sequence of solutions $(\hat{\gamma}_{\mathrm{MLE}}, \hat{\sigma}_{\mathrm{MLE}})$ that satisfies

$$\sqrt{k}\left(\hat{\gamma}_{\mathrm{MLE}} - \gamma, \frac{\hat{\sigma}_{\mathrm{MLE}}}{a(\frac{n}{k})} - 1\right) \xrightarrow{d} N(\lambda b_{\gamma,\rho}, \Sigma), \qquad (3.4.7)$$

with N standard normal,

$$b_{\gamma,\rho} := \begin{cases} \left(\frac{\gamma+1}{(1-\rho)(1+\gamma-\rho)}, \frac{-\rho}{(1-\rho)(1+\gamma-\rho)}\right), & \rho < 0 \\ (1,0), & \rho = 0, \end{cases}$$

and the matrix Σ is given by

$$\begin{pmatrix} (1+\gamma)^2 & -(1+\gamma) \\ -(1+\gamma) & 1+(1+\gamma)^2 \end{pmatrix}.$$

Moreover, for any sequence of solutions $(\hat{\gamma}^\star_{\mathrm{MLE}}, \hat{\sigma}^\star_{\mathrm{MLE}})$ for which the convergence (3.4.7) does not hold, one must have $\sqrt{k}|\hat{\gamma}^\star_{\mathrm{MLE}} - \gamma| \xrightarrow{P} \infty$ or $\sqrt{k}|\hat{\sigma}^\star_{\mathrm{MLE}}/a(k/n) - 1| \xrightarrow{P} \infty$.

Recall the relation between the parameter σ and the function f in (3.4.1): this function was first introduced in Theorem 1.1.6 (Section 1.1.3), from where it is known that $f(t)$ can be chosen as $a(1/(1-F)(t))$. Then we see that σ must be close to $a(n/k)$ as $n \to \infty$.

We now give the line of reasoning for proving Theorem 3.4.2. A detailed proof will be given only for $\gamma > 0$. The proof for the other cases is similar.

For the true $\gamma > 0$ we rewrite equations (3.4.4), which we want to solve as

$$\begin{cases} \int_0^1 \log\left(1 + \frac{\gamma'}{\sigma'_0} \frac{X_{n-[ks],n} - X_{n-k,n}}{a_0(\frac{n}{k})}\right) ds = \gamma', \\ \int_0^1 \left(1 + \frac{\gamma'}{\sigma'_0} \frac{X_{n-[ks],n} - X_{n-k,n}}{a_0(\frac{n}{k})}\right)^{-1} ds = \frac{1}{\gamma'+1}, \end{cases} \qquad (3.4.8)$$

where a_0 is a suitably chosen positive function (more specifically the one from Theorem 2.3.6) and $\sigma'_0 := \sigma'/a_0(n/k)$.

Under the second-order condition given in Theorem 3.4.2, from Corollary 2.4.6, we have that

$$\left(\frac{X_{n-[ks],n} - X_{n-k,n}}{a_0(\frac{n}{k})}\right)_{s\in(0,1]} = \left(\frac{s^{-\gamma} - 1}{\gamma} + \frac{Z_n(s)}{\sqrt{k}}\right)_{s\in(0,1]}, \qquad (3.4.9)$$

where $\{Z_n(s)\}_{s\in(0,1]}$ is an asymptotically Gaussian process with known mean and covariance function, and γ is the true parameter. Then we write

$$1 + \frac{\gamma'}{\sigma_0'}\frac{X_{n-[ks],n} - X_{n-k,n}}{a_0(\frac{k_n}{n})} = 1 + \frac{\gamma'}{\sigma_0'}\left(\frac{s^{-\gamma} - 1}{\gamma} + \frac{Z_n(s)}{\sqrt{k}}\right)$$

$$= s^{-\gamma} + \left(\frac{\gamma'}{\sigma_0'} - \gamma\right)\frac{s^{-\gamma} - 1}{\gamma} + \frac{\gamma'}{\sigma_0'}\frac{Z_n(s)}{\sqrt{k}}. \quad (3.4.10)$$

When multiplied by s^{γ}, this becomes

$$1 + \left(\frac{\gamma'}{\sigma_0'} - \gamma\right)\frac{1 - s^{\gamma}}{\gamma} + s^{\gamma}\frac{\gamma'}{\sigma_0'}\frac{Z_n(s)}{\sqrt{k}}.$$

Hence

$$\log\left(s^{\gamma}\left(1 + \frac{\gamma'}{\sigma_0'}\frac{X_{n-[ks],n} - X_{n-k,n}}{a_0(\frac{n}{k})}\right)\right) \approx \left(\frac{\gamma'}{\sigma_0'} - \gamma\right)\frac{1 - s^{\gamma}}{\gamma} + s^{\gamma}\frac{\gamma'}{\sigma_0'}\frac{Z_n(s)}{\sqrt{k}}.$$

Now

$$\gamma' = \int_0^1 \log\left(1 + \frac{\gamma'}{\sigma_0'}\frac{X_{n-[ks],n} - X_{n-k,n}}{a_0(\frac{n}{k})}\right)ds,$$

and hence

$$\gamma' - \gamma = \int_0^1 \log\left(s^{\gamma}\left(1 + \frac{\gamma'}{\sigma_0'}\frac{X_{n-[ks],n} - X_{n-k,n}}{a_0(\frac{n}{k})}\right)\right)ds$$

$$\approx \left(\frac{\gamma'}{\sigma_0'} - \gamma\right)\int_0^1 \frac{1 - s^{\gamma}}{\gamma}ds + \frac{\gamma'}{\sigma_0'}\int_0^1 s^{\gamma}\frac{Z_n(s)}{\sqrt{k}}ds$$

$$\approx \left(\frac{\gamma'}{\sigma_0'} - \gamma\right)\frac{1}{1 + \gamma} + \gamma\int_0^1 s^{\gamma}\frac{Z_n(s)}{\sqrt{k}}ds.$$

Starting again from (3.4.10), for the second equation in (3.4.8) we have

$$\left(1 + \frac{\gamma'}{\sigma_0'}\frac{X_{n-[ks],n} - X_{n-k,n}}{a_0(\frac{n}{k})}\right)^{-1} \approx s^{\gamma} - \left(\frac{\gamma'}{\sigma_0'} - \gamma\right)\frac{s^{\gamma} - s^{2\gamma}}{\gamma} - s^{2\gamma}\frac{\gamma'}{\sigma_0'}\frac{Z_n(s)}{\sqrt{k}}$$

and so

$$\frac{1}{\gamma + 1} - \frac{1}{\gamma' + 1} \approx \left(\frac{\gamma'}{\sigma_0'} - \gamma\right)\int_0^1 \frac{s^{\gamma} - s^{2\gamma}}{\gamma}ds + \frac{\gamma'}{\sigma_0'}\int_0^1 s^{2\gamma}\frac{Z_n(s)}{\sqrt{k}}ds$$

$$\approx \left(\frac{\gamma'}{\sigma_0'} - \gamma\right)\frac{1}{(\gamma + 1)(2\gamma + 1)} + \gamma\int_0^1 s^{2\gamma}\frac{Z_n(s)}{\sqrt{k}}ds.$$

Summing up, we show that equations (3.4.8) are equivalent to linear equations in the unknown parameters γ' and σ_0', which can be solved readily.

For the proof of Theorem 3.4.2 we start by proving some auxiliary results.

Lemma 3.4.3 *Assume* (3.4.5) *with* $\gamma > 0$ *and* (3.4.6). *Let* $(\gamma', \sigma_0') := (\gamma'(n), \sigma_0'(n))$ *be such that*

$$\left| \frac{\gamma'}{\sigma_0'} - \gamma \right| = O_P\left(\frac{1}{\sqrt{k}}\right) . \tag{3.4.11}$$

Then

$$P\left(1 + \frac{\gamma'}{\sigma_0'} \frac{X_{n-[ks],n} - X_{n-k,n}}{a_0(\frac{n}{k})} \geq C_n s^{-\gamma}, \quad s \in [(2k)^{-1}, 1] \right) \to 1 , \tag{3.4.12}$$

$n \to \infty$, *for some random variables* $C_n > 0$ *such that* $1/C_n = O_P(1)$.

Proof. Let $U_{k,n}$, $k = 1, \ldots, n$, denote the order statistics from an independent and identically uniform $(0, 1)$ sample of size n. By Shorack and Wellner (1986) (Chapter 10, Section 3, Inequality 2, p. 416),

$$\sup_{1/(2k) \leq s \leq 1} \frac{n U_{[ks]+1,n}}{ks} = O_P(1) , \qquad \sup_{0 \leq s \leq 1} \frac{ks}{n U_{[ks]+1,n}} = O_P(1) , \tag{3.4.13}$$

as $n \to \infty$. Combining these bounds with the bounds given in Theorem 2.3.6, for some functions $a_0(t) \sim a(t)$ and $A_0(t) \sim A(t)$, $t \to \infty$, for all $x_0 > 0$ and $\delta > 0$, we obtain

$$\sup_{s \in [1/(2k),1]} s^{\gamma+\rho+\delta} \left| \frac{\frac{U\left(\frac{1}{U_{[ks]+1,n}}\right) - U\left(\frac{n}{k}\right)}{a_0(\frac{n}{k})} - \frac{\left(\frac{k}{nU_{[ks]+1,n}}\right)^{\gamma} - 1}{\gamma}}{A_0(\frac{n}{k})} - \Psi_{\gamma,\rho}\left(\frac{k}{nU_{[ks]+1,n}}\right) \right|$$

$$= o_P(1) .$$

Next use this approximation simultaneously for $s \in [(2k)^{-1}, 1]$ and $s = 1$. Then we have

$$\frac{X_{n-[ks],n} - X_{n-k,n}}{a_0(\frac{n}{k})} \overset{d}{=} \frac{U\left(\frac{1}{U_{[ks]+1,n}}\right) - U\left(\frac{1}{U_{k+1,n}}\right)}{a_0(\frac{n}{k})}$$

$$= \frac{1}{\gamma}\left(\frac{k}{nU_{[ks]+1,n}}\right)^{\gamma} - \frac{1}{\gamma}\left(\frac{k}{nU_{k+1,n}}\right)^{\gamma} + A_0\left(\frac{n}{k}\right)\Psi_{\gamma,\rho}\left(\frac{k}{nU_{[ks]+1,n}}\right)$$

$$- A_0\left(\frac{n}{k}\right)\Psi_{\gamma,\rho}\left(\frac{k}{nU_{k+1,n}}\right) + o_P\left((s^{-\gamma-\rho-\delta} - 1)A_0\left(\frac{n}{k}\right)\right) .$$

Hence

$$1 + \frac{\gamma'}{\sigma_0'} \frac{X_{n-[ks],n} - X_{n-k,n}}{a_0(\frac{n}{k})}$$

$$\overset{d}{=} \left(1 - \left(\frac{k}{nU_{k+1,n}}\right)^{\gamma}\right) - \left(\frac{\gamma'}{\sigma_0'} - \gamma\right)\frac{1}{\gamma}\left(\frac{k}{nU_{k+1,n}}\right)^{\gamma}$$

$$+ \frac{\gamma'}{\sigma_0'} \frac{1}{\gamma} \left(\frac{k}{n U_{[ks]+1,n}} \right)^{\gamma} + \frac{\gamma'}{\sigma_0'} A_0 \left(\frac{n}{k} \right) \Psi_{\gamma,\rho} \left(\frac{k}{n U_{[ks]+1,n}} \right)$$

$$- \frac{\gamma'}{\sigma_0'} A_0 \left(\frac{n}{k} \right) \Psi_{\gamma,\rho} \left(\frac{k}{n U_{k+1,n}} \right) + o_P \left((s^{-\gamma-\rho-\delta} - 1) k^{-1/2} \right)$$

$$= I + II + III + IV + V + VI .$$

By (3.4.13), $s^\gamma III$ is bounded away from zero uniformly for $s \in [(2k_n)^{-1}, 1]$. We will show that all the other terms tend to 0 uniformly when multiplied by s^γ, so that assertion (3.4.12) follows with $C_n := \inf_{s \in [(2k)^{-1},1]} s^\gamma III - \varepsilon_n$ for a suitable sequence $\varepsilon_n \downarrow 0$.

By the asymptotic normality of intermediate order statistics (see Theorem 2.2.1), part I is $O_P(k^{-1/2})$, hence $s^\gamma I = o_P(1)$. By (3.4.13) and assumption (3.4.11), part II is $O_P(k^{-1/2})$, so that $s^\gamma II = o_P(1)$. Next note that $s^\gamma \Psi_{\gamma,\rho}(s^{-1}) = o(s^{-1/2})$ as $s \downarrow 0$. This combined with (3.4.6) and (3.4.13) gives that $s^\gamma IV$ and $s^\gamma V$ are $o_P(1)$. Finally, $s^\gamma VI = o_P(1)$, provided one chooses $\delta < \frac{1}{2}$. Hence we have proved (3.4.12). ∎

Define

$$Z_n(s) := \sqrt{k} \left(\frac{X_{n-[ks],n} - X_{n-k,n}}{a_0(\frac{n}{k})} - \frac{s^{-\gamma} - 1}{\gamma} \right) \tag{3.4.14}$$

(read $(s^{-\gamma} - 1)/\gamma$ as $-\log s$, when $\gamma = 0$). Then, from Corollary 2.4.6, for suitably chosen functions a_0 and A_0, and for all $\varepsilon > 0$,

$$Z_n(s) = s^{-\gamma-1} W_n(s) - W_n(1)$$
$$+ \sqrt{k} A_0 \left(\frac{n}{k} \right) \Psi_{\gamma,\rho}(s^{-1}) + o_P(1) s^{-\gamma-1/2-\varepsilon} , \tag{3.4.15}$$

as $n \to \infty$, where $\{W_n(s)\}_{s>0}$ is a sequence of Brownian motions and the o_P-term is uniform for $s \in (0, 1]$. Moreover, under the conditions of Theorem 3.4.2, for all $\varepsilon > 0$,

$$Z_n(s) = O_P(1) s^{-\gamma-1/2-\varepsilon} , \tag{3.4.16}$$

as $n \to \infty$, where the O_P term is uniform for $s \in (0, 1]$.

Proposition 3.4.4 *Assume condition (3.4.5) with $\gamma > 0$ and (3.4.6). Then any solution (γ', σ_0') of (3.4.8) satisfying (3.4.11) admits the approximation*

$$\sqrt{k}(\gamma' - \gamma) - \frac{(\gamma + 1)^2}{\gamma} \int_0^1 \left(s^\gamma - (2\gamma + 1)s^{2\gamma} \right) Z_n(s) \, ds = o_P(1),$$

$$\sqrt{k}(\sigma_0' - 1) - \frac{\gamma + 1}{\gamma} \int_0^1 ((\gamma + 1)(2\gamma + 1)s^{2\gamma} - s^\gamma) Z_n(s) \, ds = o_P(1) , \tag{3.4.17}$$

as $n \to \infty$. Conversely, there exists a solution of (3.4.8) that satisfies (3.4.17), and hence also (3.4.11).

Remark 3.4.5 Though we prove Proposition 3.4.4 only for $\gamma > 0$, in fact the statement is true for any $\gamma > -\frac{1}{2}$. For more details see Drees, Ferreira, and de Haan (2003).

Proof (of Proposition 3.4.4). We start by obtaining an expansion for the left-hand side of the first equation of (3.4.8). Rewrite it as

$$
\int_0^{(2k)^{-1}} \log\left(1 + \frac{\gamma'}{\sigma_0'} \frac{X_{n-[ks],n} - X_{n-k,n}}{a_0(\frac{n}{k})}\right) ds + \int_{(2k)^{-1}}^1 \log s^{-\gamma} ds
$$

$$
+ \int_{(2k)^{-1}}^1 \log\left(s^\gamma \left(1 + \frac{\gamma'}{\sigma_0'} \frac{X_{n-[ks],n} - X_{n-k,n}}{a_0(\frac{n}{k})}\right)\right) ds
$$

$$
= I_1 + \gamma(1 - O(k^{-1}\log k)) + I_2 .
$$

First we prove that I_1 is negligible. Since $X_{n-[ks],n}$ is constant when $s \in (0, (2k)^{-1}]$, from (3.4.12), with probability tending to 1,

$$
1 + \frac{\gamma'}{\sigma_0'} \frac{X_{n-[ks],n} - X_{n-k,n}}{a_0(\frac{n}{k})} = 1 + \frac{\gamma'}{\sigma_0'} \frac{X_{n,n} - X_{n-k,n}}{a_0(\frac{n}{k})} \geq (2k)^\gamma C_n \qquad (3.4.18)
$$

for all $s \in (0, (2k)^{-1}]$, so that $-I_1 \leq (2k)^{-1} O_P(\log k)$. On the other hand, from (3.4.14), (3.4.16), and (3.4.11),

$$
1 + \frac{\gamma'}{\sigma_0'} \frac{X_{n,n} - X_{n-k,n}}{a_0(\frac{n}{k})} = 1 + \left(\gamma + O_P(k^{-1/2})\right) \left(\frac{(2k)^\gamma - 1}{\gamma} + O_P(k^{\gamma+\varepsilon})\right)
$$

$$
= O_P(k^{\gamma+\varepsilon}) .
$$

Hence it follows that $I_1 = o_P(k^{-1/2})$.

Next we turn to the main term I_2. We will apply the inequality $0 \leq x - \log(1+x) \leq x^2/(2(1 \wedge (1+x)))$, valid for all $x > -1$, to

$$
x = s^\gamma \left(1 + \frac{\gamma'}{\sigma_0'} \frac{X_{n-[ks],n} - X_{n-k,n}}{a_0(\frac{n}{k})}\right) - 1
$$

$$
= \left(\frac{\gamma'}{\sigma_0'} - \gamma\right) \frac{1 - s^\gamma}{\gamma} + \frac{\gamma'}{\sigma_0'} k^{-1/2} s^\gamma Z_n(s).
$$

Then, from (3.4.12) it follows that $0 < 1/(1 \wedge (1+x)) \leq 1 \vee 1/C_n = O_P(1)$ with probability tending to one. Moreover, note that relation (3.4.16) implies

$$
\int_0^{(2k)^{-1}} s^\gamma Z_n(s) ds = O_P\left(\int_0^{(2k)^{-1}} s^{-1/2-\varepsilon} ds\right) = O_P((2k)^{-1/2+\varepsilon}) = o_P(1),
$$

for $\varepsilon \in \left(0, \frac{1}{2}\right)$. Hence from (3.4.14) and (3.4.16), as $n \to \infty$,

$$
I_2 = \int_{(2k)^{-1}}^1 \left\{s^\gamma \left(1 + \frac{\gamma'}{\sigma_0'} \frac{X_{n-[ks],n} - X_{n-k,n}}{a_0(\frac{n}{k})}\right) - 1\right\} ds
$$

$$
+ O_P\left(\int_{(2k)^{-1}}^1 \left\{s^\gamma \left(1 + \frac{\gamma'}{\sigma_0'} \frac{X_{n-[ks],n} - X_{n-k,n}}{a_0(\frac{n}{k})}\right) - 1\right\}^2 ds\right)
$$

$$= \int_{(2k)^{-1}}^{1} \left(\left(\frac{\gamma'}{\sigma_0'} - \gamma \right) \frac{1 - s^\gamma}{\gamma} + \frac{\gamma'}{\sigma_0'} k^{-1/2} s^\gamma Z_n(s) \right) ds$$

$$+ O_P \left(\int_{(2k)^{-1}}^{1} \left\{ \left(\frac{\gamma'}{\sigma_0'} - \gamma \right) \frac{1 - s^\gamma}{\gamma} + \frac{\gamma'}{\sigma_0'} k^{-1/2} s^\gamma Z_n(s) \right\}^2 ds \right)$$

$$= \left(\left(\frac{\gamma'}{\sigma_0'} - \gamma \right) \frac{1}{\gamma + 1} + O_P(k^{-1/2}(2k)^{-1}) \right)$$

$$+ \left(\frac{\gamma'}{\sigma_0'} k^{-1/2} \int_0^1 s^\gamma Z_n(s) \, ds + o_P(k^{-1/2}) \right)$$

$$+ O_P \left(k^{-1} + k^{-1}(2k)^{2\varepsilon} + k^{-1}(2k)^{-1/2+\varepsilon} \right)$$

$$= \left(\frac{\gamma'}{\sigma_0'} - \gamma \right) \frac{1}{(\gamma + 1)} + \frac{\gamma'}{\sigma_0'} k^{-1/2} \int_0^1 s^\gamma Z_n(s) \, ds + o_P(k^{-1/2}) ,$$

where for the last equality we choose $\varepsilon < \frac{1}{4}$. To sum up, we have proved that

$$\int_0^1 \log \left(1 + \frac{\gamma'}{\sigma_0'} \frac{X_{n-[ks],n} - X_{n-k,n}}{a_0(\frac{n}{k})} \right) ds$$

$$= \gamma + \left(\frac{\gamma'}{\sigma_0'} - \gamma \right) \frac{1}{(\gamma + 1)} + \frac{\gamma'}{\sigma_0'} k^{-1/2} \int_0^1 s^\gamma Z_n(s) \, ds + o_P(k^{-1/2}) .$$

This means that the first equation of (3.4.8) is equivalent to

$$\gamma' = \gamma + \left(\frac{\gamma'}{\sigma_0'} - \gamma \right) \frac{1}{\gamma + 1} + \frac{\gamma'}{\sigma_0'} k^{-1/2} \int_0^1 s^\gamma Z_n(s) \, ds + o_P(k^{-1/2}) .$$

The second equation of (3.4.8) can be treated with somewhat similar arguments.
Then one gets

$$\int_0^1 \left(1 + \frac{\gamma'}{\sigma_0'} \frac{X_{n-[ks],n} - X_{n-k,n}}{a_0(\frac{n}{k})} \right)^{-1} ds$$

$$= \frac{1}{\gamma + 1} - \left(\frac{\gamma'}{\sigma_0'} - \gamma \right) \frac{1}{(\gamma + 1)(2\gamma + 1)}$$

$$- \frac{\gamma'}{\sigma_0'} k^{-1/2} \int_0^1 s^{2\gamma} Z_n(s) \, ds + o_P(k^{-1/2}) .$$

Hence, under the given conditions, the system of likelihood equations (3.4.8) is
equivalent to

$$\gamma + \left(\frac{\gamma'}{\sigma_0'} - \gamma \right) \frac{1}{\gamma + 1} + \frac{\gamma'}{\sigma_0'} k^{-1/2} \int_0^1 s^\gamma Z_n(s) \, ds + o_P(k^{-1/2}) = \gamma',$$

$$\frac{1}{\gamma + 1} - \left(\frac{\gamma'}{\sigma_0'} - \gamma \right) \frac{1}{(\gamma + 1)(2\gamma + 1)} - \frac{\gamma'}{\sigma_0'} k^{-1/2} \int_0^1 s^{2\gamma} Z_n(s) \, ds$$

$$+ o_P(k^{-1/2}) = \frac{1}{\gamma' + 1} . \quad (3.4.19)$$

Then, in view of (3.4.11) and (3.4.16), (3.4.19) implies

$$\gamma + \left(\frac{\gamma'}{\sigma_0'} - \gamma\right)\frac{1}{\gamma + 1} + \gamma k^{-1/2}\int_0^1 s^\gamma Z_n(s)\,ds + o_P(k^{-1/2}) = \gamma',$$

$$\frac{1}{\gamma + 1} - \left(\frac{\gamma'}{\sigma_0'} - \gamma\right)\frac{1}{(\gamma + 1)(2\gamma + 1)} - \gamma k^{-1/2}\int_0^1 s^{2\gamma} Z_n(s)\,ds + o_P(k^{-1/2})$$

$$= \frac{1}{\gamma' + 1}. \quad (3.4.20)$$

The first equation and (3.4.11) show that $|\gamma' - \gamma| = O_P(k^{-1/2})$; hence $|\gamma' - \gamma|^2 = o_P(k^{-1/2})$. Therefore $1/(\gamma + 1) - 1/(\gamma' + 1) = (\gamma' - \gamma)/(\gamma + 1)^2 + o(k^{-1/2})$, and so (3.4.20) implies

$$\gamma' - \gamma - \left(\frac{\gamma'}{\sigma_0'} - \gamma\right)\frac{1}{\gamma + 1} - k^{-1/2}\gamma\int_0^1 s^\gamma Z_n(s)\,ds + o_P(k^{-1/2}) = 0,$$

$$\frac{\gamma' - \gamma}{(\gamma + 1)^2} - \left(\frac{\gamma'}{\sigma_0'} - \gamma\right)\frac{1}{(\gamma + 1)(2\gamma + 1)} - k^{-1/2}\gamma\int_0^1 s^{2\gamma} Z_n(s)\,ds + o_P(k^{-1/2})$$

$$= 0. \quad (3.4.21)$$

Straightforward calculations show that a solution of this linear system in $\gamma' - \gamma$ and $\gamma'/\sigma_0' - \gamma$ satisfies (3.4.17).

Since conversely a solution of type (3.4.17) obviously satisfies condition (3.4.11), it is easily seen that it also solves (3.4.19) and thus (3.4.8). ∎

Proof (of Theorem 3.4.2). We shall prove the theorem only for the case $\gamma > 0$. The case $-\frac{1}{2} < \gamma < 0$ requires somewhat similar arguments. The proof in the case $\gamma = 0$ requires longer expansions but the arguments are also similar. For the complete proof we refer to Drees, Ferreira, and de Haan (2003). A different proof but only for the case $\gamma > 0$ can be found in Drees (1998), and for a slightly different approach we refer to Smith (1987).

Hence suppose $\gamma > 0$. Let $a_0(n/k)$ and $A_0(n/k)$ from Theorem 2.3.6. From Proposition 3.4.4 and (3.4.15) the sequence of solutions of (3.4.4), $(\hat{\gamma}_{MLE}, \hat{\sigma}_{MLE})$ say, satisfies

$$\sqrt{k}(\hat{\gamma}_{MLE} - \gamma) - \frac{(\gamma + 1)^2}{\gamma}\sqrt{k}A_0(\frac{n}{k})\int_0^1 \left(s^\gamma - (2\gamma + 1)s^{2\gamma}\right)\Psi_{\gamma,\rho}(s^{-1})\,ds$$

$$\xrightarrow{d} \frac{(\gamma + 1)^2}{\gamma}\int_0^1 \left(s^\gamma - (2\gamma + 1)s^{2\gamma}\right)\left(s^{-\gamma-1}W(s) - W(1)\right)\,ds$$

and

$$\sqrt{k}\left(\frac{\hat{\sigma}_{\text{MLE}}}{a_0(\frac{n}{k})} - 1\right)$$

$$-\frac{\gamma+1}{\gamma}\sqrt{k}A_0\left(\frac{n}{k}\right)\int_0^1 \left((\gamma+1)(2\gamma+1)s^{2\gamma} - s^\gamma\right)\Psi_{\gamma,\rho}(s^{-1})\,ds$$

$$\xrightarrow{d} \frac{\gamma+1}{\gamma}\int_0^1\left((\gamma+1)(2\gamma+1)s^{2\gamma} - s^\gamma\right)\left(s^{-\gamma-1}W(s) - W(1)\right)\,ds,$$

as $n \to \infty$, and the convergence holds jointly with the same limiting standard Brownian motion W.

Next from Corollary 2.3.5 and Theorem 2.3.6 it follows that

$$\lim_{t\to\infty}\frac{\frac{a_0(t)}{a(t)} - 1}{A(t)} = \begin{cases} -1/\rho, & \rho < 0 \\ -1/\gamma, & \rho = 0 \neq \gamma \\ 0, & \rho = 0 = \gamma \end{cases} =: L$$

and

$$A_0(t) \sim A(t)\left(1_{\{\rho<0\}}\frac{1}{\rho} + 1_{\{\rho=0\}}\right),$$

as $t \to \infty$. Hence the above relations in terms of a and A become

$$\sqrt{k}(\hat{\gamma}_{MLE} - \gamma) - \frac{(\gamma+1)^2}{\gamma}$$

$$\times \left(1_{\{\rho<0\}}\frac{1}{\rho} + 1_{\{\rho=0\}}\right)\sqrt{k}A\left(\frac{n}{k}\right)\int_0^1\left(s^\gamma - (2\gamma+1)s^{2\gamma}\right)\Psi_{\gamma,\rho}(s^{-1})\,ds$$

$$\xrightarrow{d} \frac{(\gamma+1)^2}{\gamma}\int_0^1\left(s^\gamma - (2\gamma+1)s^{2\gamma}\right)\left(s^{-\gamma-1}W(s) - W(1)\right)\,ds$$

and

$$\sqrt{k}\left(\frac{\hat{\sigma}_{MLE}}{a(\frac{n}{k})} - 1\right) - \sqrt{k}A(\frac{n}{k})\left\{L + \frac{\gamma+1}{\gamma}\right.$$

$$\left. \times \left(1_{\{\rho<0\}}\frac{1}{\rho} + 1_{\{\rho=0\}}\right)\int_0^1\left((\gamma+1)(2\gamma+1)s^{2\gamma} - s^\gamma\right)\Psi_{\gamma,\rho}(s^{-1})\,ds\right\}$$

$$\xrightarrow{d} \frac{\gamma+1}{\gamma}\int_0^1\left((\gamma+1)(2\gamma+1)s^{2\gamma} - s^\gamma\right)\left(s^{-\gamma-1}W(s) - W(1)\right)\,ds .$$

Therefore, the components of the left-hand side of (3.4.7) minus deterministic bias terms converge to certain integrals of a Gaussian process that are normal random variables. If $\sqrt{k}A(k/n) \to \lambda$, the bias term of $\sqrt{k}(\hat{\gamma}_{MLE} - \gamma)$ tends to

$$\left(1_{\{\rho<0\}}\rho^{-1} + 1_{\{\rho=0\}}\right)\lambda\gamma^{-1}(\gamma+1)^2\int_0^1\left(s^\gamma - (2\gamma+1)s^{2\gamma}\right)\Psi_{\gamma,\rho}(s^{-1})\,ds .$$

Using the definition of $\Psi_{\gamma,\rho}$, the result follows by simple calculations. The asymptotic bias of the second component can be derived similarly.

To calculate the variance of the limiting normal random variable of $\sqrt{k}(\hat{\gamma}_{\mathrm{MLE}}-\gamma)$, let

$$X(s) := \gamma^{-1}(\gamma + 1)^2 \left(s^\gamma - (2\gamma + 1)s^{2\gamma}\right)\left(s^{-\gamma-1}W(s) - W(1)\right) .$$

Then straightforward calculations show that

$$\mathrm{Var}\left(\int_0^1 X(s)\,ds\right) = \int_0^1 \int_0^1 E\left(X(s)X(t)\right) ds\,dt = (\gamma + 1)^2 .$$

Likewise, to obtain the asymptotic covariance of $\sqrt{k}(\hat{\gamma}_{\mathrm{MLE}} - \gamma)$ with $\sqrt{k}(\hat{\sigma}_{\mathrm{MLE}}/a(n/k) - 1)$, let

$$Z(s) := \gamma^{-1}(\gamma + 1)\left((\gamma + 1)(2\gamma + 1)s^{2\gamma} - s^\gamma\right)\left(s^{-\gamma-1}W(s) - W(1)\right) .$$

Then

$$\mathrm{Cov}\left(\int_0^1 X(s)\,ds, \int_0^1 Z(s)\,ds\right) = \int_0^1 \int_0^1 E\left(X(s)Z(t)\right) ds\,dt = -1 - \gamma .$$

The limiting variance of the scale estimator is obtained similarly. ∎

3.5 A Moment Estimator ($\gamma \in \mathbb{R}$)

Next we want to develop an estimator similar to the Hill estimator but one that can be used for general $\gamma \in \mathbb{R}$, not only for $\gamma > 0$. In order to introduce the estimator let us look at the behavior of the Hill estimator for general γ. We look at a slightly more general statistic.

An immediate problem with applying the Hill estimator for the case $\gamma \leq 0$ is that $U(\infty) \leq 0$ is possible, in which case the logarithm of the observations is not defined. In order to deal with this we shall assume throughout that $U(\infty) > 0$, which can be achieved by applying a shift to the data. However, one should be aware that this shift influences the behavior of the estimator.

Lemma 3.5.1 *Let $X_1, X_2, \ldots$ be i.i.d. random variables with distribution function F and suppose $F \in \mathcal{D}(G_\gamma)$, $x^* = U(\infty) > 0$, i.e., for $x > 0$,*

$$\lim_{t \to \infty} \frac{U(tx) - U(t)}{a(t)} = \frac{x^\gamma - 1}{\gamma} . \tag{3.5.1}$$

Define for $j = 1, 2$,

$$M_n^{(j)} := \frac{1}{k}\sum_{i=0}^{k-1}\left(\log X_{n-i,n} - \log X_{n-k,n}\right)^j . \tag{3.5.2}$$

Then for $k = k(n) \to \infty$, $k/n \to 0$, $n \to \infty$,

$$\frac{M_n^{(j)}}{\left(a\left(\frac{n}{k}\right)/U\left(\frac{n}{k}\right)\right)^j} \xrightarrow{P} \prod_{i=1}^{j} \frac{i}{1-i\gamma_-} \tag{3.5.3}$$

with $\gamma_- = \min(0,\gamma)$.

Proof. For $\gamma > 0$ relation (3.5.1) simplifies (cf. Corollary 1.2.10) to

$$\lim_{t\to\infty} \frac{U(tx)}{U(t)} = x^\gamma, \quad x > 0,$$

i.e.,

$$\lim_{t\to\infty} \frac{\log U(tx) - \log U(t)}{\gamma} = \log x .$$

For $\gamma \leq 0$ we have (cf. Lemma 1.2.9) $\lim_{t\to\infty} U(tx)/U(t) = 1$ for all $x > 0$. Hence (3.5.1) is equivalent to

$$\lim_{t\to\infty} \frac{\log U(tx) - \log U(t)}{a(t)/U(t)} = \frac{x^\gamma - 1}{\gamma} .$$

Summarizing we get that relation (3.5.1) is equivalent to

$$\lim_{t\to\infty} \frac{\log U(tx) - \log U(t)}{a(t)/U(t)} = \frac{x^{\gamma_-} - 1}{\gamma_-} \tag{3.5.4}$$

with $\gamma_- := \min(0,\gamma)$ and (cf. Lemma 1.2.9)

$$\lim_{t\to\infty} \frac{a(t)}{U(t)} = \gamma_+ \tag{3.5.5}$$

with $\gamma_+ := \max(0,\gamma)$.

Next we use the inequalities of Theorem B.2.18: for each $\varepsilon > 0$ there exists t_0 such that for $t \geq t_0$, $x \geq 1$,

$$-\varepsilon x^{\gamma_-+\varepsilon} < \frac{\log U(tx) - \log U(t)}{q_0(t)} - \frac{x^{\gamma_-} - 1}{\gamma_-} < \varepsilon x^{\gamma_-+\varepsilon} , \tag{3.5.6}$$

where q_0 is a positive function satisfying $q_0(t) \sim a(t)/U(t)$, as $t \to \infty$.

Let $Y_1, Y_2, \ldots, Y_n$ be independent and identically distributed with distribution function $1 - 1/x$, $x > 1$. We apply these inequalities with $t := Y_{n-k,n}$ (tending to infinity a.s., as $n \to \infty$) and $x := Y_{n-i,n}/Y_{n-k,n}$. We then have eventually, uniformly for $0 \leq i \leq k - 1$,

$$\frac{\log U(Y_{n-i,n}) - \log U(Y_{n-k,n})}{q_0(Y_{n-k,n})} - \frac{\left(\frac{Y_{n-i,n}}{Y_{n-k,n}}\right)^{\gamma_-} - 1}{\gamma_-} < \varepsilon \left(\frac{Y_{n-i,n}}{Y_{n-k,n}}\right)^{\gamma_-+\varepsilon} .$$

It follows by adding the inequalities for $i = 0, 1, \ldots, k - 1$ that

$$\frac{\frac{1}{k}\sum_{i=0}^{k-1}\log U(Y_{n-i,n}) - \log U(Y_{n-k,n})}{q_0(Y_{n-k,n})}$$

$$< \frac{1}{k}\sum_{i=0}^{k-1}\frac{\left(\frac{Y_{n-i,n}}{Y_{n-k,n}}\right)^{\gamma_-} - 1}{\gamma_-} + \frac{\varepsilon}{k}\sum_{i=0}^{k-1}\left(\frac{Y_{n-i,n}}{Y_{n-k,n}}\right)^{\gamma_-+\varepsilon}$$

$$\stackrel{d}{=} \frac{1}{k}\sum_{i=1}^{k}\frac{(Y_i^*)^{\gamma_-} - 1}{\gamma_-} + \frac{\varepsilon}{k}\sum_{i=1}^{k}(Y_i^*)^{\gamma_-+\varepsilon}$$

with $Y_1^*, Y_2^*, \ldots, Y_k^*$ independent and identically distributed with distribution function $1 - 1/x, x > 1$, by the reasoning from the proof of Lemma 3.3.2. By the law of large numbers the right-hand side converges in probability to the mean, i.e.,

$$E\frac{Y^{\gamma_-} - 1}{\gamma_-} + \varepsilon E\, Y^{\gamma_-+\varepsilon} = \frac{1}{1-\gamma_-} + \frac{\varepsilon}{1-\gamma_- - \varepsilon}.$$

A similar lower bound applies.

Next by squaring the expansions (3.5.6) we find that

$$\left(\frac{\log U(tx) - \log U(t)}{q_0(t)}\right)^2$$

is bounded above by

$$\left(\frac{x^{\gamma_-} - 1}{\gamma_-}\right)^2 + 2\,\varepsilon\, x^{\gamma_-+\varepsilon}\,\frac{x^{\gamma_-} - 1}{\gamma_-} + \varepsilon^2 x^{2\gamma_-+2\varepsilon},$$

and a similar lower bound applies. Starting from these inequalities, we follow the same reasoning as before. This leads to (3.5.3) for $j = 2$. ∎

It follows from Lemma 3.5.1 (cf. (3.5.5)) that the Hill estimator converges to zero for $\gamma \leq 0$; hence this estimator is noninformative in this range. However, this lemma helps us to find a consistent estimator of γ for $\gamma < 0$, since under its conditions,

$$\frac{\left(M_n^{(1)}\right)^2}{M_n^{(2)}} \stackrel{P}{\to} \frac{1-2\gamma_-}{2(1-\gamma_-)}. \tag{3.5.7}$$

As mentioned before we also have

$$\hat{\gamma}_H \stackrel{P}{\to} \gamma_+. \tag{3.5.8}$$

This leads to the following combination of the Hill estimator and the statistic in (3.5.7):

$$\hat{\gamma}_M := M_n^{(1)} + 1 - \frac{1}{2}\left(1 - \frac{\left(M_n^{(1)}\right)^2}{M_n^{(2)}}\right)^{-1}, \tag{3.5.9}$$

for which we have proved the following:

Theorem 3.5.2 *Let $X_1, X_2, \ldots$ be i.i.d. random variables with distribution function F. Suppose $F \in \mathcal{D}(G_\gamma)$ and $x^* > 0$. Then*

$$\hat{\gamma}_M \xrightarrow{P} \gamma$$

for $\gamma \in \mathbb{R}$ provided the sequence k is intermediate, i.e., $k = k(n) \to \infty, k/n \to 0$ as $n \to \infty$, i.e., $\hat{\gamma}_M$ is consistent for γ.

Remark 3.5.3 The estimator $\hat{\gamma}_M$ is called *moment estimator*. The name stems from the fact that the left-hand side of (3.5.3) converges to $E(Y^{\gamma_-} - 1)^j / \gamma_-^j$, $j = 1, 2$, which is the jth moment of the limiting generalized Pareto distribution. In contrast, remember that the Pickands estimator is a quantile estimator (Remark 3.3.4).

Next we prove that $\hat{\gamma}_H$ is asymptotically normal under appropriate conditions. In particular we need a second-order condition for the function $\log U(t)$. From Lemma B.3.16 (see Appendix B) we know that under the usual second-order condition for $U(t)$,

$$\lim_{t \to \infty} \frac{\frac{U(tx)-U(t)}{a(t)} - \frac{x^\gamma - 1}{\gamma}}{A(t)} = \int_1^x s^{\gamma-1} \int_1^s u^{\rho-1} \, du \, ds \,, \tag{3.5.10}$$

for all $x > 0$, and if $\gamma \neq \rho$ and $\rho < 0$ if $\gamma > 0$, a second-order condition for $\log U(t)$ holds:

$$\lim_{t \to \infty} \frac{\frac{\log U(tx)-\log U(t)}{q(t)} - \frac{x^{\gamma_-} - 1}{\gamma_-}}{Q(t)} = \int_1^x s^{\gamma_- -1} \int_1^s u^{\rho' - 1} \, du \, ds =: H_{\gamma_-, \rho'}(x) \tag{3.5.11}$$

with $\gamma_- := \min(0, \gamma)$ as before, $q := a/U$ a positive function, and Q not changing sign eventually with $Q(t) \to 0, t \to \infty$. When $\gamma > 0$ and $\rho = 0$ the limit (3.5.11) vanishes. One possible $Q(t)$ is

$$Q(t) = \begin{cases} A(t), & \gamma < \rho \leq 0, \\ \gamma_+ - \frac{a(t)}{U(t)}, & \rho < \gamma \leq 0 \text{ or } (0 < \gamma < -\rho \text{ and } l \neq 0) \text{ or } \gamma = -\rho, \\ \frac{\rho}{\gamma+\rho} A(t), & (0 < \gamma < -\rho \text{ and } l = 0) \text{ or } \gamma > -\rho > 0, \\ A(t), & \gamma > \rho = 0, \end{cases} \tag{3.5.12}$$

where $l := \lim_{t \to \infty}(U(t) - a(t)/\gamma)$ (cf. Appendix B).

Then, according to Theorem 2.3.6 with suitably chosen Q_0,

$$q_0(t) := \begin{cases} ct^{\gamma_-}, & \rho < 0, \\ -\gamma_-(\log U(\infty) - \log U(t)), & \gamma_- < \rho = 0, \\ (\log U(t))\widehat{} + (\log U(t))\widehat{}, & \gamma_- = \rho = 0 \end{cases} \tag{3.5.13}$$

(see the theorem for the meaning of $(\log U(t))\widehat{}$ and $(\log U(t))\widehat{}$) and $c := \lim_{t \to \infty} t^{-\gamma_-} q(t) > 0$, we have that for each $\varepsilon, \delta > 0$ there exists $t_0 = t_0(\varepsilon, \delta) > 0$ such that for all $t, tx \geq t_0$,

$$\left| \frac{\frac{\log U(tx) - \log U(t)}{q_0(t)} - \frac{x^{\gamma_-} - 1}{\gamma_-}}{Q_0(t)} - \Psi_{\gamma_-,\rho}(x) \right| \le \varepsilon x^{\gamma_- + \rho} \max(x^\delta, x^{-\delta}) .$$

Theorem 3.5.4 (Dekkers, Einmahl, and de Haan (1989)) *Let $X_1, X_2, \ldots$ be i.i.d. random variables with distribution function F with $x^* > 0$. Suppose the second-order condition (3.5.10) holds with $\gamma \ne \rho$. If the sequence of integers $k = k(n)$ satisfies $k \to \infty$, $k/n \to 0$,*

$$\lim_{n \to \infty} \sqrt{k} Q\left(\frac{n}{k}\right) = \lambda \qquad (3.5.14)$$

with Q from (3.5.11) and λ finite, then

$$\sqrt{k}(\hat{\gamma}_M - \gamma) \xrightarrow{d} N(\lambda b_{\gamma,\rho}, var_\gamma) \qquad (3.5.15)$$

with N standard normal, where

$$b_{\gamma,\rho} := \begin{cases} \frac{(1-\gamma)(1-2\gamma)}{(1-\gamma-\rho)(1-2\gamma-\rho)}, & \gamma < \rho \le 0, \\ \frac{\gamma(1+\gamma)}{(1-\gamma)(1-3\gamma)}, & \rho < \gamma \le 0, \\ -\frac{\gamma}{(1+\gamma)^2}, & 0 < \gamma < -\rho \text{ and } l \ne 0, \\ \frac{\gamma - \gamma\rho + \rho}{\rho(1-\rho)^2}, & (0 < \gamma < -\rho \text{ and } l = 0) \text{ or } \gamma \ge -\rho > 0, \\ 1, & \gamma > \rho = 0, \end{cases} \qquad (3.5.16)$$

and

$$var_\gamma := \begin{cases} \gamma^2 + 1, & \gamma \ge 0, \\ \frac{(1-\gamma)^2(1-2\gamma)(1-\gamma+6\gamma^2)}{(1-3\gamma)(1-4\gamma)}, & \gamma < 0. \end{cases} \qquad (3.5.17)$$

For the proof we start with an extension of Lemma 3.5.1. Since in the proof we use the result of Theorem 2.3.6 (the uniform inequalities connected to (3.5.11)), we use the function q_0 from that theorem for the formulation of the lemma.

Lemma 3.5.5 *Assume the conditions of Theorem 3.5.4. Write $X_i = U(Y_i)$, $i = 1, 2, \ldots$, where $Y_1, Y_2, \ldots$ are i.i.d. with distribution function $1 - 1/x$, $x \ge 1$. With the notation of Lemma 3.5.1:*

1. if $\gamma \le 0$ or $\rho \ne 0$ the random vector

$$\sqrt{k}\left(\frac{M_n^{(1)}}{q_0(Y_{n-k,n})} - \frac{1}{1-\gamma_-}, \frac{M_n^{(2)}}{q_0^2(Y_{n-k,n})} - \frac{2}{(1-\gamma_-)(1-2\gamma_-)}\right) \qquad (3.5.18)$$

converges in distribution to a random vector, (P, Q) say, normally distributed with mean

$$\lambda \left(1_{\{\rho' < 0\}} \frac{1}{\rho'} + 1_{\{\rho'=0\}}\right)\left(E\Psi_{\gamma_-,\rho'}(Y), 2E\left(\frac{Y^{\gamma_-}-1}{\gamma_-}\Psi_{\gamma_-,\rho'}(Y)\right)\right)$$

$$= \begin{cases} \frac{\lambda}{\rho'}\left(\frac{1}{1-\gamma_-\rho'}, \frac{2(2-2\gamma_--\rho')}{(1-\gamma_-)(1-\gamma_--\rho')(1-2\gamma_--\rho')}\right), & \rho' < 0, \\ \lambda\left(\frac{1}{\gamma_-(1-\gamma_-)^2}, \frac{2(2-3\gamma_-)}{\gamma_-(1-\gamma_-)^2(1-2\gamma_-)^2}\right), & \rho' = 0 > \gamma_-, \\ (0,0), & \rho' = 0 = \gamma_-, \end{cases}$$

and covariance matrix

$$\frac{1}{(1-\gamma_-)^2(1-2\gamma_-)}\begin{pmatrix} 1 & \frac{4}{1-3\gamma_-} \\ \frac{4}{1-3\gamma_-} & \frac{4(5-11\gamma_-)}{(1-2\gamma_-)(1-3\gamma_-)(1-4\gamma_-)} \end{pmatrix};$$

recall (cf. (2.3.16))

$$\Psi_{\gamma_-,\rho'}(x) := \begin{cases} \frac{x^{\gamma_-+\rho'}-1}{\gamma_-+\rho'}, & \rho' < 0, \\ \frac{1}{\gamma_-}x^{\gamma_-}\log x, & \rho' = 0 > \gamma_-; \end{cases}$$

2. *if* $\gamma > 0$ *and* $\rho = 0$ *the random vector*

$$\sqrt{k}\left(\frac{M_n^{(1)}}{q(Y_{n-k,n})} - 1, \frac{M_n^{(2)}}{q^2(Y_{n-k,n})} - 2\right) \tag{3.5.19}$$

converges in distribution to a random vector, (P, Q) say, normally distributed with mean $(0, 0)$ and covariance matrix

$$\begin{pmatrix} 1 & 4 \\ 4 & 20 \end{pmatrix}.$$

Proof. The proof is somewhat similar to that of the corresponding result for the Hill estimator (Theorem 3.2.5). Theorem 2.3.6 tells us that one can choose functions $q_0 > 0$ and Q_0 such that for any $\varepsilon > 0$ there exists t_0 such that for all $t \geq t_0$ and $x \geq 1$,

$$\frac{x^{\gamma_-} - 1}{\gamma_-} + Q_0(t)\Psi_{\gamma_-,\rho'}(x) - \varepsilon|Q_0(t)|x^{\gamma_-+\rho'+\varepsilon}$$

$$\leq \frac{\log U(tx) - \log U(t)}{q_0(t)}$$

$$\leq \frac{x^{\gamma_-} - 1}{\gamma_-} + Q_0(t)\Psi_{\gamma_-,\rho'}(x) + \varepsilon|Q_0(t)|x^{\gamma_-+\rho'+\varepsilon}. \tag{3.5.20}$$

Let us concentrate on the upper inequality. We apply this inequality with t replaced by $Y_{n-k,n}$ (tending a.s. to infinity) and x replaced by $Y_{n-i,n}/Y_{n-k,n}$ for $i = 0, 1, \ldots, k - 1$. Then we eventually get that

$$\frac{M_n^{(1)}}{q_0(Y_{n-k,n})} \leq \frac{1}{k}\sum_{i=0}^{k-1}\frac{\left(\frac{Y_{n-i,n}}{Y_{n-k,n}}\right)^{\gamma_-} - 1}{\gamma_-}$$

$$+ Q_0(Y_{n-k,n})\frac{1}{k}\sum_{i=0}^{k-1}\Psi_{\gamma_-,\rho'}\left(\frac{Y_{n-i,n}}{Y_{n-k,n}}\right)$$

$$+ \varepsilon\left|Q_0(Y_{n-k,n})\right|\frac{1}{k}\sum_{i=0}^{k-1}\left(\frac{Y_{n-i,n}}{Y_{n-k,n}}\right)^{\gamma_-+\rho'+\varepsilon}.$$

As in the proof of Lemma 3.5.1, the right-hand side is equal in distribution to

$$
\frac{1}{k} \sum_{i=1}^{k} \frac{(Y_i^{\star})^{\gamma_-} - 1}{\gamma_-}
$$

$$
+ Q_0(Y_{n-k,n}) \frac{1}{k} \sum_{i=1}^{k} \Psi_{\gamma_-,\rho'}(Y_i^{\star}) + \varepsilon \left| Q_0(Y_{n-k,n}) \right| \frac{1}{k} \sum_{i=1}^{k} (Y_i^{\star})^{\gamma_- + \rho' + \varepsilon}
$$

with $Y^{\star}, Y_1^{\star}, Y_2^{\star}, \ldots, Y_k^{\star}$ independent and identically distributed with distribution function $1 - 1/x$, $x \geq 1$, and independent of $Y_{n-k,n}$. Hence

$$
\sqrt{k} \left(\frac{M_n^{(1)}}{q_0(Y_{n-k,n})} - \frac{1}{1 - \gamma_-} \right)
$$

$$
\leq \sqrt{k} \left(\frac{1}{k} \sum_{i=1}^{k} \frac{(Y_i^{\star})^{\gamma_-} - 1}{\gamma_-} - E \frac{(Y^{\star})^{\gamma_-} - 1}{\gamma_-} \right) + \sqrt{k} Q_0(Y_{n-k,n}) \frac{1}{k} \sum_{i=1}^{k} \Psi_{\gamma_-,\rho'}(Y_i^{\star})
$$

$$
+ \varepsilon \sqrt{k} \left| Q_0(Y_{n-k,n}) \right| \frac{1}{k} \sum_{i=1}^{k} (Y_i^{\star})^{\gamma_- + \rho' + \varepsilon} \ .
$$

One easily verifies that the conditions of the central limit theorem are fulfilled for the first term and the conditions of the law of large numbers for the other two terms. Then the last term vanishes in the limit. Moreover, since by Corollary 2.2.2

$$
\frac{k}{n} Y_{n-k,n} \xrightarrow{P} 1
$$

and since Q_0 is a regularly varying function, we have

$$
\frac{Q_0(Y_{n-k,n})}{Q_0 \left(\frac{n}{k} \right)} \xrightarrow{P} 1 \ .
$$

Recall from Corollary 2.3.5, Theorem 2.3.6, and assumption (3.5.14) that if $\rho' < 0$,

$$
\lim_{t \to \infty} \sqrt{k} Q_0 \left(\frac{n}{k} \right) = \lim_{t \to \infty} \sqrt{k} \frac{1}{\rho'} Q \left(\frac{n}{k} \right) = \frac{\lambda}{\rho'} ,
$$

and if $\rho' = 0$,

$$
\lim_{t \to \infty} \sqrt{k} Q_0 \left(\frac{n}{k} \right) = \lim_{t \to \infty} \sqrt{k} Q \left(\frac{n}{k} \right) = \lambda \ .
$$

Hence, since a similar lower bound applies, as $n \to \infty$,

$$
\sqrt{k} \left(\frac{M_n^{(1)}}{q_0(Y_{n-k,n})} - \frac{1}{1 - \gamma_-} \right) - \sqrt{k} \left(\frac{1}{k} \sum_{i=1}^{k} \frac{(Y_i^{\star})^{\gamma_-} - 1}{\gamma_-} - E \frac{(Y^{\star})^{\gamma_-} - 1}{\gamma_-} \right)
$$

$$
- \lambda \left(1_{\{\rho' < 0\}} \frac{1}{\rho'} + 1_{\{\rho' = 0\}} \right) E \Psi_{\gamma_-,\rho'}(Y^{\star}) \xrightarrow{P} 0 \ . \quad (3.5.21)
$$

Since in particular $M_n^{(1)}/q_0(Y_{n-k,n}) \to^P 1/(1-\gamma_-)$ and

$$\sqrt{k}\left(\frac{\left(M_n^{(1)}\right)^2}{q_0^2(Y_{n-k,n})} - \left(\frac{1}{1-\gamma_-}\right)^2\right)$$

$$= \sqrt{k}\left(\frac{M_n^{(1)}}{q_0(Y_{n-k,n})} - \frac{1}{1-\gamma_-}\right)\left(\frac{M_n^{(1)}}{q_0(Y_{n-k,n})} + \frac{1}{1-\gamma_-}\right),$$

we also have that as $n \to \infty$,

$$\sqrt{k}\left(\frac{\left(M_n^{(1)}\right)^2}{q_0^2(Y_{n-k,n})} - \frac{1}{(1-\gamma_-)^2}\right)$$

$$- \frac{2\sqrt{k}}{1-\gamma_-}\left(\frac{1}{k}\sum_{i=1}^{k}\frac{(Y_i^\star)^{\gamma_-}-1}{\gamma_-} - E\,\frac{(Y^\star)^{\gamma_-}-1}{\gamma_-}\right)$$

$$- \frac{2\lambda}{1-\gamma_-}\left(1_{\{\rho'<0\}}\frac{1}{\rho'} + 1_{\{\rho'=0\}}\right)E\,\Psi_{\gamma_-,\rho'}(Y^\star) \to^P 0 . \quad (3.5.22)$$

Now we turn to $M_n^{(2)}$. The inequalities (3.5.20) yield

$$\frac{(\log U(tx) - \log U(t))^2}{q_0^2(t)} \le \left(\frac{x^{\gamma_-}-1}{\gamma_-}\right)^2 + 2Q_0(t)\,\frac{x^{\gamma_-}-1}{\gamma_-}\,\Psi_{\gamma_-,\rho'}(x)$$

$$+2\varepsilon\,|Q_0(t)|\,\frac{x^{\gamma_-}-1}{\gamma_-}\,x^{\gamma_-+\rho'+\varepsilon}$$

$$+Q_0^2(t)\left(\Psi_{\gamma_-,\rho'}(x) + \text{sign}(Q_0)\,\varepsilon x^{\gamma_-+\rho'+\varepsilon}\right)^2 .$$

Hence eventually,

$$\frac{M_n^{(2)}}{q_0^2(Y_{n-k,n})} \le \frac{1}{k}\sum_{i=1}^{k}\left(\frac{(Y_i^\star)^{\gamma_-}-1}{\gamma_-}\right)^2$$

$$+2Q_0(Y_{n-k,n})\frac{1}{k}\sum_{i=1}^{k}\frac{(Y_i^\star)^{\gamma_-}-1}{\gamma_-}\,\Psi_{\gamma_-,\rho'}(Y_i^\star)$$

$$+2\varepsilon\,|Q_0(Y_{n-k,n})|\,\frac{1}{k}\sum_{i=1}^{k}\frac{(Y_i^\star)^{\gamma_-}-1}{\gamma_-}\,(Y_i^\star)^{\gamma_-+\rho'+\varepsilon}$$

$$+Q_0^2(Y_{n-k,n})\frac{1}{k}\sum_{i=1}^{k}\left(\Psi_{\gamma_-,\rho'}(Y_i^\star) + \text{sign}(Q_0)\,\varepsilon\,(Y_i^\star)^{\gamma_-+\rho'+\varepsilon}\right)^2 .$$

Again we can apply the central limit theorem to the first term on the right-hand side and the law of large numbers to the other terms. The last two terms vanish in the limit. A similar lower bound applies. We conclude that as $n \to \infty$,

$$\sqrt{k}\left(\frac{M_n^{(2)}}{q_0^2(Y_{n-k,n})} - \frac{2}{(1-\gamma_-)(1-2\gamma_-)}\right)$$

$$- \sqrt{k}\left(\frac{1}{k}\sum_{i=1}^{k}\left(\frac{(Y_i^\star)^{\gamma_-}-1}{\gamma_-}\right)^2 - E\left(\frac{(Y^\star)^{\gamma_-}-1}{\gamma_-}\right)^2\right)$$

$$- 2\lambda\left(1_{\{\rho'<0\}}\frac{1}{\rho'} + 1_{\{\rho'=0\}}\right)E\left(\frac{(Y^\star)^{\gamma_-}-1}{\gamma_-}\Psi_{\gamma_-,\rho'}(Y^\star)\right) \xrightarrow{P} 0. \quad (3.5.23)$$

In view of (3.5.22) and (3.5.23) it suffices for the proof to find the joint limit distribution of

$$\left\{\sqrt{k}\left(\frac{1}{k}\sum_{i=1}^{k}\frac{(Y_i^\star)^{\gamma_-}-1}{\gamma_-} - E\frac{(Y^\star)^{\gamma_-}-1}{\gamma_-}\right),\right.$$

$$\left.\sqrt{k}\left(\frac{1}{k}\sum_{i=1}^{k}\left(\frac{(Y_i^\star)^{\gamma_-}-1}{\gamma_-}\right)^2 - E\left(\frac{(Y^\star)^{\gamma_-}-1}{\gamma_-}\right)^2\right)\right\},$$

which can be found in a routine way by applying the Cramér–Wold device, Lyapunov's theorem, and the central limit theorem.

The proof of the second statement is similar. ∎

Corollary 3.5.6 *Under the conditions of Theorem 3.5.4,*

$$\sqrt{k}\left\{\left(1 - \frac{1}{2}\left(1 - \frac{\left(M_n^{(1)}\right)^2}{M_n^{(2)}}\right)^{-1}\right) - \gamma_-\right\}$$

$$= \sqrt{k}\left\{\left(1 - \frac{1}{2}\left(1 - \frac{\left(M_n^{(1)}\right)^2}{M_n^{(2)}}\right)^{-1}\right)\right.$$

$$\left. - \left(1 - \frac{1}{2}\left(1 - \frac{(1-\gamma_-)(1-2\gamma_-)}{2(1-\gamma_-)^2}\right)^{-1}\right)\right\}$$

$$\xrightarrow{d} (1-2\gamma_-)(1-\gamma_-)^2\left\{\left(\frac{1}{2}-\gamma_-\right)Q - 2P\right\},$$

where (P, Q) is the limit vector of Lemma 3.5.5. Hence the limiting random variable is normal with mean $\lambda b_{\gamma,\rho}$ where

$$b_{\gamma,\rho} := \begin{cases} \frac{(1-\gamma)(1-2\gamma)}{(1-\gamma-\rho)(1-2\gamma-\rho)}, & \gamma < \rho \le 0, \\ \frac{(1-\gamma)}{(1-3\gamma)}, & \rho < \gamma \le 0, \\ \frac{1}{(1+\gamma)^2}, & 0 < \gamma < -\rho \text{ and } l \ne 0, \\ \frac{1}{(1-\rho)^2}, & (0 < \gamma < -\rho \text{ and } l = 0) \text{ or } \gamma \ge -\rho > 0, \\ 0, & \gamma > \rho = 0, \end{cases}$$

and variance

$$var_\gamma := \begin{cases} 1, & \gamma \geq 0, \\ \frac{(1-\gamma)^2(1-2\gamma)(1-11\gamma+48\gamma^2-44\gamma^3)}{(1-3\gamma)(1-4\gamma)}, & \gamma < 0. \end{cases}$$

The bias of the limiting random variable in terms of (γ_-, ρ') is given in Exercise 3.10.

Proof. The result is straightforward from Lemma 3.5.5 and application of Cramér's delta method.   ∎

Remark 3.5.7 In subsequent chapters we shall sometimes work with

$$\hat{\gamma}_- := \hat{\gamma}_M - M_n^{(1)} = 1 - \frac{1}{2}\left(1 - \frac{\left(M_n^{(1)}\right)^2}{M_n^{(2)}}\right)^{-1}.$$

Proof (of Theorem 3.5.4). It remains to study the first part of the estimator, i.e.,

$$\sqrt{k}\left(M_n^{(1)} - \gamma_+\right).$$

We know that

$$\sqrt{k}\left(\frac{M_n^{(1)}}{q_0\left(\frac{n}{k}\right)} - \frac{1}{1-\gamma_-}\right) \xrightarrow{d} P$$

with P the first component of the limit vector from Lemma 3.5.5. Therefore

$$\sqrt{k}\left(M_n^{(1)} - \gamma_+\right) = \sqrt{k}\left\{q_0\left(\frac{n}{k}\right)\left(\frac{1}{1-\gamma_-} + \frac{P}{\sqrt{k}} + o_P\left(\frac{1}{\sqrt{k}}\right)\right) - \gamma_+\right\}$$

$$= \frac{\sqrt{k}}{1-\gamma_-}\left(q_0\left(\frac{n}{k}\right) - \gamma_+\right) + q_0\left(\frac{n}{k}\right)P + o_P(1).$$

From the relations between the functions q_0 and q given in Corollary 2.3.5 and Theorem 2.3.6, and from $q_{\gamma,\rho} = \lim_{t\to\infty}(q(t) - \gamma_+)/Q(t)$ given in (B.3.46), Appendix B, we have

$$q_{\gamma,\rho}^0 = \lim_{t\to\infty}\frac{q_0(t) - \gamma_+}{Q(t)} = \begin{cases} -1, & \rho < \gamma \leq 0, \\ 1, & \gamma > \rho = 0, \\ 0, & \text{otherwise}. \end{cases}$$

Then the limit distribution of $\sqrt{k}(\hat{\gamma}_M - \gamma)$ is the distribution of

$$\lambda\frac{q_{\gamma,\rho}^0}{1-\gamma_-} + \gamma_+P + (1 - 2\gamma_-)(1 - \gamma_-)^2\left\{\left(\frac{1}{2} - \gamma_-\right)Q - 2P\right\}. \tag{3.5.24}$$

From this and Lemma 3.5.5 one gets the asymptotic distribution by straightforward but lengthy calculations.   ∎

3.6 Other Estimators

In this section we briefly review two further estimators of γ, the probability weighted moment estimator and what we call the "negative Hill estimator."

3.6.1 The Probability-Weighted Moment Estimator ($\gamma < 1$)

First let us consider the probability-weighted moment estimator of Hosking and Wallis (1987). The starting point is the observation that if V is a random variable with a generalized Pareto distribution, i.e., with distribution function

$$H_{\gamma,\alpha}(x) := 1 - \left(1 + \frac{\gamma x}{\alpha}\right)^{-1/\gamma}, \quad 0 < x < \frac{\alpha}{0 \vee (-\gamma)},$$

where $\alpha > 0$ and γ are real parameters, then for $\gamma < 1$,

$$E\, V = \int_0^{\alpha/(0\vee(-\gamma))} \left(1 - H_{\gamma,\alpha}(x)\right)\, dx = \frac{\alpha}{1 - \gamma}. \tag{3.6.1}$$

Moreover,

$$E\left\{V\left(1 - H_{\gamma,\alpha}(V)\right)\right\} = \frac{1}{2}\int_0^{1/(0\vee(-\gamma))} \left(1 - H_{\gamma,\alpha}(x)\right)^2 dx = \frac{\alpha}{2(2 - \gamma)}, \tag{3.6.2}$$

which can be called a probability-weighted moment.

We can solve relations (3.6.1) and (3.6.2) and obtain

$$\gamma = \frac{E\, V - 4E\left\{V\left(1 - H_{\gamma,\alpha}(V)\right)\right\}}{E\, V - 2E\left\{V\left(1 - H_{\gamma,\alpha}(V)\right)\right\}} \tag{3.6.3}$$

and

$$\alpha = \frac{2E\, V\, E\left\{V\left(1 - H_{\gamma,\alpha}(V)\right)\right\}}{E\, V - 2E\left\{V\left(1 - H_{\gamma,\alpha}(V)\right)\right\}}. \tag{3.6.4}$$

A sample analogue of the right-hand sides of (3.6.3) and (3.6.4) will provide estimators for γ and α.

Consider independent and identically distributed random variables $X_1, X_2, \ldots$ with distribution function F and suppose that F is in the domain of attraction of an extreme value distribution G_γ. Then we know (e.g., Section 3.1) that for $x > 0$ and a suitably chosen positive function f,

$$\lim_{t \uparrow x^*} P\left(\frac{X_1 - t}{f(t)} > x \middle| X_1 > t\right) = \lim_{t \uparrow x^*} \frac{P\left(X_1 - t > xf(t)\right)}{P(X_1 > t)} = (1 + \gamma x)^{-1/\gamma}.$$

If we want to build analogues of (3.6.3) and (3.6.4), we can try to replace $E\, V$ by

$$\int_0^\infty \frac{1 - F(t + xf(t))}{1 - F(t)}\, dx = \frac{1}{f(t)}\int_t^\infty \frac{1 - F(u)}{1 - F(t)}\, du$$

$$= \frac{1}{f(t)}\int_t^\infty \frac{u - t}{1 - F(t)}\, dF(u) \tag{3.6.5}$$

and $E\left\{V(1 - H_{\gamma,\alpha}(V))\right\}$ by

$$\frac{1}{2}\int_0^\infty \left(\frac{1 - F(t + xf(t))}{1 - F(t)}\right)^2 dx$$

$$= \frac{1}{2f(t)}\int_t^\infty \left(\frac{1 - F(u)}{1 - F(t)}\right)^2 du$$

$$= \frac{1}{f(t)}\int_t^\infty (u - t)\frac{1 - F(u)}{(1 - F(t))^2}\, dF(u). \qquad (3.6.6)$$

Next we need sample analogues of (3.6.5) and (3.6.6). These can be obtained by replacing t with the intermediate order statistic $X_{n-k,n}$ and replacing F with the empirical distribution function F_n. Then (3.6.5) becomes after normalization (note that $1 - F_n(X_{n-k,n}) = k/n$),

$$P_n := \frac{1}{k}\sum_{i=0}^{k-1} X_{n-i,n} - X_{n-k,n}, \qquad (3.6.7)$$

and (3.6.6) becomes

$$Q_n := \frac{1}{k}\sum_{i=0}^{k-1}\frac{i}{k}\left(X_{n-i,n} - X_{n-k,n}\right). \qquad (3.6.8)$$

This leads us to consider the estimators

$$\hat{\gamma}_{PWM} := \frac{P_n - 4Q_n}{P_n - 2Q_n} = 1 - \left(\frac{P_n}{2Q_n} - 1\right)^{-1} \qquad (3.6.9)$$

and

$$\hat{\sigma}_{PWM} := \frac{2P_nQ_n}{P_n - 2Q_n} = P_n\left(\frac{P_n}{2Q_n} - 1\right)^{-1}, \qquad (3.6.10)$$

which are the *probability-weighted moment estimators*.

In fact, the definition encompasses variants whereby in the definition of Q_n the factor i/k is replaced with $(i + c)/(k + d)$, which may improve the finite sample behavior but does not affect the asymptotic behavior. Hosking and Wallis mention the choice $c = -0.35$ and $d = 0$.

In order to derive the asymptotic behavior, note that by Corollary 2.4.6, under the second-order condition for $\gamma < \frac{1}{2}$,

$$\sqrt{k}\left(\frac{P_n}{a_0\left(\frac{n}{k}\right)} - \int_0^1 \frac{s^{-\gamma} - 1}{\gamma}\, ds\right)$$

$$= \int_0^1 s^{-\gamma-1}W_n(s) - W_n(1)\, ds + \sqrt{k}A_0\left(\frac{n}{k}\right)\int_0^1 \Psi_{\gamma,\rho}(s^{-1})\, ds + o_P(1) \qquad (3.6.11)$$

with $\{W_n(s)\}$ a sequence of standard Brownian motions. Similarly,

$$\sqrt{k}\left(\frac{Q_n}{a_0\left(\frac{n}{k}\right)} - \int_0^1 s\frac{s^{-\gamma}-1}{\gamma}\,ds\right)$$
$$= \int_0^1 s^{-\gamma}W_n(s) - sW_n(1)\,ds + \sqrt{k}A_0\left(\frac{n}{k}\right)\int_0^1 s\Psi_{\gamma,\rho}(s^{-1})\,ds + o_P(1)\,.$$

$$(3.6.12)$$

For the above relations in terms of the functions a and A instead of a_0 and A_0, see the proof of Theorem 3.4.2. Then by working out the probability distribution of the corresponding right-hand sides and applying Cramér's delta method, one gets the following result:

Theorem 3.6.1 *Let $X_1, X_2, \ldots$ be i.i.d. with distribution function F.*

1. If $F \in \mathcal{D}(G_\gamma)$ with $\gamma < 1$, then

$$\hat{\gamma}_{PWM} \overset{P}{\to} \gamma \quad and \quad \frac{\hat{\sigma}_{PWM}}{a\left(\frac{n}{k}\right)} \overset{P}{\to} 1$$

provided $k = k(n) \to \infty$, $k(n)/n \to 0$, $n \to \infty$.

2. If the second-order condition (2.3.5) is fulfilled with $\gamma < \frac{1}{2}$, $k = k(n) \to \infty$, $k/n \to 0$, and $\lim_{n\to\infty}\sqrt{k}A(n/k) = \lambda$, $n \to \infty$, then

$$\sqrt{k}\left(\hat{\gamma}_{PWM} - \gamma, \frac{\hat{\sigma}_{PWM}}{a\left(\frac{n}{k}\right)} - 1\right)$$

is asymptotically normal with mean vector

$$\frac{\lambda}{(1-\gamma-\rho)(2-\gamma-\rho)}\left((1-\gamma)(2-\gamma), -\rho\right)\,, \quad \rho < 0\,,$$
$$\lambda\,(1,0)\,, \qquad\qquad\qquad \gamma \neq \rho = 0\,,$$
$$\lambda\left(1, -\tfrac{1}{2}\right)\,, \qquad\qquad\qquad \gamma = \rho = 0\,,$$

and covariance matrix

$$\begin{pmatrix} \frac{(1-\gamma)(2-\gamma)^2(1-\gamma+2\gamma^2)}{(1-2\gamma)(3-2\gamma)} & \frac{(2-\gamma)(-2+6\gamma-7\gamma^2+2\gamma^3)}{(1-2\gamma)(3-2\gamma)} \\ \frac{(2-\gamma)(-2+6\gamma-7\gamma^2+2\gamma^3)}{(1-2\gamma)(3-2\gamma)} & \frac{31-94\gamma+102\gamma^2-126\gamma^3+144\gamma^4-80\gamma^5+16\gamma^6}{(1-2\gamma)(3-2\gamma)} \end{pmatrix}\,.$$

Remark 3.6.2 For $\frac{1}{2} < \gamma < 1$ the convergence of $\hat{\gamma}_{PWM}$ to γ is slower than that for $\gamma < \frac{1}{2}$.

Remark 3.6.3 The statistic P_n is commonly called the *empirical mean excess function*. Its main use seems to be to distinguish between subexponential and superexponential distributions. It plays a role in insurance theory; see, e.g., Embrechts, Klüppelberg, and Mikosch (1997). The statistic is often discussed in books on extreme value theory; see e.g., Falk, Hüsler, and Reiss (1994) and Beirlant, Teugels, and Vynckier (1996).

3.6.2 The Negative Hill Estimator ($\gamma < -\frac{1}{2}$)

The "negative Hill estimator" was proposed by Falk (1995). It can serve as a complement, to be used when $\gamma < -\frac{1}{2}$, for the maximum likelihood estimator of Section 3.4.

One starts by observing that if the distribution of X is in the domain of attraction of G_γ with $\gamma < 0$, then the upper endpoint of the distribution of X, x^* is finite, and

$$\tilde{X} := \frac{1}{x^* - X}$$

is in the domain of attraction of $G_{-\gamma}$ (cf. Theorem 1.2.1). Hence we could apply the Hill estimator to $\tilde{X}$, but this way we do not obtain a statistic, since x^* is not known. Fortunately, for $\gamma < -\frac{1}{2}$ the endpoint x^* can be very well approximated by the largest order statistic $X_{n,n}$ (cf. Remark 4.5.5 below). This leads to what we call the *negative Hill estimator*

$$\hat{\gamma}_F := \frac{1}{k} \sum_{i=1}^{k-1} \log\left(X_{n,n} - X_{n-i,n}\right) - \log\left(X_{n,n} - X_{n-k,n}\right) . \tag{3.6.13}$$

Note that this estimator is shift and scale invariant.

For the asymptotic analysis of this estimator we once again use the theory of Section 2.4, in this case Corollary 2.4.5: for $\gamma < -\frac{1}{2}$ and $0 < s \le 1$,

$$- \gamma s^\gamma \frac{X_{n,n} - X_{n-[ks],n}}{a_0\left(\frac{n}{k}\right)}$$

$$= 1 + \frac{\gamma}{\sqrt{k}}\left(s^{-1}W_n(s) + \sqrt{k}A_0\left(\frac{n}{k}\right)s^\gamma \Psi_{\gamma,\rho}(s^{-1}) + o_P(1)\, s^{-1/2-\varepsilon}\right)$$

with $\{W_n(s)\}$ a sequence of standard Brownian motions.

One is tempted to take logarithms on both sides and then expand the right-hand side. However, the second term on the right-hand side does not go to zero uniformly in s. Nevertheless, the convergence is uniform for $s_n \le s \le 1$ with $s_n \downarrow 0$ an appropriate sequence. Then one can prove that

$$\sqrt{k} \int_0^1 \log\left(\frac{X_{n,n} - X_{n-[ks],n}}{a_0(\frac{n}{k})}\right) ds$$

$$= \gamma \int_0^1 s^{-1}W_n(s)\, ds + \gamma\sqrt{k}A_0\left(\frac{n}{k}\right)\int_0^1 s^\gamma \Psi_{\gamma,\rho}(s^{-1})\, ds + o_P(1),$$

which, together with a similar expansion with $X_{n-[ks],n}$ replaced by $X_{n-k,n}$, leads to the following result:

Theorem 3.6.4 *Let* $X_1, X_2, \ldots$ *be i.i.d. random variables with distribution function* F.

1. *If $F \in \mathcal{D}(G_\gamma)$ with $\gamma < -\frac{1}{2}$, then*

$$\hat{\gamma}_F \xrightarrow{P} \gamma$$

   provided $k = k(n) \to \infty$, $k/n \to 0$, $k^\eta / \log n \to \infty$, for some small η, as $n \to \infty$.

2. *If the second-order condition (2.3.5) is fulfilled with $-1 < \gamma < -\frac{1}{2}$, $k = k(n) \to \infty$, $k/n \to 0$, $k^\eta / \log n \to \infty$, for some small η and $\sqrt{k} A (n/k) \to \lambda$, as $n \to \infty$, then*

$$\sqrt{k} \left(\hat{\gamma}_F - \gamma \right)$$

   has asymptotically a normal distribution with variance γ^2 and mean

$$\lambda\gamma \int_0^1 s^\gamma \Psi_{\gamma,\rho}(s^{-1}) \, ds = \begin{cases} \dfrac{\lambda\gamma}{\rho(1+\gamma)(1-\rho)} , & \rho < 0, \\ \lambda , & \rho = 0 . \end{cases}$$

We give the proof of the asymptotic normality. The proof of the consistency is left to the reader (Exercise 3.11).

Proof (via the tail quantile process (Holger Drees)). We shall first consider the sum in the definition of $\hat{\gamma}_F$ for $i = 1, \ldots, j$, where $j = j(n)$ is some sequence with $1 < j(n) < k(n) - 1$. Note that

$$0 \geq \frac{1}{k} \sum_{i=1}^{j-1} \log \left(\frac{X_{n,n} - X_{n-i,n}}{X_{n,n} - X_{n-k,n}} \right) \geq \frac{j-1}{k} \log \left(\frac{X_{n,n} - X_{n-1,n}}{U(\infty) - X_{n-k,n}} \right) .$$

From Theorem 2.1.1,

$$\frac{X_{n,n} - X_{n-1,n}}{a(n)} = \frac{X_{n,n} - b_n}{a(n)} - \frac{X_{n-1,n} - b_n}{a(n)} = O_P(1)$$

for some positive sequence $a(n) \in RV_\gamma$. Consequently $\log(X_{n,n} - X_{n-1,n}) = O_P(\log n)$, $n \to \infty$. On the other hand, from Lemma 1.2.9 and Theorem 2.4.1,

$$\frac{U(\infty) - X_{n-k,n}}{a(\frac{n}{k})} = \frac{U(\infty) - U(\frac{n}{k})}{a(\frac{n}{k})} - \frac{X_{n-k,n} - U(\frac{n}{k})}{a(\frac{n}{k})} \xrightarrow{P} -\frac{1}{\gamma} ;$$

hence $\log(U(\infty) - X_{n-k,n}) = O_P(\log(n/k))$. Therefore

$$\frac{1}{k} \sum_{i=1}^{j-1} \log \left(\frac{X_{n,n} - X_{n-i,n}}{X_{n,n} - X_{n-k,n}} \right) = O_P \left(\frac{j}{k} \log n \right) . \qquad (3.6.14)$$

Next we consider

$$\frac{1}{k} \sum_{i=j}^{k-1} \log \left(\frac{X_{n,n} - X_{n-i,n}}{X_{n,n} - X_{n-k,n}} \right) = \int_{j/k}^1 \log \left(\frac{X_{n,n} - X_{n-[ks],n}}{X_{n,n} - X_{n-k,n}} \right) ds$$

$$= \int_{j/k}^1 \log \left(\frac{X_{n,n} - X_{n-[ks],n}}{a_0(\frac{n}{k})} \right) ds - \int_{j/k}^1 \log \left(\frac{X_{n,n} - X_{n-k,n}}{a_0(\frac{n}{k})} \right) ds \qquad (3.6.15)$$

and take $j = k^{-\delta+1}$, i.e., $j/k = k^{-\delta}$ with $0 < \delta < (-(2\gamma)^{-1} \wedge (1 + 2\varepsilon)^{-1})$, for some $\varepsilon < \frac{1}{2}$. Then from Corollary 2.4.5, for $k^{-\delta} \leq s \leq 1$ we have

$$\log \left(-\gamma s^\gamma \frac{X_{n,n} - X_{n-[ks],n}}{a_0(\frac{n}{k})} \right)$$

$$= \log \left(1 + \frac{\gamma}{\sqrt{k}} \left\{ s^{-1} W_n(s) + \sqrt{k} A_0 \left(\frac{n}{k} \right) s^\gamma \Psi_{\gamma,\rho}(s^{-1}) + o_P(s^{-1/2-\varepsilon}) \right\} \right)$$

$$= \frac{\gamma}{\sqrt{k}} \left\{ s^{-1} W_n(s) + \sqrt{k} A_0 \left(\frac{n}{k} \right) s^\gamma \Psi_{\gamma,\rho}(s^{-1}) + o_P(s^{-1/2-\varepsilon}) \right\} (1 + o_P(1)),$$

where $\{W_n(s)\}$ is a sequence of standard Brownian motions and the o_P-term is uniform in s.

Hence if, moreover, $\delta > \frac{1}{2}$ so that $\int_0^{k^{-\delta}} \log(-\gamma s^\gamma) \, ds = o(1/\sqrt{k})$, then

$$\int_{k^{-\delta}}^1 \log \left(\frac{X_{n,n} - X_{n-[ks],n}}{a_0(\frac{n}{k})} \right) ds$$

$$= \int_{k^{-\delta}}^1 \log \left(-\gamma s^\gamma \frac{X_{n,n} - X_{n-[ks],n}}{a_0(\frac{n}{k})} \right) ds - \int_0^1 \log(-\gamma s^\gamma) \, ds + o(\frac{1}{\sqrt{k}})$$

$$= \gamma - \log(-\gamma) + o\left(\frac{1}{\sqrt{k}} \right) + \frac{\gamma}{\sqrt{k}} \int_{k^{-\delta}}^1 \left\{ s^{-1} W_n(s) \right.$$

$$\left. + \sqrt{k} A_0 \left(\frac{n}{k} \right) s^\gamma \Psi_{\gamma,\rho}(s^{-1}) + o_P(s^{-1/2-\varepsilon}) \right\} (1 + o_P(1)) \, ds$$

$$= \frac{\gamma}{\sqrt{k}} \int_0^1 \left(s^{-1} W_n(s) + \sqrt{k} A_0 \left(\frac{n}{k} \right) s^\gamma \Psi_{\gamma,\rho}(s^{-1}) \right) ds$$

$$- \log(-\gamma) + \gamma + o\left(\frac{1}{\sqrt{k}} \right).$$

The second term in (3.6.15) is similar but simpler. Again using Corollary 2.4.5, we obtain

$$\int_{k^{-\delta}}^1 \log \left(\frac{X_{n,n} - X_{n-k,n}}{a_0(\frac{n}{k})} \right) ds$$

$$= (1 - k^{-\delta}) \left\{ \log \left(-\gamma \frac{X_{n,n} - X_{n-k,n}}{a_0(\frac{n}{k})} \right) - \log(-\gamma) \right\}$$

$$= (1 - k^{-\delta}) \left\{ \log \left(1 + \frac{\gamma}{\sqrt{k}} W_n(1) + o_P \left(\frac{1}{\sqrt{k}} \right) \right) - \log(-\gamma) \right\}$$

$$= (1 - k^{-\delta}) \left(\frac{\gamma}{\sqrt{k}} W_n(1) + o_P \left(\frac{1}{\sqrt{k}} \right) - \log(-\gamma) \right)$$

$$= \frac{\gamma}{\sqrt{k}} W_n(1) - \log(-\gamma) + o_P \left(\frac{1}{\sqrt{k}} \right).$$

Hence

$$\int_{k-\delta}^{1} \log\left(\frac{X_{n,n} - X_{n-[ks],n}}{X_{n,n} - X_{n-k,n}}\right) ds$$

$$= \gamma + \frac{\gamma}{\sqrt{k}} \int_{0}^{1} s^{-1}\left(W_n(s) - W_n(1) + \sqrt{k}A_0\left(\frac{n}{k}\right) s^{\gamma} \Psi_{\gamma,\rho}(s^{-1})\right) ds$$

$$+ o_P\left(\frac{1}{\sqrt{k}}\right). \tag{3.6.16}$$

Combining (3.6.14) with $(j/k) \log n = k^{-\delta} \log n = o(1/\sqrt{k})$, for some $\frac{1}{2} < \delta < \left(-(2\gamma)^{-1} \wedge (1+2\varepsilon)^{-1}\right)$, $\varepsilon < \frac{1}{2}$, and (3.6.16), we arrive at

$$\sqrt{k}\left(\hat{\gamma}_F - \gamma\right) - \int_{0}^{1} \gamma s^{-1} W_n(s) - \gamma W_n(1) + \sqrt{k}A_0\left(\frac{n}{k}\right)\gamma s^{\gamma} \Psi_{\gamma,\rho}(s^{-1}) ds = o_P(1) ,$$

as $n \to \infty$, from which the asymptotic normality follows. ∎

3.7 Simulations and Applications

We consider the various estimators of the extreme value index introduced and discussed in the previous sections: Hill ($\hat{\gamma}_H$), Pickands ($\hat{\gamma}_P$), maximum likelihood ($\hat{\gamma}_{MLE}$), moment ($\hat{\gamma}_M$), probability weighted moment ($\hat{\gamma}_{PWM}$) and negative Hill ($\hat{\gamma}_F$).

Recall that under appropriate conditions, the Hill estimator is consistent only for positive values of γ, the MLE is defined for $\gamma > -\frac{1}{2}$, the Pickands and moment estimators are defined and consistent for all real values of γ, the PWM is consistent for $\gamma < 1$, and the negative Hill for $\gamma < -\frac{1}{2}$. Moreover, note that the Hill and moment estimators are not shift invariant though they are scale invariant, and the Pickands, MLE, PWM and negative Hill are shift and scale invariant.

We start by comparing the asymptotic properties of these estimators. For instance, we shall see that for an important range of values of γ, the Pickands estimator has larger asymptotic variance than the others, whereas the moment and maximum likelihood estimators compete with each other when γ is around zero. Afterward, we give some simulation results for some common distributions. Finally, we apply the extreme value index estimators to the three data sets: sea level, S&P 500, and life span.

3.7.1 Asymptotic Properties

In the previous sections, under appropriate conditions we have proved the asymptotic normality of all the estimators. That is, for an independent and identically distributed sample of size n and with $k = k(n)$ an intermediate sequence,

$$\sqrt{k}(\hat{\gamma} - \gamma) \overset{d}{\approx} \sqrt{\text{var}_{\gamma}} N + \lambda b_{\gamma,\rho}$$

with N standard normal, where the constants λ, var_{γ}, and $b_{\gamma,\rho}$ are known (cf. Theorems 3.2.5, 3.3.5, 3.4.2, 3.5.4, 3.6.1, and 3.6.4).

For the asymptotic normality of the Pickands, MLE, PWM, and negative Hill estimators we require the second-order condition of Section 2.3, (2.3.5), with auxiliary second-order function $A \in RV_{\rho \le 0}$, and that the intermediate sequence k satisfy $\sqrt{k}A(n/k) \to \lambda \in \mathbb{R}$.

As is most common for the asymptotic normality of Hill's estimator, we require the second-order condition (2.3.22) of Section 2.3 with auxiliary second-order function $A^\star \in RV_{\rho^\star \le 0}$, say, and that the intermediate sequence k satisfy $\sqrt{k}A^\star(n/k) \to \lambda^\star$, say, with $\lambda^\star \in \mathbb{R}$.

Finally, for the asymptotic normality of the moment estimator we require the second-order condition (2.3.5) with $\rho \ne \gamma$. Then we have the second-order condition in terms of $\log U$; cf. (3.5.11)—recall $U := (1/(1 - F))^{\leftarrow}$ with F the underlying distribution function—with auxiliary second-order function $Q \in RV_{\rho'}$ with ρ' known (cf. Theorem B.3.1) and the intermediate sequence $k = k(n)$ satisfying $\sqrt{k}Q(n/k) \to \lambda'$, say, $\lambda' \in \mathbb{R}$.

Therefore, to compare the estimators we should first of all compare the orders of k. In Table 3.1 are the relations among the several second-order auxiliary functions and respective indices, for some combinations of γ and ρ. We have (i) for some cases the auxiliary functions are all of the same order but (ii) for some other cases $A(t) = o(A^\star(t))$ or $A(t) = o(Q(t))$, $t \to \infty$. In terms of the growth conditions for k, in case (i) they are the same for all the estimators, but in (ii), for instance $\sqrt{k}A(n/k) \to \lambda > 0$ corresponds to k of larger order than if $\sqrt{k}A^\star(n/k) \to \lambda^\star > 0$ or $\sqrt{k}Q(n/k) \to \lambda' > 0$, meaning a slower rate of convergence for the former. We shall come to this point later on, in the optimal mean square error analysis.

Table 3.1. Second-order items related to the nondegenerate behavior of the estimators; $l := \lim_{t \to \infty}(U(t) - a(t)/\gamma)$

		$\hat{\gamma}_P, \hat{\gamma}_{MLE}, \hat{\gamma}_{PWM}$	$\hat{\gamma}_H$	$\hat{\gamma}_M$	$\hat{\gamma}_F$
2nd-order condition		(2.3.5)	(2.3.22)	(3.5.11)	(2.3.5)
2nd-order auxiliary function for k's growth	$\gamma < \rho \le 0$	A	–	A	A
	$\rho < \gamma \le 0$	A	–	$\gg A$	A
	if $\quad 0 < \gamma < -\rho$ and $l \ne 0$	A	$\gg A$	$\gg A$	–
	$0 < \gamma < -\rho$ and $l = 0$	A	$\frac{\gamma A}{\gamma+\rho}$	$\frac{\rho A}{\gamma+\rho}$	–
	$\gamma > -\rho \ne 0$	A	$\frac{\gamma A}{\gamma+\rho}$	$\frac{\rho A}{\gamma+\rho}$	–
index of 2nd-order auxiliary function	$\gamma < \rho \le 0$	ρ	–	ρ	ρ
	$\rho < \gamma \le 0$	ρ	–	γ	ρ
	if $\quad 0 < \gamma < -\rho$ and $l \ne 0$	ρ	$-\gamma$	$-\gamma$	–
	$0 < \gamma < -\rho$ and $l = 0$	ρ	ρ	ρ	ρ
	$\gamma > -\rho \ne 0$	ρ	ρ	ρ	ρ

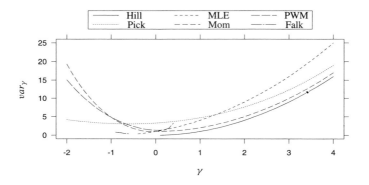

Fig. 3.2. Asymptotic variances.

Next, in Figure 3.2 we compare the asymptotic variances. The Hill estimator has systematically the smallest variance in its range of possible values. Hill's and the moment estimators have the smallest asymptotic variances for positive values of γ. The MLE and negative Hill estimators have the smallest asymptotic variances for negative values of γ.

It is more complicated to compare the bias of the estimators, since in general the bias depends on both parameters γ and ρ among other characteristics of the underlying distribution. Nonetheless, we state some general comments:

When $\gamma = 0$,

$$b_{0,\rho}(\hat{\gamma}_P) = \begin{cases} \frac{2^{-2\rho} - 2^{-\rho+1} + 1}{\rho^2 \log^2 2} & , \ \rho < 0, \\ 1 , & \rho = 0 , \end{cases}$$

$$b_{0,\rho}(\hat{\gamma}_{MLE}) = \frac{1}{(1-\rho)^2} ,$$

$$b_{0,\rho<0}(\hat{\gamma}_M) = 0 ,$$

$$b_{0,\rho}(\hat{\gamma}_{PWM}) = \begin{cases} \frac{2}{(1-\rho)(2-\rho)} & , \ \rho < 0, \\ 1 , & \rho = 0 \end{cases}$$

(see also Figure 3.3). When $\rho = 0$,

$$b_{\gamma(>0),0}(\hat{\gamma}_H) = b_{\gamma,0}(\hat{\gamma}_P) = b_{\gamma(>-1/2),0}(\hat{\gamma}_{MLE}) = b_{\gamma(\neq 0),0}(\hat{\gamma}_M)$$
$$= b_{\gamma(<1/2),0}(\hat{\gamma}_{PWM}) = b_{\gamma(<-1/2),0}(\hat{\gamma}_F) = 1 .$$

When $\rho \to -\infty$,

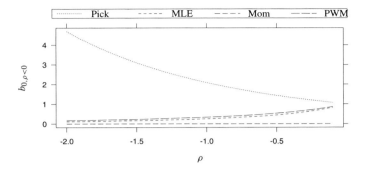

Fig. 3.3. Comparison of bias when $\rho < 0$ and $\gamma = 0$.

$$b_{\gamma(>0),-\infty}(\hat{\gamma}_H) = b_{\gamma(>-1/2),-\infty}(\hat{\gamma}_{MLE}) = b_{\gamma(<1/2),-\infty}(\hat{\gamma}_{PWM})$$
$$= b_{\gamma(<-1/2),-\infty}(\hat{\gamma}_F) = 0$$
$$b_{\gamma,-\infty}(\hat{\gamma}_P) = \infty\,,$$
$$b_{\gamma,-\infty}(\hat{\gamma}_M) = \begin{cases} \frac{\gamma(1+\gamma)}{(1-\gamma)(1-3\gamma)}\,, & \gamma \le 0 \\ -\frac{\gamma}{(1+\gamma)^2}\,, & \gamma > 0 \text{ and } l \ne 0 \\ 0\,, & \gamma > 0 \text{ and } l = 0\,, \end{cases}$$

where $l := \lim_{t\to\infty}(U(t) - a(t)/\gamma)$.

Finally, we determine an optimal sequence $k_0(n)$ following the reasoning presented in Section 3.2 for the Hill estimator. Consider the representation

$$\hat{\gamma} - \gamma \overset{d}{\approx} \sqrt{\mathrm{var}_\gamma}\,\frac{N}{\sqrt{k}} + \frac{\lambda b_{\gamma,\rho}}{\sqrt{k}} \approx \sqrt{\mathrm{var}_\gamma}\,\frac{N}{\sqrt{k}} + b_{\gamma,\rho}A\left(\frac{n}{k}\right)$$

with $k = k(n)$ an intermediate sequence such that $\sqrt{k}A(n/k) \to \lambda \in \mathbb{R}$ and N standard normal. For simplicity suppose $A(t) = ct^\rho$, for some constants c real and $\rho < 0$. We look for a sequence k for which

$$\frac{\mathrm{var}_\gamma}{k} + b_{\gamma,\rho}^2 c^2 \left(\frac{n}{k}\right)^2$$

is minimal. Similarly, as in Section 3.2 we get

$$k_0(n) = \left[\left(\frac{\mathrm{var}_\gamma}{-2\rho c^2 b_{\gamma,\rho}^2}\right)^{1/(1-2\rho)} n^{-2\rho/(1-2\rho)} \right],$$

where $[x]$ denotes the integer part of x.

Then for this choice of k we have

$$\sqrt{k_0}\,(\hat{\gamma} - \gamma) \overset{d}{\to} N\left(\frac{\mathrm{sign}(c)\sqrt{\mathrm{var}_\gamma}}{\sqrt{-2\rho}}, \mathrm{var}_\gamma \right)$$

and

$$\lim_{n\to\infty} k_0 \min_k E \left(\frac{\mathrm{var}_\gamma N}{\sqrt{k}} + \frac{A\left(\frac{n}{k}\right)}{(1-\rho)} \right)^2 = \mathrm{var}_\gamma \left(1 - \frac{1}{2\rho} \right).$$

Hence note that under the given assumptions, a comparison of this quantity for the different estimators reduces to a comparison of the asymptotic variances.

Applying this reasoning to all the estimators we have that for some cases depending mainly on γ and ρ, the order of the optimal sequence is the same for all the estimators. But for some other cases, namely when $A(t) = o(A^\star(t))$ or $A(t) = o(Q(t)), t \to \infty$, the optimal order k_0 is smaller (cf. also Table 3.1).

3.7.2 Simulations

To illustrate the finite-sample behavior of the estimators we give some simulation results for the distribution functions given in Table 3.2, namely standard Cauchy, normal, and uniform. Note that the uniform distribution satisfies the first-order extreme value condition (2.1.2) but does not satisfy the second-order condition (2.3.5) because the rate of convergence in (2.1.2) is too fast (cf. Exercise 1.3).

Table 3.2. Extreme value index and second-order parameter.

Distributions	γ	ρ
Cauchy	1	-2
Normal	0	0
Uniform	-1	$-\infty$

We generate pseudorandom independent and identically distributed samples of size $n = 1000$ and replicate them $r = 5000$ times independently. For some independent estimates $\hat\gamma_1, \ldots, \hat\gamma_r$ obtained from some estimator of γ, in the following by mean square error we mean $r^{-1} \sum_{i=1}^r (\hat\gamma_i - \gamma)^2$. Recall that all the above estimators use k upper order statistics out of the initial sample of size n.

Figures 3.4–3.6 show the so-called diagram of estimates, i.e., averages of $\hat\gamma_i$ estimates, $i = 1, \ldots, r$, for each number of upper order statistics k, and the corresponding mean square error.

The maximum likelihood estimates are obtained with numerical methods, since the maximization of the likelihood function does not have an explicit formula. We just used a naive modified Newton method with a user-supplied Hessian matrix (i.e., something that the reader can implement by itself and does not depend on any particular package). In the case of the Cauchy distribution, though very rarely, it happened that the algorithm did not reach any solution. The maximization seems to be not so straightforward for the normal distribution, so we decided to omit the results for this case.

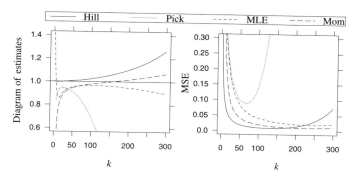

Fig. 3.4. Standard Cauchy distribution: (**a**) diagram of estimates of γ (the true value 1 is indicated by the horizontal line); (**b**) mean square error (see the text for details).

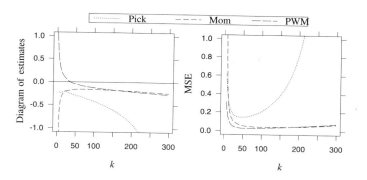

Fig. 3.5. Standard normal distribution: (**a**) diagram of estimates of γ (the true value 0 is indicated by the horizontal line); (**b**) mean square error (see the text for details).

3.7.3 Case Studies

Next we apply the estimators, Hill ($\hat{\gamma}_H$), Pickands ($\hat{\gamma}_P$), moment ($\hat{\gamma}_M$), and probability weighted moment ($\hat{\gamma}_{PWM}$) to the three case studies introduced in Section 1.1.4 and further discussed at the beginning of this chapter.

Sea Level

As described before, we have 1873 observations from the sea level (cm) at Delfzijl, the Netherlands, corresponding to winter storms during the years 1882–1991. The diagram of estimates of the extreme value index, i.e., the estimates against the number of upper order statistics k, is shown in Figure 3.7. Note the variability of the estimators for small values of k.

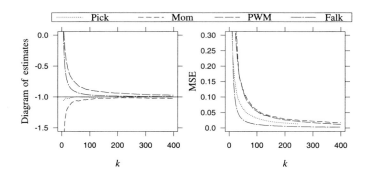

Fig. 3.6. Standard uniform distribution: (**a**) diagram of estimates of γ (the true value -1 is indicated by the horizontal line); (**b**) mean square error (see the text for details).

From the asymptotic theory one obtains the correspondent asymptotic confidence intervals. The most common approach is to assume $\sqrt{k}A(n/k) \rightarrow 0$ (or $\sqrt{k}Q(n/k) \rightarrow 0$ in case of the moment estimator), so that the limiting distribution has zero mean. This avoids the bias estimation, which generally requires the estimation of the second-order parameter ρ (for more on this we refer to Ferreira and de Vries (2004) and for the estimation of ρ see, e.g., Fraga Alves, Gomes, and de Haan (2003)). The $(1-\alpha)100\%$ approximating confidence interval is then given by

$$\hat{\gamma} - z_{\alpha/2}\sqrt{\frac{\mathrm{var}_{\hat{\gamma}}}{k}} < \gamma < \hat{\gamma} + z_{\alpha/2}\sqrt{\frac{\mathrm{var}_{\hat{\gamma}}}{k}} \,,$$

where $\mathrm{var}_{\hat{\gamma}}$ is the respective asymptotic variance with γ replaced by its estimate and $z_{\alpha/2}$ is the $1 - \alpha/2$ quantile of the standard normal distribution.

In Table 3.3 we give the 95% asymptotic confidence intervals for some values of k. The value zero belongs to all these confidence intervals, which does not contradict the hypothesis that the extreme value index is zero.

Table 3.3. Sea level data: 95% asymptotic confidence intervals for γ.

k	25	50	100
Pickands	$(-0.42,\ 1.06)$	$(-0.93,\ 0.04)$	$(-0.56,\ 0.14)$
Moment	$(-0.78,\ 0.11)$	$(-0.54,\ 0.04)$	$(-0.21,\ 0.18)$
PWM	$(-0.64,\ 0.32)$	$(-0.50,\ 0.18)$	$(-0.17,\ 0.30)$

S&P 500 Total Returns

Recall that from 5835 daily price quotes p_t (from 01/01/1980 to 14/05/2002) we analyze the log-returns $r_t = \log(p_t/p_{t-1})$. We focus on the loss log-returns comprising 2643 observations.

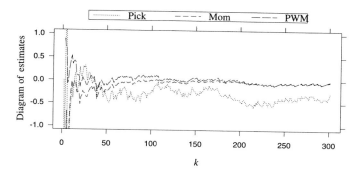

Fig. 3.7. Sea level data, diagram of estimates of γ.

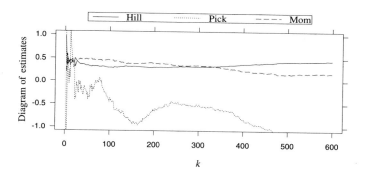

Fig. 3.8. S&P 500 data, diagram of estimates of γ.

Economists believe that this kind of financial series is heavy tailed, i.e., the underlying distribution function is in the domain of attraction of some G_γ with γ positive. Henceforth it is believed that the right endpoint of the underlying distribution is infinite. Moreover it seems that moments up to third order might exist but probably not for higher order. This means that γ should be approximately larger than $\frac{1}{3}$.

From the diagram of estimates shown in Figure 3.8, we observe a large variability of the estimates for small values of k, and then the dominance of the bias over the variance for large values of k, though more evident for the Pickands estimator. Recall

Table 3.4. S&P 500 data: 95% asymptotic confidence intervals for γ.

k	50	100	300
Hill	(0.25, 0.45)	(0.24, 0.36)	(0.28, 0.36)
Pickands	(−0.90, 0.07)	(−0.72, −0.04)	(−0.70, −0.31)
Moment	(0.16, 0.77)	(0.24, 0.67)	(0.23, 0.47)

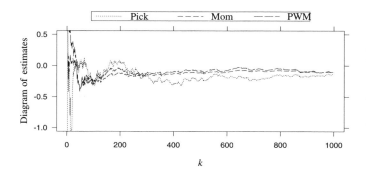

Fig. 3.9. Life span data, diagram of estimates of γ.

that the Pickands estimator has large variance for positive γ, when compared to the others. The Hill and moment estimators clearly give positive estimates for γ. The confidence intervals given in Table 3.4 also point in this direction.

Life Span

The data set consists of the total life span (in days) of all people born in the Netherlands in the years 1877–1881, still alive on January 1, 1971, and who died as resident of the Netherlands. The size of this data set is 10 391.

We want to decide whether the right endpoint of the distribution of the underlying population is finite, and for that we test whether the extreme value index is negative: $H_0 : \gamma \geq 0$ versus $H_1 : \gamma < 0$. The diagram of estimates for the Pickands, moment, and probability-weighted moment estimators is shown in Figure 3.9. The null hypothesis is rejected if for a significance level of 5%, $\hat{\gamma}_P < -2.96/\sqrt{k}$ for the Pickands estimator (cf. Theorem 3.3.5), $\hat{\gamma}_M < -1.64/\sqrt{k}$ for the moment estimator (cf. Theorem 3.5.4), and $\hat{\gamma}_{\text{PWM}} < -1.90/\sqrt{k}$ for the probability-weighted moment estimator (cf. Theorem 3.6.1). For example, for $k = 400$ the null hypothesis is rejected for the three estimators. The moment estimator is rather clear in this respect, and for practically all k the null hypothesis is rejected. The Pickands estimator is not so clear. Though the probability weighted moment has slightly larger variability than the moment estimator they behave similarly.

Exercises

3.1. Note that the generalized Pareto distributions H_γ (Section 3.1) satisfy the following property: if X is a random variable with probability distribution H_γ there exists a positive function a such that for $x > 0$ and all t with $H_\gamma(t) < 1$,

$$P\left(\frac{X - t}{a(t)} > x \,\middle|\, X > t\right) = P(X > x) .$$

Prove that this property characterizes the class of generalized Pareto distributions (cf. proof of Theorem 1.1.3).

3.2. Let $X_1, X_2, \ldots$ be independent and identically distributed random variables with distribution function in the domain of attraction of some extreme value distribution with $\gamma > 0$. Let $X_{1,n} \leq X_{2,n} \leq \cdots \leq X_{n,n}$ be the nth order statistics. Prove that if $k = k(n) \to \infty, k/n \to 0, n \to \infty$,

$$\frac{1}{k} \sum_{i=0}^{k-1} \frac{X_{n-k,n}}{X_{n-i,n}} \overset{P}{\to} \frac{1}{1+\gamma}.$$

3.3. Can you prove an asymptotic normality result for this estimator?
Hint: see Theorem 2.4.8.

3.4. Assume the conditions of Exercise 3.2. By using the methods in the proof of Theorem 3.2.2 and Lemma 3.2.3, prove that

$$\frac{\log X_{n,n} - \log X_{n-k,n}}{\log k} \overset{P}{\to} \gamma$$

and that the distribution of

$$\log X_{n,n} - \log X_{n-k,n} - \gamma \log k$$

converges to G_0; note that in the notation of the proof of Theorem 3.2.2 the distribution of $\log Y_{k,k} - \log k$ converges to $G_0(x) = \exp(-e^{-x})$. Is this estimator better or worse than the one in Exercise 2.19?

3.5. Define

$$\hat{\gamma} := 1 - 2^{-1} \left(1 - \left(m_n^{(1)} \right)^2 / m_n^{(2)} \right)^{-1}$$

with $m_n^{(j)} := k^{-1} \sum_{i=1}^{n} (X_{n-i,n} - X_{n-k,n})^j$ for $j = 1, 2$. Prove that $\hat{\gamma}$ is consistent for γ provided $\gamma < \frac{1}{2}$ and that $\sqrt{k} \, (\hat{\gamma} - \gamma)$ is asymptotically normal for $\gamma < \frac{1}{4}$ under appropriate conditions. Calculate the asymptotic variance and bias.

3.6. Prove Theorem 3.3.5 for $\gamma = 0$. Check that the variance and bias of the limiting random variable are the same as taking the limits of the given variance and bias of Theorem 3.3.5 when γ converges to zero.

3.7. Check that when $\rho = -1$ the Pickands estimator, conveniently normalized as in Theorem 3.3.5, has asymptotic bias $\lambda b_{\gamma,-1}$, where

$$b_{\gamma,-1} = \begin{cases} \frac{2\gamma(2^\gamma - 2^{\gamma-1} - 1)}{(\gamma-1)(2^\gamma - 1)\log 2}, & \gamma \neq 0, 1, \\ 2, & \gamma = 1, \\ \frac{1}{\log^2 2}, & \gamma = 0. \end{cases}$$

3.8. Let $\{W_n(s)\}_{s>0}$ be a sequence of Brownian motions.
(a) Using (3.5.11) and under the conditions of Corollary 2.4.6 show that

$$
\sup_{0<s\leq 1} \min\left(1, s^{\gamma_-+1/2+\varepsilon}\right)\left|\sqrt{k}\left(\frac{\log X_{n-[ks],n} - \log X_{n-k,n}}{q_0(\frac{n}{k})} - \frac{s^{-\gamma_-} - 1}{\gamma_-}\right)\right.
$$
$$
\left. - s^{-\gamma_- - 1}W_n(s) + W_n(1) - \sqrt{k}Q_0\left(\frac{n}{k}\right)\Psi_{\gamma_-,\rho'}(s^{-1})\right| \xrightarrow{P} 0 .
$$

(b) With the notation of Lemma 3.5.5, prove that

$$
\sqrt{k}\left(\frac{M_n^{(1)}}{q_0(\frac{n}{k})} - \frac{1}{1-\gamma_-}\right) - P(W_n) \xrightarrow{P} 0
$$

with $P(W_n) = \int_0^1 s^{-\gamma_- - 1}W_n(s) - W_n(1) + \sqrt{k}Q_0(n/k)\Psi_{\gamma_-,\rho'}(s^{-1})\,ds$.

3.9. Prove the second statement of Lemma 3.5.5.

3.10. (a) Check that the expected value of the limiting random variable of Corollary 3.5.6, as a function of (γ_-, ρ'), satisfies

$$
E\left((1-2\gamma_-)(1-\gamma_-)^2\left\{\left(\frac{1}{2}-\gamma_-\right)Q - 2P\right\}\right)
$$
$$
= \lambda\frac{(1-\gamma_-)(1-2\gamma_-)}{(1-\gamma_- - \rho')(1-2\gamma_- - \rho')},
$$

where the random variables (P, Q) are defined in Lemma 3.5.5 and the mean values and covariance matrix are given in terms of (γ_-, ρ').
(b) Use this to provide the asymptotic bias of the moment estimator in terms of γ and ρ'.

3.11. Assume the first-order regular variation condition for some $\gamma < -\frac{1}{2}$. For a sample of size n, prove the consistency of the negative Hill estimator for some intermediate sequence k with $k^{\gamma-\varepsilon}n^{2\varepsilon} \to 0$, $\varepsilon > 0$.
Hint: use the methods used in the proofs of the consistency of the Hill or the Pickands estimators.

3.12. Assume the second-order regular variation condition for some $-1 < \gamma < -\frac{1}{2}$. For a sample of size n, prove the asymptotic normality of the negative Hill estimator for some intermediate sequence k with $k^{\gamma-\varepsilon}n^{2\varepsilon} \to 0$, $\varepsilon > 0$, using the methods used, for example, in the first proof of Theorem 3.3.5.

4

Extreme Quantile and Tail Estimation

4.1 Introduction

With the sea level case study introduced in Section 1.1.4 and further discussed in Section 3.1, we illustrated the role of extreme value theory in extreme quantile estimation. In the sequel we explore this example a bit further.

The Dutch government requires the sea dikes to be so high that in a certain year a flood occurs with probability $1/10\,000$. In order to estimate the height corresponding to that probability, available are 1877 high tide water levels, monitored at the coast, one for each severe storm of a certain type, over a period of 111 years. The observations are considered as realizations of independent and identically distributed random variables. In Figure 4.1 is the empirical distribution function F_n based on these observations, i.e., we assign mass $1/n$ to each observation where n represents the sample size.

One possibility to estimate a quantile is via the empirical quantile, that is, one of the order statistics. As shown in Figure 4.1 for the 0.9 quantile, following the curve this

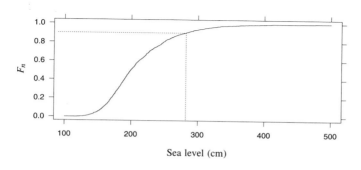

Fig. 4.1. Empirical distribution function of the sea level data.

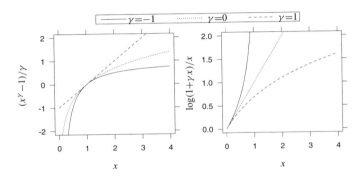

Fig. 4.2. Extrapolation function: (**a**) for $U(t)$; (**b**) for $-\log(1 - F)$.

quantile is just the level corresponding to the given probability. Now we are interested in estimating the sea level, say u, with probability $(111/1877) \times 10^{-4} \approx 6 \times 10^{-6}$ of being exceeded, that is, $1 - F(u) \approx 6 \times 10^{-6}$. So this is the probability of a flood during one windstorm of a certain (severe) type. But for the given data set the highest order statistic corresponds roughly to $F^{\leftarrow}(1 - 1/1878)$, that is, $1 - F(X_{1877,1877}) = 1/1878 \approx 5 \times 10^{-4}$. Hence in order to give a nontrivial answer one needs somehow to extrapolate beyond the range of the available data.

Recall that we want to estimate the level u such that $1 - F(u) \approx 6 \times 10^{-6}$. In terms of the function $U = (1/(1-F))^{\leftarrow}$ this means that $u \approx U(1/6 \times 10^6) \approx U(17 \times 10^4)$. Remember that (cf. (1.1.27), and also (3.1.6) and (3.1.8))

$$U\left(17 \times 10^4\right) = U(tx) \approx U(t) + a(t) \frac{x^\gamma - 1}{\gamma}. \tag{4.1.1}$$

We see that the function $(x^\gamma - 1)/\gamma$ plays a crucial role (cf. Theorem 1.1.6, Section 1.1.3). Basically, the extrapolation beyond the quantile $U(t)$ is via this function multiplied by the scale factor $a(t)$. Roughly speaking, apart from a scale factor, the function $(x^\gamma - 1)/\gamma$ approximates $U = (1/(1 - F))^{\leftarrow}$, or $\log((x^\gamma - 1)/\gamma)^{\leftarrow}$ approximates $-\log(1 - F)$. Figure 4.2 shows these functions for $\gamma = 1, 0, -1$. We see that the real parameter γ determines their shape; for instance, in the $-\log(1 - F)$ scale one gets a straight line when γ equals zero.

Hence (4.1.1) motivates the quantile estimator

$$\hat{U}(17 \times 10^4) \approx \hat{U}\left(\frac{n}{k}\right) + \hat{a}\left(\frac{n}{k}\right) \frac{\left(\frac{17 \times 10^4 \times k}{n}\right)^{\hat{\gamma}} - 1}{\hat{\gamma}},$$

where n is the sample size and k is an intermediate sequence, i.e., $k \rightarrow \infty$ and $k/n \rightarrow 0$.

Figure 4.3 displays the empirical distribution on a log-scale, i.e., the step function $-\log(1 - F_n)$ (which is a convenient scale when one is interested in the largest values

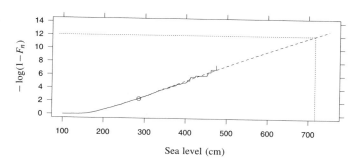

Fig. 4.3. Step function $-\log(1 - F_n)$ of the sea level sample and estimated model attached to the intermediate order statistic $X_{1699,1873}$.

of the sample) and the quantile we are interested in. Moreover, it shows one possible model fitted to the tail of the distribution, which gives the following estimate of the sea level for a failure probability of 6×10^{-6}: if we take $k = 174$,

$$\hat{U}(17 \times 10^4) = X_{1873-174,1873} + \hat{a}\left(\frac{1873}{174}\right)\frac{\left(\frac{17\times 10^4 \times 174}{1873}\right)^{\hat{\gamma}} - 1}{\hat{\gamma}}.$$

Then for instance using the moment-type estimators discussed in Section 3.5 above and Section 4.2 below, with $k = 174$, we get $\hat{\gamma} = \hat{\gamma}_M = 0.02$ and $\hat{a}(1873/174) = \hat{\sigma}_M = 40.3$, hence

$$\hat{U}(17 \times 10^4) = 286 + 40.3\frac{\left(\frac{17\times 10^4 \times 174}{1873}\right)^{0.02} - 1}{0.02} = 715.6.$$

The adjusted model in Figure 4.3 represents this formula with the quantile as a function of the given probability and with the other components fixed. It is attached to the empirical function at the intermediate order statistic $X_{1873-174,1873} = X_{1699,1873} = 286$. More details on the data analysis are given in Section 4.6.

Thus we see that a key issue in quantile estimation is the estimation of the extreme value index γ. Moreover, we have to deal with the estimation of $U(t)$ and the scale function $a(t)$. Recall that in Chapter 3 we have already discussed two estimators of the scale: the maximum likelihood estimator (cf. Section 3.4) and the probability weighted moment estimator (cf. Section 3.6). In the next section we discuss another possibility, this time related to the moment estimator of Section 3.5.

In Section 4.3 we develop some limiting theory of extreme quantile estimation. The dual problem of the estimation of tail probability is discussed in Section 4.4. A related problem is the endpoint estimation, which we also address in Section 4.5. Finally, in Section 4.6 we give some simulations and continue discussing the three case studies: sea level, S&P 500, and life span.

4.2 Scale Estimation

Suppose F is in the domain of attraction of an extreme value distribution, i.e., for sequences $a_n > 0$ and b_n, and some $\gamma \in \mathbb{R}$,

$$\lim_{n \to \infty} F^n(a_n x + b_n) = \exp{-(1 + \gamma x)^{-1/\gamma}}, \quad 1 + \gamma x > 0.$$

Next suppose, for some positive function a and A a function not changing sign and such that $A(t) \to 0, t \to \infty$, the second-order condition (2.3.5),

$$\lim_{t \to \infty} \frac{\frac{U(tx) - U(t)}{a(t)} - \frac{x^\gamma - 1}{\gamma}}{A(t)} = \frac{1}{\rho} \left(\frac{x^{\gamma + \rho} - 1}{\gamma + \rho} - \frac{x^\gamma - 1}{\gamma} \right), \quad (4.2.1)$$

for $x > 0$ with $\rho \le 0$. Then for $\gamma \ne \rho$ and $\rho < 0$ if $\gamma > 0$ we know that a second-order condition for $\log U(t)$ holds, namely

$$\lim_{t \to \infty} \frac{\frac{\log U(tx) - \log U(t)}{a(t)/U(t)} - \frac{x^{\gamma_-} - 1}{\gamma_-}}{Q(t)} = \frac{1}{\rho'} \left(\frac{x^{\gamma_- + \rho'} - 1}{\gamma_- + \rho'} - \frac{x^{\gamma_-} - 1}{\gamma_-} \right) \quad (4.2.2)$$

with $\gamma_- = \min(0, \gamma)$ and Q not changing sign eventually with $Q(t) \to 0, t \to \infty$ (cf. Lemma B.3.16 in Appendix B). When $\gamma > 0$ and $\rho = 0$ the limit (4.2.2) vanishes for all Q satisfying $A(t) = O(Q(t)), t \to \infty$.

We now study an estimator for the scale a, related to the moment estimator of γ discussed in Section 3.5. Recall the notation introduced in that section,

$$M_n^{(j)} := \frac{1}{k} \sum_{i=0}^{k-1} \left(\log X_{n-i,n} - \log X_{n-k,n} \right)^j$$

for $j = 1, 2$. Define

$$\hat{\gamma}_- := 1 - \frac{1}{2} \left(1 - \frac{\left(M_n^{(1)} \right)^2}{M_n^{(2)}} \right)^{-1}. \quad (4.2.3)$$

Note that $\hat{\gamma}_- + \hat{\gamma}_H = \hat{\gamma}_M$ with $\hat{\gamma}_M$ the moment estimator of γ. We define the estimator

$$\hat{\sigma}_M := X_{n-k,n} M_n^{(1)} (1 - \hat{\gamma}_-). \quad (4.2.4)$$

In the next theorem we give the consistency and asymptotic normality of $\hat{\sigma}_M$. Note that for the nondegenerate limit the conditions are the same as those in Theorem 3.5.4, which states the asymptotic normality of $\hat{\gamma}_M$.

Theorem 4.2.1 *Let $X_1, X_2, \ldots$ be i.i.d. random variables with distribution function F.*

1. *If $F \in \mathcal{D}(G_\gamma)$, $x^* > 0$, then*

$$\frac{\hat{\sigma}_M}{a(\frac{n}{k})} \xrightarrow{P} 1$$

   provided $k = k(n) \to \infty$, $k/n \to 0$, as $n \to \infty$.
2. *Suppose $F \in \mathcal{D}(G_\gamma)$, $x^* > 0$ and that the second-order condition (4.2.1) holds with $\gamma \neq \rho$. If the sequence of integers $k = k(n)$ satisfies $k \to \infty$, $k/n \to 0$,*

$$\lim_{n\to\infty} \sqrt{k}\, Q\left(\frac{n}{k}\right) = \lambda \tag{4.2.5}$$

   with $Q := A$ from (4.2.1) if $\gamma > 0$ and $\rho = 0$ and Q from (4.2.2) otherwise, and λ finite, then

$$\sqrt{k}\left(\frac{\hat{\sigma}_M}{a(\frac{n}{k})} - 1\right) \xrightarrow{d} N(\lambda b_{\gamma,\rho}, \mathrm{var}_\gamma) \tag{4.2.6}$$

   with N standard normal,

$$b_{\gamma,\rho} := \begin{cases} -\frac{\rho}{(1-\gamma-\rho)(1-2\gamma-\rho)}\,, & \gamma < \rho \leq 0, \\[2mm] -\frac{\gamma}{(1-2\gamma)(1-3\gamma)}\,, & \rho < \gamma \leq 0, \\[2mm] \frac{\gamma}{(1+\gamma)^2}\,, & \lim_{t\to\infty} U(t) - a(t)/\gamma \neq 0 \text{ and } 0 < \gamma < -\rho, \\[2mm] -\frac{\rho}{(1-\rho)^2}\,, & (\lim_{t\to\infty} U(t) - a(t)/\gamma = 0, \\[1mm] & \quad \text{and } 0 < \gamma < -\rho) \text{ or } \gamma \geq -\rho > 0, \\[2mm] 0\,, & \gamma > \rho = 0, \end{cases} \tag{4.2.7}$$

   and

$$\mathrm{var}_\gamma := \begin{cases} \gamma^2 + 2\,, & \gamma \geq 0, \\[2mm] \frac{2 - 16\gamma + 51\gamma^2 - 69\gamma^3 + 50\gamma^4 - 24\gamma^5}{(1-2\gamma)(1-3\gamma)(1-4\gamma)}\,, & \gamma < 0. \end{cases} \tag{4.2.8}$$

Proof. The proof of the consistency is left to the reader. Next we prove the asymptotic normality.

First observe that with $q := a/U$,

$$\frac{\hat{\sigma}_M}{a(\frac{n}{k})} \stackrel{d}{=} \frac{M_n^{(1)}}{q(Y_{n-k,n})} \frac{a(Y_{n-k,n})}{a(\frac{n}{k})}(1 - \hat{\gamma}_-)$$

$$= \frac{M_n^{(1)}}{q_0(Y_{n-k,n})} \frac{q_0(Y_{n-k,n})}{q(Y_{n-k,n})} \frac{a(Y_{n-k,n})}{a(\frac{n}{k})}(1 - \hat{\gamma}_-),$$

where the function q_0 is from Theorem 2.3.6, which gives the uniform inequalities connected with the second-order regular variation condition for the function $\log U$.

Then for the first factor we know from Lemma 3.5.5 that,

$$\sqrt{k}\left(\frac{(1-\gamma_-)M_n^{(1)}}{q_0(Y_{n-k,n})} - 1\right) \xrightarrow{d} (1 - \gamma_-)P, \tag{4.2.9}$$

where the random variable P is normally distributed. For the second factor use Corollary 2.3.5, Theorem 2.3.6, and (4.2.5) to get

$$\sqrt{k}\left(\frac{q_0(Y_{n-k,n})}{q(Y_{n-k,n})} - 1\right) \xrightarrow{P} -\frac{\lambda 1_{\{\gamma \neq 0\}}}{\rho' + \gamma 1_{\{\rho'=0\}}}.$$

For the third factor note that

$$\sqrt{k}\left(\frac{a(Y_{n-k,n})}{a(\frac{n}{k})} - 1\right)$$

$$= \sqrt{k}\left(\frac{a(Y_{n-k,n})}{a(\frac{n}{k})} - \left(\frac{k}{n}Y_{n-k,n}\right)^\gamma\right) + \sqrt{k}\left(\left(\frac{k}{n}Y_{n-k,n}\right)^\gamma - 1\right).\ (4.2.10)$$

The second term in (4.2.10) converges to γB with B a standard normal random variable (Corollary 2.2.2). By inequalities (2.3.18) of Theorem 2.3.6,

$$\sqrt{k}\left(\frac{a(Y_{n-k,n})}{a(\frac{n}{k})} - \left(\frac{k}{n}Y_{n-k,n}\right)^\gamma\right)$$

$$= \sqrt{k}A\left(\frac{k}{n}\right)(1+o(1))\left(\frac{k}{n}Y_{n-k,n}\right)^\gamma \frac{\left(\frac{k}{n}Y_{n-k,n}\right)^\rho - 1}{\rho}$$

$$+ \sqrt{k}A\left(\frac{k}{n}\right)(1+o(1))\left(\frac{k}{n}Y_{n-k,n}\right)^{\gamma+\rho}$$

$$\times \max\left(\left(\frac{k}{n}Y_{n-k,n}\right)^\delta, \left(\frac{k}{n}Y_{n-k,n}\right)^{-\delta}\right).$$

Hence, since $A = O(Q)$ by Lemma B.3.16 and $kY_{n-k,n}/n \xrightarrow{P} 1$ by Corollary 2.2.2, the first term on the right-hand side of (4.2.10) tends to zero in probability as $n \to \infty$.

Finally, for the fourth factor, from Corollary 3.5.6 we know that

$$\sqrt{k}\left(\frac{1-\hat{\gamma}_-}{1-\gamma_-} - 1\right) \xrightarrow{d} 2(1-2\gamma_-)(1-\gamma_-)P - \frac{1}{2}(1-2\gamma_-)^2(1-\gamma_-)Q,$$

where the random variables P and Q are normally distributed, and P is the same as in (4.2.9).

Hence the limiting distribution of $\sqrt{k}(\hat{\sigma}_M/a(n/k) - 1)$ is the distribution of

$$(1-\gamma_-)(3-4\gamma_-)P - \frac{1}{2}(1-\gamma_-)(1-2\gamma_-)^2Q + \gamma B - \frac{\lambda 1_{\{\gamma \neq 0 \text{ and } (\gamma < 0 \text{ or } \rho < 0)\}}}{\rho' + \gamma\ 1_{\{\rho'=0\}}},$$
$$(4.2.11)$$

where the random variables P and Q are from Lemma 3.5.5 and B is standard normal independent of P and Q (recall that $\{Y_{n-k+i,n}/Y_{n-k,n}\}_{i=0}^{k-1}$ is independent of $Y_{n-k,n}$; cf. proof of Lemma 3.2.3). ∎

The following corollary is now easy to obtain.

Corollary 4.2.2 *Under the conditions of Theorem 4.2.1(2),*

$$\sqrt{k}\left(\hat{\gamma}_M - \gamma, \frac{\hat{\sigma}_M}{a\left(\frac{n}{k}\right)} - 1, \frac{X_{n-k,n} - U\left(\frac{n}{k}\right)}{a\left(\frac{n}{k}\right)}\right) \xrightarrow{d} (R, S, T),$$

where the random vector (R, S, T) has a multivariate normal distribution with mean vector $\lambda(b_{\gamma,\rho}^{\hat{\gamma}_M}, b_{\gamma,\rho}^{\hat{\sigma}_M}, 0)$, where $b_{\gamma,\rho}^{\hat{\gamma}_M}$ and $b_{\gamma,\rho}^{\hat{\sigma}_M}$ are respectively given by (3.5.16) and (4.2.7), variances given by (3.5.17), (4.2.8), and 1, and

$$\mathrm{Cov}(R, S) = \begin{cases} \gamma - 1, & \gamma \geq 0, \\ \frac{(1-\gamma)^2(-1+4\gamma-12\gamma^2)}{(1-3\gamma)(1-4\gamma)}, & \gamma < 0, \end{cases}$$

$\mathrm{Cov}(P, T) = 0$, *and* $\mathrm{Cov}(S, T) = \gamma$.

Proof. From Theorem 2.4.1,

$$\sqrt{k}\frac{X_{n-k,n} - U\left(\frac{n}{k}\right)}{a\left(\frac{n}{k}\right)} \xrightarrow{d} B, \tag{4.2.12}$$

where B is a standard normal random variable. Moreover, from the proof of Theorem 3.5.4 we know that $\sqrt{k}(\hat{\gamma}_M - \gamma)$ has the distribution of

$$\lambda\frac{q_{\gamma,\rho}^0}{1 - \gamma_-} + \gamma_+ P + (1 - 2\gamma_-)(1 - \gamma_-)^2 \left\{ \left(\frac{1}{2} - \gamma_-\right) Q - 2P \right\}, \tag{4.2.13}$$

where the random variables P, Q, and B are the same as those in the proof of Theorem 4.2.1. Combining (4.2.11), (4.2.12), and (4.2.13), the result follows. ∎

4.3 Quantile Estimation

Suppose F is in the domain of attraction of an extreme value distribution, i.e., for sequences $a_n > 0$ and b_n, and some $\gamma \in \mathbb{R}$,

$$\lim_{n\to\infty} F^n(a_n x + b_n) = \exp\left(-(1 + \gamma x)^{-1/\gamma}\right), \quad 1 + \gamma x > 0.$$

Then for some positive function a and moreover taking $b(t) = U(t) = F^{\leftarrow}(1 - 1/t)$ we have

$$\lim_{t\to\infty} \frac{U(tx) - U(t)}{a(t)} = \frac{x^\gamma - 1}{\gamma}$$

for $x > 0$ (Theorem 1.1.6).
 Next suppose the second-order condition (2.3.5),

$$\lim_{t\to\infty} \frac{\frac{U(tx)-U(t)}{a(t)} - \frac{x^\gamma-1}{\gamma}}{A(t)} = \frac{1}{\rho}\left(\frac{x^{\gamma+\rho} - 1}{\gamma + \rho} - \frac{x^\gamma - 1}{\gamma}\right), \tag{4.3.1}$$

holds for $x > 0$ with $\rho \leq 0$ and A a function not changing sign and such that $A(t) \to 0, t \to \infty$.

Let k be an intermediate sequence, i.e., $k = k(n) \to \infty, k(n)/n \to 0 \ (n \to \infty)$. Suppose that for suitable estimators $\hat{\gamma}$, $\hat{a}(n/k)$, and $\hat{b}(n/k)$,

$$\sqrt{k} \left(\hat{\gamma} - \gamma, \frac{\hat{a}\left(\frac{n}{k}\right)}{a\left(\frac{n}{k}\right)} - 1, \frac{\hat{b}\left(\frac{n}{k}\right) - U\left(\frac{n}{k}\right)}{a\left(\frac{n}{k}\right)} \right) \xrightarrow{d} (\Gamma, \Lambda, B), \quad n \to \infty, \quad (4.3.2)$$

with (Γ, Λ, B) jointly normal random variables with known mean vector possibly depending on γ and ρ and known covariance matrix depending on γ (not on ρ).

Now we are ready to consider extreme quantile estimation. Note that in the examples we needed to estimate a $(1 - p)$ quantile on the basis of a sample of size n, where in fact p is much smaller than $1/n$. Let $x_p := U(1/p)$ be the quantile we want to estimate. We are particularly interested in the cases in which the mean number of observations above x_p, np equals a very small number. This means that we are looking for a number x_p that is to the right of all (or almost all) observations, or what is the same, we want to extrapolate outside the range of the available observations. Since this is the central issue in our problem, we want to preserve in the asymptotic analysis the fact that np should be much smaller than any positive constant. Hence we are "forced," when applying asymptotic methods, to assume that p in fact depends on n, $p = p_n$, and that

$$\lim_{n \to \infty} p_n = 0 .$$

That is, we want to estimate x_{p_n} with $1 - F(x_{p_n}) = p_n$, or equivalently, $x_{p_n} = U(1/p_n)$ with $p_n \to 0$, as $n \to \infty$.

Theorem 4.3.1 *Suppose for some function A with $A(t) \to 0, t \to \infty$, the second-order condition (4.3.1) holds. Suppose:*

1. *the second-order parameter ρ is negative, or zero with γ negative;*
2. *$k = k(n) \to \infty, n/k \to \infty$, and $\sqrt{k}A(n/k) \to \lambda \in \mathbb{R}, n \to \infty$;*
3. *condition (4.3.2) holds for suitable estimators of γ, $a(n/k)$, and $U(n/k)$;*
4. *$np_n = o(k)$ and $\log(np_n) = o(\sqrt{k}), n \to \infty$.*

Define

$$\hat{x}_{p_n} := \hat{b}\left(\frac{n}{k}\right) + \hat{a}\left(\frac{n}{k}\right) \frac{\left(\frac{k}{np_n}\right)^{\hat{\gamma}} - 1}{\hat{\gamma}} \quad and \quad x_{p_n} := U\left(\frac{1}{p_n}\right) . \quad (4.3.3)$$

Then, as $n \to \infty$,

$$\sqrt{k} \frac{\hat{x}_{p_n} - x_{p_n}}{a\left(\frac{n}{k}\right) q_\gamma(d_n)} \xrightarrow{d} \Gamma + (\gamma_-)^2 B - \gamma_- \Lambda - \lambda \frac{\gamma_-}{\gamma_- + \rho} \quad (4.3.4)$$

with $d_n := k/(np_n)$, $\gamma_- := \min(0, \gamma)$ and where for $t > 1$,

$$q_\gamma(t) := \int_1^t s^{\gamma - 1} \log s \, ds .$$

Corollary 4.3.2 *The conditions of Theorem 4.3.1 imply*

$$\frac{q_{\hat{\gamma}}(d_n)}{q_\gamma(d_n)} \xrightarrow{P} 1 \; ;$$

hence an equivalent statement is

$$\sqrt{k}\,\frac{\hat{x}_{p_n} - x_{p_n}}{\hat{a}\left(\frac{n}{k}\right)q_{\hat{\gamma}}(d_n)} \xrightarrow{d} \Gamma + (\gamma_-)^2 B - \gamma_- \Lambda - \lambda \frac{\gamma_-}{\gamma_- + \rho}\; .$$

This version is more useful for constructing an asymptotic confidence interval for $\hat{x}_{p_n}$.

Remark 4.3.3 Note that as $t \to \infty$,

$$q_\gamma(t) \sim \begin{cases} \frac{1}{\gamma}t^\gamma \log t \,, & \gamma > 0, \\ \frac{1}{2}(\log t)^2 \,, & \gamma = 0, \\ 1/\gamma^2 \,, & \gamma < 0 \,. \end{cases}$$

Moreover, from the definition of $q_\gamma(t)$ it is clear that $q_\gamma(t)$ is increasing in t and that $q_\gamma(t)$ is also an increasing function of γ when $t \geq 1$.

Remark 4.3.4 Condition $np_n = o(k)$, i.e., $p_n \ll k/n$, is quite natural since if it is not satisfied, nonparametric methods can be employed (Einmahl (1990)); see also Remark 4.3.7 below. In particular, when $np_n \to 0$ the condition $\log(np_n) = o(\sqrt{k})$, i.e., $p_n > n^{-1}e^{-\varepsilon\sqrt{k}}$ for each $\varepsilon > 0$ and sufficiently large n, means that the extrapolation cannot be pushed too far.

Note that by checking the components of the asymptotic variance in the theorem, one sees that for γ near zero the uncertainty in the estimation of x_{p_n} is to a large extent caused by the uncertainty in the estimation of γ and not so much in the estimation of $b(n/k)$ or $a(n/k)$.

For the proof we need the following lemma:

Lemma 4.3.5 *If (4.3.1) holds with $\rho < 0$ or $\rho = 0$ and $\gamma < 0$ then*

$$\lim_{\substack{t \to \infty \\ x=x(t) \to \infty}} \frac{\frac{U(tx)-U(t)}{a(t)} \frac{\gamma}{x^\gamma - 1} - 1}{A(t)} = -\frac{1}{\rho + \gamma_-}\; .$$

Proof. Use the inequalities of Theorem B.3.10 for $\rho < 0$, or $\rho = 0$ and $\gamma < 0$. ∎

Proof (of Theorem 4.3.1).

$$\hat{x}_{p_n} - x_{p_n} = \hat{b}\left(\frac{n}{k}\right) + \hat{a}\left(\frac{n}{k}\right)\frac{d_n^{\hat{\gamma}} - 1}{\hat{\gamma}} - U\left(\frac{1}{p_n}\right)$$

$$= \hat{b}\left(\frac{n}{k}\right) - U\left(\frac{n}{k}\right) + \hat{a}\left(\frac{n}{k}\right)\frac{d_n^{\hat{\gamma}} - 1}{\hat{\gamma}} - \left(U\left(\frac{1}{p_n}\right) - U\left(\frac{n}{k}\right)\right)$$

$$= \hat{b}\left(\frac{n}{k}\right) - U\left(\frac{n}{k}\right) + \hat{a}\left(\frac{n}{k}\right)\left(\frac{d_n^{\hat{\gamma}} - 1}{\hat{\gamma}} - \frac{d_n^{\gamma} - 1}{\gamma}\right)$$

$$+ \left(\hat{a}\left(\frac{n}{k}\right) - a\left(\frac{n}{k}\right)\right)\frac{d_n^{\gamma} - 1}{\gamma}$$

$$- \left\{U\left(\frac{1}{p_n}\right) - U\left(\frac{n}{k}\right) - a\left(\frac{n}{k}\right)\frac{d_n^{\gamma} - 1}{\gamma}\right\}.$$

Hence

$$\sqrt{k}\frac{\hat{x}_{p_n} - x_{p_n}}{a\left(\frac{n}{k}\right)q_{\gamma}(d_n)} =$$

I ($\hat{b}$ part)
$$\sqrt{k}\frac{\hat{b}\left(\frac{n}{k}\right) - U\left(\frac{n}{k}\right)}{a\left(\frac{n}{k}\right)}\frac{1}{q_{\gamma}(d_n)}$$

II ($\hat{\gamma}$ part)
$$+\frac{\hat{a}\left(\frac{n}{k}\right)}{a\left(\frac{n}{k}\right)}\left\{\frac{\sqrt{k}}{q_{\gamma}(d_n)}\left(\frac{d_n^{\hat{\gamma}} - 1}{\hat{\gamma}} - \frac{d_n^{\gamma} - 1}{\gamma}\right)\right\}$$

III ($\hat{a}$ part)
$$+\sqrt{k}\left(\frac{\hat{a}\left(\frac{n}{k}\right)}{a\left(\frac{n}{k}\right)} - 1\right)\frac{d_n^{\gamma} - 1}{\gamma\, q_{\gamma}(d_n)}$$

IV (nonrandom bias)
$$-\frac{\sqrt{k}}{q_{\gamma}(d_n)}\left(\frac{U\left(\frac{n}{k}d_n\right) - U\left(\frac{n}{k}\right)}{a\left(\frac{n}{k}\right)} - \frac{d_n^{\gamma} - 1}{\gamma}\right).$$

Recall Remark 4.3.3 and that as $t \to \infty$,

$$\frac{t^{\gamma} - 1}{\gamma} \sim \begin{cases} \frac{1}{\gamma}t^{\gamma}, & \gamma > 0, \\ \log t, & \gamma = 0, \\ -\frac{1}{\gamma}, & \gamma < 0. \end{cases} \tag{4.3.5}$$

Hence from (4.3.2), as $n \to \infty$, I $\to^d (\gamma_-)^2 B$ and III $\to^d -\gamma_- \Lambda$ with $\gamma_- = \min(0, \gamma)$.

Next consider II, which is basically

$$\frac{\sqrt{k}}{q_{\gamma}(d_n)}\left(\frac{d_n^{\hat{\gamma}} - 1}{\hat{\gamma}} - \frac{d_n^{\gamma} - 1}{\gamma}\right).$$

We write this as

$$\frac{\sqrt{k}(\hat{\gamma} - \gamma)}{q_{\gamma}(d_n)}\int_1^{d_n} s^{\gamma - 1}\frac{e^{(\hat{\gamma} - \gamma)\log s} - 1}{(\hat{\gamma} - \gamma)\log s}\log s\, ds.$$

Since for any $1 \le s \le d_n$,

$$|(\hat{\gamma} - \gamma)\log s| \le \left|\sqrt{k}(\hat{\gamma} - \gamma)\right|\frac{\log d_n}{\sqrt{k}} \xrightarrow{P} 0$$

by assumption, we have for $n \to \infty$,

$$\sup_{1 \le s \le d_n} \left| \frac{e^{(\hat{\gamma}-\gamma)\log s} - 1}{(\hat{\gamma} - \gamma)\log s} \right| \to 1 .$$

It follows that

$$\frac{\sqrt{k}}{q_\gamma(d_n)} \left(\frac{d_n^{\hat{\gamma}} - 1}{\hat{\gamma}} - \frac{d_n^\gamma - 1}{\gamma} \right)$$

has the same limit distribution as

$$\sqrt{k}(\hat{\gamma} - \gamma) ,$$

i.e., Γ.

Finally, we deal with part IV:

$$-\frac{\sqrt{k}}{q_\gamma(d_n)} \left(\frac{U\left(d_n \frac{n}{k}\right) - U\left(\frac{n}{k}\right)}{a\left(\frac{n}{k}\right)} - \frac{d_n^\gamma - 1}{\gamma} \right)$$

$$= -\sqrt{k} A\left(\frac{n}{k}\right) \frac{d_n^\gamma - 1}{\gamma q_\gamma(d_n)} \frac{\frac{U\left(d_n \frac{n}{k}\right) - U\left(\frac{n}{k}\right)}{a\left(\frac{n}{k}\right)} \frac{\gamma}{d_n^\gamma - 1} - 1}{A\left(\frac{n}{k}\right)} .$$

Recall (4.3.5). So part IV converges to $-\lambda \gamma_-(\gamma_- + \rho)^{-1}$ by Lemma 4.3.5 and assumption (2) of Theorem 4.3.1. ∎

Proof (of Corollary 4.3.2). Start with

$$\left| \frac{q_{\hat{\gamma}}(d_n)}{q_\gamma(d_n)} - 1 \right| = \left| \frac{\int_1^{d_n} s^{\hat{\gamma}-1} \log s \, ds}{\int_1^{d_n} s^{\gamma-1} \log s \, ds} - 1 \right| \le \frac{\int_1^{d_n} \left| s^{\hat{\gamma}-\gamma} - 1 \right| s^{\gamma-1} \log s \, ds}{\int_1^{d_n} s^{\gamma-1} \log s \, ds} .$$

$$(4.3.6)$$

We have for any $1 \le s \le d_n$,

$$\left| (\hat{\gamma} - \gamma)\log s \right| \le \left| \hat{\gamma} - \gamma \right| |\log d_n| = \left| \sqrt{k}(\hat{\gamma} - \gamma) \right| \frac{\log d_n}{\sqrt{k}} \overset{P}{\to} 0 .$$

Hence for sufficiently large n, with high probability

$$\left| s^{\hat{\gamma}-\gamma} - 1 \right| \le 2 \left| \hat{\gamma} - \gamma \right| |\log d_n| .$$

It follows that (4.3.6) is at most $2 \left| \hat{\gamma} - \gamma \right| |\log d_n|$, and hence it tends to zero. ∎

Remark 4.3.6 Note that when $\gamma < -\frac{1}{2}$, if one takes the sample maxima $X_{n,n}$ to estimate an extreme quantile ($np_n = O(1)$) one gets a better rate of convergence than that in Theorem 4.3.4 (cf. Exercise 1.15).

Remark 4.3.7 For quantile estimation in a less extreme region (e.g., if $np_n/k \to c \in (0, \infty)$) one can use the results of Section 2.4 (in particular (2.4.2); see Exercise 2.17). On the other hand, it is possible to relax the condition $np_n = o(k)$ to $np_n = O(k)$ in Theorem 4.3.1. That is, under the same conditions of Theorem 4.3.1 but with $d_n = k/(np_n) \to \tau > 0$, one can show by similar arguments that

$$\sqrt{k} \; \frac{\hat{x}_{p_n} - x_{p_n}}{a\left(\frac{n}{k}\right) q_\gamma(\tau)} \xrightarrow{d} \Gamma + \frac{B}{q_\gamma(\tau)} + \Lambda \frac{\tau^\gamma - 1}{\gamma \, q_\gamma(\tau)} - \frac{\lambda}{\rho \, q_\gamma(\tau)} \left(\frac{\tau^{\gamma+\rho} - 1}{\gamma + \rho} - \frac{\tau^\gamma - 1}{\gamma} \right),$$

as $n \to \infty$. So one could follow the approach of Theorem 2.4.2 or that of Theorem 4.3.1. The rates of convergence are the same for both cases.

A simpler version is valid when γ is positive:

Theorem 4.3.8 *Suppose for some function A, with $A(t) \to 0$, as $n \to \infty$,*

$$\lim_{t \to \infty} \frac{\frac{U(tx)}{U(t)} - x^\gamma}{A(t)} = x^\gamma \frac{x^\rho - 1}{\rho}.$$

Suppose:

1. the second-order parameter ρ is negative,
2. $k = k(n) \to \infty$, $n/k \to \infty$, and $\sqrt{k} A(n/k) \to \lambda \in \mathbb{R}$, $n \to \infty$,
3. $np_n = o(k)$ and $\log np_n = o\left(\sqrt{k}\right)$, $n \to \infty$.

Define

$$\hat{x}_{p_n} := X_{n-k,n} \left(\frac{k}{np_n}\right)^{\hat{\gamma}} \quad and \quad x_{p_n} := U\left(\frac{1}{p_n}\right).$$

Then as $n \to \infty$,

$$\frac{\sqrt{k}}{\log d_n} \left(\frac{\hat{x}_{p_n}}{x_{p_n}} - 1 \right) \xrightarrow{d} \Gamma,$$

with $d_n := k/(np_n)$.

Proof. The proof is similar to that of Theorem 4.3.1. First note that

$$\frac{\sqrt{k}}{\log d_n} \left(d_n^{\hat{\gamma}-\gamma} - 1 \right) \xrightarrow{d} \Gamma.$$

Next note (Theorem 2.4.1) that

$$\sqrt{k} \left(\frac{X_{n-k,n}}{U\left(\frac{n}{k}\right)} - 1 \right) \xrightarrow{d} \gamma B$$

and finally that (use Theorem 2.3.9)

$$\lim_{n \to \infty} \frac{\frac{U\left(\frac{n}{k}d_n\right)}{U\left(\frac{n}{k}\right)}d_n^{-\gamma} - 1}{A\left(\frac{n}{k}\right)} = -\frac{1}{\rho}.$$

Hence

$$\frac{1}{U\left(\frac{1}{p_n}\right)}\frac{\sqrt{k}}{\log d_n}\left(X_{n-k,n}d_n^{\hat{\gamma}} - U\left(\frac{1}{p_n}\right)\right)$$

$$= \frac{d_n^{\gamma}U\left(\frac{n}{k}\right)}{U\left(\frac{1}{p_n}\right)}\left(\frac{\sqrt{k}}{\log d_n}\left(\frac{X_{n-k,n}}{U\left(\frac{n}{k}\right)} - 1\right)d_n^{\hat{\gamma}-\gamma}\right.$$

$$\left. + \frac{\sqrt{k}}{\log d_n}\left(d_n^{\hat{\gamma}-\gamma} - 1\right) - \frac{\sqrt{k}A\left(\frac{n}{k}\right)}{\log d_n}\frac{\frac{U\left(\frac{1}{p_n}\right)d_n^{-\gamma}}{U\left(\frac{n}{k}\right)} - 1}{A\left(\frac{n}{k}\right)}\right). \quad \blacksquare$$

The result follows.

Corollary 4.3.9 *Under the conditions of Theorem 4.3.8,*

$$\frac{\hat{x}_{p_n}}{x_{p_n}} \xrightarrow{P} 1 \quad (n \to \infty).$$

The previous results are quite general in the sense that they are valid for any estimators of γ, $a(n/k)$, and $b(n/k)$ satisfying (4.3.2). For the estimation of the location $b(n/k) = U(n/k)$ the natural estimator is its empirical counterpart $X_{n-k,n}$. Then, from Theorem 2.4.1,

$$\sqrt{k}\,\frac{\hat{b}\left(\frac{n}{k}\right) - U\left(\frac{n}{k}\right)}{a\left(\frac{n}{k}\right)} \xrightarrow{d} B,$$

with B a standard normal random variable independent of (Γ, Λ). Next we shall find the parameters of the limit distribution in Theorem 4.3.1 for some of the estimators for γ and $a(n/k)$ introduced before. A similar exercise will be done in Sections 4.4 and 4.5.

4.3.1 Maximum Likelihood Estimators

We start with the maximum likelihood estimators of Section 3.4. Recall from Theorem 3.4.2 the joint limit distribution of (Γ, Λ) for $\gamma > -\frac{1}{2}$.

Hence if $\hat{\gamma} := \hat{\gamma}_{\text{MLE}}$ and $\hat{a}\left(\frac{n}{k}\right) := \hat{\sigma}_{\text{MLE}}$ in (4.3.3) are maximum likelihood estimators, under the conditions of Theorem 4.3.1 with $\gamma > -\frac{1}{2}$,

$$\sqrt{k}\,\frac{\hat{x}_{p_n} - x_{p_n}}{\hat{\sigma}_{\mathrm{MLE}}\,q_{\hat{\gamma}_{\mathrm{MLE}}}(d_n)}$$

converges in distribution to a normal random variable with mean

$$\begin{cases} \frac{\lambda(1+\gamma)}{(1-\rho)(\gamma-\rho+1)}\,, & \gamma \geq 0 \neq \rho, \\[2mm] \frac{\lambda\rho(1+3\gamma+2\gamma^2)}{(1+\gamma-\rho)(1-\rho)(\gamma+\rho)}\,, & \gamma < 0 \neq \rho, \\[2mm] 0\,, & \gamma < 0 = \rho, \end{cases} \tag{4.3.7}$$

and variance

$$\begin{cases} (1+\gamma)^2\,, & \gamma \geq 0, \\[2mm] 1 + 4\gamma + 5\gamma^2 + 2\gamma^3 + 2\gamma^4\,, & \gamma < 0\,. \end{cases} \tag{4.3.8}$$

4.3.2 Moment Estimators

Another possibility is to use in (4.3.3) the moment estimator of Section 3.5. To estimate the scale use $\hat{\sigma}_M = X_{n-k,n} M_n^{(1)}(1 - \hat{\gamma}_-)$, introduced earlier in Section 4.2, and take as usual $\hat{b}(n/k) = X_{n-k,n}$.

When using the moment estimator in Theorem 4.3.1 one needs to take into account that the conditions of Theorem 3.5.4 (asymptotic normality of moment estimator) and the conditions of Theorem 4.3.1 (asymptotic normality of quantile estimator) are not the same. For the asymptotic normality of the moment estimator (and for the scale as well) the extra conditions are $U(\infty) > 0$, so that the estimator is well defined, and $\gamma \neq \rho$, so that a second-order condition for $\log U$ holds, with auxiliary second-order function Q. Besides, one needs

$$\sqrt{k}\,Q\left(\frac{n}{k}\right) \to \lambda \in \mathbb{R}\,, \quad n \to \infty\,, \tag{4.3.9}$$

instead of $\sqrt{k}A(n/k) \to \lambda$ (for more details see Remark 4.3.10 below).

Consequently, under the given conditions, if $\hat{\gamma} := \hat{\gamma}_M$ and $\hat{a}\left(\frac{n}{k}\right) := \hat{\sigma}_M$ in (4.3.3) are the moment estimators, then

$$\sqrt{k}\,\frac{\hat{x}_{p_n} - x_{p_n}}{\hat{\sigma}_M q_{\hat{\gamma}_M}(d_n)} \tag{4.3.10}$$

converges in distribution to a normal random variable with mean

$$\begin{cases} \frac{\lambda\rho(1-\gamma)}{(\gamma+\rho)(1-\gamma-\rho)(1-2\gamma-\rho)}\,, & \gamma < \rho \leq 0, \\[2mm] \frac{\lambda\gamma(1-3\gamma^2)}{(1-\gamma)(1-2\gamma)(1-3\gamma)}\,, & \rho < \gamma \leq 0, \\[2mm] -\frac{\lambda\gamma}{(1+\gamma)^2}\,, & \lim_{t\to\infty} U(t) - a(t)/\gamma \neq 0 \text{ and } 0 < \gamma < -\rho, \\[2mm] \frac{\lambda(\gamma+\rho-\gamma\rho)}{\rho(1-\rho)^2}\,, & (\lim_{t\to\infty} U(t) - a(t)/\gamma = 0 \text{ and } 0 < \gamma < -\rho), \\[2mm] & \text{or } \gamma \geq -\rho > 0, \end{cases}$$

$$\tag{4.3.11}$$

and variance

$$\begin{cases} \gamma^2 + 1 , & \gamma \geq 0, \\ \frac{(1-\gamma)^2(1-3\gamma+4\gamma^2)}{(1-2\gamma)(1-3\gamma)(1-4\gamma)} , & \gamma < 0 . \end{cases} \tag{4.3.12}$$

Remark 4.3.10 From Theorems 3.5.4 and 4.2.1 and (4.2.12) we have that (Γ, Λ, B) has distribution

$$\left(-\frac{\lambda 1_{\{\rho < \gamma \leq 0\}}}{1 - \gamma_-} + \gamma_+ P + (1 - 2\gamma_-)(1 - \gamma_-)^2 \left\{ \left(\frac{1}{2} - \gamma_- \right) Q - 2P \right\} , \right.$$

$$\left. -\frac{\lambda}{\rho' + \gamma 1_{\{\rho'=0\}}} + (3 - 4\gamma_-)(1 - \gamma_-)P - \frac{1}{2}(1 - 2\gamma_-)^2(1 - \gamma_-)Q + \gamma B , B \right) ,$$

where B is a standard normal random variable and the random vector (P, Q) is the one from Lemma 3.5.5 (Section 3.5). Further, B and (P, Q) are independent.

As mentioned above, Theorem 4.3.1 is not straightforward with respect to the moment estimators. Recall that for the asymptotic normality of $\sqrt{k}(\hat{\gamma}_M - \gamma)$ and $\sqrt{k}(\hat{\sigma}_M/a(n/k) - 1)$ we require $\sqrt{k}Q(n/k) = O(1)$, as $n \to \infty$, where Q is the auxiliary second-order function in the second-order condition for $\log U$. In contrast, in Theorem 4.3.1 we require $\sqrt{k}A(n/k) = O(1)$, where A is the auxiliary second-order function in the second-order condition for U. From the proof of Theorem 4.3.1 we see that the second-order condition for U is necessary for the $\hat{b}$ part (I) and for the nonrandom bias part (IV). Hence, if one wants to use the moment estimator in quantile estimation, assume $\sqrt{k}Q(n/k) \to \lambda \in \mathbb{R}$, $n \to \infty$, and Lemma B.3.16 provides the (finite) limit of $A(t)/Q(t)$. Then the limiting distribution of (4.3.10) is

$$\Gamma + (\gamma_-)^2 B - \gamma_- \Lambda - \frac{\gamma_- \lambda 1_{\{\gamma < \rho \leq 0\}}}{\gamma_- + \rho} .$$

Use Corollary 4.2.2 to obtain the bias and the variance given in (4.3.11) and (4.3.12).

4.4 Tail Probability Estimation

On the basis of an independent and identically distributed sample of size n, now we solve the dual problem: given a large value x, how can one estimate

$$p = 1 - F(x) .$$

As in quantile estimation take (4.3.1)–(4.3.2) as our point of departure and the same considerations on $p = p_n$. Hence in particular, x should in fact depend on n, $x = x_n$, and $p_n = 1 - F(x_n) \to 0$. In the sequel an important auxiliary quantity is

$$d_n := \frac{k}{n(1 - F(x_n))},$$

where as usual, k is an intermediate sequence.

To estimate the tail probability we use (cf. (3.1.4))

$$\hat{p}_n := \frac{k}{n} \left\{ \max\left(0,\ 1 + \hat{\gamma}\,\frac{x_n - \hat{b}\left(\frac{n}{k}\right)}{\hat{a}\left(\frac{n}{k}\right)}\right) \right\}^{-1/\hat{\gamma}} \tag{4.4.1}$$

with x_n known.

Theorem 4.4.1 *Suppose for* $\gamma > -\frac{1}{2}$ *and some function A with* $A(t) \to 0, t \to \infty$, *the second-order condition (4.3.1) holds. Write as before* $d_n = k/(np_n)$. *Suppose:*

1. *the second-order parameter* ρ *is negative, or zero with* γ *negative;*
2. $k = k(n) \to \infty$, $n/k \to \infty$, *and* $\sqrt{k}A(n/k) \to \lambda \in \mathbb{R}, n \to \infty$;
3. $d_n \to \infty$ *and* $w_\gamma(d_n) = o(\sqrt{k})$, $n \to \infty$, *where for* $t > 0$,

$$w_\gamma(t) := t^{-\gamma} \int_1^t s^{\gamma-1} \log s\ ds\ ;$$

4. *condition (4.3.2) holds for some estimators of* γ, $a(n/k)$, *and* $U(n/k)$, *say* $\hat{\gamma}$, $\hat{a}(n/k)$, *and* $\hat{b}(n/k)$, *respectively.*

Then, as $n \to \infty$,

$$\frac{\sqrt{k}}{w_\gamma(d_n)}\left(\frac{\hat{p}_n}{p_n} - 1\right) \xrightarrow{d} \Gamma + (\gamma_-)^2\,B - (\gamma_-)\,\Lambda - \lambda\frac{\gamma_-}{\gamma_- + \rho} \tag{4.4.2}$$

with $\gamma_- := \min(0, \gamma)$.

Remark 4.4.2 Note that as $t \to \infty$,

$$w_\gamma(t) \sim \begin{cases} \frac{1}{\gamma}\log t\,, & \gamma > 0, \\ \frac{1}{2}(\log t)^2\,, & \gamma = 0, \\ \frac{1}{\gamma^2}t^{-\gamma}\,, & \gamma < 0\,. \end{cases}$$

Moreover, since

$$w_\gamma(t) = \int_1^t s^{-1-\gamma}(\log t - \log s)\ ds = \int_1^t \int_1^u s^{-1-\gamma}ds\ u^{-1}\ du\,,$$

it is clear that $w_\gamma(t)$ is increasing in t and that $w_\gamma(t)$ is a decreasing function of γ, when $t \geq 1$.

Remark 4.4.3 Result (4.4.2) is valid only for $\gamma > -\frac{1}{2}$: for $\gamma < 0$ condition (3) of Theorem 4.4.1 implies $k^{-1/2}d_n^{-\gamma} = k^{-1/2-\gamma}(np_n)^\gamma \to 0$, as $n \to \infty$. Hence, since $\gamma < 0$ and $np_n = O(1)$, we must have $k^{-1/2-\gamma} \to 0$. This implies $\gamma > -\frac{1}{2}$. In fact, for $\gamma < -\frac{1}{2}$ no result of this type exists.

Corollary 4.4.4 *Condition (3) of Theorem 4.4.1 implies what we may call consistency:*

$$\frac{\hat{p}_n}{p_n} \xrightarrow{P} 1 \ .$$

Corollary 4.4.5 *The conditions of Theorem 4.4.1 imply*

$$\frac{w_{\hat{\gamma}}(\hat{d}_n)}{w_{\gamma}(d_n)} \xrightarrow{P} 1, \tag{4.4.3}$$

where $\hat{d}_n := k(n\hat{p}_n)^{-1}$. *Hence an equivalent statement is*

$$\frac{\sqrt{k}}{w_{\hat{\gamma}}(\hat{d}_n)} \left(\frac{\hat{p}_n}{p_n} - 1\right) \xrightarrow{d} \Gamma + (\gamma_-)^2 B - (\gamma_-) \Lambda - \lambda \frac{\gamma_-}{\gamma_- + \rho} \ .$$

The latter form of the result is more useful for constructing a confidence interval for p_n.

Note that the limiting random variable is the same as the one in Theorem 4.3.1, as could be expected from the Bahadur–Kiefer representation. For the proper non-degenerate limit distribution of the difference of the normalized left-hand sides of (4.3.4) and (4.4.2) we refer to Einmahl (1995).

Condition $d_n \to \infty$ means that we extrapolate outside or near the boundary of the range of the available observations; cf. Remark 4.3.4 above.

Finally, note that condition (3) of Theorem 4.4.1 implies for all real γ that

$$\log d_n = o\left(\sqrt{k}\right) \ . \tag{4.4.4}$$

Proof (of Theorem 4.4.1). Define

$$\hat{x}_n := \hat{b}\left(\frac{n}{k}\right) + \hat{a}\left(\frac{n}{k}\right) \frac{d_n^{\hat{\gamma}} - 1}{\hat{\gamma}} \ .$$

Then

$$\frac{\hat{p}_n}{p_n} = d_n \left\{\max\left(0, \ 1 + \hat{\gamma}\frac{x_n - \hat{b}\left(\frac{n}{k}\right)}{\hat{a}\left(\frac{n}{k}\right)}\right)\right\}^{-1/\hat{\gamma}}$$

$$= d_n \left\{\max\left(0, \ 1 + \hat{\gamma}\left(\frac{x_n - \hat{x}_n}{\hat{a}\left(\frac{n}{k}\right)} + \frac{d_n^{\hat{\gamma}} - 1}{\hat{\gamma}}\right)\right)\right\}^{-1/\hat{\gamma}}$$

$$= d_n \left\{\max\left(0, \ d_n^{\hat{\gamma}} + \hat{\gamma}\frac{x_n - \hat{x}_n}{\hat{a}\left(\frac{n}{k}\right)}\right)\right\}^{-1/\hat{\gamma}}$$

$$= \left\{\max\left(0, \ 1 - \hat{\gamma}\frac{q_{\gamma}(d_n)}{d_n^{\hat{\gamma}}} \frac{a\left(\frac{n}{k}\right)}{\sqrt{k}} \sqrt{k}\frac{\hat{x}_n - x_n}{a\left(\frac{n}{k}\right)q_{\gamma}(d_n)}\right)\right\}^{-1/\hat{\gamma}} \ .$$

Now, as in Theorem 4.3.1,

$$\sqrt{k}\,\frac{\hat{x}_n - x_n}{a\left(\frac{n}{k}\right)q_\gamma(d_n)} \xrightarrow{d} \Gamma + (\gamma_-)^2 B - \gamma_- \Lambda - \lambda\frac{\gamma_-}{\gamma_- + \rho}\ ,$$

by assumption

$$\frac{q_\gamma(d_n)}{d_n^{\hat{\gamma}}\,\sqrt{k}} \xrightarrow{P} 0$$

and

$$\frac{a\left(\frac{n}{k}\right)}{\hat{a}\left(\frac{n}{k}\right)} \xrightarrow{P} 1\ .$$

Hence $\hat{p}_n/p_n \xrightarrow{P} 1$ and we can expand

$$\frac{\hat{p}_n}{p_n} - 1 = \frac{q_\gamma(d_n)}{d_n^{\hat{\gamma}}\,\sqrt{k}}\ \sqrt{k}\ \frac{\hat{x}_n - x_n}{a\left(\frac{n}{k}\right)q_\gamma(d_n)}\,(1 + o_P(1))\ .$$

The result follows.   ∎

Proof (of Corollary 4.4.5). Consider

$$\frac{w_{\hat{\gamma}}(\hat{d}_n)}{w_\gamma(d_n)} = \frac{w_{\hat{\gamma}}(d_n)}{w_\gamma(d_n)}\ \frac{w_{\hat{\gamma}}(\hat{d}_n)}{w_{\hat{\gamma}}(d_n)}\ .$$

The first factor converges in probability to 1 by Corollary 4.3.2 and assumption (3) of Theorem 4.4.1 (cf. (4.4.4)). The second factor also converges in probability to 1 since $\hat{d}_n/d_n \xrightarrow{P} 1$ and the function $w_{\hat{\gamma}}(t)$ is regularly varying.   ∎

Remark 4.4.6 For estimation of exceedance probabilities in a "less extreme" region (e.g., if $np_n/k \to c \in (0, \infty)$) one can use the results of Section 5.1 below (in particular (5.1.12)). On the other hand, it is possible to relax the condition $np_n = o(k)$ to $np_n = O(k)$ in Theorem 4.4.1. That is, under the same conditions of Theorem 4.4.1 but with $d_n = k/(np_n) \to \tau > 0$,

$$\frac{\sqrt{k}}{w_\gamma(\tau)}\left(\frac{\hat{p}_n}{p_n} - 1\right)$$

$$\xrightarrow{d} \Gamma + \frac{\tau^\gamma B}{w_\gamma(\tau)} + \Lambda\frac{\tau^\gamma(\tau^\gamma - 1)}{\gamma\,w_\gamma(\tau)} - \frac{\tau^\gamma \lambda}{\rho\,w_\gamma(\tau)}\left(\frac{\tau^{\gamma+\rho} - 1}{\gamma + \rho} - \frac{\tau^\gamma - 1}{\gamma}\right)\ ,$$

as $n \to \infty$. So one could follow the approach of Theorem 4.4.1 or that of Theorem 5.1.2. The rate of convergence is the same in both cases.

A simpler version is valid when γ is positive.

Theorem 4.4.7 *Suppose for some function A, with* $A(t) \to 0$, *as* $n \to \infty$,

$$\lim_{t \to \infty} \frac{\frac{U(tx)}{U(t)} - x^\gamma}{A(t)} = x^\gamma \frac{x^\rho - 1}{\rho} .$$

Suppose as well that:

1. *the second-order parameter,* ρ *is negative;*
2. $k = k(n) \to \infty$, $n/k \to \infty$, *and* $\sqrt{k} A(n/k) \to \lambda \in \mathbb{R}$, $n \to \infty$;
3. $np_n = o(k)$ *and* $\log np_n = o\left(\sqrt{k}\right)$, $n \to \infty$.

Define

$$\hat{p}_n := \frac{k}{n} \left(\frac{x_n}{X_{n-k,n}} \right)^{-1/\hat{\gamma}} \quad and \quad x_{p_n} := U\left(\frac{1}{p_n} \right) .$$

Then, as $n \to \infty$,

$$\frac{\gamma \sqrt{k}}{\log d_n} \left(\frac{\hat{p}_n}{p_n} - 1 \right) \xrightarrow{d} \Gamma ,$$

with $d_n := k/(np_n)$.

The proof of the theorem is left to the reader.

4.4.1 Maximum Likelihood Estimators

Recall that the limiting distributions of the suitably normalized quantile and tail probability estimators, i.e., the left-hand sides of (4.3.4) and (4.4.2) respectively, are the same. Therefore if the maximum likelihood estimators $\hat{\gamma}_{\text{MLE}}$ and $\hat{\sigma}_{\text{MLE}}$ are used in (4.4.1), then under the conditions of Theorem 4.4.1 the limiting random variable (4.4.2) is normal with mean (4.3.7) and variance (4.3.8).

4.4.2 Moment Estimators

When using the moment estimators (cf. Sections 3.5 and 4.2) one needs extra conditions; the same considerations of Section 4.3.2 apply here. Then, if the moment estimators $\hat{\gamma}_M$ and $\hat{\sigma}_M$ are used in (4.4.1) the limiting random variable (4.4.2) is normal with mean (4.3.11) and variance (4.3.12).

4.5 Endpoint Estimation

Next we turn to the problem of estimating the endpoint of the distribution function F. We assume $F \in \mathcal{D}(G_\gamma)$ for some negative γ, since in this case the endpoint x^* is known to be finite (cf. Lemma 1.2.9).

An estimator of x^* can be motivated from the quantile estimator. Replacing p_n by zero in (4.3.3) we get

$$\hat{x}^* := \hat{b}\left(\frac{n}{k}\right) - \frac{\hat{a}\left(\frac{n}{k}\right)}{\hat{\gamma}} . \tag{4.5.1}$$

As in the previous sections, we assume in the following that $(\hat{\gamma}, \hat{a}, \hat{b})$ when suitably normalized are asymptotically normal. Denote the limiting random vector by (Γ, Λ, B) (cf. (4.3.2)).

Theorem 4.5.1 *Suppose that for some function $A(t) \to 0$, $t \to \infty$, the second-order condition (4.3.1) holds with γ negative. Suppose $k = k(n) \to \infty$, $n/k \to \infty$, and $\sqrt{k}A(n/k) \to \lambda \in \mathbb{R}$, $n \to \infty$. Then,*

$$\sqrt{k}\gamma^2 \frac{\hat{x}^* - x^*}{a\left(\frac{n}{k}\right)} \xrightarrow{d} \Gamma + \gamma^2 B - \gamma \Lambda - \lambda \frac{\gamma}{\gamma + \rho} . \tag{4.5.2}$$

Corollary 4.5.2 *Under the given assumptions an equivalent statement is*

$$\sqrt{k}\hat{\gamma}^2 \frac{\hat{x}^* - x^*}{\hat{a}\left(\frac{n}{k}\right)} \xrightarrow{d} \Gamma + \gamma^2 B - \gamma \Lambda - \lambda \frac{\gamma}{\gamma + \rho} .$$

This version is more useful for constructing asymptotic confidence intervals for $\hat{x}^*$.

Corollary 4.5.3 *Under the conditions of Theorem 4.5.1,*

$$\hat{x}^* \xrightarrow{P} x^* .$$

For the proof of Theorem 4.5.1 we need the following lemma:

Lemma 4.5.4 *If the second-order condition (4.3.1) holds with γ negative then*

$$\lim_{t \to \infty} \frac{\frac{U(\infty) - U(t)}{a(t)} + \frac{1}{\gamma}}{A(t)} = \frac{1}{\gamma(\rho + \gamma)} .$$

Proof. From the second-order condition (4.3.1) and the second-order condition (2.3.7) for the function a, it follows that

$$\lim_{t \to \infty} \frac{\left(U(tx) - \frac{a(tx)}{\gamma}\right) - \left(U(t) - \frac{a(t)}{\gamma}\right)}{a(t)A(t)/\gamma}$$

$$= \lim_{t \to \infty} \frac{\frac{U(tx) - U(t)}{a(t)} - \frac{x^\gamma - 1}{\gamma}}{A(t)/\gamma} - \lim_{t \to \infty} \frac{\frac{a(tx)}{a(t)} - x^\gamma}{A(t)}$$

$$= -\frac{x^{\gamma + \rho} - 1}{\gamma + \rho} .$$

Now Lemma 1.2.9(2) implies that $\lim_{t \to \infty} (U(t) - a(t)/\gamma)$ exists. This limit must be $U(\infty)$, since the function a is regularly varying with negative index and hence tends to zero. Lemma 1.2.9(2) also implies

$$\lim_{t \to \infty} \frac{U(\infty) - \left(U(t) - \frac{a(t)}{\gamma}\right)}{-a(t)A(t)/\gamma} = -\frac{1}{\gamma + \rho} ,$$

hence the result.   ∎

Proof (of Theorem 4.5.1).

$$\hat{x}^* - x^* = \hat{b}\left(\frac{n}{k}\right) - U\left(\frac{n}{k}\right) - \hat{a}\left(\frac{n}{k}\right)\left(\frac{1}{\hat{\gamma}} - \frac{1}{\gamma}\right)$$

$$- \left(\hat{a}\left(\frac{n}{k}\right) - a\left(\frac{n}{k}\right)\right)\frac{1}{\gamma} - \left(U(\infty) - U\left(\frac{n}{k}\right) + a\left(\frac{n}{k}\right)\frac{1}{\gamma}\right).$$

Hence by (4.3.2) and Lemma 4.5.4,

$$\sqrt{k}\gamma^2 \frac{\hat{x}^* - x^*}{a\left(\frac{n}{k}\right)} = \gamma^2 \sqrt{k}\frac{\hat{b}\left(\frac{n}{k}\right) - U\left(\frac{n}{k}\right)}{a\left(\frac{n}{k}\right)} - \gamma^2 \frac{\hat{a}\left(\frac{n}{k}\right)}{a\left(\frac{n}{k}\right)}\sqrt{k}\left(\frac{1}{\hat{\gamma}} - \frac{1}{\gamma}\right)$$

$$- \gamma\sqrt{k}\left(\frac{\hat{a}\left(\frac{n}{k}\right)}{a\left(\frac{n}{k}\right)} - 1\right) - \gamma^2\sqrt{k}A\left(\frac{n}{k}\right)\frac{\frac{U(\infty)-U\left(\frac{n}{k}\right)}{a\left(\frac{n}{k}\right)} + \frac{1}{\gamma}}{A\left(\frac{n}{k}\right)}$$

$$\overset{d}{\to} \gamma^2 B + \Gamma - \gamma\Lambda - \lambda\frac{\gamma}{\gamma + \rho}. \qquad \blacksquare$$

Remark 4.5.5 As pointed out in Aarssen and de Haan (1994), for estimating x^* when $\gamma < -\frac{1}{2}$ it is more efficient to use different k's for the various estimators involved in $\hat{x}^*$. For instance, the authors suggest, with k_1 fixed and $k = k(n) \to \infty$, $k/n \to 0$ as usual, to take

$$\hat{x}^* := \hat{b}\left(\frac{n}{k_1}\right) - \frac{\hat{a}\left(\frac{n}{k_1}\right)}{\hat{\gamma}_M} \quad \text{with} \quad \hat{a}\left(\frac{n}{k_1}\right) := X_{n-k_1,n}M_{n,k_1}^{(1)}(1 - \hat{\gamma}_M), \quad (4.5.3)$$

where $M_{n,k_1}^{(1)}$ is like $M_n^{(1)}$ but with k fixed equal to k_1, $\hat{b}(n/k_1) = X_{n-k_1,n}$, and $\hat{\gamma}_M$ is the moment estimator of Section 3.5 with the intermediate sequence $k(n)$. When $\gamma < -\frac{1}{2}$ this estimator converges more quickly than the one from Theorem 4.5.1 (Exercise 4.6).

Another way to estimate x^* when $\gamma < -\frac{1}{2}$ is simply to use the sample maxima $X_{n,n}$, similarly as in Remark 4.3.6. Under the natural conditions the rates of convergence of $\hat{x}^*$ in (4.5.3) and $X_{n,n}$ are the same (cf. Corollary 1.2.4).

4.5.1 Maximum Likelihood Estimators

Note that for γ negative the limiting distribution in (4.5.2) is still the same as that of Theorem 4.3.1 on quantile estimation. Hence if the maximum likelihood estimators $\hat{\gamma}_{\text{MLE}}$ and $\hat{\sigma}_{\text{MLE}}$ are used in (4.5.1), then under the conditions of Theorem 4.5.1 with $\gamma > -\frac{1}{2}$ the limiting random variable (4.5.2) is normal with mean (4.3.7) and variance (4.3.8).

4.5.2 Moment Estimators

When using the moment estimators (cf. Sections 3.5 and 4.2) one needs extra conditions; the same considerations of Section 4.3.2 apply here. Then, if the moment estimators $\hat{\gamma}_M$ and $\hat{\sigma}_M$ are used in (4.5.1) the limiting random variable (4.4.2) is normal with mean (4.3.11) and variance (4.3.12).

Another option would be to use $\hat{\gamma}_-$, i.e., (4.2.3), to estimate γ since we assume the latter negative, but it turns out that this is no better than using $\hat{\gamma}_M$. The extra conditions needed are the same as for the moment estimator. Then, if $\hat{\gamma} := \hat{\gamma}_-$ and $\hat{a}(n/k) := \hat{\sigma}_M$ are used in (4.5.1),

$$\sqrt{k}\,\hat{\gamma}_-^2\,\frac{\hat{x}^* - x^*}{\hat{\sigma}_M} \tag{4.5.4}$$

converges in distribution to a normal random variable with mean

$$\begin{cases} \lambda\frac{\rho(1-\gamma)}{(\gamma+\rho)(1-\gamma-\rho)(1-2\gamma-\rho)}\,, & \gamma < \rho \le 0, \\ \lambda\frac{\gamma(1-3\gamma+3\gamma^2)}{(1-2\gamma)(1-3\gamma)}\,, & \rho < \gamma \le 0, \end{cases} \tag{4.5.5}$$

and variance

$$\frac{(1-\gamma)^2(1-3\gamma+4\gamma^2)}{(1-2\gamma)(1-3\gamma)(1-4\gamma)}\,. \tag{4.5.6}$$

The variance of the limiting variable is the same as when one uses the moment estimator $\hat{\gamma}_M$. The bias is also the same for $\gamma < \rho < 0$; otherwise, the bias is larger (in absolute value) than the one with $\hat{\gamma}_M$. Therefore this latter option of using $\hat{\gamma}_-$ shows no advantage.

4.6 Simulations and Applications

4.6.1 Simulations

In what follows we show simulation results for quantiles corresponding to $p_n = 1/(n \log n) = 1.086 \times 10^{-4}$, from independent and identically distributed samples of size $n = 1000$, with a number of independent replications $r = 5000$, for the Cauchy, normal, and uniform distributions (cf. Table 3.2). For some independent estimates $\hat{x}_{p_n,1}, \dots, \hat{x}_{p_n,r}$ obtained from some estimator of the quantile, in the following by mean square error we mean $r^{-1}\sum_{i=1}^r(\hat{x}_{p_n,i} - x_{p_n})^2$.

We use the following notation: Hill means the quantile estimator restricted to positive γ (as in Theorem 4.3.8) with the Hill estimator, Mom means the quantile estimator from Theorem 4.3.1 with the moment estimator and the scale estimator from Section 4.2, PWM means again the quantile estimator from Theorem 4.3.1 but with the probability-weighted moment estimators (3.6.9)–(3.6.10), and finally Falk means the quantile estimator from Theorem 4.3.1 with the negative Hill estimator (3.6.13) and the scale estimator the same as for Mom. Recall that all the estimators use k upper order statistics out of the initial sample of size n. In Figures 4.4–4.6 are shown the diagram of estimates of the quantile, i.e., averages of the 5000 quantile estimates for each number of upper order statistics k, and the corresponding mean square error.

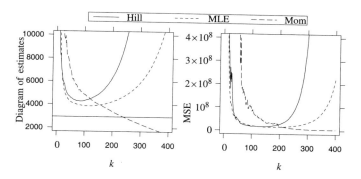

Fig. 4.4. Standard Cauchy distribution: (**a**) diagram of estimates of the quantile (the true quantile 2931.7 is indicated by the horizontal line); (**b**) mean square error (see the text for details).

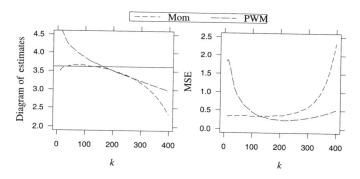

Fig. 4.5. Standard normal distribution: (**a**) diagram of estimates of the quantile (the true quantile 3.62 is indicated by the horizontal line); (**b**) mean square error (see the text for details).

4.6.2 Case Studies

Sea Level

We continue the data analysis of Section 3.7. Recall that we want to estimate the quantile corresponding to a tail probability of $1/17 \times 10^{-4}$, on the basis of 1873 observations of the sea level during severe storms. We give results for Mom (the quantile estimator from Theorem 4.3.1 with the moment estimator and the scale estimator from Section 4.2) and PWM (the quantile estimator from Theorem 4.3.1 with the probability-weighted moment estimators (3.6.9)–(3.6.10)). Moreover, from Section 3.7 we know that γ is close to zero. Then one can consider the following options: use the quantile estimator as given in Theorem 4.3.1 or assume $\gamma = 0$ and use

$$\hat{x}_{p_n} = \hat{b}\left(\frac{n}{k}\right) + \hat{a}\left(\frac{n}{k}\right)\log\left(\frac{k}{np_n}\right) .$$

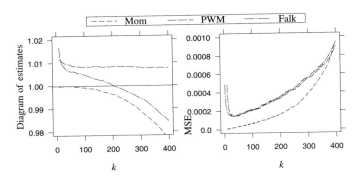

Fig. 4.6. Standard uniform distribution: (**a**) diagram of estimates of the quantile (the true quantile .99985 is indicated by the horizontal line); (**b**) mean square error (see the text for details).

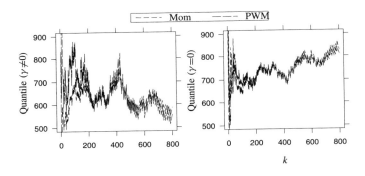

Fig. 4.7. Sea level data, diagram of estimates of the quantile (cm).

The correspondent diagram of estimates for both options are shown in Figure 4.7. As expected, one finds less volatility when γ is fixed to 0.

In any case one has to estimate the scale. The corresponding diagram of estimates is shown in Figure 4.8.

Under the conditions of Theorem 4.3.1 with $\lambda = 0$ we have

$$\sqrt{k}\,\frac{\hat{x}_{p_n} - x_{p_n}}{\hat{a}\left(\frac{n}{k}\right)q_{\hat{\gamma}}(d_n)} \overset{d}{\approx} \sqrt{\mathrm{var}_\gamma}\,N$$

with N standard normal. The approximating $(1-\alpha)100\%$ confidence intervals are given by

$$\hat{x}_{p_n} - z_{\alpha/2}\,\hat{a}\left(\frac{n}{k}\right)q_{\hat{\gamma}}(d_n)\sqrt{\frac{\mathrm{var}_{\hat{\gamma}}}{k}} < x_{p_n} < \hat{x}_{p_n} + z_{\alpha/2}\,\hat{a}\left(\frac{n}{k}\right)q_{\hat{\gamma}}(d_n)\sqrt{\frac{\mathrm{var}_{\hat{\gamma}}}{k}},$$

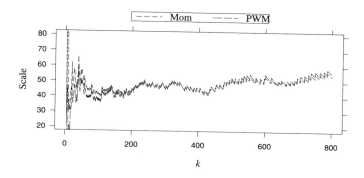

Fig. 4.8. Sea level data, diagram of estimates of the scale.

where $\text{var}_{\hat{\gamma}}$ is the respective asymptotic variance with γ replaced by its estimate and $z_{\alpha/2}$ is the $1 - \alpha/2$ quantile of the standard normal distribution. In Table 4.1 are confidence intervals for the quantile corresponding to a tail probability of $1/17 \times 10^{-4}$, for some values of k and considering $\hat{\gamma} = 0$ (cf. Exercise 4.7 for PWM).

Table 4.1. Sea level data: 95% asymptotic confidence intervals for quantile.

k	100	200	300
Moment	(615., 764.)	(656., 778.)	(669., 773.)
PWM	(579., 737.)	(641., 779.)	(655., 774.)

S&P 500

We continue the S&P 500 data analysis of Section 3.7. Recall that we focus on the log-loss returns comprising 2643 observations. A short summary of the largest observations is in Table 4.2.

Table 4.2. S&P 500 data.

3rd Quantile	$X_{n-1,n}$	$X_{n,n}$
0.009	0.086	0.228

Next we show estimates of the probability that the log-loss return exceeds the value 0.20, using the quantile estimator from Theorem 4.4.7 with the Hill estimator (we simply call it Hill) and the quantile estimator from Theorem 4.3.1 with the moment

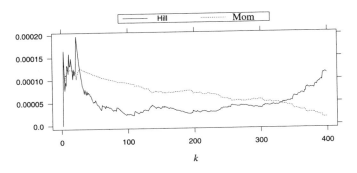

Fig. 4.9. S&P 500 data, diagram of estimates of the tail probability.

estimator and the scale estimator from Section 4.2 (we call it Mom). The diagram of estimates is in Figure 4.9.

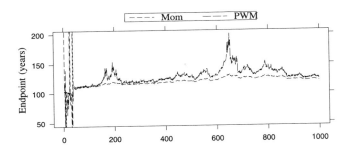

Fig. 4.10. Life span data, diagram of estimates of the right endpoint.

Life Span

We continue the life span data analysis of Section 3.7. The data set consists of the total life span (in days) of 10 391 residents of the Netherlands, and we are interested in the estimation of the right endpoint of the underlying distribution. In Section 3.7 we analyzed its existence by not rejecting the hypothesis that the underlying extreme value index is negative. So now we assume that the right endpoint exists and proceed with its estimation.

We give results for Mom (the quantile estimator from Theorem 4.3.1 with the moment estimator and the scale estimator from Section 4.2) and PWM (the quantile estimator from Theorem 4.3.1 with the probability weighted moment estimators (3.6.9)–(3.6.10)). In Figure 4.10 is the diagram of estimates of the right endpoint.

Table 4.3. Life span data: upper limit of 95% asymptotic confidence intervals for the endpoint (in years).

k	100	200	400
Moment	115.	130.	125.
PWM	121.	208.	141.

Similarly as before, the approximating $(1-\alpha)100\%$ one-sided confidence interval is given by

$$x^* < \hat{x}^* + z_\alpha \frac{\hat{a}\left(\frac{n}{k}\right)}{\hat{\gamma}^2} \sqrt{\frac{\mathrm{var}_{\hat{\gamma}}}{k}},$$

where $\mathrm{var}_{\hat{\gamma}}$ is the respective asymptotic variance with γ replaced by its estimate and z_α is the $1-\alpha$ quantile of the standard normal distribution. In Table 4.3 we give the one-sided confidence intervals for the endpoint for $k = 100, 200, 400$ (cf. Exercise 4.7 for PWM).

Exercises

4.1. Prove the consistency of $\hat{\sigma}_M$, that is, the first part of Theorem 4.2.1(1).

4.2. If one replaces p_n by cp_n $(c > 0)$ in Theorem 4.3.1, how does the limit result change?

4.3. Prove that $w_\gamma(t) = t^{-\gamma}q_\gamma(t) = H_{\gamma,0}(t)$ with q_γ from Theorem 4.3.1, w_γ from Theorem 4.4.1, and $H_{\gamma,\rho}$ from Corollary 2.3.4.

4.4. Let $\gamma \leq 0$ or $\rho \neq 0$. Verify that the expected value of the limiting random variable of Theorem 4.2.1(2), $(1 - \gamma_-)(3 - 4\gamma_-)P - 2^{-1}(1 - \gamma_-)(1 - 2\gamma_-)^2 Q - \lambda 1_{\{\gamma \neq 0 \text{ and } (\gamma < 0 \text{ or } \rho < 0)\}}(\rho' + \gamma 1_{\{\rho'=0\}})^{-1}$, as a function of (γ_-, ρ'), equals $-\lambda\rho'(1 - \gamma_- - \rho')^{-1}(1 - 2\gamma_- - \rho')^{-1}$ (consider the random variables (P, Q) of Lemma 3.5.5 and recall that in the statement of this lemma the mean values and covariance matrix are in terms of γ_- and ρ').

4.5. Prove Theorem 4.4.7.

4.6. (Aarssen and de Haan (1994)) Let $X_1, X_2, \ldots$ be i.i.d. random variables with distribution function F. Suppose $U(\infty) > 0$ and (4.2.2), i.e., the second-order condition for $\log U$, with γ negative and auxiliary second-order function Q. Let k_1 be fixed and $k = k(n) \to \infty$, $k/n \to 0$, and $\sqrt{k}Q(n/k) \to 0$, as $n \to \infty$. Then, with the notation of Remark 4.5.5,

$$\frac{\hat{x}^* - x^*}{X_{n-k_1,n} M_{n,k_1}^{(1)}}$$

converges in distribution to a random variable of the form

$$1 - \frac{1}{\gamma} + \left\{ \frac{1}{k_1} \sum_{i=0}^{k_1-1} \exp\left(\gamma \sum_{j=i+1}^{k_1} \frac{Z_j}{j} \right) - 1 \right\}^{-1},$$

where $Z_1, Z_2, \ldots$ are i.i.d. random variables with a standard exponential distribution.

4.7. Consider the quantile estimator of Theorem 4.3.1 with $\hat{\gamma}_{\text{PWM}}$ and $\hat{a}(n/k) = \hat{\sigma}_{\text{PWM}}$ from Section 3.6.1. Check that the variance of the limiting random variable in (4.3.4) is given by

$$\begin{cases} \frac{(1-\gamma)(2-\gamma)^2(1-\gamma+2\gamma^2)}{(1-2\gamma)(3-2\gamma)}, & \gamma \geq 0, \\ 2\frac{2-2\gamma+12\gamma^2-38\gamma^3+47\gamma^4-66\gamma^5+74\gamma^6-40\gamma^7+8\gamma^8}{(1-2\gamma)(3-2\gamma)}, & \gamma < 0. \end{cases}$$

5

Advanced Topics

Chapters 1–4 constitute the basic probabilistic and statistical theory of one-dimensional extremes. In this chapter we shall present additional material that can be skipped at first reading. It is not used in the rest of the book.

Section 5.1 is a mirror image of Sections 2.3 and 2.4: it offers an expansion of the tail empirical distribution function rather than the tail quantile function as in Section 2.4.

Section 5.2 offers various ways to check the extreme value condition in facing a data set. Some procedures use the tail quantile function and others the tail empirical distribution function.

Section 5.3 uses an expansion for the tail distribution function (not empirical distribution function) developed in Section 5.1 in order to obtain uniform speed of convergence results in the convergence of maxima toward the limit distribution. This also leads to a large deviation result.

Some classical results are presented in Sections 5.3.1 and 5.4: convergence of moments and weak (in probability) and strong (a.s.) behavior of the sequence of maxima.

In Sections 5.5 and 5.6 the conditions "independent and identically distributed" that we used throughout Chapters 1–4 are relaxed: in Section 5.5 the assumption of independence of the initial random variables is relaxed and in Section 5.6 the assumption of stationarity is relaxed.

5.1 Expansion of the Tail Distribution Function and Tail Empirical Process

Let us go back to the approximation of the inverse tail empirical process of Theorem 2.4.2. This theorem implies in particular (for simplicity consider $\gamma \geq -\frac{1}{2}$ and k large enough so that the bias component vanishes)

$$\sqrt{k} \left(\frac{X_{n-[ks],n} - U(\frac{n}{k})}{a_0(\frac{n}{k})} - \frac{s^{-\gamma} - 1}{\gamma} \right) \xrightarrow{d} s^{-\gamma-1} W(s) \qquad (5.1.1)$$

in $D(0, 1]$ with a_0 a suitable positive function and $\{W(s)\}_{0<s\leq 1}$ standard Brownian motion. We are going to translate this result into a result for the empirical distribution function. Let us invoke a Skorohod construction and pretend that (5.1.1) holds almost surely in $D(0, 1]$.

This result has the following structure: for some nonincreasing functions $H_n(s)$ (i.e., $(X_{n-[ks],n} - U(n/k))/a_0(n/k)$) and some function M (i.e., $(s^{-\gamma} - 1)/\gamma$) with continuous negative derivative we have

$$\lim_{n\to\infty} \frac{H_n(s) - M(s)}{\delta_n} = P(s) \tag{5.1.2}$$

locally uniformly in $(0, 1]$ where P is a continuous function (i.e., $s^{-\gamma-1}W(s)$) and δ_n a positive sequence tending to zero (i.e., $\delta_n = 1/\sqrt{k}$). From (5.1.2) and Lemma A.0.2 (Vervaat's lemma) we get

$$\lim_{n\to\infty} \frac{H_n^{\leftarrow}(x) - M^{\leftarrow}(x)}{\delta_n} = -(M^{\leftarrow})'(x)P(M^{\leftarrow}(x)) \ .$$

Specializing to the mentioned random function, this implies

$$\sqrt{k}\left(\frac{1}{k}\sum_{i=1}^{n} 1_{\{(X_i-U(n/k))/a_0(n/k)>x\}} - (1+\gamma x)^{-1/\gamma}\right) \to W\left((1+\gamma x)^{-1/\gamma}\right) ,$$

a.s. locally uniformly for those x for which $1 + \gamma x > 0$. The conclusion is that

$$\sqrt{k}\left(\frac{1}{k}\sum_{i=1}^{n} 1_{\{(X_i-U(n/k))/a_0(n/k)>x\}} - (1+\gamma x)^{-1/\gamma}\right) \xrightarrow{d} W\left((1+\gamma x)^{-1/\gamma}\right) \tag{5.1.3}$$

in $D(0, 1/(\max(0, -\gamma)))$.

The present section aims at obtaining a weighted uniform version of (5.1.3). That is, we discuss expansions for the (empirical) distribution function similar to those of Section 2.4. Since the proofs are quite technical and lengthy, some parts will be sketched rather than carried out in detail. For a full account of the theory and full proofs, see Drees, de Haan, and Li (2006). We begin with an expansion of the tail of the distribution function analogous to Section 2.3.

Theorem 5.1.1 *Let F be a probability distribution function. Suppose the second-order condition (2.3.5) holds for the function U, the inverse of $1/(1 - F)$. Let a_0, b_0 and, A_0 be the functions defined in Corollary 2.3.7.*

1. If not $\gamma = \rho = 0$, for all $c, \delta > 0$, then

$$\lim_{t\to\infty} \sup_{x\in\widetilde{D}_{t,\rho,\delta,c}} \left((1+\gamma x)^{-1/\gamma}\right)^{\rho-1} \exp\left(-\varepsilon\left|\log(1+\gamma x)^{-1/\gamma}\right|\right)$$

$$\times \left|\frac{t(1 - F(b_0(t) + xa_0(t))) - (1+\gamma x)^{-1/\gamma}}{A_0(t)}\right.$$

$$\left. - (1+\gamma x)^{-1/\gamma-1}\overline{\Psi}_{\gamma,\rho}((1+\gamma x)^{1/\gamma})\right| = 0 \tag{5.1.4}$$

with

$$\overline{\Psi}_{\gamma,\rho}(x) := \begin{cases} \frac{x^{\gamma+\rho}}{\gamma+\rho} , & \rho < 0 \neq \gamma + \rho, \\ \log x , & \rho < 0 = \gamma + \rho, \\ \frac{1}{\gamma} x^{\gamma} \log x , & \rho = 0 \neq \gamma, \end{cases}$$

and

$$\widetilde{D}_{t,\rho,\delta,c} := \begin{cases} \{x : (1 + \gamma x)^{-1/\gamma} \leq ct^{-\delta+1}\} , & \rho < 0, \\ \{x : (1 + \gamma x)^{-1/\gamma} \leq |A_0(t)|^{-c}\} , & \rho = 0 . \end{cases}$$

2. *If* $\gamma = \rho = 0$, *for all* $c, \delta > 0$, *then*

$$\lim_{t \to \infty} \sup_{x \in \widetilde{D}_{t,\rho,\delta,c}} \min \left(\frac{e^{-\varepsilon|\log(t\{1-F(b_0(t)+xa_0(t))\})|}}{t\{1 - F(b_0(t) + xa_0(t))\}}, e^{x-\varepsilon|x|} \right)$$

$$\times \left| \frac{t(1 - F(b_0(t) + xa_0(t))) - e^{-x}}{A_0(t)} - \frac{x^2}{2} e^{-x} \right| = 0 . \quad (5.1.5)$$

Proof. Here is a proof for $\rho < 0$ and F continuous and strictly increasing. As usual, we revert to the properties of U rather than F to obtain the necessary approximations. Hence we define

$$y := \frac{1}{t\{1 - F(b_0(t) + xa_0(t))\}} ,$$

so that

$$x = \frac{U(ty) - b_0(t)}{a_0(t)} .$$

This leads to the following expansion, where the notation $g(x) := (1 + \gamma x)^{-1/\gamma}$ and $q_t(x) := (U(tx) - b_0(t))/a_0(t) - (x^{\gamma} - 1)/\gamma$ is used:

$$t(1 - F(b_0(t) + xa_0(t))) - (1 + \gamma x)^{-1/\gamma}$$

$$= - \left\{ \left(1 + \gamma \frac{U(ty) - b_0(t)}{a_0(t)} \right)^{-1/\gamma} - \left(1 + \gamma \frac{y^{\gamma} - 1}{\gamma} \right)^{-1/\gamma} \right\}$$

$$= - \left\{ g \left(\frac{U(ty) - b_0(t)}{a_0(t)} \right) - g \left(\frac{y^{\gamma} - 1}{\gamma} \right) \right\}$$

$$= q_t(y) \left(-g' \left(\frac{y^{\gamma} - 1}{\gamma} \right) \right) - \int_0^{q_t(y)} \int_0^s g'' \left(\frac{y^{\gamma} - 1}{\gamma} + u \right) du \, ds$$

$$= y^{-\gamma-1} q_t(y) - (1 + \gamma) \int_0^{q_t(y)} \int_0^s \left(1 + \gamma \left(\frac{y^{\gamma} - 1}{\gamma} + u \right) \right)^{-1/\gamma-2} du \, ds .$$

Now, the integrand function $(1 + \gamma((y^{\gamma} - 1)/\gamma + u))^{-1/\gamma-2}$ always lies between its value for $u = 0$ and for $u = q_t(y)$, i.e., it lies between $(1 + \gamma(y^{\gamma} - 1)/\gamma)^{-1/\gamma-2} = y^{-1-2\gamma}$ and $(1 + \gamma((y^{\gamma} - 1)/\gamma + q_t(y)))^{-1/\gamma-2} = y^{-1-2\gamma}(1 + \gamma y^{-\gamma} q_t(y))^{-1/\gamma-2}$.

By examining the uniform inequalities of Corollary 2.3.7 for the quantile function one sees that for any $c, \delta > 0$ (recall $\rho < 0$),

$$\lim_{t \to \infty} \sup_{y \geq ct^{\delta-1}} y^{-\gamma} |q_t(y)| = 0 . \tag{5.1.6}$$

Hence

$$\left(1 + \gamma \left(\frac{y^\gamma - 1}{\gamma} + u\right)\right)^{-1/\gamma - 2} \leq 2y^{-1-2\gamma}$$

for $y \geq ct^{\delta-1}$ and t sufficiently large, and we obtain

$$\left| t(1 - F(b_0(t) + xa_0(t)) - (1 + \gamma x)^{-1/\gamma} - y^{-1-\gamma} q_t(y) \right|$$
$$\leq 2|1 + \gamma| y^{-1-2\gamma} q_t^2(y) \tag{5.1.7}$$

for all $y \geq ct^{\delta-1}$ and t sufficiently large. Upon multiplying the left- and right-hand sides in (5.1.7) by y we see that these two statements easily lead to

$$\lim_{t \to \infty} \sup_{y \geq ct^{\delta-1}} \left| y(1 + \gamma x)^{-1/\gamma} - 1 \right| = 0 \tag{5.1.8}$$

and hence (by checking the definition of $\overline{\Psi}_{\gamma,\rho}$) to

$$\lim_{t \to \infty} \sup_{y \geq ct^{\delta-1}} y^{1-\rho} \exp(-\varepsilon |\log y|)$$
$$\times \left| (1 + \gamma x)^{-(1+1/\gamma)} \overline{\Psi}_{\gamma,\rho} \left((1 + \gamma x)^{1/\gamma} \right) - y^{-(1+\gamma)} \overline{\Psi}_{\gamma,\rho}(y) \right| = 0 . \tag{5.1.9}$$

The rest is straightforward: by (5.1.8),

$$\left((1 + \gamma x)^{-1/\gamma} \right)^{\rho-1} \exp \left(-\varepsilon \left| \log(1 + \gamma x)^{-1/\gamma} \right| \right)$$
$$\times \left| \frac{t\{1 - F(b_0(t) + xa_0(t))\} - (1 + \gamma x)^{-1/\gamma}}{A_0(t)} \right.$$
$$\left. - (1 + \gamma x)^{-1/\gamma - 1} \overline{\Psi}_{\gamma,\rho} \left((1 + \gamma x)^{1/\gamma} \right) \right|$$

$$\leq \frac{y^{1-\rho} e^{-\varepsilon |\log y|}}{|A_0(t)|} \left| t(1 - F(b_0(t) + xa_0(t))) - (1 + \gamma x)^{-1/\gamma} - y^{-1-\gamma} q_t(y) \right|$$
$$+ y^{1-\rho} e^{-\varepsilon |\log y|} \left| \frac{y^{-(1+\gamma)} q_t(y)}{A_0(t)} - y^{-(1+\gamma)} \overline{\Psi}_{\gamma,\rho}(y) \right|$$
$$+ y^{1-\rho} e^{-\varepsilon |\log y|} \left| y^{-(1+\gamma)} \overline{\Psi}_{\gamma,\rho}(y) - (1 + \gamma x)^{-1/\gamma - 1} \overline{\Psi}_{\gamma,\rho}((1 + \gamma x)^{1/\gamma}) \right| . \tag{5.1.10}$$

By (5.1.7) the first term on the right-hand side is at most

$$2\,|1+\gamma|\left(y^{-(\gamma+\rho)}e^{-\varepsilon|\log y|}\left|\frac{q_t(y)}{A_0(t)}\right|\right)\left\{y^{-\gamma}q_t(y)\right\}.$$

For the last factor we use (5.1.6). The uniform boundedness of the other factor on $y \geq ct^{\delta-1}$ is again a result of the uniform inequalities of Corollary 2.3.7 for the quantile function.

The second term on the right side of (5.1.10) converges to zero uniformly on $y \geq ct^{\delta-1}$: this is just Corollary 2.3.7.

The third term on the right side of (5.1.10) converges to zero uniformly on $y \geq ct^{\delta-1}$ by (5.1.9).

Finally, note that $y \geq ct^{\delta-1}$ if and only if $x \in \widetilde{D}_{t,\rho,\delta,c}$ for $\rho < 0$.   ∎

This result will be used later in the chapter but also for the proof of the next result: an expansion of the tail *empirical* distribution function, that is, the tail empirical process.

Theorem 5.1.2 *Let $X_1, X_2, \ldots$ be i.i.d. random variables with distribution function F. Let F_n be the empirical distribution function, based on $X_1, X_2, \ldots, X_n$. Suppose that the function U, the inverse of $1/(1-F)$, satisfies the second-order condition (2.3.5) with $\gamma \in \mathbb{R}$, $\rho \leq 0$. Let $k = k(n)$ be a sequence of integers such that $k \to \infty$ and $\sqrt{k}A_0(n/k)$ is bounded, as $n \to \infty$, with A_0 from Corollary 2.3.7. We also use a_0 and b_0 from that corollary. Then the underlying sample space can be enlarged to include a sequence of Brownian motions W_n such that for all x_0 larger than the lower endpoint of the limiting extreme value distribution $-1/(\gamma \vee 0)$:*

1. *If not $\gamma = \rho = 0$, then as $n \to \infty$,*

$$\sup_{x_0 \leq x < 1/((-\gamma)\vee 0)} \left((1+\gamma x)^{-1/\gamma}\right)^{-1/2+\varepsilon}$$

$$\left|\sqrt{k}\left\{\frac{n}{k}\left(1 - F_n\left(b_0\left(\frac{n}{k}\right) + xa_0\left(\frac{n}{k}\right)\right)\right) - (1+\gamma x)^{-1/\gamma}\right\}\right.$$

$$- W_n\left((1+\gamma x)^{-1/\gamma}\right)$$

$$\left. - \sqrt{k}A_0\left(\frac{n}{k}\right)(1+\gamma x)^{-1/\gamma-1}\overline{\Psi}_{\gamma,\rho}\left((1+\gamma x)^{1/\gamma}\right)\right| \xrightarrow{P} 0.$$

2. *If $\gamma = \rho = 0$, then as $n \to \infty$ for all $\tau > 0$,*

$$\sup_{x_0 \leq x < \infty} \max(1,x)^{\tau}\left|\sqrt{k}\left\{\frac{n}{k}\left(1 - F_n\left(b_0\left(\frac{n}{k}\right) + xa_0\left(\frac{n}{k}\right)\right)\right) - e^{-x}\right\}\right.$$

$$\left. - W_n\left(e^{-x}\right) - \sqrt{k}A_0\left(\frac{n}{k}\right)\frac{x^2}{2}e^{-x}\right| \xrightarrow{P} 0.$$

Proof. The result is well known in the case of the standard uniform distribution (see, e.g., Einmahl (1997), Corollary 3.3) and this result will be our point of departure. Let

U_n be the uniform empirical distribution function. Then the underlying sample space can be enlarged to include a sequence of Brownian motions W_n,

$$\sup_{t>0} t^{-1/2} e^{-\varepsilon|\log t|} \left| \sqrt{k} \left(\frac{n}{k} U_n \left(\frac{kt}{n} \right) - t \right) - W_n(t) \right| \overset{P}{\to} 0, \tag{5.1.11}$$

as $n \to \infty$ with $k \to \infty$, $k/n \to 0$. By the well-known quantile transformation, $1-F_n$ has the same distribution as $U_n(1-F)$. Hence by (5.1.11) for suitable versions of F_n,

$$\sup_{\{x:z_n(x)>0\}} (z_n(x))^{-1/2} e^{-\varepsilon|\log z_n(x)|}$$
$$\times \left| \sqrt{k} \left\{ \frac{n}{k} \left(1 - F_n \left(b_0 \left(\frac{n}{k} \right) + x a_0 \left(\frac{n}{k} \right) \right) \right) - z_n(x) \right\} - W_n(z_n(x)) \right| \overset{P}{\to} 0 \tag{5.1.12}$$

with

$$z_n(x) := \frac{n}{k} \left(1 - F \left(b_0 \left(\frac{n}{k} \right) + x a_0 \left(\frac{n}{k} \right) \right) \right).$$

In order to get the result of the theorem, we are going to replace $z_n(x)$ by $(1+\gamma x)^{-1/\gamma}$ in (5.1.12) and we shall see how this can be done for $\rho < 0$.

First note that by (5.1.8) for $0 < \delta < 1$ and sufficiently large t with $x_0 > -1/(\gamma \vee 0)$ and $c > 0$,

$$\left[x_0, \frac{1}{(-\gamma) \vee 0} \right) = \left\{ x : 0 < (1+\gamma x)^{-1/\gamma} \le q \text{ with } q := (1+\gamma x_0)^{-1/\gamma} < \infty \right\}$$
$$\subset \left\{ x : (1+\gamma x)^{-1/\gamma} \le ct^{-\delta+1} \right\} = \widetilde{D}_{t,\rho,\delta,c} \tag{5.1.13}$$

with $\widetilde{D}$ as in Theorem 5.1.1.

It follows that we can replace the supremum over $\{x : z_n(x) > 0\}$ in (5.1.12) by the supremum over $\{x : x_0 \le x < 1/((-\gamma) \vee 0)\}$ and that we can use the inequalities of Theorem 5.1.1 in this range of x-values.

Also, (5.1.8) allows us to replace the factor $(z_n(x))^{-1/2} \exp(-\varepsilon|\log z_n(x)|)$ by $((1+\gamma x)^{-1/\gamma})^{-1/2+\varepsilon}$ (since we know that $(1+\gamma x)^{-1/\gamma}$ is bounded). The result of Theorem 5.1.1 along with the condition "$\sqrt{k} A_0(n/k)$ bounded" allows us to replace the term $z_n(x)$ in the curly brackets by $(1+\gamma x)^{-1/\gamma} + A_0(n/k)(1+\gamma x)^{-1/\gamma-1} \overline{\Psi}_{\gamma,\rho}((1+\gamma x)^{1/\gamma})$.

Finally, we need to prove that as $n \to \infty$,

$$\sup_{x_0 \le x < 1/((-\gamma)\vee 0)} \left((1+\gamma x)^{-1/\gamma} \right)^{-1/2+\varepsilon}$$
$$\times \left| W_n \left\{ \frac{n}{k} \left(1 - F \left(b_0 \left(\frac{n}{k} \right) + x a_0 \left(\frac{n}{k} \right) \right) \right) \right\} - W_n \left((1+\gamma x)^{-1/\gamma} \right) \right| \overset{P}{\to} 0. \tag{5.1.14}$$

First note that by the law of the iterated logarithm, $\lim_{t\to\infty} t^{-\varepsilon-1/2} W(t) = 0$ a.s. for $\varepsilon > 0$. Hence by time reversal for $\varepsilon > 0$,

$$\lim_{s \downarrow 0} s^{\varepsilon - 1/2} W(s) = 0 \quad \text{a.s.} \tag{5.1.15}$$

Now by (5.1.8) and (5.1.13) we have

$$\lim_{n \to \infty} \sup_{x_0 < x \le 1/((-\gamma) \vee 0)} \left| (1 + \gamma x)^{1/\gamma} \frac{n}{k} \left(1 - F \left(b_0 \left(\frac{n}{k} \right) + x a_0 \left(\frac{n}{k} \right) \right) \right) - 1 \right| = 0 .$$

Hence for (5.1.14) it is sufficient to prove that if

$$\lim_{n \to \infty} \sup_{0 < s \le s_0} \left| \frac{t_n(s)}{s} - 1 \right| = 0 , \tag{5.1.16}$$

then

$$\lim_{n \to \infty} \sup_{0 < s \le s_0} \left| \frac{W(t_n(s)) - W(s)}{(t_n(s))^{1/2 - \varepsilon}} \right| = 0 \quad \text{a.s.} \tag{5.1.17}$$

Take a sequence $s_n \to s_0 \ge 0$, $n \to \infty$. Then by (5.1.16), also $t_n(s_n) \to s_0$, $n \to \infty$. For $s_0 > 0$ by continuity of Brownian motion (5.1.17) is true. For $s_0 = 0$ by (5.1.15) and (5.1.16) both $W(t_n(s_n))/(t_n(s))^{1/2 - \varepsilon}$ and $W(s)/(t_n(s))^{1/2 - \varepsilon}$ converge to zero and (5.1.17) follows. ∎

Remark 5.1.3 The Brownian motions in Theorem 5.1.2 (on tail empirical distribution functions) are the same as the Brownian motions in Theorem 2.4.2 (on tail empirical quantile functions). This can be seen most easily by applying Vervaat's lemma (see Appendix A) to the functions in Theorem 5.1.2, restricted to a compact interval.

The result of Theorem 5.1.2 can be simplified in the case $\gamma > 0$ and reads as follows.

Theorem 5.1.4 *Let $X_1, X_2, \ldots$ be i.i.d. random variables with distribution functions F. Let F_n be the empirical distribution function based on $X_1, X_2, \ldots, X_n$. Suppose that the function U, the inverse of $1/(1 - F)$, satisfies the second-order condition of Theorem 2.3.9, hence in particular $\gamma > 0$. Let $k = k(n)$ be a sequence of integers such that $k \to \infty$, $\sqrt{k} A_0(n/k)$ bounded, $n \to \infty$ with A_0 from this theorem. Then the underlying sample space can be enlarged to include a sequence of Brownian motions W_n such that for all $x_0 > 0$,*

$$\sup_{x \ge x_0} x^{(1/2 - \varepsilon)/\gamma} \left| \sqrt{k} \left\{ \frac{n}{k} \left(1 - F_n \left(x \, U \left(\frac{n}{k} \right) \right) \right) - x^{-1/\gamma} \right\} \right.$$

$$\left. - W_n \left(x^{-1/\gamma} \right) - \sqrt{k} A_0 \left(\frac{n}{k} \right) x^{-1/\gamma} \frac{x^{\rho/\gamma} - 1}{\gamma \rho} \right| \xrightarrow{P} 0 , \tag{5.1.18}$$

as $n \to \infty$.

Proof. The proof follows the line of the proof of Theorem 5.1.2 but now we use the inequalities

$$\left| \frac{\frac{1 - F(tx)}{1 - F(t)} - x^{-1/\gamma}}{\alpha(t)} - x^{-1/\gamma} \frac{x^{\gamma \rho} - 1}{\rho/\gamma} \right| \le \varepsilon x^{-1/\gamma + \rho/\gamma} \max(x^\delta, x^{-\delta})$$

(cf. Theorem 2.3.9) and the result of Theorem 2.4.8. ∎

Example 5.1.5 As an example let us apply this result to get another proof of the asymptotic normality of the Hill estimator:

$$
\begin{aligned}
\hat{\gamma}_H &:= \frac{1}{k} \sum_{i=0}^{k-1} \log X_{n-i,n} - \log X_{n-k,n} \\
&= \frac{n}{k} \int_{X_{n-k,n}}^{\infty} \left(\log s - \log X_{n-k,n} \right) d F_n(s) \\
&= \int_{X_{n-k,n}}^{\infty} \frac{n}{k} \left(1 - F_n(s) \right) \frac{ds}{s} \\
&= \int_{X_{n-k,n}/U(n/k)}^{\infty} \frac{n}{k} \left\{ 1 - F_n \left(s\, U \left(\tfrac{n}{k} \right) \right) \right\} \frac{ds}{s} .
\end{aligned}
$$

Hence

$$
\begin{aligned}
\sqrt{k} \left(\hat{\gamma}_H - \gamma \right) &= \sqrt{k} \int_{X_{n-k,n}/U(n/k)}^{1} \frac{n}{k} \left\{ 1 - F_n \left(s U \left(\tfrac{n}{k} \right) \right) \right\} \frac{ds}{s} \\
&\quad + \sqrt{k} \int_{1}^{\infty} \left(\frac{n}{k} \left\{ 1 - F_n \left(s\, U \left(\tfrac{n}{k} \right) \right) \right\} - s^{-1/\gamma} \right) \frac{ds}{s} \\
&=: \mathrm{I} + \mathrm{II} .
\end{aligned}
$$

For part I note that by Theorem 2.4.8,

$$
\sqrt{k} \left(\frac{X_{n-k,n}}{U \left(\tfrac{n}{k} \right)} - 1 \right) - \gamma W_n(1) \overset{P}{\to} 0 . \tag{5.1.19}
$$

Hence $X_{n-k,n}/U(n/k) \to^P 1$. Using the approximation of Theorem 5.1.4 for the integrand, we see that

$$
\sqrt{k} \int_{X_{n-k,n}/U(n/k)}^{1} \left(\frac{n}{k} \left\{ 1 - F_n \left(s\, U \left(\tfrac{n}{k} \right) \right) \right\} - s^{-1/\gamma} \right) \frac{ds}{s} \overset{P}{\to} 0 . \tag{5.1.20}
$$

When we combine (5.1.19) and (5.1.20), we get

$$
\sqrt{k} \int_{X_{n-k,n}/U(n/k)}^{1} \frac{n}{k} \left\{ 1 - F_n \left(s\, U \left(\tfrac{n}{k} \right) \right) \right\} \frac{ds}{s} + \gamma W_n(1) \overset{P}{\to} 0 . \tag{5.1.21}
$$

Next we consider II. By the uniform convergence in (5.1.18) we can interchange limit and integral when applying (5.1.18) to II. The result is

$$
\begin{aligned}
&\sqrt{k} \int_{1}^{\infty} \left\{ \frac{n}{k} \left[1 - F_n \left(s U \left(\tfrac{n}{k} \right) \right) \right] - s^{-1/\gamma} \right\} \frac{ds}{s} \\
&= \int_{1}^{\infty} W_n \left(s^{-1/\gamma} \right) \frac{ds}{s} + \sqrt{k} A_0 \left(\tfrac{n}{k} \right) \int_{1}^{\infty} s^{-1/\gamma} \frac{s^{\rho/\gamma} - 1}{\gamma \rho} \frac{ds}{s} + o_p(1) .
\end{aligned}
$$

It follows that

$$\sqrt{k}\left(\hat{\gamma}_H - \gamma\right) = -\gamma W_n(1) + \int_1^\infty W_n\left(s^{-1/\gamma}\right)\frac{ds}{s}$$
$$+ \sqrt{k}A_0\left(\frac{n}{k}\right)\int_1^\infty \gamma s^{-1/\gamma}\frac{s^{\gamma\rho}-1}{\rho/\gamma}\frac{ds}{s} + o_p(1)\ .$$

Now

$$-\gamma W_n(1) + \int_1^\infty W_n\left(s^{-1/\gamma}\right)\frac{ds}{s} = -\gamma\left(W_n(1) - \int_0^1 W_n(u)\frac{du}{u}\right)$$

has a normal distribution with mean zero, and its variance is

$$\gamma^2 E\left(-W(1) + \int_0^1 W(u)\frac{du}{u}\right)\left(-W(1) + \int_0^1 W(v)\frac{dv}{v}\right)$$
$$= \gamma^2\left(EW^2(1) + 2\int_0^1\int_v^1 EW(u)W(v)\frac{du}{u}\frac{dv}{v} - 2\int_0^1 EW(1)W(v)\frac{dv}{v}\right)$$
$$= \gamma^2\left(1 + 2\int_0^1\int_v^1 v\frac{du}{u}\frac{dv}{v} - 2\int_0^1 v\frac{dv}{v}\right) = \gamma^2\ .$$

5.2 Checking the Extreme Value Condition

The extreme value condition is the only realistic framework for estimating quantiles and distribution tails outside the range of the available data if one does not have sufficient previous knowledge of the underlying distribution function. However, the condition is not always fulfilled. Hence it is useful to develop a test to see whether the tail of the empirical distribution function really looks like one of the generalized Pareto distributions (GP) that we want to use for extrapolation. Such a goodness-of-fit criterion will also be useful to guide us in deciding *which part* of the tail empirical distribution function (or tail empirical quantile process) can be well approximated by the corresponding distribution function or quantile function of a GP distribution. That is, it helps us to decide from which intermediate order statistic onward the approximation can be trusted.

We shall discuss four tests, three based on the quantile function and one on the distribution function. We start with the former.

Let $X_1, X_2, X_3, \ldots$ be independent and identically distributed random variables with distribution function F and let $X_{1,n} \leq X_{2,n} \leq \cdots \leq X_{n,n}$ be the nth order statistics. Suppose F is in the domain of attraction of an extreme value distribution $G_\gamma, \gamma \in \mathbb{R}$.

Recall the representation of the tail quantile process via a special construction (Corollary 2.4.6) that holds under the second-order condition: for $\varepsilon > 0$,

$$\frac{X_{n-[ks],n} - X_{n-k,n}}{a_0\left(\frac{n}{k}\right)} - \frac{s^{-\gamma}-1}{\gamma} = \frac{1}{\sqrt{k}}\left\{s^{-\gamma-1}W_n(s) - W_n(1)\right.$$

$$\left. + \sqrt{k}A_0\left(\frac{n}{k}\right)\left(\overline{\Psi}_{\gamma,\rho}\left(s^{-1}\right) - \overline{\Psi}_{\gamma,\rho}(1)\right) + o_P(1)\max\left(1, s^{-\gamma-1/2-\varepsilon}\right)\right\}, \quad (5.2.1)$$

where $o_P(1)$ tends to zero uniformly for $0 < s \le 1$. Since the left-hand side is small uniformly in s, we use it for the test. However, it contains the two unknown quantities γ and a_0. We replace these by estimators $\hat{\gamma}$ and $\hat{a}(n/k)$. The test statistic becomes

$$I_{k,n} := \int_0^1 \left(\frac{X_{n-[ks],n} - X_{n-k,n}}{\hat{a}\left(\frac{n}{k}\right)} - \frac{s^{-\hat{\gamma}}-1}{\hat{\gamma}}\right)^2 s^{2\hat{\gamma}_+ +1} ds\,, \quad (5.2.2)$$

with $\hat{\gamma}_+ := \max(0, \hat{\gamma})$. The weight function is necessary to ensure that all integrals converge.

We are going to see that $I_{k,n} \to^P 0$ under the extreme value condition, so that it can be used as a goodness-of-fit criterion, and that $kI_{k,n}$ has a nondegenerate limit distribution under the second-order condition.

Theorem 5.2.1 *Let $X_1, X_2, \ldots$ be i.i.d. random variables and suppose that their distribution function F is in the domain of attraction of some extreme value distribution.*

1. *Let $k = k(n) \to \infty, k/n \to 0, as\, n \to \infty$. Assume $\hat{\gamma}_n \to^P \gamma$ and $\hat{a}(n/k)/a(n/k)$ $\to^P 1$. Then*

$$I_{k,n} \xrightarrow{P} 0\,.$$

2. *Suppose moreover that the second-order condition (2.3.5) holds and that $\sqrt{k}A(n/k) \to 0$, as $n \to \infty$, where A is the second-order auxiliary function. Further assume that*

$$\sqrt{k}\left(\hat{\gamma} - \gamma, \frac{\hat{a}\left(\frac{n}{k}\right)}{a\left(\frac{n}{k}\right)} - 1\right) - (\Gamma(W_n), \alpha(W_n)) \xrightarrow{P} 0 \quad (5.2.3)$$

where Γ and α are measurable real valued functionals of the Brownian motion from (5.2.1). Then

$$kI_{k,n} \xrightarrow{d} I_\gamma$$

where

$$I_\gamma := \int_0^1 \left(s^{-\gamma-1}W(s) - W(1) - \alpha(W_n)\frac{s^{-\gamma}-1}{\gamma}\right.$$

$$\left. + \Gamma(W_n)\int_s^1 u^{-\gamma-1}\log u\, du\right)^2 s^{2\gamma_+ +1} ds$$

with W Brownian motion.

Remark 5.2.2 The reader may want to check that (5.2.3) holds for all the estimators discussed in Chapter 3.

For the proof we need some lemmas. We have seen the following one previously:

Lemma 5.2.3 *For $\varepsilon > 0$,*

$$\lim_{\theta \downarrow 0} \sup_{0 < s \le \theta} s^{-1/2+\varepsilon} W(s) = 0 \quad a.s.$$

with $W(s)$ Brownian motion.

Lemma 5.2.4 *If $\hat{\gamma}$ is any consistent estimator of γ, then*

$$\frac{s^{-\hat{\gamma}} - 1}{\hat{\gamma}} - \frac{s^{-\gamma} - 1}{\gamma} = -(\hat{\gamma} - \gamma) \int_s^1 u^{-\gamma-1} \log u \, du$$

$$+ |\hat{\gamma} - \gamma| \, O_P(1) \left(s^{-|\hat{\gamma}-\gamma|} - 1 \right) \int_s^1 u^{-\gamma-1} |\log u| \, du,$$

where the $O_P(1)$ term is bounded uniformly for $s \in (0, 1]$.

Proof. Since for $x \in \mathbb{R}$,

$$\left| \frac{e^x - 1}{x} - 1 \right| \le e^{|x|} - 1$$

(check by series expansion), we have

$$\left| \frac{s^{-\hat{\gamma}} - 1}{\hat{\gamma}} - \frac{s^{-\gamma} - 1}{\gamma} + (\hat{\gamma} - \gamma) \int_s^1 u^{-\gamma-1} \log u \, du \right|$$

$$= \left| \int_s^1 u^{-\gamma-1} \left(u^{\gamma-\hat{\gamma}} - 1 + (\hat{\gamma} - \gamma) \log u \right) du \right|$$

$$= |\hat{\gamma} - \gamma| \left| \int_s^1 \left(\frac{e^{(\gamma-\hat{\gamma}) \log u} - 1}{(\gamma - \hat{\gamma}) \log u} - 1 \right) u^{-\gamma-1} \log u \, du \right|$$

$$\le |\hat{\gamma} - \gamma| \int_s^1 \left(u^{-|\hat{\gamma}-\gamma|} - 1 \right) u^{-\gamma-1} |\log u| \, du$$

$$\le |\hat{\gamma} - \gamma| \left(s^{-|\hat{\gamma}-\gamma|} - 1 \right) \int_s^1 u^{-\gamma-1} |\log u| \, du. \qquad \blacksquare$$

Lemma 5.2.5 *Under the assumptions of Theorem 5.2.1(1), for $\varepsilon > 0$,*

$$\frac{X_{n-[ks],n} - X_{n-k,n}}{\hat{a}\left(\frac{n}{k}\right)} - \frac{s^{-\hat{\gamma}} - 1}{\hat{\gamma}} = o_P(1) \left(s^{-\gamma-1/2-\varepsilon} \vee 1 \right),$$

where the $o_P(1)$ term tends to zero uniformly for $0 < s \le 1$.

Proof.

$$\frac{X_{n-[ks],n} - X_{n-k,n}}{\hat{a}\left(\frac{n}{k}\right)} - \frac{s^{-\hat{\gamma}} - 1}{\hat{\gamma}} \tag{5.2.4}$$

$$= \frac{a_0\left(\frac{n}{k}\right)}{\hat{a}\left(\frac{n}{k}\right)} \left(\frac{X_{n-[ks],n} - X_{n-k,n}}{a_0\left(\frac{n}{k}\right)} - \frac{s^{-\gamma} - 1}{\gamma} \right) \tag{5.2.5}$$

$$- \frac{a_0\left(\frac{n}{k}\right)}{\hat{a}\left(\frac{n}{k}\right)} \left(\frac{\hat{a}\left(\frac{n}{k}\right)}{a_0\left(\frac{n}{k}\right)} - 1 \right) \frac{s^{-\gamma} - 1}{\gamma} \tag{5.2.6}$$

$$- \left(\frac{s^{-\hat{\gamma}} - 1}{\hat{\gamma}} - \frac{s^{-\gamma} - 1}{\gamma} \right). \tag{5.2.7}$$

For (5.2.5) we proceed as follows: for convenience we replace $X_1, X_2, \ldots$ by $U(Y_1), U(Y_2), \ldots$, where $U := (1/(1 - F))^{\leftarrow}$ and $Y_1, Y_2, \ldots$ are independent and identically distributed $1 - 1/x$, $x > 1$. Then $(X_{1,n}, \ldots, X_{n,n}) =^d (U(Y_{1,n}), \ldots, (U(Y_{n,n}))$. We write

$$\frac{U(Y_{n-[ks],n}) - U(Y_{n-k,n})}{a_0(Y_{n-k,n})} - \frac{s^{-\gamma} - 1}{\gamma}$$

$$= \left\{ \frac{U(Y_{n-[ks],n} - U(Y_{n-k,n})}{a_0(Y_{n-k,n})} - \frac{\left(\frac{Y_{n-[ks],n}}{Y_{n-k,n}}\right)^{\gamma} - 1}{\gamma} \right\}$$

$$+ \left\{ \frac{\left(\frac{Y_{n-[ks],n}}{Y_{n-k,n}}\right)^{\gamma} - 1}{\gamma} - \frac{s^{-\gamma} - 1}{\gamma} \right\}. \tag{5.2.8}$$

We start with the second term on the right side of (5.2.8) and use Lemma 2.4.10. As in the proof of Corollary 2.4.10 one sees that the supremum in Lemma 2.4.10 can be taken over $0 < s \le 1$. By combining the expansions for $Y_{n-[ks],n}$ and $Y_{n-k,n}$ in Lemma 2.4.10 we get

$$\sup_{0<s\le 1} s^{\gamma+1/2+\varepsilon} \left| \sqrt{k} \left(\frac{\left(\frac{Y_{n-[ks],n}}{Y_{n-k,n}}\right)^{\gamma} - 1}{\gamma} - \frac{s^{-\gamma} - 1}{\gamma} \right) \right.$$

$$\left. - s^{-\gamma-1} W_n(s) + s^{-\gamma} W_n(1) \right| = o_P(1) \tag{5.2.9}$$

(check separately for $\gamma > 0$, $\gamma < 0$, and $\gamma = 0$). It follows by Lemma 5.2.3 that

$$\sup_{0<s\le 1} s^{\gamma+1/2+\varepsilon} \left(\frac{\left(\frac{Y_{n-[ks],n}}{Y_{n-k,n}}\right)^{\gamma} - 1}{\gamma} - \frac{s^{-\gamma} - 1}{\gamma} \right) = o_P(1). \tag{5.2.10}$$

For the first part of the right-hand side of (5.2.8) we use the uniform inequalities for extended regularly varying functions (Theorem B.2.18): there exists $a_0(t) \sim a(t)$, $t \to \infty$, such that for all $\varepsilon > 0$ there exists $t_0(\varepsilon)$ such that for $t \geq t_0$ and $x \geq 1$,

$$x^{-\gamma-\varepsilon} \left| \frac{U(tx) - U(t)}{a_0(t)} - \frac{x^\gamma - 1}{\gamma} \right| < \varepsilon .$$

We apply this with $t := Y_{n-k,n}$ ($\to \infty$ a.s., as $n \to \infty$, cf. Lemma 3.2.1) and $x := Y_{n-[ks],n}/Y_{n-k,n}$ and get

$$\frac{U(Y_{n-[ks],n}) - U(Y_{n-k,n})}{a_0(Y_{n-k,n})} - \frac{\left(\frac{Y_{n-[ks],n}}{Y_{n-k,n}}\right)^\gamma - 1}{\gamma} = o_P(1) \left(\frac{Y_{n-[ks],n}}{Y_{n-k,n}}\right)^{\gamma+\varepsilon}$$

with the $o_P(1)$ term tending to zero uniformly for $0 < s \leq 1$. Next we apply (5.2.10) to get for $\beta \in \mathbb{R}$,

$$\left(\frac{Y_{n-[ks],n}}{Y_{n-k,n}}\right)^\beta = s^{-\beta} + o_P(1)s^{-\beta-1/2-\varepsilon} .$$

We have

$$\frac{U(Y_{n-[ks],n}) - U(Y_{n-k,n})}{a_0(Y_{n-k,n})} - \frac{\left(\frac{Y_{n-[ks],n}}{Y_{n-k,n}}\right)^\gamma - 1}{\gamma} = o_P(1)s^{-\gamma-1/2-\varepsilon}.$$

For (5.2.6) use $\hat{a}(n/k)/a(n/k) \sim \hat{a}(n/k)/a_0(n/k) \to^P 1$. For (5.2.7) use Lemma 5.2.4. We obtain

$$\frac{X_{n-[ks],n} - X_{n-k,n}}{\hat{a}\left(\frac{n}{k}\right)} - \frac{s^{-\hat{\gamma}} - 1}{\hat{\gamma}}$$

$$= (1 + o_P(1)) o_P(1)s^{-\gamma-1/2-\varepsilon} + (1 + o_P(1)) o_P(1)\frac{s^{-\gamma} - 1}{\gamma}$$

$$+ o_P(1)s^{-\varepsilon}s^{-\gamma}|\log s|. \qquad \blacksquare$$

Lemma 5.2.6 *Under the assumptions of Theorem 5.2.1(2), for $\varepsilon > 0$,*

$$\sqrt{k}\left(\frac{X_{n-[ks],n} - X_{n-k,n}}{\hat{a}\left(\frac{n}{k}\right)} - \frac{s^{-\hat{\gamma}} - 1}{\hat{\gamma}}\right)$$

$$= s^{-\gamma-1}W_n(s) - W_n(1) - \alpha(W_n)\frac{s^{-\gamma} - 1}{\gamma} + \Gamma(W_n)\int_s^1 u^{-\gamma-1}\log u \, du$$

$$+ o_P(1)s^{-\gamma+-1/2-\varepsilon}, \qquad (5.2.11)$$

where the $o_P(1)$ term tends to zero uniformly for $0 < s \leq 1$.

Proof. First note that since (2.3.5) holds with a replaced with a_0, we have $a_0(t) \sim a(t)$ and even $a_0(t)/a(t) - 1 = o(A_0(t))$, $t \to \infty$. Hence (5.2.3) implies

$$\sqrt{k}\left(\frac{\hat{a}\left(\frac{n}{k}\right)}{a_0\left(\frac{n}{k}\right)}-1\right)-\alpha\left(W_n\right)\xrightarrow{P}0\,.$$

The left-hand side of (5.2.11) is, according to (5.2.1) and Lemma 5.2.4,

$$\frac{a_0\left(\frac{n}{k}\right)}{\hat{a}\left(\frac{n}{k}\right)}\sqrt{k}\left(\frac{X_{n-[ks],n}-X_{n-k,n}}{a_0\left(\frac{n}{k}\right)}-\frac{s^{-\gamma}-1}{\gamma}\right)\tag{5.2.12}$$

$$-\frac{a_0\left(\frac{n}{k}\right)}{\hat{a}\left(\frac{n}{k}\right)}\sqrt{k}\left(\frac{\hat{a}\left(\frac{n}{k}\right)}{a_0\left(\frac{n}{k}\right)}-1\right)\frac{s^{-\gamma}-1}{\gamma}\tag{5.2.13}$$

$$-\sqrt{k}\left(\frac{s^{-\hat{\gamma}}-1}{\hat{\gamma}}-\frac{s^{-\gamma}-1}{\gamma}\right)\tag{5.2.14}$$

$$=(1+O_P(1))\left\{s^{-\gamma-1}W_n(s)-W_n(1)+\sqrt{k}A_0\left(\frac{n}{k}\right)\right.$$
$$\left.\times\left(\overline{\Psi}_{\gamma,\rho}(s^{-1})-\overline{\Psi}_{\gamma,\rho}(1)\right)+o_P(1)s^{-\gamma-1/2-\varepsilon}\right\}$$
$$-(1+o_P(1))\,\alpha(W_n)\int_s^1 u^{-\gamma-1}\log u\,du$$
$$+O_P(1)\left(s^{-|\hat{\gamma}-\gamma|}-1\right)\int_s^1 u^{-\gamma-1}|\log u|\,du\,.$$

By Lemma 5.2.3 the error term connected with (5.2.12) is $o_P(1)s^{-\gamma-1/2-\varepsilon}$. The error term connected with (5.2.13) is $o_P(1)(s^{-\gamma}-1)/\gamma$. The error term connected with (5.2.14) is

$$O_P(1)\left(s^{-|\hat{\gamma}-\gamma|}-1\right)\int_s^1 u^{-\gamma-1}|\log u|\,du\,.$$

Now

$$s^{-|\hat{\gamma}-\gamma|}-1=|\hat{\gamma}-\gamma|\int_s^1 u^{-|\hat{\gamma}-\gamma|-1}\,du\le|\hat{\gamma}-\gamma|s^{-|\hat{\gamma}-\gamma|}|\log s|,$$

i.e., for $\varepsilon>0$,

$$s^{-|\hat{\gamma}-\gamma|}-1=o_P(1)\,s^{-\varepsilon}\,.$$

Moreover, for $\gamma>0$, $\gamma<0$, and $\gamma=0$ one checks that

$$\int_s^1 u^{-\gamma-1}|\log u|\,du\le c\,(\log s)^2\,s^{-\gamma_+}\,.$$

Hence the combined error term is $o_P(1)\,s^{-\gamma_++1/2-\varepsilon}$. ∎

Proof (of Theorem 5.2.1). (1) Lemma 5.2.5 implies

$$I_{k,n}=o_P(1)\int_0^1\left(s^{-\gamma-1/2-\varepsilon}\vee1\right)^2 s^{2\hat{\gamma}_++1}\,ds$$
$$=o_P(1)\int_0^1\left(s^{-\gamma-1/2-\varepsilon}\vee1\right)^2 s^{-\varepsilon}s^{2\gamma_++1}\,ds.$$

Since the integral is finite $I_{k,n} \to^P 0$.

(2) We write the right-hand side of (5.2.11) as $\alpha(W_n) + o_P(1) \, s^{-\gamma_+ - 1/2 - \varepsilon}$. Hence

$$k \left(\frac{X_{n-[ks],n} - X_{n-k,n}}{\hat{a}\left(\frac{n}{k}\right)} - \frac{s^{-\hat{\gamma}} - 1}{\hat{\gamma}} \right)^2 s^{2\hat{\gamma}_+ + 1}$$

$$\leq \left(2 \left(\alpha(W_n) \right)^2 + 2\varepsilon s^{-2\gamma_+ - 1 - 2\varepsilon} \right) s^{-\varepsilon} s^{2\gamma_+ + 1},$$

which is integrable (and note that the distribution of $\alpha(W_n)$ does not depend on n). Hence by Lebesgue's theorem on dominated convergence (and Skorohod construction), $k I_{k,n} \to^d I_\gamma$, as $n \to \infty$.   ∎

Simulations seem to tell us (cf. Hüsler and Li (2005)) that this quite natural test does not perform as well as a similar one involving the logarithms of the observations (Dietrich, de Haan, and Hüsler (2002)). The background is the following. The domain of attraction condition

$$\lim_{t \to \infty} \frac{U(tx) - U(t)}{a(t)} = \frac{x^\gamma - 1}{\gamma}, \quad x > 0,$$

is easily seen to be equivalent (provided that $U(\infty) > 0$) to

$$\lim_{t \to \infty} \frac{\log U(tx) - \log U(t)}{a(t)/U(t)} = \frac{x^{\gamma_-} - 1}{\gamma_-}$$

and

$$\lim_{t \to \infty} \frac{a(t)}{U(t)} = \gamma_+ .$$

Now the moment estimator (Section 3.5) provides separate estimators $\hat{\gamma}_H$ and $\hat{\gamma}_-$ for γ_+ and γ_-. Lemma 3.5.1 states that for $\gamma \in \mathbb{R}$, the estimator $\hat{\gamma}_H$, which in fact is the Hill estimator from Section 3.2, satisfies

$$\frac{\hat{\gamma}_H}{a\left(\frac{n}{k}\right)/U\left(\frac{n}{k}\right)} \xrightarrow{P} \frac{1}{1 - \gamma_-}$$

provided $k = k(n) \to \infty$, $k/n \to 0$ as $n \to \infty$. This suggests that one use the following test statistic:

$$D_{k,n} := \int_0^1 \left(\frac{\log X_{n-[ks],n} - \log X_{n-k,n}}{\hat{\gamma}_H} - \frac{s^{-\hat{\gamma}_-} - 1}{\hat{\gamma}_-} \left(1 - \hat{\gamma}_- \right) \right)^2 s^2 \, ds,$$

where $\hat{\gamma}_H$ and $\hat{\gamma}_-$ are as in Section 3.5, with $\hat{\gamma}_H$ the Hill estimator and $\hat{\gamma}_-$ the one in Remark 3.5.7.

We have the following result.

Theorem 5.2.7 *Let $X_1, X_2, \ldots$ be i.i.d. random variables and suppose that their distribution function F is in the domain of attraction of some extreme value distribution.*

1. *Let $k = k(n) \to \infty$, $k/n \to 0$ as $n \to \infty$. Then*

$$D_{k,n} \overset{P}{\to} 0 .$$

2. *Suppose that moreover, the second-order condition for $\log U$ (3.5.11) (cf. also Lemma B.3.16) holds. Let $k = k(n)$ be such that*

$$\lim_{n \to \infty} \sqrt{k}\, Q\left(\frac{n}{k}\right) = 0 .$$

Then $k D_{k,n}$ converges in distribution to

$$D_\gamma := \int_0^1 \left\{ (1 - \gamma_-)\left(s^{-\gamma_- - 1} W(s) - W(1)\right) - (1 - \gamma_-)^2 \frac{s^{-\gamma_-} - 1}{\gamma_-} P(W) \right.$$
$$\left. + \frac{s^{-\gamma_-} - 1}{\gamma_-} R(W) + (1 - \gamma_-)\, R(W) \int_s^1 u^{-\gamma_- - 1} \log u\; du \right\}^2 s^2\, ds$$

with W Brownian motion and

$$P(W) := \int_0^1 s^{-\gamma_- - 1} W(s) - W(1)\; ds,$$

$$Q(W) := 2 \int_0^1 \frac{s^{-\gamma_-} - 1}{\gamma_-} \left(s^{-\gamma_- - 1} W(s) - W(1)\right) ds,$$

$$R(W) := (1 - \gamma_-)^2 (1 - 2\gamma_-) \left\{\left(\frac{1}{2} - \gamma_-\right) Q(W) - 2 P(W)\right\} .$$

Table 5.1. Quantiles of the asymptotic test statistic $k D_{k,n}$.

γ	.10	0.30	0.50	0.70	0.90	0.95	0.9750	0.99
≥ 0	.028	.042	.057	.078	.122	.150	.181	.222
-0.1	.027	.041	.054	.074	.116	.144	.174	.213
-0.2	.027	.040	.053	.072	.114	.141	.169	.208
-0.3	.027	.040	.054	.073	.113	.140	.168	.206
-0.4	.027	.040	.054	.073	.114	.141	.169	.207
-0.5	.027	.040	.054	.073	.115	.141	.169	.208
-0.6	.027	.040	.054	.074	.116	.144	.173	.212
-0.7	.028	.041	.055	.074	.118	.147	.176	.218

For the proof we need some auxiliary results.

Lemma 5.2.8 *Under the conditions of Theorem 5.2.7(2), for each $\varepsilon > 0$*

$$\sup_{0<s\leq 1} \min\left(1, s^{\gamma_-+1/2+\varepsilon}\right) \left| \sqrt{k}\left(\frac{\log X_{n-[ks],n} - \log X_{n-k,n}}{q_0\left(\frac{n}{k}\right)} - \frac{s^{-\gamma_-}-1}{\gamma_-}\right) \right.$$

$$\left. - s^{-\gamma_- -1} W_n(s) + W_n(1) \right| \xrightarrow{P} 0 ,$$

with q_0 from (3.5.13).

Proof. Apply Corollary 2.4.6 with the random variables $X_1, \ldots, X_n$ replaced with $\log X_1, \ldots, \log X_n$. ∎

Lemma 5.2.9 *Under the assumptions of Theorem 5.2.7(2),*

$$\sqrt{k}\left(\frac{\hat{\gamma}_H}{a\left(\frac{n}{k}\right)/U\left(\frac{n}{k}\right)} - \frac{1}{1-\gamma_-}\right) - P(W_n) \xrightarrow{P} 0,$$

$$\sqrt{k}\left(\hat{\gamma}_- - \gamma_-\right) - R(W_n) \xrightarrow{P} 0,$$

with

$$P(W_n) := \int_0^1 s^{-\gamma_- -1} W_n(s) - W_n(1)\, ds,$$

$$R(W_n) := (1-\gamma_-)^2(1-2\gamma_-)$$

$$\times \left\{-2P_n + (1-2\gamma_-)\int_0^1 \frac{s^{-\gamma_-}-1}{\gamma_-}\left(s^{-\gamma_- -1} W_n(s) - W_n(1)\right)\, ds\right\}$$

and W_n is Brownian motion for all n.

Proof. We use Lemma 5.2.8 with $0 < \varepsilon < \frac{1}{2}$:

$$\frac{\hat{\gamma}_H}{q_0\left(\frac{n}{k}\right)} = \int_0^1 \frac{\log X_{n-[ks],n} - \log X_{n-k,n}}{q_0\left(\frac{n}{k}\right)}\, ds$$

$$= \int_0^1 \frac{s^{-\gamma_-}-1}{\gamma_-}\, ds + \frac{1}{\sqrt{k}}\left\{\int_0^1 s^{-\gamma_- -1} W_n(s) - W_n(1)\, ds\right.$$

$$\left. + o_P(1)\int_0^1 s^{-\gamma_- -1/2-\varepsilon}\, ds\right\} .$$

It follows that

$$\sqrt{k}\left(\frac{\hat{\gamma}_H}{q_0\left(\frac{n}{k}\right)} - \frac{1}{1-\gamma_-}\right) - P(W_n) \xrightarrow{P} 0 . \qquad (5.2.15)$$

Next note that by Remark B.3.2, (3.5.11) holds with q replaced by q_0, since

$$\frac{q_0\left(\frac{n}{k}\right)}{q\left(\frac{n}{k}\right)} - 1 = o\left(Q\left(\frac{n}{k}\right)\right) .$$

Then (5.2.15) follows with q_0 replaced by q.

Similarly,

$$
\begin{aligned}
\frac{M_n^{(2)}}{\left(q_0\left(\frac{n}{k}\right)\right)^2} &= \int_0^1 \left(\frac{\log X_{n-[ks],n} - \log X_{n-k,n}}{q_0\left(\frac{n}{k}\right)}\right)^2 ds \\
&= \int_0^1 \left(\frac{s^{-\gamma_-} - 1}{\gamma_-}\right)^2 ds \\
&\quad + \frac{2}{\sqrt{k}} \int_0^1 \frac{s^{-\gamma_-} - 1}{\gamma_-} \left(s^{-\gamma_- - 1} W_n(s) - W_n(1)\right) ds + \frac{o_P(1)}{\sqrt{k}} .
\end{aligned}
$$

Hence

$$
\begin{aligned}
\sqrt{k} &\left(\frac{M_n^{(2)}}{\left(q_0\left(\frac{n}{k}\right)\right)^2} - \frac{2}{(1-\gamma_-)(1-2\gamma_-)}\right) \\
&- 2\int_0^1 \frac{s^{-\gamma_-} - 1}{\gamma_-} \left(s^{-\gamma_- - 1} W(s) - W_n(1)\right) ds \xrightarrow{P} 0 .
\end{aligned}
$$

Since $\hat{\gamma}_- = 1 - 2^{-1}(1 - \hat{\gamma}_H^2/M_n^{(2)})^{-1}$ (cf. Remark 3.5.7), Cramér's delta method finishes the proof. ∎

Lemma 5.2.10 *Under the conditions of Theorem 5.2.7(1),*

$$
\begin{aligned}
\frac{s^{-\hat{\gamma}_-} - 1}{\hat{\gamma}_-}(1 - \hat{\gamma}_-) &- \frac{s^{-\gamma_-} - 1}{\gamma_-}(1 - \gamma_-) \\
&= -(\hat{\gamma}_- - \gamma_-)(1 - \gamma_-)\int_s^1 u^{-\gamma_- - 1} \log u \, du \\
&\quad + \left|\hat{\gamma}_- - \gamma_-\right| O_P(1)(\log s)^2 \left(s^{-|\hat{\gamma}_- - \gamma_-|} - 1\right) ,
\end{aligned}
$$

where the $O_P(1)$ term is bounded uniformly for $s \in (0, 1]$.

Proof. Apply Lemmas 5.2.4 and 5.2.9. ∎

Proof (of Theorem 5.2.7). The proof is similar to that proof of Theorem 5.2.1, now with the use of Lemmas 5.2.8–5.2.10. It is left to the reader. ∎

Remark 5.2.11 Here and in the next theorem a similar result can be proved when one replaces $s^2 ds$ in the definition of $D_{k,n}$ with $s^\eta ds$ as long as $\eta > 0$. Hüsler and Li (2005) recommended the value $\eta = 2$.

In the special case that only positive values of gamma are possible (for example if the distribution is not bounded above), a simpler test can be used.

Theorem 5.2.12 *Let $X_1, X_2, \ldots$ be i.i.d. random variables with distribution function F. Suppose F is in the domain of attraction of an extreme value distribution G_γ with $\gamma > 0$. Define*

$$S_{k,n} := \int_0^1 \left(\frac{\log X_{n-[ks],n} - \log X_{n-k,n}}{\hat{\gamma}_H} + \log s \right)^2 s^\eta \, ds,$$

where $\hat{\gamma}_H$ is the Hill estimator from Section 3.2.

1. If $k = k(n) \to \infty$, $k/n \to 0$, $n \to \infty$, then

$$S_{k,n} \overset{P}{\to} 0 .$$

2. Suppose in addition that a second-order condition holds: for some positive or negative function α and all $x > 0$ with $\rho \le 0$,

$$\lim_{t\to\infty} \frac{\frac{1-F(tx)}{1-F(t)} - x^{-1/\gamma}}{\alpha(t)} = x^{-1/\gamma} \frac{x^{\rho/\gamma} - 1}{\gamma\rho} . \tag{5.2.16}$$

Then for sequences $k = k(n) \to \infty$, $\sqrt{k}\alpha(U(n/k)) \to 0$, where U is the inverse function of $1/(1 - F)$,

$$k S_{k,n} \overset{d}{\to} S$$

as $n \to \infty$, where

$$S := \int_0^1 \left(B(s) + s \log s \int_0^1 u^{-1} B(u) \, du \right)^2 s^{\eta-1} ds$$

with B Brownian bridge.

Proof. The proof of part (1) is as before. Next note that the second-order relation (5.2.16) implies (cf. Theorem 2.4.8)

$$\limsup_{0<s\le 1} s^{1/2+\varepsilon} \left| \sqrt{k} \left(\frac{\log X_{n-[ks],n} - \log X_{n-k,n}}{\gamma} + \log s \right) \right.$$
$$\left. -s^{-1} W_n(s) + W_n(1) - \sqrt{k}A_0 \left(\frac{n}{k}\right) \frac{1}{\gamma} \frac{s^{-\rho} - 1}{\rho} \right| \overset{P}{\to} 0. \tag{5.2.17}$$

The rest of the proof is similar to that of the previous theorem. ∎

According to Hüsler and Li (2005) this test does not perform so well as the others.

Next we discuss the behavior of the test of Theorem 5.2.12 under two types of alternatives.

Example 5.2.13 (Super-heavy tails) Let $F(x) = 1 - (\log x)^{-\beta}$ for $x \ge e$ and β a positive parameter. Note that $\log X = Y^{1/\beta}$, where Y has distribution function $1 - 1/x$, $x \ge 1$. Then with $\alpha = 1/\beta$ (cf. proof of Lemma 3.2.3),

$$S_{k,n} = \int_0^1 \left(\frac{(Y_{n-[ks],n}/Y_{n-k,n})^\alpha - 1}{\frac{1}{k}\sum_{i=0}^{k-1} (Y_{n-i,n}/Y_{n-k,n})^\alpha - 1} + \log s \right)^2 s^2 \, ds.$$

$$\stackrel{d}{=} \int_0^1 \left(\frac{Y_{k-[ks],k}^{\alpha} - 1}{\frac{1}{k} \sum_{i=0}^{k-1} Y_{k-i,k}^{\alpha} - 1} + \log s \right)^2 s^2 \, ds \,.$$

For $0 < \alpha < 1$,

$$\frac{1}{k} \sum_{i=0}^{k-1} Y_{k-i,k}^{\alpha} \stackrel{d}{=} \frac{1}{k} \sum_{i=1}^{k} (Y_i^{\star})^{\alpha}$$

with $Y_i^{\star}$ independent and identically distributed, $1 - 1/x$, $x \geq 1$; hence

$$\frac{1}{k} \sum_{i=0}^{k-1} Y_{k-i,k}^{\alpha} \stackrel{P}{\to} \frac{1}{1-\alpha} \,.$$

Further, by the proof of Lemma 2.4.10, lines 4–5,

$$Y_{k-[ks],k}^{\alpha} = s^{-\alpha} \left(1 + O_P \left(\frac{1}{\sqrt{k}} \right) \right)$$

uniformly on $\left[k^{-1}, 1 \right]$. Hence

$$S_{k,n} \stackrel{P}{\to} \int_0^1 \left((s^{-\alpha} - 1)(1 - \alpha) + \log s \right)^2 s^2 \, ds > 0 \,.$$

Example 5.2.14 (Periodicity) Let $E_1, E_2, E_3, \dots$ be independent and identically distributed random variables with standard exponential distribution. Then the distribution of $\exp(E_1)$ is in the domain of attraction of some extreme value distribution. But the distribution of

$$X := \exp[E_1]$$

(the exponent of the integer part) is not in any domain of attraction. We show that the statistic $S_{k,n}$ does not tend to zero as $n \to \infty$ for all sequences $k = k(n) \to \infty$, $k/n \to 0$. Let $X_{1,n} \leq X_{2,n} \leq \cdots \leq X_{n,n}$ be the order statistics of a sample from this distribution. Then $\{X_{i,n}\}_{i=1}^n \stackrel{d}{=} \{\exp[E_{i,n}]\}_{i=1}^n$, where the $E_{i,n}$ are the order statistics for the exponential distribution. Hence

$$\log X_{n-[ks],n} - \log X_{n-k,n} + \hat{\gamma}_H \log s$$

$$= \left[E_{n-[ks],n} \right] - \left[E_{n-k,n} \right] + (\log s) \frac{1}{k} \sum_{i=0}^{k-1} \left(\left[E_{n-i,n} \right] - \left[E_{n-k,n} \right] \right)$$

$$= \left[(E_{n-[ks],n} - E_{n-k,n} + \log s) + E_{n-k,n} - \left[E_{n-k,n} \right] - \log s \right]$$

$$+ (\log s) \frac{1}{k} \sum_{i=0}^{k-1} \left[(E_{n-i,n} - E_{n-k,n}) + E_{n-k,n} - \left[E_{n-k,n} \right] \right] \,.$$

Now let us take a subsequence $k = k(n)$ such that

$$E_{n-k,n} - [E_{n-k,n}] \approx \log \frac{n}{k} - \left[\log \frac{n}{k}\right] \to 0$$

in probability as $n \to \infty$. Moreover, apply Lemma 2.4.10. Then the above becomes

$$\left(\frac{1}{\sqrt{k}} s^{-1/2-\varepsilon} O_P(1) + o_P(1) - \log s\right) + (\log s)\frac{1}{k}\sum_{i=0}^{k-1}[E_i + o_P(1)] ,$$

with $E_1, E_2, E_3, \ldots$ as before. Note that the expectation of $[E_1 + p]$ is $e^p/(e-1)$. Hence, as $n \to \infty$, and for this sequence $k = k(n)$,

$$S_{k,n} \xrightarrow{P} \int_0^1 \left([-\log s] + (e-1)^{-1}\log s\right)^2 s^2 \, ds > 0 .$$

Simulation results can be found in Hüsler and Li (2005).

We conclude with a version of the test using the tail empirical distribution function rather than the tail empirical quantile function. This is closer to the usual goodness-of-fit tests of Cramér–von Mises and Andersen–Darling. The proof is omitted (cf. Drees, de Haan, and Li (2006)); it is similar to but more complicated than that of Theorem 5.1.2, now using the asymptotic theory of Section 5.1.

Theorem 5.2.15 *Under the conditions and with the notation of Theorem 5.2.1(2), with*

$$T_{k,n} := \int_0^1 \left\{ \frac{n}{k}\left\{1 - F_n\left(\hat{a}\left(\frac{n}{k}\right)\frac{x^{-\hat{\gamma}} - 1}{\hat{\gamma}} + X_{n-k,n}\right)\right\} - x\right\}^2 x^{\eta-2} \, dx$$

we have

$$kT_{k,n} - \int_0^1 \left(W_n(x) + L_n^{(\gamma)}(x)\right)^2 x^{\eta-2} dx \xrightarrow{P} 0$$

for all $\eta > 0$ if not $\gamma = \rho = 0$, and all $\eta \geq 1$ if $\gamma = \rho = 0$. Here

$$L_n^{(\gamma)}(x) := \begin{cases} \frac{1}{\gamma}x\left(\frac{1}{\gamma}\Gamma(W_n) - \alpha(W_n)\right) + \frac{1}{\gamma}\Gamma(W_n)x\log x \\ \qquad - \frac{1}{\gamma}x^{1+\gamma}\left(\gamma W_n(1) + \frac{1}{\gamma}\Gamma(W_n) - \alpha(W_n)\right) , & \gamma \neq 0, \\ x\left(-W_n(1) - \frac{1}{2}\Gamma(W_n)\log^2 x + \alpha(W_n)\log x\right) , & \gamma = 0 . \end{cases}$$

Hüsler and Li recommend the value $\eta = 1$.

For more information about these tests we refer to Hüsler and Li (2005). In particular, for applications they suggest: (i) to estimate the extreme value index γ based on both estimators maximum likelihood and moment; (ii) if the extreme value index can be believed to be positive (for example, both estimators of γ are larger than 0.05), then it might be better to use the test statistic $kT_{k,n}$; otherwise use the test $kD_{k,n}$.

Table 5.2. Quantiles of the asymptotic test statistic $kT_{k,n}$ with $\eta = 1$ and the maximum likelihood estimators.

γ	.10	0.30	0.50	0.70	0.90	0.95	0.9750	0.99
4	.086	.123	.161	.212	.322	.393	.462	.558
3	.085	.120	.156	.205	.307	.372	.440	.532
2	.083	.116	.150	.195	.286	.344	.402	.489
1.5	.082	.115	.148	.192	.282	.340	.400	.480
1	.082	.114	.146	.189	.276	.330	.388	.466
0.5	.083	.116	.149	.194	.285	.343	.404	.481
0.25	.085	.119	.153	.120	.295	.355	.415	.499
0	.089	.126	.163	.213	.319	.388	.455	.542
-0.1	.091	.129	.168	.221	.330	.400	.471	.569
-0.2	.093	.133	.174	.231	.350	.425	.500	.604
-0.3	.096	.139	.183	.242	.369	.449	.531	.653
-0.4	.100	.145	.192	.256	.393	.484	.576	.690
-0.45	.103	.150	.199	.320	.416	.511	.605	.735
-0.499	.107	.157	.210	.338	.439	.546	.652	.799

5.3 Convergence of Moments, Speed of Convergence, and Large Deviations

In this section we shall see that the domain of attraction conditions also imply convergence of the relevant moments of normalized sample maxima. Next we shall prove that the second-order condition allows for a precise uniform speed of convergence result, which in turn will imply a large deviations result.

Let $X_1, X_2, \ldots$ be independent and identically distributed random variables with distribution function F and assume that F is in the domain of attraction of an extreme value distribution G_γ, $\gamma \in \mathbb{R}$. Let $X_{n,n} := \max(X_1, X_2, \ldots, X_n)$, $n = 1, 2, \ldots$.

5.3.1 Convergence of Moments

We have seen in Chapter 1 (Exercise 1.16) that if $F \in D(G_\gamma)$ with $\gamma \le 0$, then $E|X|^\alpha 1_{\{X>x\}}$ is finite for all $\alpha > 0$ and all $x < U(\infty)$, where $U(\infty)$ is the right endpoint of the distribution (recall that $U := (1/(1-F))^{\leftarrow}$). However, if $F \in D(G_\gamma)$, $\gamma > 0$, then

$$E|X|^\alpha 1_{\{X>x\}} < \infty \tag{5.3.1}$$

for $0 < \alpha < 1/\gamma$, but is infinite for $\alpha > 1/\gamma$. This defines the scope of possible convergence of moments in extreme value theory.

Recall that under the domain of attraction condition for some positive function a,

$$\lim_{t \to \infty} \frac{U(tx) - U(t)}{a(t)} = \frac{x^\gamma - 1}{\gamma}$$

for all $x > 0$. Then

$$\lim_{n \to \infty} P\left(\frac{X_{n,n} - U(n)}{a(n)} \leq x\right) = G_\gamma(x) \tag{5.3.2}$$

for all x with $1 + \gamma x > 0$.

Theorem 5.3.1 *Let X be F distributed and assume $F \in D(G_\gamma)$, $\gamma \in \mathbb{R}$. Let k be an integer with $0 < k < 1/\gamma_+$, where $\gamma_+ := \max(0, \gamma)$. Suppose $E|X|^k$ is finite, which in view of (5.3.1) is implied by $E|X|^k 1_{\{X \leq x\}}$ finite for some x for which $0 < F(x) < 1$. Then*

$$\lim_{n \to \infty} E\left(\frac{X_{n,n} - U(n)}{a(n)}\right)^k = \int_{-\infty}^{\infty} x^k \, dG_\gamma(x) . \tag{5.3.3}$$

Proof. Let Z be a random variable with distribution function $\exp(-1/x)$, $x > 0$. By comparing the distribution functions of the left- and right-hand sides of the equation it is easily seen that

$$\frac{X_{n,n} - V(n)}{a(n)} \stackrel{d}{=} \frac{V(nZ) - V(n)}{a(n)}$$

with $V := (1/(-\log F))^{\leftarrow}$.

Next note that (5.3.2) is equivalent to

$$\lim_{n \to \infty} n \left(-\log F(U(n) + xa(n))\right) = (1 + \gamma x)^{-1/\gamma}$$

for all x with $1 + \gamma x > 0$. Application of Lemma 1.1.1, Theorem 1.1.2, and Lemma 1.2.12 gives

$$\lim_{t \to \infty} \frac{V(tx) - V(t)}{a(t)} = \frac{x^\gamma - 1}{\gamma}$$

for all $x > 0$. By Theorem B.2.18 for $\varepsilon, \varepsilon' > 0$ there exists t_0 such that for $t, tx \geq t_0$,

$$x^{-\gamma} e^{-\varepsilon'|\log x|} \left| \frac{V(tx) - V(t)}{a_0(t)} - \frac{x^\gamma - 1}{\gamma} \right| < \varepsilon \tag{5.3.4}$$

for some a_0 satisfying $a_0(t) \sim a(t)$, $t \to \infty$. We write for $n \geq t_0$,

$$E\left(\frac{V(nZ) - V(n)}{a_0(n)}\right)^k$$

$$= E\left(\frac{V(nZ) - V(n)}{a_0(n)}\right)^k 1_{\{nZ \geq t_0\}} + E\left(\frac{V(nZ) - V(n)}{a_0(n)}\right)^k 1_{\{nZ < t_0\}}$$

$$=: \mathrm{I} + \mathrm{II}.$$

For I we use inequalities (5.3.4). The upper bound is (note that $|a + b|^k \leq 2^k(|a|^k + |b|^k)$)

$$2^k \int_0^\infty \left|\frac{x^\gamma - 1}{\gamma}\right|^k d\left(e^{-1/x}\right) + (2\varepsilon)^k \int_0^\infty x^{k\gamma} e^{\varepsilon'k|\log x|} d\left(e^{-1/x}\right),$$

which is finite. Hence by Lebesgue's theorem on dominated convergence the limit of part I is

$$E\left(\frac{Z^\gamma - 1}{\gamma}\right)^k = \int_{-\infty}^{\infty} x^k \, dG_\gamma(x) .$$

For II we proceed as follows. It is bounded by

$$\int_0^{t_0/n} \left|\frac{V(nx) - V(n)}{a_0(n)}\right|^k d\left(e^{-1/x}\right)$$

$$= \int_0^{t_0} \left|\frac{V(x) - V(n)}{a_0(n)}\right|^k d\left(e^{-n/x}\right)$$

$$= \int_0^{t_0} \left|\frac{V(x) - V(n)}{a_0(n)}\right|^k ne^{-(n-1)/x} \, d\left(e^{-1/x}\right)$$

$$\leq ne^{-(n-1)/t_0} \, 2^k \left(\int_0^{\infty} \left|\frac{V(x)}{a_0(n)}\right|^k d\left(e^{-1/x}\right) + \left|\frac{V(n)}{a_0(n)}\right|^k \int_0^{\infty} d\left(e^{-1/x}\right)\right)$$

$$\leq \left(ne^{-(n-1)/t_0}\right) 2^k \frac{E|x|^k + |V(n)|^k}{|a_0(n)|^k} .$$

Since the sequences $V(n)$ and $a(n)$ are of polynomial growth and since the first factor tends to zero exponentially fast, part II tends to zero.

Since $(-\log F(x))/(1-F(x)) \to 1$, as $x \uparrow U(\infty)$, we get $(U(n)-V(n))/a_0(n))^k \to 0$, $n \to \infty$ (Lemma 1.2.12, Chapter 1). Hence $E((X_{n,n} - U(n))/a_0(n))^k = E((X_{n,n} - V(n))/a_0(n) + \varepsilon_n)^k$ with $\varepsilon_n \to 0$, $n \to \infty$. By going through the proof again with this modification, it is easy to see that also $E((X_{n,n} - V(n))/a_0(n) + \varepsilon_n)^k \to \int_{-\infty}^{\infty} x^k G_\gamma(x)$ for any $\varepsilon_n \to 0$. It follows that (5.3.3) holds with $U(n)$ replaced by $V(n)$. Finally, note that changing from a_0 to a does not affect the result. This finishes the proof. ■

For $\gamma \neq 0$ we have somewhat simpler results. The proof is very similar to the proof of Theorem 5.3.1 and it is omitted.

Theorem 5.3.2 *Suppose that the conditions of Theorem 5.3.1 hold.*

1. *If $\gamma > 0$,*

$$\lim_{n\to\infty} E\left(\frac{X_{n,n}}{U(n)}\right)^k = \int_0^{\infty} x^k \, d\left\{\exp\left(-x^{-1/\gamma}\right)\right\}.$$

2. *If $\gamma < 0$ (note that $U(\infty) := \lim_{x\to\infty} U(x)$ is finite in this case),*

$$\lim_{n\to\infty} E\left(\frac{X_{n,n} - U(\infty)}{U(\infty) - U(n)}\right)^k = \int_{-\infty}^{0} x^k \, d\left\{\exp\left(-(-x)^{1/\gamma}\right)\right\}.$$

5.3.2 Speed of Convergence; Large Deviations

Let $X_1, X_2, \ldots$ be a sequence of independent and identically F distributed random variables. The basic convergence in extreme value theory is that if the extreme value conditions holds, then for some sequence of constants $a_n > 0$ and b_n real, $n = 1, 2, \ldots$, and some $\gamma \in \mathbb{R}$,

$$\lim_{n \to \infty} P\left(\frac{\max(X_1, X_2, \ldots, X_n) - b_n}{a_n} \leq x\right) = \exp\left(-(1 + \gamma x)^{-1/\gamma}\right) =: G_\gamma(x)$$

for all x with $1 + \gamma x > 0$. Then by the continuity of the limit distribution function,

$$\lim_{n \to \infty} \sup_{x \in \mathbb{R}} \left|F^n(a_n x + b_n) - G_\gamma(x)\right| = 0 . \tag{5.3.5}$$

The speed of convergence in (5.3.5) is not the same for all distributions in some domain of attraction. For example, the convergence rate for the exponential distribution is of order n^{-1} (Hall and Wellner (1979)), but for the normal distribution it is of order $(\log n)^{-1}$ (de Haan and Resnick (1996)). Rootzén (1984) proves that if the convergence rate is faster than exponential, the initial distribution must be an extreme value distribution. In fact, the convergence rate depends on the *second-order* behavior. This came out for the first time in a paper by Smith (1982). As we shall see, the second-order condition is sufficient for a uniformly weighted version of a second-order expansion for F^n.

We are going to assume that the function $V := (1/(-\log F))^{\leftarrow} = F^{\leftarrow}(e^{-1/t})$ satisfies the second-order condition of Section 2.3. Consequently, for some $\gamma \in \mathbb{R}$, $\rho \leq 0$, and all $\varepsilon, \delta > 0$ there exists $t_0 = t_0(\varepsilon, \delta) > 0$ such that for all $t, tx \geq t_0$,

$$x^{-(\gamma+\rho)} e^{-\delta|\log x|} \left| \frac{\frac{V(tx)-V(t)}{a_0(t)} - \frac{x^\gamma - 1}{\gamma}}{A_0(t)} - \Psi_{\gamma,\rho}(x) \right| < \varepsilon \tag{5.3.6}$$

with

$$\Psi_{\gamma,\rho}(x) := \begin{cases} \frac{x^{\gamma+\rho}-1}{\gamma+\rho} , & \rho < 0, \\ \gamma^{-1} x^\gamma \log x , & \rho = 0 \neq \gamma, \\ 2^{-1} (\log x)^2 , & \rho = 0 = \gamma, \end{cases} \tag{5.3.7}$$

and where the functions a_0 and $A_0(t)$ are from Theorem B.3.10.

In order to get a uniform rate of convergence in (5.3.5) we have to choose the normalizing constants a_n and b_n in a special way. For the first take $a_0(n)$ with the function a_0 from (5.3.6), and for the second,

$$b_0(n) = \begin{cases} V(n) , & \gamma \geq 0, \\ V(\infty) + \gamma^{-1} a_0(n) , & \gamma < 0 = \rho, \\ V(\infty) + \gamma^{-1} a_0(n) + (\gamma + \rho)^{-1} a_0(n) A_0(n) , & \gamma < 0 , \rho < 0 , \end{cases}$$

with $A_0(n)$ again from (5.3.6). Further, define the function $\overline{\overline{\Psi}}_{\gamma,\rho}$ by

$$\overline{\overline{\Psi}}_{\gamma,\rho}(x) := \begin{cases} \Psi_{\gamma,\rho}(x) + (\gamma + \rho)^{-1} , & \rho < 0 \neq \gamma + \rho, \\ \Psi_{\gamma,\rho}(x) , & \text{otherwise} , \end{cases}$$

Theorem 5.3.3 *Suppose the function* $V := (1/(-\log F))^{\leftarrow}$ *satisfies the second-order condition, so that (5.3.6) holds. Then*

$$\lim_{n\to\infty} \frac{F^n(a_0(n)x + b_0(n)) - G_\gamma(x)}{A_0(n)} = -J\left(w_\gamma(x)\right) \tag{5.3.8}$$

uniformly for $x \in \mathbb{R}$, *where for* $x \in \mathbb{R}$,

$$w_\gamma(x) = ((1 + \gamma x) \vee 0)^{1/\gamma} ,$$

and for $x > 0$,

$$J(x) = x^{-(1+\gamma)} e^{-1/x} \overline{\overline{\Psi}}_{\gamma,\rho}(x),$$

and $J(0)$ *and* $J(\infty)$ *are defined by continuity, i.e.,* $J(0) = J(\infty) = 0$.

Remark 5.3.4 The second-order condition for this theorem is imposed on the function $V := (1/(-\log F))^{\leftarrow}$, whereas in Theorem 5.1.1, for example, it is imposed on the function $U := (1/(1 - F))^{\leftarrow}$. The relation between these two conditions is discussed in Drees, de Haan, and Li (2003).

For the proof we need some lemmas.

Lemma 5.3.5 *Assume the conditions of Theorem 5.3.3. For any* $\varepsilon, \delta > 0$, *there exists* $n_0 > 0$ *such that*

$$\left| \frac{p_{n,\gamma}(x)}{A_0(n)} - \overline{\overline{\Psi}}_{\gamma,\rho}(x) \right| \leq \varepsilon \max\left(x^{\gamma+\rho+\delta}, x^{\gamma+\rho-\delta}\right) \tag{5.3.9}$$

for all $n, nx > n_0$ *with*

$$p_{n,\gamma}(x) = \frac{V(nx) - b_0(n)}{a_0(n)} - \frac{x^\gamma - 1}{\gamma} ,$$

for all $x > 0$.

Proof. It follows from (5.3.6) that for any $\varepsilon, \delta > 0$, there exists $n_0 = n_0(\varepsilon, \delta) > 0$ such that for all $n, nx \geq n_0$,

$$\left| \frac{\frac{V(nx)-V(n)}{a_0(n)} - \frac{x^\gamma-1}{\gamma}}{A_0(n)} - \Psi_{\gamma,\rho}(x) \right| \leq \varepsilon \max\left(x^{\gamma+\rho+\delta}, x^{\gamma+\rho-\delta}\right) .$$

This gives (5.3.9) for $\gamma \geq 0$. For $\gamma < 0$, it is easily checked by the definitions of $a_0(n)$, $A_0(n)$, and $b_0(n)$ that

$$\frac{V(n) - b_0(n)}{a_0(n) A_0(n)} = \begin{cases} 0, & \rho = 0, \\ -(\gamma + \rho)^{-1}, & \rho < 0, \end{cases}$$

holds for all n, so that (5.3.9) follows also.   ∎

Lemma 5.3.6 *Under the conditions of Theorem 5.3.3,*

$$\lim_{n\to\infty} \sup_{\alpha_n \leq s \leq \beta_n} x^{-(1+\gamma)} e^{-1/x} \left| \frac{p_{n,\gamma}(x)}{A_0(n)} - \overline{\overline{\Psi}}_{\gamma,\rho}(x) \right| = 0 , \tag{5.3.10}$$

$$\lim_{n\to\infty} \sup_{\alpha_n \leq s \leq \beta_n} \frac{p_{n,\gamma}^2(x)}{A_0(n)\, x^{2\gamma}} = 0 , \tag{5.3.11}$$

where $\alpha_n = -\log^{-1}(A_0(n))^2$ and $\beta_n = (A_0(n))^{-2}$.

Proof. Note that $|A_0| \in RV_\rho$ and therefore $A_0^2 \in RV_{2\rho}$ with $\rho \leq 0$. Hence (Proposition B.1.9(5)) there exist a constant $C > 0$ and an integer $n_0 > 0$ such that $A_0^2 \geq C\, n^{2\rho-1}$ for all $n \geq n_0$. Hence $n\,\alpha_n \geq -n((2\rho - 1)\log n + \log C)^{-1} \to \infty$, as $n \to \infty$. By Lemma 5.3.5, the latter implies that (5.3.9) holds for all $x \in [\alpha_n, \beta_n]$ and $n \geq n_0$. Thus we have for $\delta \in (0, 1)$,

$$\sup_{\alpha_n \leq x \leq \beta_n} x^{-(1+\gamma)} e^{-1/x} \left| \frac{p_{n,\gamma}(x)}{A_0(n)} - \overline{\overline{\Psi}}_{\gamma,\rho}(x) \right|$$

$$\leq \varepsilon \sup_{x>0} x^{-(1+\gamma)} e^{-1/x} \max \left(x^{\gamma+\rho+\delta} x^{\gamma+\rho-\delta} \right)$$

$$= \varepsilon \sup_{x>0} e^{-1/x} \max \left(x^{-1+\rho+\delta}, x^{-1+\rho-\delta} \right) ,$$

so that (5.3.10) holds by noting that $\sup_{x>0} e^{-1/x} \max \left(x^{-1+\rho+\delta}, x^{-1+\rho-\delta} \right) < \infty$. Choosing $\delta \in \left(0, \frac{1}{4} \right)$, we have from (5.3.9)

$$I_{n,1} := A_0(n) \sup_{\alpha_n \leq x \leq \beta_n} x^{-2\gamma} \left| \frac{p_{n,\gamma}(x)}{A_0(n)} - \overline{\overline{\Psi}}_{\gamma,\rho}(x) \right|^2$$

$$\leq A_0(n) \sup_{\alpha_n \leq x \leq \beta_n} \max \left(x^{2(\rho+\delta)}, x^{2(\rho-\delta)} \right)$$

$$\leq A_0(n) \max \left((A_0(n))^{-4\delta}, \left(-\log^{-1}(A_0(n))^2 \right)^{2(\rho-\delta)} \right) \to 0 .$$

On the other hand, it is easily seen that

$$I_{n,2} := A_0(n) \sup_{\alpha_n \leq x \leq \beta_n} x^{-2\gamma} \overline{\overline{\Psi}}_{\gamma,\rho}^2(x) \to 0 , \quad n \to \infty .$$

Hence (5.3.11) follows from the inequality

$$\sup_{\alpha_n \leq x \leq \beta_n} \frac{p_{\gamma,\rho}^2(x)}{A_0(n)\, x^{2\gamma}} \leq 2\,(I_{n,1} + I_{n,2}) .$$

Reading $\gamma^{-1} \log(1 + \gamma x) = x$, for all $x \in \mathbb{R}$ and $\gamma = 0$, we denote

$$J_n(x) = G_0\left(\frac{1}{\gamma} \log\left(1 + \gamma \frac{V(nx) - b_0(n)}{a_0(n)}\right) - G_0(\log x)\right), \quad x > 0.$$

Moreover, for any function f on (a, b) with $-\infty \leq a < b \leq \infty$, define $f(a) := \lim_{t \to a} f(t)$ and $f(b) := \lim_{t \to b} f(t)$ if the limits exist, e.g., $J(0) = J(\infty) = 0$. ∎

Lemma 5.3.7 *Under the conditions of Theorem 5.3.3,*

$$\lim_{n \to \infty} \sup_{0 \leq x \leq \infty} \left| \frac{J_n(x)}{A_0(n)} - J(x) \right| = 0. \tag{5.3.12}$$

Proof. We shall prove (5.3.12) only for the case that A_0 is positive near infinity, because the proof for the other case is similar.

Since for every positive integer n and $x > 0$, there exists $\theta = \theta(n, x) \in [0, 1]$ such that

$$J_n(x) = G_0(\log x + q_n(x)) - G_0(\log x) = q_n(x) G_0'(\log x + \theta q_n(x))$$

with $q_n(x) = \gamma^{-1} \log\{1 + \gamma(V(nx) - b_0(n))/a_0(n)\} - \log x$, we have

$$\left| (A_0(n))^{-1} J_n(x) - J(x) \right|$$

$$= \left| (A_0(n))^{-1} q_n(x) G_0'(\log x + \theta q_n(x)) - J(x) \right|$$

$$\leq (A_0(n))^{-1} \left| q_n(x) - x^{-\gamma} p_{n,\gamma}(x) \right| G_0'(\log x + \theta q_n(x))$$

$$+ (A_0(n))^{-1} x^{-\gamma} p_{n,\gamma}(x) \left| G_0'(\log x + \theta q_n(x)) - G_0'(\log x) \right|$$

$$+ x^{-\gamma} G_0'(\log x) \left| (A_0(n))^{-1} p_{n,\gamma}(x) - \overline{\overline{\Psi}}_{\gamma,\rho}(x) \right|$$

$$=: J_{n,1}(x) + J_{n,2}(x) + J_{n,3}(x).$$

Note that for some $\theta_0 \in [0, 1]$,

$$q_n(x) = \gamma^{-1} \log\left\{1 + \gamma (a_0(n))^{-1} (V(nx) - b_0(n))\right\} - \gamma^{-1} \log\left\{1 + \gamma\gamma^{-1}(x^\gamma - 1)\right\}$$

$$= \frac{p_{n,\gamma}(x)}{x^\gamma + \theta_0 \gamma p_{n,\gamma}(x)} = \frac{x^{-\gamma} p_{n,\gamma}(x)}{1 + \theta_0 \gamma x^{-\gamma} p_{n,\gamma}(x)}.$$

Letting $M = \max(\sup_{x>0} G_0'(\log x), \sup_{x>0} G_0''(\log x))$, we have from (5.3.11) that

$$\sup_{\alpha_n \leq x \leq \beta_n} J_{n,1}(x) \leq M \sup_{\alpha_n \leq x \leq \beta_n} \left| (A_0(n))^{-1} \left(q_n(x) - x^{-\gamma} p_{n,\gamma}(x) \right) \right|$$

$$= M |\gamma| \sup_{\alpha_n \leq x \leq \beta_n} \frac{p_{n,\gamma}^2(x)}{x^{2\gamma} A_0(n) \left| 1 + \theta_0 \gamma x^{-\gamma} p_{n,\gamma}(x) \right|} \to 0$$

and that for some $\tilde{\theta} \in [0, 1]$

$$\sup_{\alpha_n \le x \le \beta_n} J_{n,2}(x)$$

$$= \sup_{\alpha_n \le x \le \beta_n} (A_0(n))^{-1} x^{-\gamma} p_{n,\gamma}(x) \left| G_0'' \left(\log x + \tilde{\theta}\theta q_n(x) \right) \theta q_n(x) \right|$$

$$\le M \sup_{\alpha_n \le x \le \beta_n} (A_0(n))^{-1} x^{-2\gamma} p_{n,\gamma}^2(x) \left[1 + \theta_0\gamma x^{-\gamma} p_{n,\gamma}(x) \right]^{-1} \to 0 .$$

On the other hand, (5.3.10) gives $\lim_{n\to\infty} \sup_{\alpha_n \le x \le \beta_n} J_{n,3}(x) = 0$. So we obtain

$$\lim_{n\to\infty} \sup_{\alpha_n \le x \le \beta_n} |J_n(x) - J(x)| = 0 . \tag{5.3.13}$$

It follows from (5.3.13) that

$$(A_0(n))^{-1} \sup_{\beta_n \le x \le \infty} |J_n(x)|$$

$$\le (A_0(n))^{-1} \sup_{\beta_n \le x \le \infty} \left(1 - G_0 \left(\gamma^{-1} \log \left\{ 1 + \gamma a_n^{-1} (V(nx) - b_0(n)) \right\} \right) \right)$$

$$+ (A_0(n))^{-1} \sup_{\beta_n \le x \le \infty} (1 - G_0(\log x))$$

$$\le (A_0(n))^{-1} \left(1 - G_0 \left(\gamma^{-1} \log \left\{ 1 + \gamma a_n^{-1} (V(n\beta_n) - b_0(n)) \right\} \right) \right)$$

$$+ (A_0(n))^{-1} (1 - G_0(\log \beta_n))$$

$$\le (A_0(n))^{-1} \left\{ G_0 \left(\gamma^{-1} \log \left\{ 1 + \gamma a_n^{-1} (V(n\beta_n) - b_0(n)) \right\} \right) - G_0(\log \beta_n) \right\}$$

$$+ 2 (A_0(n))^{-1} (1 - G_0(\log \beta_n)) \to 0 , \quad n \to \infty .$$

Noting that $J(\infty) = 0$, we have $\lim_{n\to\infty} \sup_{\beta_n \le x \le \infty} |J_n(x) - J(x)| = 0$. Similarly, it may be shown that $\lim_{n\to\infty} \sup_{0 \le x \le \alpha_n} |J_n(x) - J(x)| = 0$, completing the proof of the lemma. ∎

Proof (of Theorem 5.3.3). Let $x_n(u) = \{-n \log F(a_0(n)u + b_0(n))\}^{-1}$, for all $u \in \mathbb{R}$. We have

$$K_n(u) := \frac{P\left(X_{n,n} \le a_0(n)u + b_0(n) \right) - G_\gamma(u)}{A_0(n)} + J(\omega_\gamma(u))$$

$$= \frac{F^n(a_0(n)u + b_0(n)) - G_\gamma(u)}{A_0(n)} + J(\omega_\gamma(u))$$

$$= \frac{G_0(\log x_n(u)) - G_0(\log \omega_\gamma(u))}{A_0(n)} + J(\omega_\gamma(x_n(u)))$$

$$+ \left\{ J(\omega_\gamma(u)) - J(\omega_\gamma(x_n(u))) \right\}$$

$$=: K_{n,1}(u) + K_{n,2}(u) .$$

In order to establish (5.3.8) we need only to prove

$$\lim_{n\to\infty} \sup_{0 < F(a_0(n)u + b_0(n)) < 1} |K_{n,i}(u)| = 0 , \quad i = 1, 2 , \tag{5.3.14}$$

$$\lim_{n\to\infty} \sup_{F(a_0(n)u + b_0(n)) = 0} , |K_n(u)| = 0, \tag{5.3.15}$$

and

$$\lim_{n \to \infty} \sup_{F(a_0(n)u+b_0(n))=1} |K_n(u)| = 0 . \qquad (5.3.16)$$

It follows from the definition of V that if $0 < F(a_0(n)u + b_0(n)) < 1$, then

$$\frac{V(nx_n(u)) - b_0(n)}{a_0(n)} \le u \le \frac{V^+(nx_n(u)) - b_0(n)}{a_0(n)} , \quad n = 1, 2, \ldots$$

(recall that V^+ is the right-continuous version of V). Therefore for $u > 0$ such that $0 < F(a_0(n)u + b_0(n)) < 1$, we have

$$\frac{G_0(\log x_n(u)) - G_0 \left(\frac{1}{\gamma} \log \left\{ 1 + \gamma \frac{V^+(nx_n(u)) - b_0(n)}{a_0(n)} \right\} \right)}{A_0(n)} + J(\omega_\gamma(x_n(u)))$$

$$\le K_{n,1}(u)$$

$$\le \frac{G_0(\log x_n(u)) - G_0 \left(\frac{1}{\gamma} \log \left\{ 1 + \gamma \frac{V(nx_n(u)) - b_0(n)}{a_0(n)} \right\} \right)}{A_0(n)} + J(\omega_\gamma(x_n(u))) .$$

Combining this with (5.3.12), we obtain (5.3.14) for $i = 1$. Since $J(\cdot)$ is continuous on $(0, \infty)$ and $J(0) = J(\infty) = 0$, it is easily seen that (5.3.14) for $i = 2$ is also true.

Since $F(a_0(n)u + b_0(n)) = 0$ implies $u \le (V(0) - b_0(n))/a_0(n)$, using (5.3.12) once again, we have

$$\lim_{n \to \infty} \sup_{F(a_0(n)u+b_0(n))=0} \frac{G_\gamma(u)}{A_0(n)} \le \lim_{n \to \infty} \frac{G_\gamma \left(\frac{V(0) - b_0(n)}{a_0(n)} \right)}{A_0(n)}$$

$$= \lim_{n \to \infty} |J_n(0) - J(0)| = 0 .$$

Note that (5.3.6) implies for $x > 0$,

$$\lim_{t \to \infty} \frac{V(tx) - V(t)}{a(t)} = \frac{x^\gamma - 1}{\gamma}.$$

For any $\varepsilon > 0$, there exists n_0 such that $(V(0) - b_0(n))/a_0(n) < \omega_\gamma^{\leftarrow}(\varepsilon)$ for all $n \ge n_0$ and therefore

$$\lim_{n \to \infty} \sup_{F(a_0(n)u+b_0(n))=0} \left| J(\omega_\gamma(u)) \right| \le \sup_{u \le \omega_\gamma^{\leftarrow}(\varepsilon)} \left| J(\omega_\gamma(u)) \right| = \sup_{0 < x \le \varepsilon} |J(x)| .$$

This then gives

$$\lim_{n \to \infty} \sup_{F(a_0(n)u+b_0(n))=0} |K_n(u)|$$

$$\le \lim_{n \to \infty} \sup_{F(a_0(n)u+b_0(n))=0} \frac{G_\gamma(u)}{A_0(n)} + \lim_{n \to \infty} \sup_{F(a_0(n)u+b_0(n))=0} \left| J(\omega_\gamma(u)) \right|$$

$$\le \sup_{0 < x \le \varepsilon} |J(x)| .$$

Hence (5.3.15) is obtained by letting $\varepsilon \to 0$ in the above. Similarly we may prove (5.3.16), completing the proof of the theorem. ∎

Remark 5.3.8 The uniform limit (5.3.8) gives an Edgeworth expansion as follows:

$$P\left(X_{n,n} \leq a_0(n)u + b_0(n)\right)$$
$$= G_\gamma(u) - A_0(n)G_\gamma(u)\left(-\log^{\gamma+1} G_\gamma(u)\right)\overline{\overline{\Psi}}_{\gamma,\rho}\left(-\log^{-1} G_\gamma(u)\right) + o\left(A_0(n)\right)$$

holds uniformly on $\mathbb{R}$.

Remark 5.3.9 The uniform limit (5.3.8) also gives a rate for convergence, that is,

$$\lim_{n\to\infty}\sup_{u\in\mathbb{R}}\left|\frac{P\left(X_{n,n} \leq a_0(n)u + b_0(n)\right) - G_\gamma(u)}{A_0(n)}\right| = \sup_{x>0} x^{-(\gamma+1)}e^{-1/x}\left|\overline{\overline{\Psi}}_{\gamma,\rho}(x)\right|.$$

The surprising part is that under a weak extra condition the converse of Theorem 5.3.3 holds.

Theorem 5.3.10 (Cheng and Jiang (2001)) *If there exist sequences $a_n > 0$, b_n real, and $A_n > 0$ satisfying*

$$\lim_{n\to\infty} A_n = 0 \quad and \quad \lim_{n\to\infty}\frac{A_{n+1}}{A_n} = 1$$

and a function K such that

$$\lim_{n\to\infty}\frac{P\left(X_{n,n} \leq a_n x + b_n\right) - G_\gamma(x)}{A_n} = K(x) \tag{5.3.17}$$

holds locally uniformly on $\mathbb{R}$, and K is not a multiple of $x\,(\omega_\gamma(x))^{-\gamma-1}G_\gamma(x)$, then V satisfies the second-order condition for some $\rho \leq 0$.

For the proof we need the following lemma.

Lemma 5.3.11 *Suppose that f is a positive measurable function. If there exist sequences $a_n > 0$, b_n real, and $A_n > 0$ satisfying*

$$\lim_{n\to\infty} A_n = 0 \quad and \quad \lim_{n\to\infty}\frac{A_{n+1}}{A_n} = 1 \tag{5.3.18}$$

and a function K such that

$$\lim_{n\to\infty}\frac{\frac{f(nx)-b_n}{a_n} - \frac{x^\gamma-1}{\gamma}}{A_n} = K(x) \tag{5.3.19}$$

holds locally uniformly on $\mathbb{R}$, with K not a multiple of $(x^\gamma - 1)/\gamma$, then f satisfies the second-order condition.

Proof. Let $[t]$ be the integer part of $t \in \mathbb{R}$ and set

$$\alpha(t) = \begin{cases} \frac{1-\{t/([t]+1)\}^{\gamma}}{(t/[t])^{\gamma}-\{t/([t]+1)\}^{\gamma}} \,, & \gamma \neq 0, \\[2ex] \frac{\log([t]+1)-\log t}{\log([t]+1)-\log[t]} \,, & \gamma = 0, \end{cases}$$

$$\beta(t) = \begin{cases} \frac{(t/[t])^{\gamma}-1}{(t/[t])^{\gamma}-\{t/([t]+1)\}^{\gamma}} \,, & \gamma \neq 0, \\[2ex] \frac{\log t-\log[t]}{\log([t]+1)-\log[t]} \,, & \gamma = 0, \end{cases}$$

$$a(t) = \left[\frac{\alpha(t)}{a_{[t]}} + \frac{\beta(t)}{a_{[t]+1}}\right]^{-1}, \quad b(t) = a(t)\left[\frac{b_{[t]}}{a_{[t]}}\alpha(t) + \frac{b_{[t]+1}}{a_{[t]+1}}\beta(t)\right],$$

and $A(t) = A_{[t]}$. It follows from (5.3.18) and local uniformly for (5.3.19) that

$$\frac{\frac{f(tx)-b(t)}{a(t)} - \frac{x^{\gamma}-1}{\gamma}}{A(t)} - K(x)$$

$$= \alpha(t)\left\{\frac{\frac{f(tx)-b([t])}{a_{[t]}} - \frac{(tx/[t])^{\gamma}-1}{\gamma}}{A_{[t]}} - K(x)\right\}$$

$$+ \beta(t)\left\{\frac{\frac{f(tx)-b([t]+1)}{a_{[t]+1}} - \frac{\{tx/([t]+1)\}^{\gamma}-1}{\gamma}}{A_{[t]}} - K(x)\right\}$$

$$\to 0, \quad t \to \infty.$$

This then gives

$$\lim_{t\to\infty} \frac{\frac{f(tx)-f(t)}{a(t)} - \frac{x^{\gamma}-1}{\gamma}}{A(t)} = K(x),$$

completing the proof of the lemma. ∎

Proof (of Theorem 5.3.10). If (5.3.17) holds locally uniformly on $\mathbb{R}$, then

$$\frac{n\log F(a_n x + b_n) - \log G_{\gamma}(x)}{A_n}$$

$$= \frac{F^n(a_n x + b_n) - G_{\gamma}(x)}{A_n} \frac{n\log F(a_n x + b_n) - \log G_{\gamma}(x)}{F^n(a_n x + b_n) - G_{\gamma}(x)}$$

$$\to \frac{K(x)}{G_{\gamma}(x)}$$

holds locally uniformly on $\mathbb{R}$ and therefore

$$\frac{1}{A_n}\left\{-\frac{1}{n\log F(a_n x + b_n)} + \frac{1}{\log G_{\gamma}(x)}\right\}$$

$$= \frac{n\log F(a_n x + b_n) - \log G_{\gamma}(x)}{A_n} \frac{-(n\log F(a_n x + b_n))^{-1} + \log^{-1} G_{\gamma}(x)}{n\log F(a_n x + b_n) - \log G_{\gamma}(x)}$$

$$\to \frac{\omega_{\gamma}^2(x) K(x)}{G_{\gamma}(x)}$$

holds locally uniformly on $\mathbb{R}$. By Vervaat's lemma (Appendix A), the above gives

$$\lim_{n\to\infty} \frac{\frac{V(nx)-b_n}{a_n} - \frac{x^\gamma-1}{\gamma}}{A_n} = -x^{\gamma+1}e^{1/x}K\left(\frac{x^\gamma-1}{\gamma}\right).$$

This implies that V satisfies the second-order condition by Lemma 5.3.11, completing the proof of the theorem. ∎

Theorem 5.3.3 gives a uniform convergence rate. In fact there is a somewhat sharper result: a weighted approximation of the left-hand side of (5.3.8) by the right-hand side, restricted to the right tail. We are not going to prove this result (cf. Drees, de Haan, and Li (2006)), which is a consequence of the weighted approximation of Theorem 5.1.1. We only prove the following large deviations result.

Theorem 5.3.12 (Large deviations) *Suppose that the second-order relation for* $U :=$ $(1/(1-F))^{\leftarrow}$ *holds (cf. Section 2.3) with* $\rho < 0$. *Then*

$$\lim_{n\to\infty} \frac{1-F^n(a_n x_n + b_n)}{1-G_\gamma(x_n)} = 1$$

for any sequence $x_n \uparrow 1/((-\gamma)\vee 0)$.

Proof. Note that $1/((-\gamma)\vee 0)$ is the right endpoint of the distribution function G_γ. Statement (5.1.4) of Theorem 5.1.1 implies

$$\lim_{\substack{t\to\infty \\ x(t)\uparrow 1/((-\gamma)\vee 0)}} \left((1+\gamma x(t))^{-1/\gamma}\right)^{\rho+\varepsilon}$$

$$\times \left\{ \left(\frac{t\,(1-F(b_0(t)+x(t)a_0(t)))}{(1+\gamma x(t))^{-1/\gamma}} - 1 \right) \Big/ A_0(t) \right.$$

$$\left. - \frac{1}{1+\gamma x(t)}\overline{\Psi}_{\gamma,\rho}\left((1+\gamma x(t))^{1/\gamma}\right) \right\} = 0.$$

Note that $((1+\gamma x(t))^{-1/\gamma})^{\rho+\varepsilon} \to \infty$ if $\varepsilon < -\rho$ and $\overline{\Psi}_{\gamma,\rho}((1+\gamma x(t))^{1/\gamma})/(1+\gamma x(t)) \to 0$, as $t \to \infty$. Hence

$$\lim_{t\to\infty} \frac{t\,\{1-F\,(b_0(t)+x(t)a_0(t))\}}{-\log G_\gamma(x(t))} = 1.$$

Note that for $x_n \uparrow 1/((-\gamma)\vee 0)$,

$$1-F^n(a_0(n)x_n+b_0(n)) \sim -n\log F(a_0(n)x_n+b_0(n))$$
$$\sim n\{1-F(a_0(n)x_n+b_0(n))\}.$$

Finally, it is clear that we can replace $a_0(n), b_0(n)$ by any set of normalizing constants a_n, b_n. ∎

5.4 Weak and Strong Laws of Large Numbers and Law of the Iterated Logarithm

Let $X_1, X_2, X_3, \ldots$ be independent and identically distributed random variables with distribution function F. Define $X_{n,n} := \max(X_1, X_2, \ldots, X_n)$ for $n = 1, 2, \ldots$. We are going to discuss analogues for partial maxima, of the weak and strong laws of large numbers, and the law of the iterated logarithm for partial sums.

Whereas in the partial sum case existence of moments is the right thing to consider, for partial maxima conditions of regular variation type turn out to play an important role. We start with a weak law of large numbers.

Theorem 5.4.1 *Suppose $F(x) < 1$ for all real x. The following assertions are equivalent:*

1. *There exists a sequence a_n of positive numbers such that*

$$\frac{X_{n,n}}{a_n} \overset{P}{\to} 1 . \tag{5.4.1}$$

2. *With $b_n := U(n) = (1/(1-F))^{\leftarrow}(n)$,*

$$\frac{X_{n,n}}{b_n} \overset{P}{\to} 1 . \tag{5.4.2}$$

3. *For all $x > 1$,*

$$\lim_{t \to \infty} \frac{1 - F(tx)}{1 - F(t)} = 0 . \tag{5.4.3}$$

4. *$\int_0^\infty s \, dF(s)$ is finite and*

$$\lim_{t \to \infty} \frac{\int_t^\infty s \, dF(s)}{t(1 - F(t))} = 1 . \tag{5.4.4}$$

Proof. Relation (5.4.1) can be expressed as

$$\lim_{n \to \infty} F^n(a_n x) = \begin{cases} 1 , & x > 1, \\ 0 , & 0 < x < 1 . \end{cases}$$

We first show that this is equivalent to

$$\lim_{n \to \infty} n \, (1 - F(a_n x)) = \begin{cases} 0 , & x > 1, \\ \infty , & 0 < x < 1 . \end{cases} \tag{5.4.5}$$

We use the following inequalities: for $0 < t < 1$,

$$t < -\log(1 - t) < \frac{t}{1 - t} . \tag{5.4.6}$$

Let first $x > 1$. Then

$$0 \le n(1 - F(a_n x)) < -n \log F(a_n x) \to 0 , \quad n \to \infty .$$

Next let $0 < x < 1$. Then

$$n(1 - F(a_n x)) > F(a_n x)(-n \log F(a_n x)) \to \infty ,$$

$n \to \infty$.

Next we prove that (5.4.1) implies (5.4.2). Let U be the (generalized) inverse function of $1/(1 - F)$. By inversion (5.4.5) is equivalent to

$$\lim_{n \to \infty} \frac{U(nx)}{a_n} = 1 \text{ for } x > 0 .$$

It follows that $a_n \sim U(n), n \to \infty$.

We continue with the proof of the equivalence between (5.4.2) and (5.4.3). Assume (5.4.2). For some $x > 1$ take $0 < b < 1 < d < \infty$ such that $x = d/b$. Let $n = n(t) := \min\{m : a_{m+1} > t\}$. Then

$$0 \le \lim_{t \to \infty} \frac{1 - F(tx)}{1 - F(t)} = \lim_{t \to \infty} \frac{1 - F(td/b)}{1 - F(t)}$$
$$= \lim_{t \to \infty} \frac{1 - F(td)}{1 - F(tb)} \le \lim_{n \to \infty} \frac{n+1}{n} \frac{n(1 - F(a_n d))}{(n + 1)(1 - F(a_{n+1}b))} = 0 .$$

For $0 < x < 1$ one proceeds in a similar way.

To go from (5.4.3) to (5.4.2) recall the inequalities

$$1 - F^+(U(t)) \le t^{-1} \le 1 - F^-(U(t)) \tag{5.4.7}$$

with U the inverse function of $1/(1 - F)$. For $x > 1$ by (5.4.6) and (5.4.7),

$$\limsup_{n \to \infty} - \log F^n(b_n x) = \limsup_{n \to \infty} -n \log F(b_n x)$$
$$\le \limsup_{n \to \infty} \frac{n(1 - F(b_n x))}{F(b_n x)}$$
$$\le 2 \limsup_{n \to \infty} \frac{1 - F(b_n x)}{1 - F^+(b_n)}$$
$$= 0 .$$

For $0 < x < 1$ one proceeds in a similar way.

Next we prove the equivalence of (5.4.3) and (5.4.4). First assume (5.4.3). For $\lambda > 1$ there exists $x_0(\lambda)$ such that for $x \ge x_0(\lambda)$,

$$1 - F(\lambda x) < 2^{-1}(1 - F(x)) ;$$

hence by repeated application,

$$1 - F(\lambda^n x) < 2^{-n}(1 - F(x))$$

and

$$
\begin{aligned}
\int_x^\infty (1 - F(t))\, dt &= \sum_{n=0}^\infty \int_{x\lambda^n}^{x\lambda^{n+1}} (1 - F(t))\, dt \\
&= \sum_{n=0}^\infty x\lambda^n \int_1^\lambda (1 - F(tx\lambda^n))\, dt \le \sum_{n=0}^\infty x\lambda^n (1 - F(x\lambda^n)) \int_1^\lambda dt \\
&\le x\,(1 - F(x))(\lambda - 1) \sum_{n=0}^\infty \left(\frac{\lambda}{2}\right)^n.
\end{aligned}
$$

By letting λ approach 1 we see that

$$\limsup_{x\to\infty} \frac{\int_x^\infty (1 - F(t))\, dt}{x(1 - F(x))} = 0, \tag{5.4.8}$$

which gives (5.4.4) by partial integration.

Finally we prove that (5.4.4) implies (5.4.3) by contradiction: suppose for some $x_0 > 1$ and sequence $t_n \to \infty$,

$$\liminf_{n\to\infty} \frac{1 - F(t_n x_0)}{1 - F(t_n)} \ge c > 0.$$

Then by Fatou's lemma,

$$\liminf_{n\to\infty} \int_1^\infty \frac{1 - F(t_n s)}{1 - F(t_n)}\, ds \ge \int_1^{x_0} \liminf_{n\to\infty} \frac{1 - F(t_n s)}{1 - F(t_n)}\, ds \ge c\,(x_0 - 1) > 0,$$

which contradicts (5.4.8) and hence also (5.4.4). ∎

Corollary 5.4.2 *If F is in the domain of attraction of the Gumbel distribution $G_0(x)$ and $F(x) < 1$ for all x, the result of Theorem 5.4.1 holds.*

Proof. By Theorem 1.2.5,

$$\lim_{t\to\infty} \frac{1 - F(t + xf(t))}{1 - F(t)} = e^{-x}$$

for $x \in \mathbb{R}$ and the function f is such that

$$\lim_{t\to\infty} \frac{f(t)}{t} = 0.$$

Hence (5.4.3) is applicable. ∎

Remark 5.4.3 Clearly if F is in the domain of attraction of $G_\gamma(x)$ for some $\gamma > 0$, relation (5.4.3) does not hold (cf. Theorem 1.2.1(1)), hence (5.4.1) cannot hold.

Now we turn to strong (a.s.) laws. The validity of the strong law of large numbers depends on the finiteness of a certain integral. For the law of the iterated logarithm the second-order conditions are basically sufficient.

We provide proofs for all the results except for the necessity of the condition for the strong law, which is lengthy and complicated. We need the following lemma.

Lemma 5.4.4 *Let c_n be a sequence of positive constants and $b_n := (1/(1-F))^{\leftarrow}(n)$. Suppose that $b_n + xc_n$ is an ultimately nondecreasing sequence for all real $x > -1$.*

1. *For each distribution function F we have almost surely*

$$\liminf_{n\to\infty} \frac{X_{n,n} - b_n}{c_n} \leq 0 .$$

2. *Let c be a finite constant. We have almost surely*

$$\limsup_{n\to\infty} \frac{X_{n,n} - b_n}{c_n} = c$$

   if and only if

$$\sum_{n=1}^{\infty}(1 - F(c_n x + b_n)) \tag{5.4.9}$$

   converges for all $x > c$ and diverges for all $x < c$.
3. *If for all $-1 < x < 0$,*

$$\sum_{n=1}^{\infty}(1 - F(c_n x + b_n)) \exp(-n(1 - F(c_n x + b_n))) < \infty , \tag{5.4.10}$$

   then almost surely

$$\liminf_{n\to\infty} \frac{X_{n,n} - b_n}{c_n} \geq 0. \tag{5.4.11}$$

Proof. (1) Note that

$$P(X_{n,n} \leq b_n \text{ infinitely often}) \geq \limsup_{n\to\infty} P(X_{n,n} \leq b_n)$$

$$= \limsup_{n\to\infty} F^n(b_n) \geq \lim_{n\to\infty}(1 - 1/n)^n = e^{-1} > 0 .$$

Since $\{X_{n,n} \leq b_n \text{ infinitely often}\}$ is a tail event, we have

$$P(X_{n,n}/c_n \leq b_n/c_n \text{ infinitely often}) = P(X_{n,n} \leq b_n \text{ infinitely often}) = 1 .$$

(2) Since $c_n x + b_n$ is a nondecreasing sequence for all real $x > -1$, we have $X_{n,n} > c_n x + b_n$ infinitely often if and only if $X_n > c_n x + b_n$ infinitely often. Since the X_n are independent, part (2) is a direct consequence of the Borel–Cantelli lemmas.

(3) Since $\sum_{n=1}^{\infty}(1 - F(b_n)) = \infty$, we have almost surely $X_{n,n} > b_n$ infinitely often. Hence also $X_{n,n} > c_n x + b_n$ infinitely often for all $x < 0$. So to prove (5.4.11) it is sufficient to show that

$$P(X_{n,n} \le c_n x + b_n \text{ and } X_{n+1,n+1} > c_{n+1}x + b_{n+1} \text{ finitely often}) = 1 ,$$

or equivalently (since $c_n x + b_n$ is a nondecreasing sequence for $x > -1$),

$$P(X_{n,n} \le c_n x + b_n \text{ and } X_{n+1} > c_{n+1}x + b_{n+1} \text{ finitely often}) = 1 .$$

By the first Borel–Cantelli lemma this is true if

$$\sum_{n=1}^{\infty} P(X_{n,n} \le c_n x + b_n \text{ and } X_{n+1} > c_{n+1}x + b_{n+1})$$

$$= \sum_{n=1}^{\infty} (1 - F(c_{n+1}x + b_{n+1})) \ F^n(c_n x + b_n) \qquad (5.4.12)$$

converges. Now

$$1 - F(c_{n+1}x + b_{n+1}) \le 1 - F(c_n x + b_n)$$

and

$$F^n(c_n x + b_n) = \exp(n \log F(c_n x + b_n)) \le \exp\{-n(1 - F(c_n x + b_n))\} ;$$

hence the convergence of (5.4.12) is implied by (5.4.10). ∎

Theorem 5.4.5 *Let $F(x) < 1$ for all real x. Equivalent are:*

1. *For some sequence b_n,*

$$\frac{X_{n,n}}{b_n} \to 1 \ a.s. \qquad (5.4.13)$$

2. *For all $0 < v < 1$,*

$$\int_1^{\infty} \frac{dF(x)}{1 - F(vx)} < \infty . \qquad (5.4.14)$$

Proof. We prove that (5.4.14) implies (5.4.13). First note that (5.4.14) implies

$$\lim_{x \to \infty} \frac{1 - F(x)}{1 - F(vx)} = 0$$

for $0 < v < 1$. Hence (5.4.14) implies

$$\int_1^{\infty} (1 - F(vx)) \exp\left(-\frac{1 - F(vx)}{1 - F(x)}\right) d\left(\frac{1}{1 - F(x)}\right)$$

$$= \int_1^{\infty} \left(\frac{1 - F(vx)}{1 - F(x)}\right)^2 \exp\left(-\frac{1 - F(vx)}{1 - F(x)}\right) \frac{dF(x)}{1 - F(vx)} < \infty .$$

With $U := (1/(1 - F))^{\leftarrow}$ this gives

$$\int_1^\infty \{1 - F(vU(t))\} \exp\left(-t\{1 - F(vU(t))\}\right) \, dt < \infty .$$

By applying $b_n \leq U(t) \leq b_{n+1}$ for $n \leq t \leq n + 1$ this entails

$$\sum_{n=1}^\infty (1 - F(vb_{n+1})) \exp\{-(n + 1)(1 - F(vb_n))\} < \infty,$$

which is easily seen to lead to

$$\sum_{n=1}^\infty (1 - F(vb_n)) \exp\{-n(1 - F(vb_n))\} < \infty$$

for $0 < v < 1$. This is the condition of Lemma 5.4.4(3) with $c_n = b_n$. Hence $\liminf_{n\to\infty} X_{n,n}/b_n \geq 1$ a.s. In order to get $\limsup_{n\to\infty} X_{n,n}/b_n = 1$ we need to prove that the sum in (5.4.9) with $c_n = b_n$ converges for $x > 0$ and diverges for $x < 0$. That is, $\sum_{n=1}^\infty 1 - F(vb_n)$ is finite for $v > 1$ and infinite for $v < 1$. First note that $\sum_{n=1}^\infty 1 - F(vb_n)$ is finite if and only if $\int_1^\infty 1 - F(vU(t)) \, dt$ is finite. Clearly by the definition of $U(t)$ this integral is infinite for $v < 1$. For $v > 1$ by partial integration

$$\int_1^\infty 1 - F(vU(t)) \, dt = \int_1^\infty \frac{dF(s)}{1 - F(vs)},$$

which is finite by assumption.

We conclude from the result of Lemma 5.4.4 with $c_n = b_n$ and $c = 0$ that

$$\limsup_{n\to\infty} \frac{X_{n,n}}{b_n} = 1 \quad \text{and} \quad \liminf_{n\to\infty} \frac{X_{n,n}}{b_n} = 1 \quad \text{a.s.}$$

We omit the proof that (5.4.13) implies (5.4.14), which can be found in Barndorff–Nielsen (1963).   ∎

For the law of the iterated logarithm we have conditions that are believed to be new (cf. Pickands (1967) and de Haan and Hordijk (1972)).

Theorem 5.4.6 *Let $F(x) < 1$ for all x. Define $V(x) := U(e^x)$ for real x. Suppose there is a positive function p such that for all real x and some real β,*

$$\lim_{t\to\infty} \frac{V(t + x \log t) - V(t)}{p(t)} = \frac{e^{\beta x} - 1}{\beta} . \tag{5.4.15}$$

Then almost surely

$$\limsup_{n\to\infty} \frac{X_{n,n} - V(\log n)}{p(\log n)} = \frac{e^\beta - 1}{\beta}$$

and

$$\liminf_{n\to\infty} \frac{X_{n,n} - V(\log n)}{p(\log n)} = 0 .$$

Remark 5.4.7 By writing $V(x) = Q(\int_1^x (\log s)^{-1} \, ds)$ for $x > 1$ and noticing that $\lim_{t \to \infty} \int_t^{t+x \log t} (\log s)^{-1} \, ds = x$ for real x, one sees that $\lim_{t \to \infty} (Q(t + x) - Q(t))/p(t)$ must exist for real x. Hence the limit function in (5.4.15) is (except for shift and scale constants) the only possible nonconstant one.

Proof. Let $E_{n,n}$ be the maximum of n independent standard exponential random variables. It is easy to see that in this case the conditions of Lemma 5.4.4 are fulfilled with $c = 1$, $b_n = \log n$, and $c_n = \log \log n$.
   Hence, with

$$Q_n := \frac{E_{n,n} - \log n}{\log \log n}$$

we have almost surely

$$\limsup_{n \to \infty} Q_n = 1 \quad \text{and} \quad \liminf_{n \to \infty} Q_n = 0 \, .$$

Clearly $\{X_{n,n}\}_{n=1}^\infty =^d \{V(E_{n,n})\}_{n=1}^\infty$ and

$$\frac{X_{n,n} - V(\log n)}{p(\log n)} = \frac{V(\log n + Q_n \log \log n) - V(\log n)}{p(\log n)} \, .$$

The result follows from (5.4.15). ∎

Corollary 5.4.8 *If for some positive function a, a distribution function satisfies the second-order condition of Section 2.3 with $\gamma = 0$ and $\rho < 0$, then the result of Theorem 5.4.6 is true with $b_n = U(n)$ and $c_n = a(n) \log \log n$.*

Proof. The uniform inequalities of Theorem 2.3.6 imply for $x > 0$,

$$\frac{V(t + x) - V(t)}{a_0(e^t)} = x + A_0(e^t) \Psi_{0,\rho}(e^x) + o(1) A_0(e^t) e^{(\rho + \varepsilon)x},$$

where the $o(1)$ term is uniform for $x > 0$, as $t \to \infty$. Hence

$$\frac{V(t + x \log t) - V(t)}{a_0(e^t) \log t} = x + \frac{A_0(e^t)}{\log t} \Psi_{0,\rho}(t^x) + o(1) \frac{A_0(e^t)}{\log t} t^{(\rho + \varepsilon)x} \, .$$

It is easily seen that the last two terms tend to zero as $t \to \infty$; hence (5.4.15) holds. ∎

   If the second-order condition holds with $\gamma = 0$ and $\rho = 0$, the result of Theorem 5.4.6 still holds in many cases. We give an example.

Example 5.4.9 (Normal distribution) Since

$$1 - \Phi(t) = (2\pi)^{-1/2} e^{-t^2/2} \left(1/t - 1/t^3 + o(1/t^3) \right)$$

as $t \to \infty$, with Φ the standard normal distribution function, we have, with $V^\leftarrow$ the inverse function of V,

$$V^{\leftarrow}(t) = \log\left(\frac{1}{1-\Phi(t)}\right) = \frac{t^2}{2} + \log t + \frac{1}{2}\log(2\pi) + o(1) .$$

From this one sees that $\log V^{\leftarrow}(t) \sim 2\log t$ and

$$\lim_{t\to\infty} \frac{V^{\leftarrow}(t + x(\log t)/t) - V^{\leftarrow}(t)}{\log V^{\leftarrow}(t)} = \frac{x}{2} .$$

By inversion one gets relation (5.4.15) with $p(t) = (\log t)/\sqrt{2t}$.

Finally, we mention a result of a somewhat different nature due to Klass (1985).

Theorem 5.4.10 *Let* $X, X_1, X_2, \ldots$ *be i.i.d. random variables. Let* b_n *be a non-decreasing sequence,* $P(X > b_n) \to 0$, $nP(X > b_n) \to \infty$, *as* $n \to \infty$. *Then*

$$P\left(X_{n,n} \leq b_n \ i.o.\right) = \begin{cases} 1 , \ if \ \displaystyle\sum_{n=1}^{\infty} P(X > b_n)\exp\left(-nP(X > b_n)\right) = \infty, \\[2ex] 0 , \ if \ \displaystyle\sum_{n=1}^{\infty} P(X > b_n)\exp\left(-nP(X > b_n)\right) < \infty . \end{cases}$$

5.5 Weak "Temporal" Dependence

Up to now we have considered partial maxima of a sequence of independent and identically distributed random variables. In many practical situations the assumption of independence in particular is not realistic. In this section we consider a random sequence $\{X_i\}_{i=1}^{\infty} = X_1, X_2, \ldots$ that is strictly stationary, i.e., $\{X_{n+k}\}_{n=1}^{\infty}$ has the same distribution as $\{X_n\}_{n=1}^{\infty}$ for all positive integers k. Now not only the marginal distribution function but also the dependence structure come into play.

We call this type of dependence "temporal," since we imagine the observations obtained one after the other as time progresses. In the next chapter we shall consider a sequence of vector-valued observations. In that situation the dependence between the components of the vector will be the object of study. The latter kind of dependence will be called "spatial." Of course in practical situations these concepts may not be connected with real time or space.

Since the topic of dependent observations is vast and rather specialized, and since excellent books on the subject are available (for example Leadbetter, Lindgren, and Rootzén (1983), in the sequel LLR), the present section contains hardly any proofs. We intend to give only a flavor of what is possible. We start with some examples.

Example 5.5.1 Let X be a random variable and consider the sequence $X, X, X, \ldots$. The partial maximum is just X and no limit theory is involved.

Example 5.5.2 Let $Y_1, Y_2, \ldots$ be independent and identically distributed with distribution function $\exp(-1/x)$, $x > 0$. Consider the random sequence

$$X_1, X_2, X_3, \ldots := \begin{cases} Y_1, Y_2, Y_2, Y_2, Y_3, Y_3, Y_3, \ldots, & \text{with probability } \frac{1}{3}, \\ Y_1, Y_1, Y_2, Y_2, Y_2, Y_3, Y_3, \ldots, & \text{with probability } \frac{1}{3}, \\ Y_1, Y_1, Y_1, Y_2, Y_2, Y_2, Y_3, \ldots, & \text{with probability } \frac{1}{3}. \end{cases}$$

Hence $\max(X_1, \ldots, X_{3n-2})$ behaves like $\max(Y_1, \ldots, Y_n)$.

Example 5.5.3 Let $Y_0, Y_1, Y_2, \ldots$ be independent and identically distributed with distribution function $\exp(-1/x)$, $x > 0$, and define for $n = 1, 2, \ldots$,

$$X_n := \frac{1}{2} \max(Y_{n-1}, Y_n) .$$

This is an example of a max-moving average process. Note that the sequence $X_1, X_2, \ldots$ is stationary and the marginal distribution is the same as that of Y. Further,

$$\max(X_1, \ldots, X_n) = \frac{1}{2} \max(Y_0, Y_1, \ldots, Y_n) ,$$

so that

$$P(\max(X_1, \ldots, X_{2n-1}) \le x) = P(\max(Y_0, \ldots, Y_n) \le x) .$$

Hence $\max(X_1, \ldots, X_n)$ behaves approximately like the maximum of $n/2$ independent and identically distributed random variables Y_i.

The situation is more or less the same if $\{X_n\}_{n=1}^{\infty}$ is an *infinite-order* "max-moving average" process.

Example 5.5.4 Let $Y_1, Y_2, \ldots$ be as in the previous example and let V be some random variable. Define for $n = 1, 2, \ldots$,

$$X_n := Y_n + V .$$

Then $\max(X_1, \ldots, X_n) = V + \max(Y_1, \ldots, Y_n)$ and

$$P(n^{-1} \max(X_1, \ldots, X_n) \le x) = P(n^{-1}V + n^{-1} \max(Y_1, \ldots, Y_n) \le x),$$

which converges to $\exp(-1/x)$, $x > 0$, as $n \to \infty$.

Example 5.5.5 Let $U_1, U_2, \ldots$ be independent and identically distributed uniform $(0, 1)$ random variables and V some random variable. Define for $n = 1, 2, \ldots$,

$$X_n := U_n + V .$$

Then $P(\max(X_1, \ldots, X_n) \le x)$ converges to $P(V + 1 \le x)$, as $n \to \infty$.

Example 5.5.6 Let $E_1, E_2, \ldots$ be independent and identically distributed standard exponential random variables and V some independent random variable. Define for $n = 1, 2, \ldots$,

$$X_n := E_n + V .$$

Then

$$P(\max(X_1, X_2, \ldots, X_n) - \log n \le x) = P(\max(E_1, \ldots, E_n) - \log n + V \le x)$$

converges to $P(M + V \le x)$, where M is a random variable with extreme value distribution G_0 from Theorem 1.1.3 independent of V.

Example 5.5.7 Let $X_1, X_2, \ldots$ be a stationary Gaussian sequence and let $E X_1 = 0$, $E X_1^2 = 1$. Let r_n be the correlation function $r_n = E(X_1, X_{n+1})$. Consider also a sequence $Y_1, Y_2, \ldots$ of independent and identically distributed standard normal random variables. Then we know from Example 1.1.7 that sequences $a_n > 0$ and b_n exist such that

$$\lim_{n \to \infty} P\left(\frac{\max (Y_1, \ldots, Y_n) - b_n}{a_n} \le x\right) = \exp\left(-e^{-x}\right)$$

for all x. If

$$\lim_{n \to \infty} r_n \log n = \gamma \ge 0 ,$$

then

$$\lim_{n \to \infty} P\left(\frac{\max (X_1, X_2, \ldots, X_n) - b_n}{a_n} \le x\right) = P\left(M + N\sqrt{2\gamma} - \gamma \le x\right),$$

where M and N are independent, M has distribution function $\exp\left(-e^{-x}\right)$, and N is standard normal.

If $r_n \log n \to \infty$, under certain conditions with different normalizing constants a normal limit distribution is obtained (Pickands (1967)).

One aim of extreme value theory for stationary dependent sequences is to formulate general conditions that allow most of the theory for sequences of independent and identically distributed random variables to go through.

The Conditions D and D'

The direction followed in the basic book of LLR is to formulate mixing conditions, as weak as possible, for probabilities of events connected with large values of the random variables.

Suppose $X_1, X_2, X_3, \ldots$ is a stationary stochastic process and the distribution function F of X_1 is in the domain of attraction of some extreme value distribution; that is, if $Y_1, Y_2, \ldots$ is an independent and identically distributed sequence with the same marginal distribution, there are sequences $a_n > 0$ and b_n such that for all x and some $\gamma \in \mathbb{R}$,

$$\lim_{n \to \infty} P\left(\frac{\max (Y_1, Y_2, \ldots, Y_n) - b_n}{a_n} \le x\right) = G_\gamma(x) . \qquad (5.5.1)$$

For fixed x define $u_n := b_n + a_n x$. The mixing conditions for the process $\{X_n\}$ will be concerned with events above the level u_n, for fixed x and sample size n.

Let l and p be positive integers. For any random vector $(X_1, \ldots, X_p)$ let $F_{1,\ldots,p}$ denote its joint distribution function. The condition $D(u_n)$ will be said to hold if for any integers

$$1 \le i_1 < \cdots < i_p < j_1 < \cdots < j_p \le n$$

for which $j_1 - i_p \ge l$, we have

$$\left| F_{i_1,\ldots,i_p,j_1,\ldots,j_p}(u_n) - F_{i_1,\ldots,i_p}(u_n)F_{j_1,\ldots,j_p}(u_n) \right| \leq x_{n,l} , \qquad (5.5.2)$$

where $x_{n,l} \to 0$, $n \to \infty$, for some sequence $l = l_n = o(n)$.

The condition $D'(u_n)$ will be said to hold for the stationary sequence $\{X_j\}$ and the sequence of constants u_n if

$$\limsup_{n \to \infty} n \sum_{j=2}^{[n/k]} P(X_1 > u_n, X_j > u_n) \to 0 , \qquad (5.5.3)$$

as $k \to \infty$.

Theorem 5.5.8 (LLR, Theorem 3.4.1) *Let $X_1, X_2, \ldots$ be a stationary sequence and let $Y_1, Y_2, \ldots$ be an independent and identically distributed sequence with the same marginal distribution for which (5.5.1) holds. Assume that the conditions $D(u_n)$ and $D'(u_n)$ hold with $u_n = b_n + a_n x$. Then*

$$\lim_{n \to \infty} P \left(\frac{\max(X_1, X_2, \ldots, X_n) - b_n}{a_n} \leq x \right) = G_\gamma(x) .$$

Next we describe a situation in which the normalizing constants are still the same as under independence but the limit distribution is slightly different. This is the situation of Examples 5.5.2 and 5.5.3.

Definition 5.5.9 Let $X_1, X_2, \ldots$ be a stationary sequence. Let $Y_1, Y_2, \ldots$ be an independent and identically distributed sequence with the same marginal distribution for which (5.5.1) holds. If for some $0 < \theta \leq 1$,

$$\lim_{n \to \infty} P \left(\frac{\max(X_1, \ldots, X_n) - b_n}{a_n} \leq x \right) = G_\gamma^\theta(x)$$

for all x, then the sequence $X_1, X_2, \ldots$ is said to have *extremal index* θ.

Note that our definition is slightly more specific than the definition in LLR, p. 67. An example of a sequence with arbitrary extremal index $\theta \in (0, 1]$ is the process

$$X_1 := Y_1,$$
$$X_{n+1} := \max((1 - \theta)X_n, \theta Y_{n+1}) \text{ for } n \geq 1,$$

where $Y_1, Y_2, \ldots$ are independent and identically distributed with distribution function $\exp(-1/x)$, $x > 0$. It is easy to see that $P(X_n \leq x) = \exp(-1/x)$ and that θ is the extremal index of the sequence.

Many processes have an extremal index between zero and one: well-known examples are a moving average process of stable random variables (LLR, Section 3.8) and a process satisfying a stochastic difference equation (de Haan, Resnick, Rootzén, and de Vries (1989)). The point process convergence of Section 2.1 can be generalized to this case: the epochs of the points in the limiting point process still form a homogeneous Poisson process. However the two-dimensional points are now arranged on vertical lines where the mean number of points on a vertical line is $1/\theta$.

Several estimators for the extremal index have been developed based on this interpretation of θ. A general form for those estimators is given by (cf. Ancona-Navarrete and Tawn (2000))

$$\hat{\theta} := \frac{C(u_n)}{N(u_n)},$$
(5.5.4)

where $N(u_n)$ is the number of exceedances of a high threshold u_n and $C(u_n)$ is the number of clusters. There are two general ways of identifying clusters. The *runs estimator* for θ is, for $1 < l < n$,

$$\hat{\theta}_R := \frac{1}{N(u_n)} \sum_{i=1}^{n-l} 1_{\{X_i > u_n\}} 1_{\{X_{i+1} \leq u_n\}} \cdots 1_{\{X_{i+l} \leq u_n\}}.$$

It recognizes two different clusters of exceedances when there are at least l consecutive observations below the threshold between them.

The second kind of estimator for θ is called a *blocks estimator*. By first dividing the sample into k blocks of length m, so n approximately equals km, the number of clusters $C(u_n)$ in (5.5.4) is estimated as the number of blocks in which at least one exceedance of u_n occurs.

This approach is mainly connected with the behavior of extreme order statistics and point process convergence (as in Section 2.1). For results concerning intermediate order statistics, convergence to Gaussian processes and asymptotic behavior of estimators we need other conditions than the conditions D and D', and the existence of the extremal index. That is what we discuss next.

Mixing Conditions

A different way to deal with dependence has been explored by Rootzén (1995) and developed by Drees. The aim is to formulate rather general but quite specific conditions under which the approximation of the tail empirical quantile process by Brownian motion of Section 2.4 can be generalized. As before, this approximation can serve as a basis for a wealth of statistical results. It can be proved that many specific stochastic processes used in applications satisfy the conditions.

We consider a version of the conditions and the ensuing result. Let $\{X_n\}$ be a stationary sequence. The common distribution function is denoted by F and the inverse function of $1/(1 - F)$ is denoted by U. We assume that F is in the domain of attraction of an extreme value distribution G_γ.

Further, we assume that the sequence $\{X_n\}$ is β-*mixing*, i.e.,

$$\beta(l) := \sup_{m \in \mathbb{N}} E\left(\sup_{A \in \mathcal{B}_{m+l+1}^\infty} |P(A | \mathcal{B}_1^m) - P(A)| \right) \to 0, \quad l \to \infty,$$

where $\mathcal{B}_1^m$ and $\mathcal{B}_{m+l+1}^\infty$ denote the σ-fields generated by $\{X_i\}_{1 \leq i \leq m}$ and $\{X_i\}_{i \geq m+l+1}$, respectively.

The other conditions are the following:

Condition 5.5.10

$$\lim_{n \to \infty} \frac{\beta(l_n)}{l_n} n + l_n k_n^{-1/2} \log^2 k_n = 0,$$

where l_n and k_n are sequences of integers, l_n, $k_n \to \infty$, $n \to \infty$. The growth of the sequence k_n has to be restricted by Condition 5.5.13 below.

Condition 5.5.11 There exist $\varepsilon > 0$ and functions c_m, $m = 1, 2, \ldots$, such that

$$\lim_{n \to \infty} \frac{n}{k_n} P\left(X_1 > U\left(\frac{n}{k_n x}\right), X_{m+1} > U\left(\frac{n}{k_n y}\right)\right) = c_m(x, y)$$

for all $x, y \in (0, 1 + \varepsilon)$ and each m.

Condition 5.5.12 There exist a constant $D_1 \geq 0$ and a sequence $\tilde{\rho}_m$, with $\sum_{m=1}^{\infty} \tilde{\rho}_m < \infty$, such that

$$\frac{n}{k_n} P\left(X_1 \in I_n(x, y), X_{m+1} \in I_n(x, y)\right) \leq (y - x)\left(\tilde{\rho}_m + D_1 \frac{k_n}{n}\right)$$

for all $x, y \in (0, 1 + \varepsilon)$ and each m, where

$$I_n(x, y) := \left[U\left(\frac{n}{k_n y}\right), U\left(\frac{n}{k_n x}\right)\right].$$

Condition 5.5.13

$$\lim_{n \to \infty} \sqrt{k_n} \sup_{0 < s \leq 1 + \varepsilon} \frac{s^{\gamma + 1/2}}{\sqrt{1 + |\log s|}} \left| \frac{U\left(\frac{n}{k_n s}\right) - U\left(\frac{n}{k_n}\right)}{a\left(\frac{n}{k_n}\right)} - \frac{s^{-\gamma} - 1}{\gamma} \right| = 0,$$

where $a > 0$ is a suitable version of the auxiliary function of Theorem 1.1.6.

Theorem 5.5.14 Let $X_1, X_2, \ldots$ be a β-mixing stationary random sequence with common marginal distribution function $F \in \mathcal{D}(G_\gamma)$, for some real γ. Under Conditions 5.5.10–5.5.13, for some sequence $l_n = o(n/k_n)$, there exist versions of the tail quantile process, denoted by $\{X_{n-[k_n s],n}\}_{0 < s \leq 1}$, and a centered Gaussian process E with covariance function c defined by

$$c(x, y) := \min(x, y) + \sum_{m=1}^{\infty} (c_m(x, y) + c_m(y, x)),$$

such that

$$\sup_{0 < s \leq 1} s^{\gamma + 1/2} (1 + |\log s|)^{-1/2}$$

$$\times \left| \sqrt{k_n} \left(\frac{X_{n-[k_n s],n} - D_n}{a\left(\frac{n}{k_n}\right)} - \frac{s^{-\gamma} - 1}{\gamma} \right) - s^{-(\gamma + 1)} E(s) \right| \xrightarrow{P} 0, \quad n \to \infty.$$

Here $D_1, D_2, \ldots$ is a sequence of random variables which for $\gamma \geq \frac{1}{2}$ can be chosen as $U(n/k_n)$ (hence nonrandom).

Under very general conditions, estimators of γ, $a(n/k_n)$, large quantiles, etc. can be expressed as functionals of the tail empirical quantile function, and an appropriate invariance theorem entails asymptotic normality of such statistics, very much in the same way as for the approximation developed in Section 2.4. For details we refer to Drees (2000, 2002) and in particular to Drees (2003), where the case $\gamma > 0$ receives special attention. It is shown that under certain conditions a stochastic difference equation (hence also the ARCH process) satisfies the stated conditions, as well as a moving average process.

5.6 Mejzler's Theorem

Instead of dropping the assumption of independence in the usual requirement of independent and identically distributed random variables, one can also drop the assumption that the random variables have all the same distribution. The most general and interesting result in this area is due to Mejzler (1956). It is somewhat similar to Lévy's class L in the theory of partial sums of random variables.

Theorem 5.6.1 *Suppose $X_1, X_2, \ldots$ are independent random variables with distribution functions $F_1, F_2, \ldots$ respectively. Suppose there exist sequences $a_n > 0$ and b_n such that $(\max(X_1, X_2, \ldots, X_n) - b_n)/a_n$ has a nondegenerate limit distribution, which we call G. Suppose that as $n \to \infty$,*

$$|\log a_n| + |b_n| \to \infty \tag{5.6.1}$$

and

$$\begin{cases} a_{n+1}/a_n \to 1, \\ (b_{n+1} - b_n)/a_n \to 0 . \end{cases} \tag{5.6.2}$$

Then

$$\begin{cases} -\log G(x) \text{ is convex} & \text{if } x^*(G) = \infty , \\ -\log G\left(x^* - e^{-x}\right) \text{ is convex} & \text{if } x^*(G) < \infty . \end{cases} \tag{5.6.3}$$

Here $x^*(G) := \sup\{x : G(x) < 1\}$.

Conversely, any distribution function G satisfying (5.6.3) occurs as a limit in the given set-up.

Remark 5.6.2 The conditions of Mejzler's version of the theorem are the same except for (5.6.1) and (5.6.2). Mejzler's condition is that for any x with $G(x) > 0$ and any sequence $k = k(n)$ of integers with $1 \leq k \leq n$,

$$\lim_{n\to\infty} F_k(a_n x + b_n) = 1 . \tag{5.6.4}$$

By taking k fixed in (5.6.4) one sees that (5.6.1) follows. Next take $k(n) = n$ in (5.6.4). This implies $\lim_{n\to\infty} \prod_{k=1}^{n-1} F_k(a_n x + b_n) = G(x)$ for continuity points x of G. Since we also have $\lim_{n\to\infty} \prod_{k=1}^{n} F_k(a_n x + b_n) = G(x)$ (5.6.2), follows by the convergence of types theorem.

Proof (Balkema, de Haan, and Karandikar (1993)). First the direct statement. Define

$$\alpha_n x := a_n x + b_n \; ,$$

$$M_n(x) := \sum_{k=1}^{n} -\log F_k(\alpha_n x) \; ,$$

$$\|\alpha_n\| := |\log a_n| + |b_n| \; .$$

We know that $M_n \to M := -\log G, n \to \infty$, weakly. Further, M_n is a "tail function" for any n, i.e., M_n is nonincreasing and $\lim_{x \to \infty} M_n(x) = 0$. Take $\varepsilon > 0$. By virtue of (5.6.1) and (5.6.2) we can construct a sequence of integers $k(n)$ such that

$$\lim_{n \to \infty} \left\| \frac{\alpha_n}{\alpha_{n+k(n)}} \right\| = \varepsilon \; .$$

Then along a subsequence n' we have convergence of $\alpha_{n'+k(n')}^{-1} \alpha_{n'}$ to α_ε, say. Now

$$\frac{M_{n+k}}{\alpha_{n+k}} - \frac{M_n}{\alpha_n} = \sum_{r=n+1}^{n+k} -\log F_r$$

is a tail function for $n, k = 1, 2, \ldots$. Also,

$$\frac{M_{n'+k(n')} \alpha_{n'}}{\alpha_{n'+k(n')}} - M_{n'} = \left(\frac{M_{n'+k(n')}}{\alpha_{n'+k(n')}} - \frac{M_{n'}}{\alpha_{n'}} \right) \alpha_{n'} \xrightarrow{d} -\log G \, \alpha_\varepsilon + \log G \; .$$

So $-\log G \, \alpha_\varepsilon + \log G$ is a tail function. Let

$$A := \{ \alpha \; : \; M\alpha - M \text{ is a tail function} \} \; .$$

We have proved that for all $\varepsilon > 0$ there exists $\alpha \in A$ with $\|\alpha\| = \varepsilon$. We shall prove that there is a $\beta \in A$ such that $\beta^t \in A$ for all $t > 0$ (note that if $\beta x =: ax + b$, then $\beta^t x =: a^t x + b \int_0^t a^s \, ds$).

For the proof choose $\alpha_n \in A$, $n = 1, 2, \ldots$, such that $\|\alpha_n\| = 1/n$. Choose $r(n) \in \mathbb{N}, r(n) \to \infty$, such that $\|\alpha_n^{r(n)}\| \to 1, n \to \infty$. Note that $\alpha_n^{r(n)} \in A$ for all n. Then along a subsequence n' we have convergence: $\alpha_{n'}^{r(n')} \to \beta$, say. Note that β is not the identity and $\beta \in A$. Now $\alpha_{n'}^{[tr(n')]} \in A$ for $t > 0$ and $\lim_{n \to \infty} \alpha_n^{[tr(n)]} \alpha_n^{-tr(n)}$ is the identity, since α_n converges to the identity. Since $\alpha_{n'}^{tr(n')} \to \beta^t$, also $\alpha_{n'}^{[tr(n')]} \to \beta^t$ and hence $\beta^t \in A$ for all $t > 0$. This proves the statement. Now, as we have seen, β^t has a specific form:

$$\beta^t x = a^t x + b \int_0^t a^s \, ds \; , \tag{5.6.5}$$

for some $a > 0$ and $b \in \mathbb{R}$. Hence $M(\beta^t x) - M(x)$ is a tail function for all $t > 0$, i.e., either

$$M(a^t(x - x_0) + x_0) - M(x) \qquad \text{(with } a \neq 1\text{)} \tag{5.6.6}$$

or

$$M(x - bt) - M(x) \tag{5.6.7}$$

is a tail function for all $t > 0$. Since $M(\beta^t x) - M(x)$ must be nonnegative, we must have

$$\beta^t x \leq x \text{ for all } x < x^* := x^*(G) . \tag{5.6.8}$$

Let $_*x := \inf\{x : G(x) > 0\}$. Then if (5.6.6) holds with $a > 1$, we have $x_0 \geq x^*$; if (5.6.6) holds with $a < 1$, then $x_0 \leq_* x$; if (5.6.7) holds, then $b > 0$.

Let us consider the case $a > 1$. Then $x^* < \infty$. We know that for all $t > 0$ the function

$$M(a^t(x - x_0) + x_0) - M((x - x_0) + x_0)$$

is nonnegative and nonincreasing for $x < x^*$. That is, for all $t > 0$,

$$M(x_0 - e^t e^y) - M(x_0 - e^y)$$

is nonnegative and nonincreasing for $e^y > x_0 - x^*$. It follows that $M(x_0 - e^{-y})$ is nondecreasing and convex for $e^y > x_0 - x^*$, i.e., (with M' the right one-sided derivative of M) $-e^y M'(x_0 - e^y)$ is nonnegative and nondecreasing for $e^y > x_0 - x^*$. Write $x_0 - e^y = x^* - e^x$; then also $-e^x M'(x^* - e^x)$ is nonnegative and nondecreasing for $x \in \mathbb{R}$, since $1 + e^x(x_0 - x^*)$ is nonnegative and nonincreasing. It follows that $M(x^* - e^x)$ is a convex function and hence also $M(x^* - e^{-x})$. Similar reasoning applies in the other two cases.

Conclusions: if (5.6.6) holds with $a > 1$, then $x^* < \infty$ and $M(x^* - e^{-x})$ is convex; if (5.6.6) holds with $0 < a < 1$, then $_*x > -\infty$ and $M(x_0 + e^x)$ is convex for some $x_0 \leq_* x$; if (5.6.7) holds, then $x^* = \infty$ and $M(x)$ is convex. Some reflection shows that in fact the convexity of $M(x_0 + e^x)$ for some $x_0 \leq_* x$ implies the convexity of $M(x)$. This proves the direct statement of the theorem.

Conversely, suppose that G satisfies (5.6.3). We shall consider only the case $-\log G(x)$ convex. The other case is similar. Relation (5.6.7) implies that the function

$$F_n(x) := \begin{cases} \dfrac{G(x - \log(n+1))}{G(x - \log n)} , & x \geq_* x(G) + \log(n + 1), \\ 0 , & x <_* x(G) + \log(n + 1), \end{cases}$$

is a distribution function for $n = 1, 2, \ldots$. Moreover,

$$\prod_{k=1}^{n} F_k(x + \log n) = \frac{G(x)}{G(x + \log n)} \to G(x) ,$$

as $n \to \infty$, for all x. Note that (5.6.1) and (5.6.2) apply. ∎

Corollary 5.6.3 *If (5.6.6) holds with $a > 1$, then the same relation holds with x_0 replaced by any x_1 satisfying $x^* \leq x_1 < x_0$. If (5.6.7) holds, then (5.6.6) holds for $a > 1$ and $x_0 \geq x^*$. If (5.6.6) holds with $0 < a < 1$, then the same relation holds with x_0 replaced by any $x_1 < x_0$; moreover, (5.6.7) holds.*

Proof. Let us consider the case $a > 1$. The previous proof shows that then (5.6.6) holds with x_0 replaced by x^*. The same proof also gives (5.6.6) with x_0 replaced by any x_1 satisfying $x^* \leq x_1 < x_0$. Similar reasoning applies in the other two cases. ∎

Exercises

5.1. Let F be twice differentiable and write $r(t) := \left((1 - F)/F'\right)'(t)$. Recall that $\lim_{t\to\infty} r(t) = 0$ implies that $F \in \mathcal{D}(G_0)$. Prove that if $r(t) \log \log 1/(1 - F(t)) \to 0$, then the condition of Theorem 5.4.6 holds.

5.2. Prove that Theorem 5.4.6 holds for any gamma distribution.

5.3. Prove that under the conditions of Theorem 5.1.1 with $\rho < 0$,

$$\lim_{n\to\infty} \sup_{x_0 < x < 1/((-\gamma)\vee 0)} \frac{n\,(1 - F(a_0(n)x + b_0(n)))}{-\log G_\gamma(x)} = 1$$

for any $x_0 > -1/(\gamma \vee 0)$.

5.4. Suppose for a sequence of i.i.d. random variables $X_1, X_2, \ldots$ we have that $(\max_{1 \le i \le n} X_i - b_n)/a_n$ converges to an extreme value distribution $G_\gamma(x)$ with $G_\gamma(x) < 1$ for all x (hence $\gamma \ge 0$). Consider now maxima in the presence of a trend, i.e., $\max_{1 \le i \le n}(X_i + cb_i)$ with, say, $c > 0$. Prove that

$$P\left(\max_{1 \le i \le n} \frac{X_i + cb_i - (1 + c)b_n}{a_n} \le x\right)$$

converges to

$$P\left(\sup_{i \ge 1} \frac{W_i^\gamma - 1}{\gamma} + c\frac{T_i^\gamma - 1}{\gamma} \le x\right),$$

where $\{(W_i, T_i)\}_{i=1}^\infty$ is an enumeration of the points of a Poisson point process on $\mathbb{R}_+ \times [0, 1]$ with mean measure ν defined by $\nu\{(x, \infty) \times (0, t)\} = t/x$. Hence

$$\lim_{n\to\infty} P\left(\max_{1 \le i \le n} \frac{X_i + cb_i - (1 + c)b_n}{a_n} \le x\right)$$

$$= \exp\left\{-\int_0^1 \left(1 + \gamma\left(x - c\frac{t^\gamma - 1}{\gamma}\right)\right)^{-1/\gamma} dt\right\}.$$

Hint: Use the point process convergence of Theorem 2.1.2, the interpretation of point process convergence given at the end of Section 2.1, and Theorem 1.1.6(2). Note that for a Poisson point process the probability that a certain set is empty (contains no points) equals e^{-q}, where q is the mean measure of that set.

5.5. Let $E_1, E_2, \ldots$ be i.i.d. standard exponential random variables. Define for $x > 1$,

$$N := \min\{n : E_k < x \log k \text{ for } k \ge n\}.$$

Show that $N < \infty$ a.s. For what values of x does EN exist?

5.6. Let $X_1, X_2, \ldots$ be i.i.d. positive random variables with distribution function F. Prove that $\max(X_1, X_2, \ldots, X_n)/\sqrt{n} \to^P 0$ if and only if $\lim_{x\to\infty} x^2(1 - F(x)) = 0$.

5.7. Let $X_1, X_2, \ldots$ be i.i.d. positive random variables. Prove that $\max(X_1, X_2, \ldots, X_n)/n \to 0$ a.s. if and only if $EX < \infty$.

Part II

Finite-Dimensional Observations

6

Basic Theory

6.1 Limit Laws

6.1.1 Introduction: An Example

In order to see the usefulness of developing multivariate extremes we start with a problem for which multivariate extreme value theory seems to be relevant.

Consider the situation of the first case study in Section 1.1.4. We want to assess the safety of a seawall near the town of Petten, in the province of North Holland, the Netherlands. The seawall is called Pettemer Zeewering or Hondsbossche Zeewering. The seawall could be in danger of overflowing and eventual collapse in case of high levels of the water of the North Sea combined with high waves (both mainly due to wind storm activity). High-tide seawater levels, i.e., the water level with the waves filtered out, that is, the average level taken over a few minutes, have been monitored over a long period of time. However, these measure still water levels and do not take into account wave activity since it involves a kind of time averaging. Hence it makes sense to introduce the waves as a second variable, thus completing the picture. Since monitoring waves is much more problematic than that of still water levels, less data are available.

The wave height (HmO) and still water level (SWL) have been recorded during 828 storm events that are relevant for the Pettemer Zeewering. Also, engineers of RIKZ (Institute for Coastal and Marine Management) have determined *failure conditions*, that is, those combinations of HmO and SWL that result in overtopping the seawall, thus creating a dangerous situation. The set of those combinations forms a failure set C. Figure 6.1 shows the observations as well as a simplified failure set C, which is

$$\{(\text{HmO}, \text{SWL}) \ : \ 0.3\,\text{HmO} + \text{SWL} > 7.6\} \ .$$

Clearly the observations stay clear of C, indicating that there has been no dangerous situation during the observation period.

The question is, what is the probability that during a wind storm of the considered type, we get an observation (HmO, SWL) in the set C, i.e., what is the probability of failure?

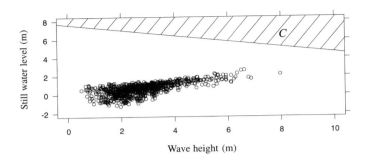

Fig. 6.1. Observations and failure set C.

This problem is clearly similar to estimation of an exceedance probability considered in Section 4.4, but we need a multivariate extension of the theory that leads to the estimator used there.

We shall proceed as follows. After stating the definition of multivariate extremes, we shall identify all possible limit distributions as in the one-dimensional case and then determine their domains of attraction. Then we go on to discuss the statistical issues.

6.1.2 The Limit Distribution; Standardization

In order to make the results more accessible, we shall mainly focus on the two-dimensional case. Generalization to higher-dimensional spaces is usually obvious. When needed, we shall mention the extension explicitly.

It is not obvious how one should define the maximum of a set of vectors. In fact, there are many possibilities, but it turns out that a very naive definition leads to a rich enough theory to tackle problems in applications. We just take the maximum of each coordinate separately and then we assemble these maxima into a vector. Note that the resulting vector maximum will usually not be one of the constituent vectors, but this will not disturb us.

Suppose $(X_1, Y_1), (X_2, Y_2), \ldots$ are independent and identically distributed random vectors with distribution function F. Suppose that there exist sequences of constants $a_n, c_n > 0$, b_n, and d_n real and a distribution function G with nondegenerate marginals such that for all continuity points (x, y) of G,

$$\lim_{n \to \infty} P \left(\frac{\max(X_1, X_2, \ldots, X_n) - b_n}{a_n} \leq x, \frac{\max(Y_1, Y_2, \ldots, Y_n) - d_n}{c_n} \leq y \right)$$
$$= G(x, y) . \tag{6.1.1}$$

Any limit distribution function G in (6.1.1) with nondegenerate marginals is called a *multivariate extreme value distribution*.

In this section we are going to determine the class of all possible limit distributions G. In doing so we will heavily rely on the theory developed in Chapter 1. Since (6.1.1) implies convergence of the one-dimensional two marginal distributions, we have

$$\lim_{n\to\infty} P\left(\frac{\max(X_1, X_2, \ldots, X_n) - b_n}{a_n} \leq x\right) = G(x, \infty) \qquad (6.1.2)$$

and

$$\lim_{n\to\infty} P\left(\frac{\max(Y_1, Y_2, \ldots, Y_n) - d_n}{c_n} \leq y\right) = G(\infty, y) . \qquad (6.1.3)$$

Now we choose the constants a_n, c_n, b_n, and d_n such that (cf. Theorem 1.1.3) for some $\gamma_1, \gamma_2 \in \mathbb{R}$,

$$G(x, \infty) = \exp\left(-(1 + \gamma_1 x)^{-1/\gamma_1}\right) \qquad (6.1.4)$$

and

$$G(\infty, y) = \exp\left(-(1 + \gamma_2 y)^{-1/\gamma_2}\right) . \qquad (6.1.5)$$

We note in passing that since the two marginal distributions of G are continuous, G must be continuous as well.

Next we are going to use the results of Lemma 1.2.9 and Corollary 1.2.10. Let F_i, $i = 1, 2$, be the marginal distribution functions of F. Define $U_i(t) := F_i^{\leftarrow}(1 - 1/t)$, $t > 1$, for $i = 1, 2$. Then according to Theorem 1.1.6 there are positive functions $a_i(t)$, $i = 1, 2$, such that

$$\lim_{t\to\infty} \frac{U_i(tx) - U_i(t)}{a_i(t)} = \frac{x^{\gamma_i} - 1}{\gamma_i} \qquad (6.1.6)$$

and

$$\lim_{t\to\infty} \frac{a_i(tx)}{a_i(t)} = x^{\gamma_i}$$

for all $x > 0$. Moreover, (6.1.2)–(6.1.5) hold with

$$b_n := U_1([n]) ,$$
$$d_n := U_2([n]) ,$$
$$a_n := a_1([n]) ,$$
$$c_n := a_2([n])$$

(cf. proof of Theorem 1.1.2). It follows from (6.1.6) that for $x, y > 0$,

$$\lim_{n\to\infty} \frac{U_1(nx) - b_n}{a_n} = \frac{x^{\gamma_1} - 1}{\gamma_1} ,$$

$$\lim_{n\to\infty} \frac{U_2(ny) - d_n}{c_n} = \frac{y^{\gamma_2} - 1}{\gamma_2} . \qquad (6.1.7)$$

Now we return to (6.1.1), which can be written as

$$\lim_{n\to\infty} F^n(a_n x + b_n, c_n y + d_n) = G(x, y) \,. \tag{6.1.8}$$

Note that if $x_n \to u$, $y_n \to v$, then by the continuity of G and the monotonicity of F,

$$\lim_{n\to\infty} F^n(a_n x_n + b_n, c_n y_n + d_n) = G(u, v) \,. \tag{6.1.9}$$

We apply (6.1.9) with

$$x_n := \frac{U_1(nx) - b_n}{a_n} \,,$$

$$y_n := \frac{U_2(ny) - d_n}{c_n} \,,$$

$x, y > 0$, and get by (6.1.7)

$$\lim_{n\to\infty} F^n\left(U_1(nx), U_2(ny)\right) = G\left(\frac{x^{\gamma_1} - 1}{\gamma_1}, \frac{y^{\gamma_2} - 1}{\gamma_2}\right) \,.$$

We have proved the following theorem:

Theorem 6.1.1 *Suppose that there are real constants $a_n, c_n > 0$, b_n, and d_n such that*

$$\lim_{n\to\infty} F^n(a_n x + b_n, c_n y + d_n) = G(x, y)$$

for all (x, y) of G, and the marginals of G are standardized as in (6.1.4) and (6.1.5). Then with $F_1(x) := F(x, \infty)$, $F_2(y) := F(\infty, y)$, and $U_i(x) := F_i^{\leftarrow}(1 - 1/x)$, $i = 1, 2$,

$$\lim_{n\to\infty} F^n\left(U_1(nx), U_2(ny)\right) = G_0(x, y) \tag{6.1.10}$$

for all $x, y > 0$, where

$$G_0(x, y) := G\left(\frac{x^{\gamma_1} - 1}{\gamma_1}, \frac{y^{\gamma_2} - 1}{\gamma_2}\right)$$

and γ_1, γ_2 are the marginal extreme value indices from (6.1.2)–(6.1.5).

Remark 6.1.2 In case F has continuous marginal distribution functions F_1 and F_2, relation (6.1.10) can be formulated simply as

$$\lim_{n\to\infty} P\left\{\max\left(\frac{1}{1 - F_1(X_1)}, \ldots, \frac{1}{1 - F_1(X_n)}\right) \leq nx \,, \right.$$
$$\left. \max\left(\frac{1}{1 - F_2(Y_1)}, \ldots, \frac{1}{1 - F_2(Y_n)}\right) \leq ny\right\}$$
$$= G_0(x, y)$$

for $x, y > 0$, i.e., after a transformation of the marginal distributions to a standard distribution, namely $F(x) := 1 - 1/x$, $x \geq 1$, a simplified limit relation applies. This means that we have reformulated the problem of identifying the limit distribution in such a way that the marginal distributions no longer play a role. From now on we can focus solely on the dependence structure.

Corollary 6.1.3 *For any* (x, y) *for which* $0 < G_0(x, y) < 1$,

$$\lim_{n \to \infty} n \{1 - F(U_1(nx), U_2(ny))\} = - \log G_0(x, y) .$$ (6.1.11)

Proof. Taking logarithms to the left and to the right of (6.1.10), we get

$$\lim_{n \to \infty} -n \log \{F(U_1(nx), U_2(ny))\} = - \log G_0(x, y) .$$ (6.1.12)

Note that (6.1.12) implies $F(U_1(nx), U_2(ny)) \to 1$; hence

$$\frac{- \log F(U_1(nx), U_2(ny))}{1 - F(U_1(nx), U_2(ny))} \to 1,$$

and (with (6.1.12)) relation (6.1.11) follows. ∎

We shall also use the following slight extension.

Corollary 6.1.4 *For any* (x, y) *for which* $0 < G_0(x, y) < 1$,

$$\lim_{t \to \infty} t \{1 - F(U_1(tx), U_2(ty))\} = - \log G_0(x, y) ,$$ (6.1.13)

where t runs through the real numbers.

Proof. By applying inequalities as in the proof of Theorem 1.1.2 to relation (6.1.11) one sees then, that (6.1.13) also holds. ∎

6.1.3 The Exponent Measure

Next we take (6.1.11) as our point of departure. Take $a > 0$ and define for $(x, y) \in \mathbb{R}_+^2 := [0, \infty)^2$ with $\max(x, y) > a$ and $n = 1, 2, \ldots$,

$$H_{n,a}(x, y) := 1 - \frac{1 - F(U_1(nx), U_2(ny))}{1 - F(U_1(na), U_2(na))} .$$

Clearly $H_{n,a}$ is the distribution function of a probability measure, $P_{n,a}$ say, on $\mathbb{R}_+^2 \setminus [0, a]^2$ for all n, and by (6.1.11),

$$\lim_{n \to \infty} H_{n,a}(x, y) =: H_a(x, y)$$

exists for all x, y with $\max(x, y) > a$. Hence by Billingsley (1979), Theorem 29.1, we have that H_a is the distribution function of a probability measure, P_a say, and it follows that

$$\lim_{n \to \infty} P_{n,a}(A) = P_a(A)$$

for all Borel sets $A \subset \mathbb{R}_+^2 \setminus [0, a]^2$ with $P_a(\partial A) = 0$. Now clearly

$$\nu_n := n \{1 - F(U_1(na), U_2(na))\} P_{n,a}$$

is a measure on $\mathbb{R}_+^2 \setminus [0, a]^2$, *not depending on a* and such that for all Borel sets $A \subset \mathbb{R}_+^2 \setminus [0, a]^2$,

$$\lim_{n \to \infty} \nu_n(A) = \nu(A)$$

with

$$\nu := -\log G_0(a, a) \, P_a \, .$$

Note that since $a > 0$ is arbitrary, $\nu_n(A)$ and $\nu(A)$ are defined for all Borel sets A with

$$\inf_{(x,y) \in A} \max(x, y) > 0 \tag{6.1.14}$$

and that for x, y with $\max(x, y) > 0$,

$$\nu_n \left\{ (s, t) \in \mathbb{R}_+^2 \; : \; s > x \text{ or } t > y \right\} = n \left\{ 1 - F(U_1(nx), U_2(ny)) \right\},$$
$$\nu \left\{ (s, t) \in \mathbb{R}_+^2 \; : \; s > x \text{ or } t > y \right\} = -\log G_0(x, y) \, .$$

Finally, for all Borel sets A such that (6.1.14) holds we have $\lim_{n \to \infty} \nu_n(A) = \nu(A)$. We formulate these results in the following theorem.

Theorem 6.1.5 *Let F and G_0 be probability distribution functions for which (6.1.11) holds, i.e., for x, $y > 0$ with $0 < G_0(x, y) < 1$,*

$$\lim_{n \to \infty} n \left\{ 1 - F(U_1(nx), U_2(ny)) \right\} = -\log G_0(x, y),$$

where $U_i(1/(1 - x))$ is the inverse function of the ith marginal distribution, $i = 1, 2$. Then there are set functions $\nu, \nu_1, \nu_2, \ldots$ defined for all Borel sets $A \subset \mathbb{R}_+^2$ with

$$\inf_{(x,y) \in A} \max(x, y) > 0$$

such that:

1.

$$\nu_n \left\{ (s, t) \in \mathbb{R}_+^2 \; : \; s > x \text{ or } t > y \right\} = n \left\{ 1 - F(U_1(nx), U_2(ny)) \right\},$$
$$\nu \left\{ (s, t) \in \mathbb{R}_+^2 \; : \; s > x \text{ or } t > y \right\} = -\log G_0(x, y) \, ;$$

2. for all $a > 0$ the set functions $\nu, \nu_1, \nu_2, \ldots$ are finite measures on $\mathbb{R}_+^2 \setminus [0, a]^2$;
3. for each Borel set $A \subset \mathbb{R}_+^2$ with $\inf_{(x,y) \in A} \max(x, y) > 0$ and $\nu(\partial A) = 0$,

$$\lim_{n \to \infty} \nu_n(A) = \nu(A) \, .$$

Remark 6.1.6 One can consider this result in the framework of convergence of measures on a metric space as follows. Consider the space

$$\mathbb{R}_+^{2\star} := \mathbb{R}_+^2 \setminus \{(0, 0)\} \, .$$

One can write this space as a product space by using the transformation

$$(x, y) \rightarrow \left(\max(x, y), \frac{(x, y)}{\max(x, y)}\right) .$$

Then

$$\mathbb{R}_+^{2\star} := (0, \infty) \times Q,$$

where

$$Q := \left\{(s, t) \in \mathbb{R}_+^2 \ : \ s, t \geq 0 \ , \ \max(s, t) = 1\right\} .$$

Next we extend the space $\mathbb{R}_+^{2\star}$ to

$$\overline{Q} := (0, \infty] \times Q$$

(although there is never any mass at infinity) and we change the Euclidean metric on $(0, \infty]$ to the metric $\varrho(x, y) := |1/x - 1/y|$. With this change $(0, \infty]$ is a complete separable metric space (CSMS) and so is $(0, \infty] \times Q$. The set functions $\nu, \nu_1, \nu_2, \dots$ are boundedly finite measures on this CSMS, i.e., $\nu(A), \nu_i(A) < \infty$ for each bounded Borel set, for $i = 1, 2, \dots$. Moreover, $\nu_n(A) \rightarrow \nu(A)$ for each bounded Borel set A with $\nu(\partial A) = 0$, i.e., ν_n converges weakly to ν on this CSMS. For details about this type of convergence see the appendix in Daley and Vere-Jones (1988). Compare also the somewhat analogous results of Theorem 9.3.1 in an infinite-dimensional setting.

Definition 6.1.7 The measure ν from Theorem 6.1.5 is sometimes called the *exponent measure* of the extreme value distribution G_0, since

$$G_0(x, y) = \exp\left(-\nu(A_{x,y})\right) \tag{6.1.15}$$

with

$$A_{x,y} := \{(s, t) \in \mathbb{R}_+^2 \ : \ s > x \text{ or } t > y\} . \tag{6.1.16}$$

Remark 6.1.8 Relation (6.1.15) does not hold for all Borel sets $A_{x,y}$.

The characterizing property of the exponent measure is the following homogeneity relation.

Theorem 6.1.9 *For any Borel set $A \subset \mathbb{R}_+^2$ with $\inf_{(x,y) \in A} \max(x, y) > 0$ and $\nu(\partial A) = 0$, and any $a > 0$,*

$$\nu(aA) = a^{-1}\nu(A) , \tag{6.1.17}$$

where aA is the set obtained by multiplying all elements of A by a.

Proof. Taking $t_n = na$ for some $a > 0$ in (6.1.13) we obtain

$$\lim_{n \to \infty} n\{1 - F(U_1(nax), U_2(nay))\} = -a^{-1} \log G_0(x, y) .$$

On the other hand, by direct application of (6.1.11),

$$\lim_{n \to \infty} n\{1 - F(U_1(nax), U_2(nay))\} = - \log G_0(ax, ay) . \tag{6.1.18}$$

Hence

$$-a^{-1} \log G_0(x, y) = - \log G_0(ax, ay), \qquad (6.1.19)$$

and the statement of the theorem holds for all sets $A_{x,y}$ defined by (6.1.16). It is then clear that this relation must also hold for the generated σ-field. ∎

Remark 6.1.10 Relation (6.1.17) implies that $v(A)$ is finite for all sets A with positive distance from the origin, but v is not bounded. Note in particular that

$$G_0(ax, ay) = G_0^{1/a}(x, y) , \quad \text{for} \quad a, x, y > 0 . \qquad (6.1.20)$$

A nice intuitive background for the role of the exponent measure is provided by the following theorem. The proof is very similar to the proof of Theorem 2.1.2 and is omitted.

Theorem 6.1.11 *Let $(X_1, Y_1), (X_2, Y_2) \ldots$, be i.i.d random vectors with distribution function F. Suppose (6.1.1) holds with a_n, b_n, c_n, and d_n as in (6.1.2)–(6.1.5), i.e.,*

$$\lim_{n \to \infty} nP \left(\left(1 + \gamma_1 \frac{X - b_n}{a_n} \right)^{1/\gamma_1} > x \text{ or } \left(1 + \gamma_2 \frac{Y - d_n}{c_n} \right)^{1/\gamma_2} > y \right)$$
$$= - \log G_0(x, y),$$

that is, more generally for each Borel set A of $\overline{Q}$ (for the definition of $\overline{Q}$ see Remark 6.1.6),

$$\lim_{n \to \infty} nP \left\{ \left(\left(1 + \gamma_1 \frac{X - b_n}{a_n} \right)^{1/\gamma_1}, \left(1 + \gamma_2 \frac{Y - d_n}{c_n} \right)^{1/\gamma_2} \right) \in A \right\} = v(A) .$$

Define the point process N_n as follows: for each Borel set $B \in \mathbb{R}_+ \times \overline{Q}$,

$$N_n(B) = \sum_{i=1}^{\infty} 1 \left\{ \left(i/n, (1+\gamma_1(X_i-b_n)/a_n)^{1/\gamma_1}, (1+\gamma_2(Y_i-d_n)/c_n)^{1/\gamma_2} \right) \in B \right\} .$$

Define also a Poisson point process N on the same space with mean measure $\lambda \times v$ with λ Lebesgue measure and v the measure defined in Theorem 6.1.5. Then N_n converges in distribution to N, i.e., for Borel sets $B_1, B_2, \ldots, B_r \in \mathbb{R}_+ \times \overline{Q}$ with $(\lambda \times v)(\partial B_i) = 0, i = 1, \ldots, r$,

$$(N_n(B_1), \ldots, N_n(B_r)) \overset{d}{\to} (N(B_1), \ldots, N(B_r)) .$$

This theorem opens the way to estimating the measure v by just counting the number of observations in certain sets, as we shall see later on (Sections 7.2 and 8.2).

6.1.4 The Spectral Measure

The homogeneity property (6.1.17) of the exponent measure v suggests a coordinate transformation in order to capitalize on that. Recall $\mathbb{R}_+^{2\star}$ from Remark 6.1.6. Take any one-to-one transformation $\mathbb{R}_+^{2\star} \to (0, \infty) \times [0, c]$ for some $c > 0$,

$$\begin{cases} r = r(x, y), \\ d = d(x, y), \end{cases}$$

with the property that for all $a, x, y > 0$,

$$\begin{cases} r(ax, ay) = a r(x, y), \\ d(ax, ay) = d(x, y) . \end{cases}$$

One can think of r as a radius and d as a direction. Examples are

$$\begin{cases} r(x, y) = \sqrt{x^2 + y^2}, \\ d(x, y) = \arctan \dfrac{y}{x}, \end{cases} \qquad \begin{cases} r(x, y) = x + y, \\ d(x, y) = \dfrac{x}{x + y}, \end{cases}$$

and

$$\begin{cases} r(x, y) = x \vee y, \\ d(x, y) = \arctan \dfrac{x}{y} . \end{cases} \tag{6.1.21}$$

It will turn out that the measure ν has a simple structure when expressed in the new coordinates.

Let us start with the first transformation. Define for constants $r > 0$ and $\theta \in [0, \pi/2]$ the set

$$B_{r,\theta} := \left\{ (x, y) \in \mathbb{R}_+^{2\star} : \sqrt{x^2 + y^2} > r \text{ and } \arctan \frac{y}{x} \le \theta \right\} .$$

Clearly

$$B_{r,\theta} = r B_{1,\theta} ,$$

and hence by (6.1.17),

$$\nu \left(B_{r,\theta} \right) := r^{-1} \nu \left(B_{1,\theta} \right) . \tag{6.1.22}$$

This relation means that after transformation to the new coordinates $r(x, y)$ and $d(x, y)$ the measure ν becomes a product measure. Set for $0 \le \theta \le \pi/2$,

$$\Psi(\theta) := \nu \left(B_{1,\theta} \right) . \tag{6.1.23}$$

Clearly Ψ is the distribution function of a finite measure on $[0, \pi/2]$. This finite measure is called the *spectral measure* of the limit distribution G. The spectral measure determines the distribution function G in the following way. Write $s = r \cos \theta$, $t = r \sin \theta$. Take $x, y > 0$,

$$- \log G_0(x, y) = \nu \{(s, t) : s > x \text{ or } t > y\}$$

$$= \nu \{(s, t) : r \cos \theta > x \text{ or } r \sin \theta > y\}$$

$$= \nu \left\{ (s, t) : r > \frac{x}{\cos \theta} \wedge \frac{y}{\sin \theta} \right\} . \tag{6.1.24}$$

We consider two subsets in order to evaluate. First the subset where $x/\cos\theta < y/\sin\theta$. Then $r > \min(x/\cos\theta, y/\sin\theta)$ translates into $r > x/\cos\theta$. Hence by (6.1.22) and (6.1.23) the ν measure of this set is the integral

$$\int_{x/(\cos\theta)<y/(\sin\theta)} \int_{r>x/(\cos\theta)} \frac{dr}{r^2}\, \Psi(d\theta) = \int_{(\cos\theta)/x>(\sin\theta)/y} \frac{\cos\theta}{x}\, \Psi(d\theta)\,.$$

The integral over the other subset, namely where $x/(\cos\theta) > y/(\sin\theta)$, can be evaluated similarly and yields

$$\int_{(\cos\theta)/x<(\sin\theta)/y} \frac{\sin\theta}{y}\, \Psi(d\theta)\,.$$

Combination of the two integrals gives that (6.1.24) equals

$$\int_0^{\pi/2} \left(\frac{\cos\theta}{x} \vee \frac{\sin\theta}{y} \right) \Psi(d\theta)\,.$$

The term "spectral" can be seen as an analogue to the light spectrum, which highlights the contribution of each color separately. Here the spectral measure highlights the contribution of each direction separately. The terminology comes from corresponding results in the theory of partial sums rather than partial maxima, see, e.g., Breiman (1968), Section 11.6.

We have proved the direct statement of the following.

Proposition 6.1.12 *For any extreme value distribution function G from (6.1.1) with (6.1.4) and (6.1.5) there exists a finite measure on the set $[0, \pi/2]$, called spectral measure, with the property that if Ψ is the distribution function of this measure, for $x, y > 0$,*

$$G\left(\frac{x^{\gamma_1} - 1}{\gamma_1}, \frac{y^{\gamma_2} - 1}{\gamma_2} \right) = G_0(x, y) = \exp\left(-\int_0^{\pi/2} \left(\frac{\cos\theta}{x} \vee \frac{\sin\theta}{y} \right) \Psi(d\theta) \right),$$

(6.1.25)

where γ_1 and γ_2 are the extreme value indices of the marginal distributions of G. Moreover, we have the side conditions

$$\int_0^{\pi/2} \cos\theta\ \Psi(d\theta) = \int_0^{\pi/2} \sin\theta\ \Psi(d\theta) = 1\,. \qquad (6.1.26)$$

Conversely, any finite measure represented by its distribution function Ψ gives rise to a limit distribution function G in (6.1.1) via (6.1.25) provided the side conditions (6.1.26) are fulfilled.

Proof. We have already proved the direct statement. The side conditions (6.1.26) stem from the fact that $G_0(x, \infty) = G_0(\infty, x) = \exp(-1/x)$ for $x > 0$.

For the converse we first prove that G_0 defined by (6.1.25) is the distribution function of a probability measure.

Clearly $\exp-((\cos\theta)/x \vee (\sin\theta)/y)$ is the distribution function of the random vector $(V\cos\theta, V\sin\theta)$, where V has the distribution function $\exp(-1/x)$, $x > 0$. Further, if F_i is the distribution function of the random vector (V_i, W_i), $i = 1, 2$, and (V_1, W_1) and (V_2, W_2) are independent, then $F_1 F_2$ is the distribution function of $(\max(V_1, V_2), \max(W_1, W_2))$. Hence any product of distribution functions is a distribution function. It follows that

$$\exp\left(-\sum_{i=1}^{n} \Psi_i \left(\frac{\cos\theta_i}{x} \vee \frac{\sin\theta_i}{y}\right)\right) \tag{6.1.27}$$

is a distribution function for any $0 \le \theta_1 \le \cdots \le \theta_n \le \pi/2$ and $\Psi_i > 0$, $i = 1, 2, \ldots, n$. Now the expression on the right-hand side in (6.1.25) can be approximated by a sequence of type (6.1.27). This proves that G_0 is a distribution function.

Next we prove that G can serve as a limit distribution in (6.1.1). Note that for all $x, y > 0$ and $n = 1, 2, \ldots,$

$$G_0^n(nx, ny) = G_0(x, y) .$$

Hence for all x, y with $1 + \gamma_1 x > 0, 1 + \gamma_2 y > 0$,

$$G^n\left(\frac{n^{\gamma_1} - 1}{\gamma_1} + n^{\gamma_1}x, \frac{n^{\gamma_2} - 1}{\gamma_2} + n^{\gamma_2}y\right)$$

$$= G_0^n\left(n(1 + \gamma_1 x)^{1/\gamma_1}, n(1 + \gamma_2 y)^{1/\gamma_2}\right)$$

$$= G_0\left((1 + \gamma_1 x)^{1/\gamma_1}, (1 + \gamma_2 y)^{1/\gamma_2}\right)$$

$$= G(x, y) . \tag{6.1.28}$$

Hence (6.1.8) holds with $F = G$, $a_n = n^{\gamma_1}$, $c_n = n^{\gamma_2}$, $b_n = (n^{\gamma_1} - 1)/\gamma_1$, $d_n = (n^{\gamma_2} - 1)/\gamma_2$. It follows that any distribution function G satisfying (6.1.25) can occur as a limit function in (6.1.1). ∎

Definition 6.1.13 We call the class of limit distribution functions G in (6.1.1) the class of *max-stable distributions*, as suggested by relation (6.1.28). Hence any extreme value distribution is max-stable and vice versa. The class of limit distribution functions G_0 in (6.1.10) is called the class of *simple max-stable distributions*, "simple" meaning that the marginal distributions are fixed as follows: $G_0(x, \infty) = G_0(\infty, x) = \exp(-1/x)$, $x > 0$.

So far we have considered only the first transformation from (6.1.21). A similar analysis of the other transformations yields the following result. The proof is left to the reader.

Theorem 6.1.14 *For each limit distribution G from (6.1.1), (6.1.4), and (6.1.5) there exist:*

1. *A finite measure (denoted by the distribution function Ψ) on $[0, \pi/2]$ such that for $x, y > 0$,*

$$G\left(\frac{x^{\gamma_1} - 1}{\gamma_1}, \frac{y^{\gamma_2} - 1}{\gamma_2}\right) = G_0(x, y)$$

$$= \exp\left(-\int_0^{\pi/2} \left(\frac{\cos\theta}{x} \vee \frac{\sin\theta}{y}\right) \Psi(d\theta)\right) \quad (6.1.29)$$

with the side conditions

$$\int_0^{\pi/2} \cos\theta \; \Psi(d\theta) = \int_0^{\pi/2} \sin\theta \; \Psi(d\theta) = 1 .$$

2. *A probability distribution (denoted by the distribution function H) concentrated on $[0, 1]$ with mean $\frac{1}{2}$ such that for $x, y > 0$,*

$$G\left(\frac{x^{\gamma_1} - 1}{\gamma_1}, \frac{y^{\gamma_2} - 1}{\gamma_2}\right) = G_0(x, y)$$

$$= \exp\left(-2\int_0^1 \left(\frac{w}{x} \vee \frac{1-w}{y}\right) H(dw)\right) . \quad (6.1.30)$$

3. *A finite measure (denoted by the distribution function Φ) on $[0, \pi/2]$ such that for $x, y > 0$,*

$$G\left(\frac{x^{\gamma_1} - 1}{\gamma_1}, \frac{y^{\gamma_2} - 1}{\gamma_2}\right) = G_0(x, y)$$

$$= \exp\left(-\int_0^{\pi/2} \left(\frac{1 \wedge \tan\theta}{x} \vee \frac{1 \wedge \cot\theta}{y}\right) \Phi(d\theta)\right) \quad (6.1.31)$$

with the side conditions

$$\int_0^{\pi/2} (1 \wedge \tan\theta) \; \Phi(d\theta) = \int_0^{\pi/2} (1 \wedge \cot\theta) \; \Phi(d\theta) = 1 .$$

The parameters γ_1 and γ_2 are the extreme value indices of the marginal distributions of G.

Conversely, any finite measures represented by the distribution function Ψ, H, or Φ gives rise to a limit distribution function G in (6.1.1) via (6.1.29), (6.1.30), and (6.1.31) respectively, provided that the stated side conditions are fulfilled.

Corollary 6.1.15 *Convergence in distribution for a sequence of simple max-stable distributions $G_0^{(n)}$ is equivalent to convergence in distribution of their spectral measures. In particular, the class of simple max-stable distributions is closed under weak convergence.*

Representations (6.1.29)–(6.1.31) mean that the limit distributions in (6.1.1) are characterized by just the spectral measure and the extreme value indices of the marginal distributions.

Clearly the list of possible versions of the spectral measure can be extended *ad infinitum*. The spectral measures Ψ, H, and Φ are the most common ones. Clearly one can transform one into the other. Which one is more convenient depends on the particular situation at hand.

For example, the relation between H and Φ follows from

$$
-\log G_0(x, y) = 2 \int_0^1 \left(\frac{w}{x} \vee \frac{1-w}{y} \right) H(dw)
$$

$$
= 2 \int_0^{\pi/2} \left\{ \frac{1}{x\,(1+\cot\theta)} \vee \left(\frac{1}{y} \left(1 - \frac{1}{1+\cot\theta} \right) \right) \right\} H \left(d\,\frac{1}{1+\cot\theta} \right)
$$

$$
= 2 \int_0^{\pi/2} \left(\frac{\sin\theta}{x} \vee \frac{\cos\theta}{y} \right) \left(\frac{1}{\sin\theta} \wedge \frac{1}{\cos\theta} \right) \frac{H\left(d\,\frac{1}{1+\cot\theta} \right)}{(\sin\theta + \cos\theta)\left(\frac{1}{\sin\theta} \wedge \frac{1}{\cos\theta} \right)}
$$

$$
= 2 \int_0^{\pi/2} \left(\frac{1 \wedge \tan\theta}{x} \vee \frac{1 \wedge \cot\theta}{y} \right)
$$

$$
\times \left\{ \frac{1}{1+\cot\theta} \vee \left(1 - \frac{1}{1+\cot\theta} \right) \right\} H \left(d\,\frac{1}{1+\cot\theta} \right)
$$

$$
=: \int_0^{\pi/2} \left(\frac{1 \wedge \tan\theta}{x} \vee \frac{1 \wedge \cot\theta}{y} \right) \Phi(d\theta).
$$

That is,

$$
\Phi(\theta) = 2 \int_0^{1/(1+\cot\theta)} (w \vee (1-w))\, H(dw) .
$$

Remark 6.1.16 We discuss two extreme cases of spectral measures. For simplicity we formulate only the results for H. We consider

$$
G \left(\frac{x_1^{\gamma_1} - 1}{\gamma_1}, \ldots, \frac{x_d^{\gamma_d} - 1}{\gamma_d} \right)
$$

$$
= \exp \left(-d \int \cdots \int_{w_1 + \cdots + w_d = 1} \left(\frac{w_1}{x_1} \vee \cdots \vee \frac{w_d}{x_d} \right) H(dw) \right),
$$

where H is the distribution function of a probability measure on

$$
\{ \mathbf{w} = (w_1, \ldots, w_d) \;:\; w_1 + \cdots + w_d = 1,\; w_i \geq 0,\; i = 1, 2, \ldots, d \} \quad (6.1.32)
$$

with

$$
\int \cdots \int_{w_1 + \cdots + w_d = 1} w_i\, H(d\mathbf{w}) = \frac{1}{d}
$$

for $i = 1, 2, \ldots, d$.

1. Let the spectral measure be concentrated at the point $(1/d, 1/d, \ldots, 1/d)$ with mass 1. If $(X_1, X_2, \ldots, X_d)$ is a random vector with distribution function $G\left((x_1^{\gamma_1} - 1)/\gamma_1, \ldots, (x_d^{\gamma_d} - 1)/\gamma_d\right)$, then $X_1 = X_2 = \cdots = X_d$ a.s.

2. Let the spectral measure be concentrated at the extreme points of the set (6.1.32), i.e., the d points $(1, 0, 0, \ldots, 0)$, $(0, 1, 0, \ldots, 0)$, $\ldots$, $(0, 0, \ldots, 0, 1)$ with masses $1/d$. If $(X_1, X_2, \ldots, X_d)$ is a random vector with distribution function $G\left((x_1^{\gamma_1} - 1)/\gamma_1, \ldots, (x_d^{\gamma_d} - 1)/\gamma_d\right)$, then $X_1, X_2, \ldots, X_d$ are independent.

Let us consider some examples of limit distributions and their spectral measures. It is useful to note first that the transformation theory for integrals implies that if, for example, the spectral measure has density Ψ', then $-\log G_0(x, y)$ has density $q(x, y)$, say. For instance, for a Borel set $A \subset \mathbb{R}_+^2$ with $\inf_{(x,y)\in A} \max(x, y) > 0$ and $\nu(\partial A) = 0$, and with $r = \sqrt{x^2 + y^2}$ and $\theta = \arctan(y/x)$,

$$\nu(A) = \int_A q(x, y)dx\, dy = \int_{(r\cos\theta, r\sin\theta)\in A} q(r\cos\theta, r\sin\theta)r\, dr\, d\theta \, ;$$

hence

$$\Psi'(\theta) = r^3 q(r\cos\theta, r\sin\theta) \, .$$

Similarly for the other spectral measures (cf. Exercises 6.6 and 6.7).

Example 6.1.17 (Geffroy (1958)) Take $\Psi(\theta) = \theta$ for $0 \leq \theta \leq \pi/2$. Then the side conditions are fulfilled and

$$G_0(x, y) = \exp\left\{-\left(x^{-2} + y^{-2}\right)^{1/2}\right\}, \qquad x > 0, y > 0 \, .$$

Generalization: for $0 \leq a \leq 1$,

$$G_0(x, y) = \exp\left\{-\left(x^{-1/a} + y^{-1/a}\right)^a\right\}, \qquad x > 0, y > 0 \, .$$

Note that $a = 1$ corresponds to independence of the coordinates, and $a = 0$ (defined as a limit) corresponds to full dependence.

Example 6.1.18 (Sibuya (1960)) Take $H(w) = w$ for $0 \leq w \leq 1$. Then the side conditions are fulfilled and

$$G_0(x, y) = \exp\left\{-\left(x^{-1} + y^{-1} - (x + y)^{-1}\right)\right\}, \qquad x > 0, y > 0 \, .$$

Generalization: for $k \geq 0$,

$$G_0(x, y) = \exp\left\{-\left(x^{-1} + y^{-1} - k(x + y)^{-1}\right)\right\}, \qquad x > 0, y > 0 \, .$$

Note that $k = 0$ corresponds to independence of the coordinates.

Example 6.1.19 Take $\Phi(\theta) = (\pi/4 + 2^{-1}\log 2)^{-1}\theta$ for $0 \leq \theta \leq \pi/2$. Then the side conditions are fulfilled and

$$\left(\frac{\pi}{4} + \frac{1}{2}\log 2\right)(-\log G_0(x, y)) = \frac{1}{x}\left\{\frac{\pi}{4} \wedge \arctan\frac{y}{x} + 0 \vee \log\left(\frac{y\sqrt{2}}{\sqrt{x^2 + y^2}}\right)\right\}$$

$$+ \frac{1}{y}\left\{\frac{\pi}{4} \wedge \arctan\frac{x}{y} + 0 \vee \log\left(\frac{x\sqrt{2}}{\sqrt{x^2 + y^2}}\right)\right\}.$$

To see this note that

$$\int_0^{\pi/2}\left(\frac{1 \wedge \tan\theta}{x} \vee \frac{1 \wedge \cot\theta}{y}\right) d\theta$$

$$= \int_{\arctan(x/y)}^{\pi/2} \frac{1 \wedge \tan\theta}{x} d\theta + \int_0^{\arctan(x/y)} \frac{1 \wedge \cot\theta}{y} d\theta.$$

Now we have for example

$$\int_{\arctan(x/y)}^{\pi/2} \frac{1 \wedge \tan\theta}{x} d\theta$$

$$= \frac{1}{x}\int_{(\pi/4)\wedge\arctan(x/y)}^{\pi/4} \tan\theta \, d\theta + \frac{1}{x}\int_{(\pi/4)\vee\arctan(x/y)}^{\pi/2} d\theta.$$

Finally note that $\int_0^z \tan\theta \, d\theta = -\log\cos z$, for $0 \leq z < \pi/2$.

Example 6.1.20 This example is based on the normal distribution: for $c > 0$,

$$-\log G_0(x, y) = E\left(\frac{1}{x} \vee \left(\frac{1}{y}e^{cN - c^2/2}\right)\right)$$

with N a standard normal random variable. Then

$$G_0(x, y) = \exp\left\{-\left(\frac{1}{x}F\left(\frac{c}{2} + \frac{1}{c}\log\frac{y}{x}\right) + \frac{1}{y}F\left(\frac{c}{2} + \frac{1}{c}\log\frac{x}{y}\right)\right)\right\},$$

where F is the standard normal distribution function. Again, when $c \to \infty$ we have independence and when $c \downarrow 0$ we have full dependence. This distribution function can be obtained in several other ways (Eddy and Gale (1981), Hüsler and Reiss (1989), and de Haan and Pereira (2005)).

6.1.5 The Sets Q_c and the Functions L, χ, and A

Finally, we discuss a few other ways to characterize the max-stable distributions. Since the dependence structure is quite general, one could describe the dependence using copulas. If F is the distribution function of the random vector (X, Y), the *copula* C associated with F is a distribution function that satisfies $F(x, y) = C(F_1(x), F_2(y))$ with $F_1(x) := F(x, \infty)$ and $F_2(y) := F(\infty, y)$. It contains complete information about the joint distribution of F apart from the marginal distributions (for more details see Nelsen (1998) and Joe (1997)).

Define for $0 < x, y < 1$,

$$C(x, y) := G_0\left(-1/\log x, -1/\log y\right) .$$

Then C is a copula and relation (6.1.20) translates into the following: for $0 < x, y < 1$, $a > 0$,

$$C\left(x^a, y^a\right) = C^a(x, y) .$$

Since this relation is not very tractable for analysis, it is usual to consider instead the function L defined by

$$L(x, y) := -\log G_0\left(1/x, 1/y\right)$$

for $x, y > 0$. We can express the function L in terms of the exponent measure v (cf. Section 6.1.3) as

$$L(x, y) = v\left\{(s, t) \in \mathbb{R}_+^2 \; : \; s > 1/x \text{ or } t > 1/y\right\} . \tag{6.1.33}$$

Proposition 6.1.21 (Properties of the function L)

1. Homogeneity of order 1: $L(ax, ay) = aL(x, y)$, for all $a, x, y > 0$.
2. $L(x, 0) = L(0, x) = x$, for all $x > 0$.
3. $x \vee y \leq L(x, y) \leq x + y$, for all $x, y > 0$.
4. Let (X, Y) be a random vector with distribution function G_0. If X and Y are independent, then $L(x, y) = x + y$, for $x, y > 0$. If on the other hand X and Y are completely positive dependent, i.e., $X = Y$ a.s., then $L(x, y) = \max(x, y)$ for $x, y > 0$.
5. L is continuous.
6. $L(x, y)$ is a convex function: $L(\lambda(x_1, y_1) + (1 - \lambda)(x_2, y_2)) \leq \lambda L(x_1, y_1) + (1 - \lambda)L(x_2, y_2)$, for all $x_1, y_1, x_2, y_2 > 0$ and $\lambda \in [0, 1]$.

Proof. (1) Direct consequence of Theorem 6.1.9.

(2) We have

$$L(x, 0) = v\{(s, t) \in \mathbb{R}_+^2 \; : \; s > 1/x\}$$
$$= -\log G_0\left(\frac{1}{x}, \infty\right) = -\log G\left(\frac{x^{-\gamma_1} - 1}{\gamma_1}, \infty\right) = x ,$$

for all $x > 0$. Similarly it follows for $L(0, x)$.

(3) We have

$$L(x, y) = v\{(s, t) \in \mathbb{R}_+^2 : s > 1/x \text{ or } t > 1/y\}$$
$$\leq v\{(s, t) \in \mathbb{R}_+^2 : s > 1/x\} + v\{(s, t) \in \mathbb{R}_+^2 : t > 1/y\}$$
$$= L(x, 0) + L(0, y)$$
$$= x + y ,$$

for all $x, y > 0$. On the other hand,

$$L(x, y) \geq v\{(s, t) \in \mathbb{R}_+^2 : s > 1/x\} \vee v\{(s, t) \in \mathbb{R}_+^2 : t > 1/y\}$$
$$= L(x, 0) \vee L(0, y) = x \vee y ,$$

for all $x, y > 0$.

(4) If X and Y are independent, $G_0(1/x, 1/y) = G_0(1/x, \infty) \, G_0(\infty, 1/y) = e^{-x} e^{-y}$. Hence $L(x, y) = -\log G_0(1/x, 1/y) = x + y$. If $X = Y$ a.s., $G_0(x, y) = P(X \leq x, Y \leq y) = P(X \leq \min(x, y)) = \exp(-1/\min(x, y))$. Hence $L(x, y) = -\log G_0(1/x, 1/y) = \max(x, y)$.

(5) The statement is an immediate consequence of the continuity of G, which follows by Lebesgue's theorem on dominated convergence using, for example, representation (6.1.29).

(6) The function $(ax) \vee (by)$ is convex for $x, y > 0$, provided $a, b \geq 0$. Also, positive linear combinations of such functions are convex. Hence $L(x, y) = -\log G_0(1/x, 1/y)$, which can be written as

$$-\log G_0 (1/x, 1/y) = \int_0^{\pi/2} ((x \cos \theta) \vee (y \sin \theta)) \; \Psi(d\theta) ,$$

is convex. ∎

The function L leads to a characterization of the limit distribution in the following way. For $c > 0$ define the level sets Q_c by

$$Q_c := \{(x, y) \in \mathbb{R}_+^2 \; : \; L(x, y) \leq c\} .$$

The sets Q_c have the following properties:

1. Q_c is a closed convex set.
2. The points $(0, 0)$, $(c, 0)$, and $(0, c)$ are extreme points.
3. $Q_c = c \, Q_1$.

The convexity of the level set Q_1 is characteristic for a limit distribution, as the following theorem shows.

Theorem 6.1.22 *For any simple max-stable distribution G_0 define the set Q_1 by*

$$Q_1 := \left\{(x, y) \in \mathbb{R}_+^2 : -\log G_0 (1/x, 1/y) \leq 1\right\} . \qquad (6.1.34)$$

The set Q_1 is closed convex and the points $(0, 0)$, $(0, 1)$, and $(1, 0)$ are vertices. Conversely, any closed convex set Q_1 with vertices $(0, 0)$, $(0, 1)$, and $(1, 0)$ gives rise to a limit distribution G_0 for which (6.1.34) holds. The mapping is one-to-one.

Proof. Let G_0 be a simple max-stable distribution. The convexity of Q_1 is an immediate consequence of the convexity of the function L (cf. Proposition 6.1.21 (6)). The statement about the vertices follows from the side conditions for Ψ.

Conversely, let Q_1 satisfy the stated properties. The closed convex set Q_1 can be approximated from below and above by sequences of sets $Q_L^{(n)}$ and $Q_U^{(n)}$ that satisfy the properties and have a polygonal boundary, i.e., a boundary on $\mathbb{R}_+^2$ satisfying

$\sum_1^n \max(a_i x, b_i y) = 1$ with $a_i \geq 0$, $b_i \geq 0$, $i = 1, 2, \ldots, n$, being constants satisfying $\sum_1^n a_i = \sum_1^n b_i = 1$. Then we have

$$Q_L^{(n)} \subset Q_1 \subset Q_U^{(n)} \tag{6.1.35}$$

and

$$Q_U^{(n)} \setminus Q_L^{(n)} \downarrow 0, \quad n \to \infty . \tag{6.1.36}$$

Note that the sets $Q_L^{(n)}$ and $Q_U^{(n)}$ can be written as follows:

$$Q_L^{(n)} = \left\{ (x, y) \in \mathbb{R}_+^2 : \sum_{i=1}^{m(n)} ((A_i x) \vee (B_i y)) \leq 1 \right\},$$

$$Q_U^{(n)} = \left\{ (x, y) \in \mathbb{R}_+^2 : \sum_{i=1}^{r(n)} ((C_i x) \vee (D_i y)) \leq 1 \right\},$$

for some sequences $m(n)$, $r(n)$ and positive constants A_i, B_i, C_i, $D_i > 0$ satisfying $\sum_1^{m(n)} A_i = \sum_1^{m(n)} B_i = \sum_1^{r(n)} C_i = \sum_1^{r(n)} D_i = 1$.

Define distribution functions $G_L^{(n)}$ and $G_U^{(n)}$ by

$$G_L^{(n)}(x, y) := \exp\left(-\sum_{i=1}^{m(n)} \left(\frac{A_i}{x} \vee \frac{B_i}{y} \right) \right),$$

$$G_U^{(n)}(x, y) := \exp\left(-\sum_{i=1}^{r(n)} \left(\frac{C_i}{x} \vee \frac{D_i}{y} \right) \right).$$

Clearly $G_L^{(n)}$ and $G_U^{(n)}$ are simple max-stable distributions and there exist discrete spectral measures $\Psi_L^{(n)}$ and $\Psi_U^{(n)}$ with

$$G_L^{(n)}(x, y) = \exp\left(-\int_0^{\pi/2} \left(\frac{\cos\theta}{x} \vee \frac{\sin\theta}{y} \right) \Psi_L^{(n)}(d\theta) \right),$$

$$G_U^{(n)}(x, y) = \exp\left(-\int_0^{\pi/2} \left(\frac{\cos\theta}{x} \vee \frac{\sin\theta}{y} \right) \Psi_U^{(n)}(d\theta) \right).$$

The inclusions (6.1.35) and relation (6.1.36) imply

$$0 \leq \sum_{i=1}^{m(n)} ((A_i x) \vee (B_i y)) - \sum_{i=1}^{r(n)} ((C_i x) \vee (D_i y)) \to 0 , \quad n \to \infty ,$$

for $x, y > 0$. It follows that

$$0 \leq G_U^{(n)}(x, y) - G_L^{(n)}(x, y) \to 0 , \quad n \to \infty,$$

for $x, y > 0$ and hence there is a distribution function, G_0 say, such that for $x, y > 0$,

$$\lim_{n\to\infty} G_U^{(n)}(x, y) = \lim_{n\to\infty} G_L^{(n)}(x, y) = G_0(x, y) .$$

By Corollary 6.1.15 we have also

$$\lim_{n\to\infty} \Psi_L^{(n)}(\theta) = \lim_{n\to\infty} \Psi_U^{(n)}(\theta) = \Psi(\theta)$$

for $0 \le \theta \le \pi/2$. Clearly for

$$Q^\star := \left\{ (x, y) \in \mathbb{R}_+^2 : \int_0^{\pi/2} ((x \cos \theta) \vee (y \sin \theta)) \; \Psi(d\theta) \le 1 \right\}$$

we have

$$Q_L^{(n)} \subset Q^\star \subset Q_U^{(n)}$$

for all n. It follows that $Q^\star = Q_1$. Hence

$$Q_1 = \left\{ (x, y) \in \mathbb{R}_+^2 \; : \; -\log G_0 \, (1/x, 1/y) \le 1 \right\}$$

with

$$G_0(x, y) := \exp \left(-\int_0^{\pi/2} \left(\frac{\cos \theta}{x} \vee \frac{\sin \theta}{y} \right) \Psi(d\theta) \right) . \quad\blacksquare$$

Since the convexity of the level set Q_1 is typical for a limit distribution, this property can be used to check whether the tail of a given distribution function resembles a limit distribution. Details will be given later.

A related function to the function L (or the measure ν) is the function R:

$$R(x, y) := x + y - L(x, y) = \nu \left\{ (s, t) \in \mathbb{R}_+^2 : s > 1/x \text{ and } t > 1/y \right\} .$$

Note that the function R is the distribution function of a measure.

Finally, we review two other ways of characterizing the limit distribution G_0 in the two-dimensional context.

Sibuya (1960), see also Geffroy (1958), introduced for $t > 0$,

$$\chi(t) := -\log G_0 \, (1/t, 1) + \log G_0 \, (1/t, \infty) + \log G_0 \, (\infty, 1)$$
$$= L(t, 1) - L(t, 0) - L(0, 1) = -R(t, 1) . \qquad (6.1.37)$$

By the homogeneity of the function $-\log G_0$, the function χ determines the function G_0. The determining properties for the function χ are as follows:

1. χ is convex,
2. $((-t) \vee (-1)) \le \chi(t) \le 0$ for $t > 0$.

Pickands (1981) introduced for $0 \le t \le 1$,

$$A(t) := -\log G_0 \left(\frac{1}{1-t}, \frac{1}{t} \right) = L(1-t, t) . \tag{6.1.38}$$

By the homogeneity of the function $-\log G_0$, the function A determines the function G_0. The determining properties of the function A are the following:

1. A is convex,
2. $A(0) = A(1) = 1$,
3. $((1-t) \vee t) \leq A(t) \leq 1$.

Any function A satisfying Properties (1)–(3) leads to a unique limit function G_0. The convexity of A can be proved using

$$A(t) = 2 \int_0^1 (w(1-t) \vee ((1-w)t) H(dw)$$

with H as in Theorem 6.1.14: in case H has a density, break the integral into two integrals according to $w(1-t) > (1-w)t$ or $w(1-t) < (1-w)t$ and differentiate. If H does not have a density, one needs to approximate the measure H.

For the characterization of multivariate max-stable distributions one can use the distribution function of the spectral measure or any of the functions χ and A. Since the functions χ and A are more complicated objects (convex functions rather than monotone functions), we concentrate on the use of the spectral measure, which also has a simple intuitive meaning. Moreover, the generalization to higher-dimensional spaces is straightforward in that case.

6.2 Domains of Attraction; Asymptotic Independence

As stated in Section 6.1, the class of limiting distributions in (6.1.1) is characterized by three objects: the extreme value indices γ_1 and γ_2 of the marginal distributions and the spectral measure that governs the dependence structure. This is reflected in the domain of attraction conditions, which concern the marginal distribution and, separately, the dependence structure.

Let $G : \mathbb{R}^2 \to \mathbb{R}_+$ be a max-stable distribution function. A distribution function F is said to be in its *(max-) domain of attraction* (notation $F \in \mathcal{D}(G)$) if for sequences of constants $a_n, c_n > 0$ and b_n, d_n real,

$$\lim_{n \to \infty} F^n(a_n x + b_n, c_n y + d_n) = G(x, y)$$

for all $x, y \in \mathbb{R}$. This is the same as convergence in distribution since any max-stable distribution is continuous.

Theorem 6.2.1 *Let G be a max-stable distribution. Let the marginal distribution functions be $\exp\left(-(1 + \gamma_i x)^{-1/\gamma_i}\right)$ for $i = 1, 2$ and let Ψ or H or Φ be its spectral measure according to the representations of Theorem 6.1.14.*

1. *If the distribution function F of the random vector (X, Y) with continuous marginal distribution functions F_1 and F_2 is in the domain of attraction of G, then the following equivalent conditions are fulfilled:*

(a) *With* $U_i := (1/(1 - F_i))^\leftarrow, i = 1, 2, \text{for } x, y > 0,$

$$\lim_{t \to \infty} \frac{1 - F(U_1(tx), U_2(ty))}{1 - F(U_1(t), U_2(t))} = S(x, y) \qquad (6.2.1)$$

with $S(x, y) := \log G((x^{\gamma_1} - 1)/\gamma_1, (y^{\gamma_2} - 1)/\gamma_2) / \log G(0, 0).$

(b) *(Via the circle) For all $r > 1$ and all $\theta \in [0, \pi/2]$ that are continuity points of* $\Psi,$

$$\lim_{t \to \infty} P\left(V^2 + W^2 > t^2 r^2 \text{ and } \frac{W}{V} \leq \tan \theta \,\Big|\, V^2 + W^2 > t^2\right)$$
$$= r^{-1} \frac{\Psi(\theta)}{\Psi\left(\frac{\pi}{2}\right)}, \qquad (6.2.2)$$

where $V := 1/(1 - F_1(X))$ *and* $W := 1/(1 - F_2(Y)).$

(c) *(Via the triangle) For all $r > 1$ and all $s \in [0, 1]$ that are continuity points of* $H,$

$$\lim_{t \to \infty} P\left(V + W > tr \text{ and } \frac{V}{V + W} \leq s \,\Big|\, V + W > t\right)$$
$$= r^{-1} H(s), \qquad (6.2.3)$$

where $V := 1/(1 - F_1(X))$ *and* $W := 1/(1 - F_2(Y)).$

(d) *(Via the square) For all $r > 1$ and all $\theta \in [0, \pi/2]$ that are continuity points of* $\Phi,$

$$\lim_{t \to \infty} P\left(V \vee W > tr \text{ and } \frac{V}{W} \leq \tan \theta \,\Big|\, V \vee W > t\right)$$
$$= r^{-1} \frac{\Phi(\theta)}{\Phi\left(\frac{\pi}{2}\right)}, \qquad (6.2.4)$$

where $V := 1/(1 - F_1(X))$ *and* $W := 1/(1 - F_2(Y)).$

2. *Conversely, if the continuous marginal distribution functions F_i are in the domain of attraction of* $\exp\left(-(1 + \gamma_i x)^{-1/\gamma_i}\right),$ *for $i = 1, 2$, and any limit relation* (6.2.1)–(6.2.4) *holds for some positive function S or some bounded distribution function Ψ, H, or Φ, then F is in the domain of attraction of G.*

Proof. The direct statements follow immediately from the results of Section 6.1.

For the converse statement assume for example (6.2.4). For any $a, t > 0$ we define a probability measure $P_{a,t}$ on $\mathbb{R}_+^2 \setminus [0, a]^2$ by

$$P_{a,t}(B) := P\left((V, W) \in tB \mid V \vee W > ta\right)$$

for Borel sets $B \subset \mathbb{R}_+^2 \setminus [0, a]^2$. Note that by (6.2.4),

$$\lim_{t \to \infty} P_{a,t}(A) = P_a(A),$$

where P_a is a probability measure on $\mathbb{R}_+^2 \setminus [0, a]^2$ and A a P_a-continuity set of the form

$$\{(x, y) \ : \ x \vee y > r \text{ and } x/y \le \tan \theta\}$$

with $r > a$ and $0 \le \theta \le \pi/2$. The finite unions of sets of this form constitute a family that is closed under finite intersections and such that each open set in $\mathbb{R}_+^2 \setminus [0, a]^2$ is a countable union of sets in the family. It follows (Billingsley (1968), Theorem 2.2) that

$$\lim_{t \to \infty} P_{a,t}(B) = P_a(B)$$

for P_a-continuity Borel sets B in $\mathbb{R}_+^2 \setminus [0, a]^2$. Since this is true for all $a > 0$, in particular (6.2.1) holds. Hence the statements (1a)–(1d) are equivalent. We proceed with statement (1a).

Since the function S is homogeneous of order -1, the statement implies that the function $1 - F(U_1(t), U_2(t))$ is regularly varying with index -1. Hence there exists a sequence $a_n > 0$, $a_n \to \infty$ as $n \to \infty$, with

$$\lim_{n \to \infty} n \{1 - F(U_1(a_n), U_2(a_n))\} = - \log G(0, 0) .$$

It then follows from (6.2.1) that

$$\lim_{n \to \infty} n \{(1 - F(U_1(a_n x), U_2(a_n y)))\} = - \log G \left(\frac{x^{\gamma_1} - 1}{\gamma_1}, \frac{y^{\gamma_2} - 1}{\gamma_2} \right)$$

and hence

$$\lim_{n \to \infty} F^n (U_1(a_n x), U_2(a_n y)) = G \left(\frac{x^{\gamma_1} - 1}{\gamma_1}, \frac{y^{\gamma_2} - 1}{\gamma_2} \right) .$$

In particular, the marginal distribution converges:

$$\lim_{n \to \infty} F^n (U_1(a_n x), \infty) = G \left(\frac{x^{\gamma_1} - 1}{\gamma_1}, \infty \right) = \exp \left(-\frac{1}{x} \right) .$$

Since the distribution function $F(U_1(x), \infty)$ is in fact $1 - 1/x$, $x \ge 1$, we have also

$$\lim_{n \to \infty} F^n (U_1(nx), \infty) = \exp \left(-\frac{1}{x} \right) .$$

It follows that $\lim_{n \to \infty} a_n/n = 1$, i.e.,

$$\lim_{n \to \infty} F^n (U_1(nx), U_2(ny)) = G \left(\frac{x^{\gamma_1} - 1}{\gamma_1}, \frac{y^{\gamma_2} - 1}{\gamma_2} \right) . \tag{6.2.5}$$

We now proceed as in the proof of Theorem 6.1.1. The convergence of the marginal distributions implies, for $a_n, c_n > 0$,

$$\lim_{n \to \infty} \frac{U_1(nx) - U_1(n)}{a_n} = \frac{x^{\gamma_1} - 1}{\gamma_1},$$

$$\lim_{n \to \infty} \frac{U_2(ny) - U_2(n)}{c_n} = \frac{y^{\gamma_2} - 1}{\gamma_2} . \tag{6.2.6}$$

Combining (6.2.5) and (6.2.6) as in the proof of Theorem 6.1.1 we get

$$\lim_{n\to\infty} F^n \left(a_n x + U_1(n), c_n y + U_2(n)\right) = G(x, y) . \qquad \blacksquare$$

Remark 6.2.2 In fact if the marginal distributions are in some domains of attraction, if for all x, y,

$$\lim_{t\to\infty} \frac{1 - F(U_1(tx), U_2(ty))}{1 - F(U_1(t), U_2(t))}$$

exists and is positive and if the regularly varying function $1 - F(U_1(t), U_2(t))$ has index -1, then F is in the domain of attraction of some max-stable distribution.

A particular case is the domain of attraction of a max-stable distribution with independent components, i.e., one that is the product of its marginal distributions. A random vector $(X_1, X_2, \ldots, X_d)$ whose distribution is in the domain of attraction of such a max-stable distribution is said to have the property of *asymptotic independence*. A simple criterion for this to happen is given in the next theorem.

Theorem 6.2.3 *Let $F : \mathbb{R}^d \to \mathbb{R}_+$ be a probability distribution function. Suppose that its marginal distribution functions $F_i : \mathbb{R} \to \mathbb{R}_+$ satisfy*

$$\lim_{n\to\infty} F_i^n \left(a_n^{(i)} x + b_n^{(i)}\right) = \exp\left(-(1 + \gamma_i x)^{-1/\gamma_i}\right)$$

for all x for which $1 + \gamma_i x > 0$ and where $a_n^{(i)} > 0$ and $b_n^{(i)}$ are sequences of real constants, $i = 1, 2, \ldots, d$. Let $(X_1, X_2, \ldots, X_d)$ be a random vector with distribution function F. If

$$\lim_{t\to\infty} \frac{P\left(X_i > U_i(t), X_j > U_j(t)\right)}{P\left(X_i > U_i(t)\right)} = 0 \qquad (6.2.7)$$

for all $1 \leq i < j \leq d$, then

$$\lim_{n\to\infty} F^n \left(a_n^{(1)} x_1 + b_n^{(1)}, \ldots, a_n^{(d)} x_d + b_n^{(d)}\right) = \exp\left(- \sum_{i=1}^{d} (1 + \gamma_i x_i)^{-1/\gamma_i}\right)$$

for $1 + \gamma_i x_i > 0$, $i = 1, 2, \ldots, d$. Hence the components of $(X_1, X_2, \ldots, X_d)$ are asymptotically independent.

Remark 6.2.4 It is clear that conversely, asymptotic independence entails (6.2.7).

Remark 6.2.5 The result means in particular that pairwise asymptotic independence implies joint asymptotic independence.

Proof (of Theorem 6.2.3). Relation (6.2.7) means that for the exponent measure ν (Section 6.1.3) we have

$$\nu \left\{(s_1, s_2, \ldots, s_d) \in \mathbb{R}_+^d : s_i > 0 \text{ and } s_j > 0\right\} = 0 .$$

Since this is true for all pairs (i, j), the exponent measure must be concentrated on the lines

$$l_i = \left\{ (s_1, s_2, \ldots, s_d) \in \mathbb{R}_+^d : s_i > 0 \text{ and } s_j = 0 \quad \text{for} \quad i \neq j \right\}.$$

This is the same as saying that the spectral measure is concentrated on the extreme points, i.e., the limit distribution has independent components (cf. Remark 6.1.16). ∎

Example 6.2.6 (Sibuya (1960)) Consider the random vector (X, Y), normally distributed with mean zero, variances one, and correlation coefficient $\rho < 1$. We shall prove asymptotic independence in this case, i.e.,

$$\lim_{n \to \infty} n P (X > b_n, Y > b_n) = 0$$

with b_n chosen in such a way that

$$\lim_{n \to \infty} n P (X > b_n) = \lim_{n \to \infty} n (1 - F(b_n)) = 1$$

(cf. Example 1.1.7). Note that

$$n P (X > b_n, Y > b_n) \leq n P \left(\frac{X + Y}{2} > b_n \right).$$

Now, $(X + Y)/2$ has a normal distribution with variance $(1 + \rho)/2$. If $\rho = -1$ the result is immediate. If $|\rho| < 1$,

$$\lim_{n \to \infty} n P \left(\frac{X + Y}{2} > b_n \right)$$

$$= \lim_{n \to \infty} n P \left(X > \sqrt{\frac{2}{1 + \rho}} b_n \right) = \lim_{n \to \infty} \frac{P \left(X > \sqrt{\frac{2}{1+\rho}} b_n \right)}{P (X > b_n)},$$

and this limit is zero by Corollary 5.4.2.

Note that if $\rho = 1$ the limit is one.

In the two-dimensional situation, when we have asymptotic independence, a quite natural submodel can be defined. Since the model appears most naturally in a statistical context, we postpone the discussion to Sections 7.5 and 7.6.

Exercises

6.1. Let A_1, A_2 be positive random variables with $E A_i = 1$ for $i = 1, 2$. Prove that

$$G(x, y) := \exp \left(-E \left(\frac{A_1}{x} \vee \frac{A_2}{y} \right) \right)$$

is a distribution function that is simple max-stable.

6.2. Let F_n be two-dimensional normal distribution functions with means zero, variances one, and covariances ρ_n. Define $a_n := (2\log n)^{1/2}$ and $b_n := (2\log n - \log\log n - \log(4\pi))^{1/2}$. Then we have (Example 1.1.7) $\lim_{n\to\infty} n(1 - F(a_n x + b_n)) = e^{-x}$, for $x \in \mathbb{R}$, where F is the standard normal distribution function. Take ρ_n such that $a_n^2/(1-\rho_n) \to \lambda > 0, n \to \infty$. Prove that $n(\partial^2/(\partial x \partial y))(1 - F_n(a_n x + b_n, a_n y + b_n))$ converges to $2^{-1}\log(\lambda/(4\pi)) - (4\lambda)^{-1} - \lambda 4^{-1}(x-y)^2 - 2^{-1}(x+y)$ for $x, y \in \mathbb{R}$. Conclude that

$$\lim_{n\to\infty} n\left(1 - F_n(a_n x + b_n, a_n y + b_n)\right) = -\log G_0(x, y)$$

with G_0 form Theorem 6.1.1. Cf. Example 6.1.20.

6.3. Prove that if (X, Y) are random variables with distribution function F with continuous marginals then the following are equivalent:
(a) F is in the domain of attraction of some max-stable distribution G.
(b) For any Borel set $A \subset \mathbb{R}^2_+$ with $\inf_{(x,y)\in A} \max(x, y) > 0$ and $\nu(\partial A) = 0$,

$$\lim_{n\to\infty} nP\left\{\left(\left(1 + \gamma_1 \frac{X - b_n}{a_n}\right)^{1/\gamma_1}, \left(1 + \gamma_2 \frac{Y - d_n}{c_n}\right)^{1/\gamma_2}\right) \in A\right\} = \nu(A),$$

where the sequences $a_n, c_n > 0, b_n, d_n \in \mathbb{R}$ are chosen so that $G(x, \infty)$ is as (6.1.4), $G(\infty, y)$ is as (6.1.5), and ν is the exponent measure defined in Section 6.1.3.
(c) For any Borel set $A \subset \mathbb{R}^2_+$ with $\inf_{(x,y)\in A} \max(x, y) > 0$ and $\nu(\partial A) = 0$,

$$\lim_{n\to\infty} nP\left\{\left(\frac{1}{1 - F_1(X)}, \frac{1}{1 - F_2(Y)}\right) \in nA\right\} = \nu(A),$$

where $F_i : \mathbb{R} \to \mathbb{R}_+, i = 1, 2$, are the marginal distribution functions of F and satisfy, for some sequences $a_n, c_n > 0, b_n, d_n \in \mathbb{R}$, $\lim_{n\to\infty} F_1^n(a_n x + b_n) = \exp\left(-(1 + \gamma_1 x)^{-1/\gamma_1}\right)$ and $\lim_{n\to\infty} F_2^n(c_n x + d_n) = \exp\left(-(1 + \gamma_2 x)^{-1/\gamma_2}\right)$ for all x for which $1 + \gamma_i x > 0, i = 1, 2$.
(d) For any Borel set $A \subset \mathbb{R}^2_+$ with $\inf_{(x,y)\in A} \max(x, y) > 0$ and $\nu(\partial A) = 0$,

$$\lim_{t\downarrow 0} t^{-1} P\left\{(1 - F_1(X), 1 - F_2(Y)) \in tA^{-1}\right\} = \nu(A)$$

with $F_i, i = 1, 2$, as before.

6.4. Prove Theorem 6.1.14.

6.5. A distribution function F in the d-dimensional space is called *max-infinitely divisible* if for all n there is a distribution function F_n with $F_n^n = F$, i.e., for each n the random vector can be written as the maximum of n independent and identically distributed random vectors. Using the method of Section 6.1.3 prove that F is max-infinitely divisible if and only if

$$-\log F(x_1, x_2, \ldots, x_d)$$
$$= \nu\{(s_1, s_2, \ldots, s_n) : s_i > x_i \text{ for at least one } i \text{ with } i = 1, 2, \ldots, d\}$$

for all $(x_1, x_2, \ldots, x_d)$ with $0 < F(x_1, x_2, \ldots, x_d) < 1$, where ν is a measure (not necessarily homogeneous).

6.6. With H' the density of the spectral measure H and $q(x, y)$ the density of $-\log G_0(x, y)$, verify that for $r = x + y$ and $\theta = x/(x + y)$, $H'(\theta) = r^3 q(\theta r, r(1 - \theta))$.

6.7. With Φ' the density of the spectral measure Φ and $q(x, y)$ the density of $-\log G_0(x, y)$, verify that for $r = \max(x, y)$ and $\theta = \arctan(x/y)$,

$$\Phi'(\theta) = \begin{cases} r^3 q(r, r \tan \theta)/\cos^2 \theta, & 0 \le \theta \le \pi/4, \\ r^3 q\,(r, r/\tan \theta)\,/\sin^2 \theta, & \pi/4 \le \theta \le \pi/2 . \end{cases}$$

6.8. If (X, Y) is a random vector with some simple max-stable distribution function, then $L(x, y) = x + y$ for all $x, y > 0$ if and only if (X, Y) are independent.

6.9. Discuss properties of the function R (cf. Proposition 6.1.21).

6.10. Prove that $R(x, y)$ is positive for all $x, y > 0$ or $R(x, y) = 0$ for all $x, y > 0$.

6.11. Let $(V_1, V_2, \ldots, V_d)$ be independent and identically distributed random variables with distribution function $\exp -(1/x)$, $x > 0$. Let $\{r_{i,j}\}_{i,j=1}^d$ be a matrix with positive entries. Show that the random vector $(\vee_{j=1}^d r_{1,j} V_j, \ldots, \vee_{j=1}^d r_{d,j} V_j)$ has a simple max-stable distribution. Find the distribution function. Show that any two-dimensional simple max-stable distribution function can be obtained as a limit of elements in this class.

6.12. Let X, Y be independent positive random variables with distribution function F. Suppose $\lim_{n \to \infty} n P(X > x\, a(n)) = x^{-\alpha}$ for some $\alpha > 0$ and all $x > 0$. Show that for $\lambda_1, \lambda_2, \nu_1, \nu_2$ positive, the random vector $(\lambda_1 X_1 + \lambda_2 X_2, \nu_1 X_1 + \nu_2 X_2)$ is in the domain of attraction of the extreme value distribution

$$\exp - \left(\left(\frac{x}{\lambda_1} \vee \frac{y}{\nu_1} \right)^{-\alpha} + \left(\frac{x}{\lambda_2} \vee \frac{y}{\nu_2} \right)^{-\alpha} \right),$$

the spectral measure of which is purely discrete, with two atoms.
Hint: Apply Theorem 6.1.5 to $\nu_n(\cdot) := n P((X, Y) \in a(n)\cdot)$.

6.13. Let X, Y be independent random variables with distribution function F. Suppose that $\lim_{n \to \infty} n P(-X < x\, a(n)) = x^{-\alpha}$ for some $\alpha > 0$ and all $x > 0$. That is, F is in the domain of attraction of an extreme value distribution G_γ with $\gamma = -1/\alpha < 0$. Show that as $n \to \infty$, for $\lambda_1, \lambda_2, \nu_1, \nu_2$ positive and all $x > 0$,

$$\lim_{n \to \infty} n^2 P \left(-(\lambda_1 X_1 + \lambda_2 X_2) < x\, a(n) \text{ or } -(\nu_1 X_1 + \nu_2 X_2) < y\, a(n) \right)$$

$$= \alpha^2 \iint_B s^{-\alpha-1} t^{-\alpha-1} \, ds\, dt$$

where $B := \{(s, t) \in \mathbb{R}_+^2 : \lambda_1 s + \lambda_2 t \leq x, \nu_1 s + \nu_2 t \leq y\}$. Conclude that the distribution of $(\lambda_1 X_1 + \lambda_2 X_2, \nu_1 X_1 + \nu_2 X_2)$ is in the domain of attraction of the extreme value distribution with nondiscrete spectral measure. The marginal distributions have extreme value index 2α.

Hint: Apply Theorem 6.1.5 to $\nu_n(\cdot) := n^2 P((X, Y) \in a(n)\cdot)$.

7

Estimation of the Dependence Structure

7.1 Introduction

In Chapter 6 we have seen that a multivariate extreme value distribution is characterized by the marginal extreme value indices plus a homogeneous exponent measure
or alternatively a spectral measure. In particular, there is no finite parametrization
for extreme value distributions. This suggests the use of nonparametric methods for
estimating the dependence structure, and in fact we are going to emphasize those
methods.

In Sections 7.2 and 7.3 we shall consider estimation of the exponent measure ν
exemplified by the function L and the sets Q_c of Section 6.1.5, as well as estimation
of the spectral measure introduced in Section 6.1.4.

Further, in Section 7.4 we shall discuss a simple coefficient that summarizes the
amount of dependence between components of the random vector.

Finally, in Sections 7.5–7.6, for the case of asymptotic independence of the components, we shall discuss a submodel that allows for a more precise analysis in that case.

7.2 Estimation of the Function L and the Sets Q_c

Recall that for any extreme value distribution function G for which the marginal distribution functions have the standard von Mises form we have defined the distribution
function G_0 by (cf. Section 6.1.2)

$$G_0(x, y) := G\left(\frac{x^{\gamma_1} - 1}{\gamma_1}, \frac{y^{\gamma_2} - 1}{\gamma_2}\right) ,$$

where γ_1, γ_2 are the extreme value indices of the marginal distributions. The relation between G_0 and the exponent measure ν from Section 6.1.3 is (cf. Theorem 6.1.5)

$$G_0(x, y) = \exp\left(-\nu\left\{(s, t) \in \mathbb{R}_+^2 : s > x \text{ or } t > y)\right\}\right) , \quad x, y > 0 .$$

Next we defined the function L by (cf. Section 6.1.5)

$$L(x, y) := -\log G_0 \left(\frac{1}{x}, \frac{1}{y} \right)$$

for $x, y > 0$. In fact, L is connected to the exponent measure ν as follows:

$$L(x, y) := \nu \left\{ (s, t) \in \mathbb{R}_+^2 \; : \; s > 1/x \text{ or } t > 1/y \right\} .$$

It is clear that L determines G_0 and ν. So in estimating the function L in fact we estimate the dependence structure of the extreme value distribution.

Since it is not realistic to assume that the available observations have been taken from the extreme value distribution itself, we assume instead that we have independent and identically distributed observations $(X_1, Y_1), (X_2, Y_2), \ldots, (X_n, Y_n)$ from a distribution function F that is in the domain of attraction of an extreme value distribution. Suppose that the marginal distribution functions of F are continuous. The domain of attraction condition can be expressed as (cf. Corollary 6.1.4, Section 6.1.2)

$$\lim_{t \to \infty} t \{1 - F (U_1(tx), U_2(ty))\} = -\log G_0(x, y),$$

where t runs through the reals, i.e.,

$$\lim_{t \to \infty} t \left\{ 1 - F \left(U_1 \left(\frac{t}{x} \right), U_2 \left(\frac{t}{y} \right) \right) \right\} = L(x, y) . \tag{7.2.1}$$

We want to use relation (7.2.1) to obtain an estimator for L by replacing F by its empirical measure and also U_1 and U_2 by their empirical counterparts (we shall use the empirical left-continuous versions). The empirical counterpart of $U_1(t/x)$ is $X_{n-[nx/t]+1,n}$. In particular, for $U_1(t)$ and $U_2(t)$, two basic quantities, we get $X_{n-[n/t]+1,n}$ and $Y_{n-[n/t]+1,n}$. Since this vector plays an anchor role we want to call them $X_{n-k+1,n}$ and $Y_{n-k+1,n}$ respectively. That means that we read (7.2.1) as

$$\lim_{n \to \infty} \frac{n}{k} \left\{ 1 - F \left(U_1 \left(\frac{n}{kx} \right), U_2 \left(\frac{n}{ky} \right) \right) \right\} = L(x, y), \tag{7.2.2}$$

where k may depend on n but we need to have $k = o(n)$. We shall see that in order to get consistency for our estimators we also need to assume $k = k(n) \to \infty, n \to \infty$. Replacing F by its empirical distribution function, $U_1(n/(kx))$ by $X_{n-[kx]+1,n}$, and $U_2(n/(ky))$ by $Y_{n-[ky]+1,n}$ in the left-hand side of (7.2.2), we get

$$\hat{L}(x, y) := \frac{1}{k} \sum_{i=1}^{n} 1_{\{X_i \geq X_{n-[kx]+1,n} \text{ or } Y_i \geq Y_{n-[ky]+1,n}\}} \tag{7.2.3}$$

$$= \frac{1}{k} \sum_{i=1}^{n} 1_{\{R(X_i) \geq n-kx+1 \text{ or } R(Y_i) \geq n-ky+1\}}, \tag{7.2.4}$$

where $R(X_i)$ is the *rank* of X_i among $(X_1, X_2, \ldots, X_n), i = 1, \ldots, n$, i.e.,

$$R(X_i) := \sum_{j=1}^{n} 1_{\{X_j \leq X_i\}} \, ,$$

and $R(Y_i)$ is the rank of Y_i among $(Y_1, Y_2, \ldots, Y_n)$.

Indeed, this estimator is invariant under monotone transformations of the components of the random vector; hence it does not depend on the marginal distributions.

We will establish consistency and asymptotic normality for $\hat{L}$. We start with the consistency.

Theorem 7.2.1 *Let $(X_1, Y_1), (X_2, Y_2), \ldots$ be i.i.d. random vectors with distribution function F. Suppose F is in the domain of attraction of an extreme value distribution G. Define*

$$L(x, y) := -\log G\left(\frac{x^{-\gamma_1} - 1}{\gamma_1}, \frac{y^{-\gamma_2} - 1}{\gamma_2}\right)$$

for $x, y > 0$, where γ_1, γ_2 are the marginal extreme value indices. Let G be such that $L(x, 0) = L(0, x) = x$ (this means that the marginal distribution functions of G are exactly $\exp\left(-(1 + \gamma_i x)^{-1/\gamma_i}\right)$ for $i = 1, 2$). Then for $T > 0$, as $n \to \infty$, $k = k(n) \to \infty$, $k/n \to 0$,

$$\sup_{0 \leq x, y \leq T} \left|\hat{L}(x, y) - L(x, y)\right| \xrightarrow{P} 0 \, .$$

Proof. First we show that it is sufficient to prove pointwise convergence. Fix $\varepsilon > 0$. Select

$$(0, 0) = (x_0, y_0); (x_1, y_1), \ldots, (x_r, y_r) \in [0, T] \times [0, T]; (x_{r+1}, y_{r+1}) = (T, T) \, ,$$

such that for $i = 0, 1, 2, \ldots, r$,

$$0 < L(x_{i+1}, y_{i+1}) - L(x_i, y_i) < \frac{\varepsilon}{3} \, .$$

This is possible since L is bounded on finite rectangles, continuous, and monotone (cf. Proposition 6.1.21). Suppose

$$\hat{L}(x_i, y_i) \xrightarrow{P} L(x_i, y_i)$$

for $i = 0, 1, 2, \ldots, r + 1$. Then as $n \to \infty$,

$$P\left(\sup_{0 \leq i \leq r+1} \left|\hat{L}(x_i, y_i) - L(x_i, y_i)\right| > \frac{\varepsilon}{3}\right) \to 0 \, . \tag{7.2.5}$$

By the monotonicity of $\hat{L}$ and L,

$$P\left(\sup_{(0,0) \leq (x,y) \leq (T,T)} \left|\hat{L}(x, y) - L(x, y)\right| > \varepsilon\right)$$

is at most the left-hand side of (7.2.5).

Note that the result is quite general: it holds for any function L bounded on finite rectangles, continuous and monotone, and $\hat{L}$ monotone.

Next we prove pointwise convergence. Consider fixed $x, y > 0$. Since $\hat{L}(x, y)$ is invariant under monotone transformations of the components of the vector, we can write, with $U_i := 1 - F_1(X_i)$ and $W_i := 1 - F_2(Y_i)$, for $i = 1, 2, \ldots, n$,

$$\hat{L}(x, y) := \frac{1}{k} \sum_{i=1}^{n} 1_{\{U_i \leq U_{[kx],n} \text{ or } W_i \leq W_{[ky],n}\}} .$$

We deal with this in two steps. First we consider

$$V_{n,k}(x, y) := \frac{1}{k} \sum_{i=1}^{n} 1_{\{U_i \leq kx/n \text{ or } W_i \leq ky/n\}} . \qquad (7.2.6)$$

The characteristic function of $V_{n,k}(x, y)$ is

$$\left\{ (1 - p_{n,k}) + p_{n,k} e^{it/k} \right\}^n = \left\{ 1 - p_{n,k} \left(1 - e^{it/k} \right) \right\}^n$$
$$= \left\{ 1 - \frac{1}{n} \left(\frac{n}{k} p_{n,k} k \left(1 - e^{it/k} \right) \right) \right\}^n$$

with $p_{n,k} := P(U_i \leq kx/n \text{ or } W_i \leq ky/n)$. We know by (7.2.2) that $n p_{n,k}/k \to L(x, y), n \to \infty$. It follows that the characteristic function converges to $\exp(it L(x, y))$, i.e.,

$$V_{n,k}(x, y) \overset{P}{\to} L(x, y) .$$

Again by continuity and monotonicity the convergence is locally uniform. Next note that

$$\hat{L}(x, y) = V_{n,k} \left(\frac{n}{k} U_{[kx],n}, \frac{n}{k} W_{[ky],n} \right)$$

(the random objects $U_{[kx],n}$, $W_{[ky],n}$, and $V_{n,k}$ are dependent but this is not relevant for the present consistency proof).

Since

$$\frac{n}{k} U_{[kx],n} \overset{P}{\to} x \quad \text{and} \quad \frac{n}{k} W_{[ky],n} \overset{P}{\to} y$$

(cf. Lemma 2.4.11), we have proved

$$\hat{L}(x, y) \overset{P}{\to} L(x, y) . \qquad \blacksquare$$

Next we move to the more complicated matter of proving asymptotic normality for $\hat{L}$.

For the formulation of the theorem it is useful to introduce a measure μ that is closely related to the measure ν from Section 6.1.3 (cf. Remark 6.1.6) as follows: for $x, y > 0$,

$$\mu\{(s, t) \in [0, \infty]^2 \setminus \{(\infty, \infty)\} \ : \ s < x \text{ or } t < y\}$$
$$:= \nu\{(s, t) \in [0, \infty]^2 \setminus \{(0, 0)\} \ : \ s > 1/x \text{ or } t > 1/y\} . \quad (7.2.7)$$

So the two measures are the same modulo the marginal transformations $x \mapsto 1/x$ and $y \mapsto 1/y$.

Let $D([0, T] \times [0, T])$ be the space of functions in $[0, T] \times [0, T]$ that are right-continuous and have left-hand limits.

Theorem 7.2.2 *Let $(X_1, Y_1), (X_2, Y_2), \ldots$ be i.i.d random vectors with distribution function F. Suppose F is in the domain of attraction of some extreme value distribution G. Suppose also that the marginal distribution functions F_1 and F_2 are continuous. Let the marginal distributions of G be standard, i.e., the function L defined by*

$$L(x, y) := -\log G \left(\frac{x^{-\gamma_1} - 1}{\gamma_1}, \frac{y^{-\gamma_2} - 1}{\gamma_2} \right)$$

(with γ_1, γ_2 the marginal extreme value indices) satisfies $L(x, 0) = L(0, x) = x$. Suppose that for some $\alpha > 0$ and for all $x, y > 0$,

$$t \left\{ 1 - F \left(U_1 \left(\frac{t}{x} \right), U_2 \left(\frac{t}{y} \right) \right) \right\} = L(x, y) + O(t^{-\alpha}) , \quad (7.2.8)$$

$t \to \infty$, *where $U_i := (1/(1 - F_i))^{\leftarrow}$, $i = 1, 2$, holds uniformly on the set*

$$\left\{ x^2 + y^2 = 1, x \geq 0, y \geq 0 \right\} .$$

Suppose further that the function L has continuous first-order partial derivatives

$$L_1(x, y) := \frac{\partial}{\partial x} L(x, y) \quad \text{and} \quad L_2(x, y) := \frac{\partial}{\partial y} L(x, y)$$

for $x, y \geq 0$. Then for $k = k(n) \to \infty$, $k(n) = o \left(n^{2\alpha/(1+2\alpha)} \right)$, as $n \to \infty$,

$$\sqrt{k} \left(\hat{L}(x, y) - L(x, y) \right) \xrightarrow{d} B(x, y)$$

in $D([0, T] \times [0, T])$, for every $T > 0$, where

$$B(x, y) = W(x, y) - L_1(x, y)W(x, 0) - L_2(x, y)W(0, y)$$

and W is a continuous mean-zero Gaussian process with covariance structure

$$EW(x_1, y_1)W(x_2, y_2) = \mu \left(R(x_1, y_1) \cap R(x_2, y_2) \right)$$

with

$$R(x, y) := \left\{ (u, v) \in \mathbb{R}_+^2 \ : \ 0 \leq u \leq x \text{ or } 0 \leq v \leq y \right\} .$$

It is useful to prove a similar statement for $V_{n,k}$ (see the proof of Theorem 7.2.1) before giving the proof of the theorem.

Proposition 7.2.3 *Let* $(X_1, Y_1), (X_2, Y_2), \ldots$ *be i.i.d random vectors with distribution function* F. *Suppose that for* $x, y \geq 0$,

$$\lim_{t \to \infty} t \left\{ 1 - F\left(U_1\left(\frac{t}{x}\right), U_2\left(\frac{t}{y}\right)\right) \right\} = L(x, y) .$$

Then, provided $k = k(n) \to \infty$, $k/n \to 0$ *as* $n \to \infty$,

$$\sqrt{k}\left(V_{n,k}(x, y) - \frac{n}{k}\left\{1 - F\left(U_1\left(\frac{n}{kx}\right), U_2\left(\frac{n}{ky}\right)\right)\right\}\right) \overset{d}{\to} W(x, y)$$

in $D([0, T] \times [0, T])$, *for every* $T > 0$.

Proof. We prove convergence of finite-dimensional distributions plus tightness (cf. Billingsley (1968), Theorem 15.1). For ease of writing we take $T = 1$. The changes in the proof for $T \neq 1$ are obvious. For the convergence of finite-dimensional distributions it is sufficient to prove (Cramér–Wold device) that

$$\sum_{r=1}^{d} t_r \sqrt{k}\left(V_{n,k}(x_r, y_r) - \frac{n}{k}\left\{1 - F\left(U_1\left(\frac{n}{kx_r}\right), U_2\left(\frac{n}{ky_r}\right)\right)\right\}\right)$$

$$\overset{d}{\to} \sum_{r=1}^{d} t_r W(x_r, y_r)$$

for $d = 1, 2, \ldots$, all real numbers $t_1, \ldots, t_d$, and all $(x_1, y_1), \ldots, (x_d, y_d)$. This can be done conveniently by applying Lyapunov's form of the central limit theorem (Chung (1974), Theorem 7.1.2).

In order to establish tightness we define subrectangles

$$I_{ij} := \left[\frac{i}{m}, \frac{i+1}{m}\right] \times \left[\frac{j}{m}, \frac{j+1}{m}\right]$$

for $i, j = 0, 1, 2, \ldots, m - 1$ and define

$$W_n(x, y) := \sqrt{k}\left(V_{n,k}(x, y) - \frac{n}{k}\left\{1 - F\left(U_1\left(\frac{n}{kx}\right), U_2\left(\frac{n}{ky}\right)\right)\right\}\right)$$

for $0 \leq x, y \leq 1$.

The main tool is an inequality from Einmahl (1987), which for our purposes can be written as follows. Define for a rectangle $S \in [0, 1]^2$,

$$V_{n,k}(S) := \frac{1}{k}\sum_{i=1}^{n} 1_{\{n(1-F_1(X_i), 1-F_2(Y_i))/k \in S\}}$$

and

$$v_{n,k}(S) := \frac{n}{k} P\left(\frac{n}{k}(1 - F_1(X), 1 - F_2(Y)) \in S\right) .$$

Einmahl's inequality: Let R be a rectangle in $[0, 1]^2$ with

$$0 < \frac{n}{k} P\left(\frac{n}{k}\left(1 - F_1(X), 1 - F_2(Y)\right) \in R\right) \le \frac{1}{2}.$$

Then there exists a constant $C > 0$ such that

$$P\left(\sup_{S \subset R} \left|\sqrt{k}\left(V_{n,k}(S) - v_{n,k}(S)\right)\right| \ge \lambda\right) \le C \exp\left(\frac{-\lambda^2}{4v_{n,k}(R)}\psi\left(\frac{\lambda}{\sqrt{k}v_{n,k}(R)}\right)\right)$$

for $\lambda \ge 0$. The function ψ satisfies the following conditions: $\psi(x)$ is continuous and decreasing, $x\psi(x)$ is increasing, and $\psi(0) = 1$. In particular, this implies that $\psi(x) \ge 0$ for $x \ge 0$.

We shall apply the inequality to the rectangles

$$J_{i,j} := \left[0, \frac{i+1}{m}\right] \times \left[\frac{j}{m}, \frac{j+1}{m}\right],$$

$$K_{i,j} := \left[\frac{i}{m}, \frac{i+1}{m}\right] \times \left[0, \frac{j+1}{m}\right],$$

$i, j = 1, 2, \ldots, m - 1$.

First note that if for some $\mathbf{x} = (x_1, x_2)$ and $\mathbf{y} = (y_1, y_2)$ with $|\mathbf{x} - \mathbf{y}| < \delta$ (consider the Euclidean norm) we have $|W_n(\mathbf{x}) - W_n(\mathbf{y})| > \varepsilon$, then there exist i, j such that $\mathbf{x}$ and $\mathbf{y}$ are in I_{ij} with $m = \lceil \sqrt{2}/\delta \rceil$ and

$$\left|W_n(\mathbf{x}) - W_n\left(\frac{i}{m}, \frac{j}{m}\right)\right| > \frac{\varepsilon}{2} \quad \text{or} \quad \left|W_n(\mathbf{y}) - W_n\left(\frac{i}{m}, \frac{j}{m}\right)\right| > \frac{\varepsilon}{2}.$$

Hence

$$P\left(\sup_{|\mathbf{x}-\mathbf{y}|<\delta/2} |W_n(\mathbf{x}) - W_n(\mathbf{y})| > 4\varepsilon\right)$$

$$\le P\left(\max_{i,j=0,1,2,\ldots,m-1} \sup_{\mathbf{x} \in I_{ij}} \left|W_n(\mathbf{x}) - W_n\left(\frac{i}{m}, \frac{j}{m}\right)\right| > 2\varepsilon\right)$$

$$\le \sum_{i=0}^{m-1}\sum_{j=0}^{m-1} P\left(\sup_{\mathbf{x} \in I_{ij}} \left|W_n(\mathbf{x}) - W_n\left(\frac{i}{m}, \frac{j}{m}\right)\right| > 2\varepsilon\right)$$

$$\le \sum_{i=0}^{m-1}\sum_{j=0}^{m-1} P\left(\sup_{\mathbf{x} \in I_{ij}} \left|W_n(\mathbf{x}) - W_n\left(x_1, \frac{j}{m}\right)\right| > \varepsilon\right)$$

$$+ P\left(\sup_{\mathbf{x} \in I_{ij}} \left|W_n\left(x_1, \frac{j}{m}\right) - W_n\left(\frac{i}{m}, \frac{j}{m}\right)\right| > \varepsilon\right)$$

$$\le 2\sum_{i=0}^{m-1}\sum_{j=0}^{m-1} C \exp\left(-\frac{\varepsilon^2}{2v_{n,k}(J_{ij})}\psi\left(\frac{\varepsilon}{\sqrt{k}v_{n,k}(J_{ij})}\right)\right)$$

$$+ C \exp\left(-\frac{\varepsilon^2}{2v_{n,k}(K_{ij})}\psi\left(\frac{\varepsilon}{\sqrt{k}v_{n,k}(K_{ij})}\right)\right), \tag{7.2.9}$$

where for the last inequality we apply Einmahl's inequality with R replaced with J_{ij} and K_{ij}. Note that $W_{x_1,0} = W_{0,x_2} = 0$ for $x_1, x_2 \geq 0$.

Next note that

$$v_{n,k}(J_{ij}) \leq \frac{n}{k} P\left\{ \frac{n}{k}\left(1 - F_2(Y) \in \left[\frac{j}{m}, \frac{j+1}{m} \right] \right) \right\} = \frac{1}{m}$$

and

$$v_{n,k}(K_{ij}) \leq \frac{n}{k} P\left\{ \frac{n}{k}\left(1 - F_1(X) \in \left[\frac{i}{m}, \frac{i+1}{m} \right] \right) \right\} = \frac{1}{m}.$$

Hence by the monotonicity of $x\psi(x)$ expression (7.2.9) is at most

$$\sum_{i=0}^{m-1} \sum_{j=0}^{m-1} 2C \exp\left(-\frac{m\varepsilon^2}{4} \psi\left(\frac{\varepsilon m}{2\sqrt{k}} \right) \right).$$

Since $\psi(0) = 1$, this clearly converges to zero as $k \to \infty$, $m \to \infty$, $m = o(\sqrt{k})$. ∎

Corollary 7.2.4 *If moreover (7.2.8) holds, $k(n) \to \infty$ and $k(n) = o\left(n^{2\alpha/(1+2\alpha)} \right)$ as $n \to \infty$, then*

$$\sqrt{k}\left(V_{n,k}(x, y) - L(x, y) \right) \xrightarrow{d} W(x, y)$$

in $D([0, T] \times [0, T])$, for every $T > 0$.

Proof. Invoking a Skorohod construction we can start from

$$\sup_{0 \leq x, y \leq T} \left| \sqrt{k}\left\{ V_{n,k}(x, y) - \frac{k}{n}\left[1 - F\left(U_1\left(\frac{n}{kx} \right), U_2\left(\frac{n}{ky} \right) \right) \right] \right\} - W(x, y) \right|$$
$$\to 0 \qquad \text{a.s.}$$

Then

$$\sup_{0 \leq x, y \leq T} \left| \sqrt{k}\left(V_{n,k}(x, y) - L(x, y) \right) - W(x, y) \right|$$
$$\leq \sup_{0 \leq x, y \leq T} \left| \sqrt{k}\left\{ V_{n,k}(x, y) - \frac{k}{n}\left[1 - F\left(U_1\left(\frac{n}{kx} \right), U_2\left(\frac{n}{ky} \right) \right) \right] \right\} - W(x, y) \right|$$
$$+ \sup_{0 \leq x, y \leq T} \sqrt{k}\left| \frac{n}{k}\left[1 - F\left(U_1\left(\frac{n}{kx} \right), U_2\left(\frac{n}{ky} \right) \right) \right] - L(x, y) \right|.$$

In view of the condition $k(n) = o\left(n^{2\alpha/(1+2\alpha)} \right)$ it is now sufficient to prove that

$$\sup_{0 \leq x_1^2 + x_2^2 \leq T} t^{-\alpha}\left| t^{-1}\left(1 - F\left(U_1\left(\frac{1}{tx_1} \right), U_2\left(\frac{1}{tx_2} \right) \right) \right) - L(x_1, x_2) \right| \qquad (7.2.10)$$

is bounded as $t \downarrow 0$. We shall prove this for $T = 1$. Now,

$$t^{-\alpha}\left|t^{-1}\left(1 - F\left(U_1\left(\frac{1}{tx_1}\right), U_2\left(\frac{1}{tx_2}\right)\right)\right) - L(x_1, x_2)\right|$$

$$= |\mathbf{x}|^{\alpha}\,(t|\mathbf{x}|)^{-\alpha}\left||\mathbf{x}|\,(t|\mathbf{x}|)^{-1}\left[1 - F\left(U_1\left(\frac{1}{t|\mathbf{x}|x_1/|\mathbf{x}|}\right), U_2\left(\frac{1}{t|\mathbf{x}|x_2/|\mathbf{x}|}\right)\right)\right]\right.$$

$$\left. - |\mathbf{x}|L\left(\frac{x_1}{|\mathbf{x}|}, \frac{x_2}{|\mathbf{x}|}\right)\right|$$

$$= |\mathbf{x}|^{1+\alpha}\left\{(t|\mathbf{x}|)^{-\alpha}\left|(t|\mathbf{x}|)^{-1}\left[1 - F\left(U_1\left(\frac{1}{t|\mathbf{x}|x_1/|\mathbf{x}|}\right), U_2\left(\frac{1}{t|\mathbf{x}|x_2/|\mathbf{x}|}\right)\right)\right]\right.\right.$$

$$\left.\left. - L\left(\frac{x_1}{|\mathbf{x}|}, \frac{x_2}{|\mathbf{x}|}\right)\right|\right\}.$$

This expression remains bounded uniformly for $|\mathbf{x}| \leq 1$ as $t \downarrow 0$ since $|\mathbf{x}|^{1+\alpha} \leq 1$ and since by (7.2.8) the second factor remains bounded uniformly for $|\mathbf{x}| \leq 1$. ∎

Proof (of Theorem 7.2.2). Once again it is sufficient to prove the result for $T = 1$. Again we invoke a Skorohod construction (but keep the same notation) and we start from

$$\sup_{0 \leq x, y \leq 1} \left|\sqrt{k}\left(V_{n,k}(x, y) - L(x, y)\right) - W(x, y)\right| \to 0 \qquad \text{a.s.,}$$

which implies by setting $y = 0$,

$$\sup_{0 \leq x \leq 1} \left|\sqrt{k}\left(V_{n,k}(x, 0) - x\right) - W(x, 0)\right| \to 0 \qquad \text{a.s.} \qquad (7.2.11)$$

with

$$V_{n,k}(x, 0) = \frac{1}{k}\sum_{i=1}^{n} 1_{\{1 - F_1(X_i) \leq kx/n\}}.$$

Let $U_i := 1 - F_1(X_i), i = 1, 2, \ldots, n$. The function $V_{n,k}(x, 0)$ is a nondecreasing function. Its inverse function is $(n/k)U_{[kx],n}$, the $[kx]$th order statistic from $U_1, \ldots, U_n$. Vervaat's lemma (Appendix A) allows us to invert relation (7.2.11) and we get

$$\sup_{0 \leq x \leq 1} \left|\sqrt{k}\left(\frac{n}{k}U_{[kx],n} - x\right) + W(x, 0)\right| \to 0 \qquad \text{a.s.} \qquad (7.2.12)$$

Similarly we get

$$\sup_{0 \leq y \leq 1} \left|\sqrt{k}\left(\frac{n}{k}W_{[ky],n} - x\right) + W(0, y)\right| \to 0 \qquad \text{a.s.} \qquad (7.2.13)$$

with $W_i := 1 - F_2(Y_i), i = 1, 2, \ldots, n$.

Since we have uniform convergence in

$$\sqrt{k}\left(V_{n,k}(x, y) - L(x, y)\right) - W(x, y) \to 0$$

and since by (7.2.12) and (7.2.13)

$$\sup_{0\le x\le 1}\left|\frac{n}{k}U_{[kx],n}-x\right|\to 0 \quad\text{and}\quad \sup_{0\le y\le 1}\left|\frac{n}{k}W_{[ky],n}-y\right|\to 0 \qquad\text{a.s.}, \qquad (7.2.14)$$

we have

$$\sqrt{k}\left\{V_{n,k}\left(\frac{n}{k}U_{[kx],n},\frac{n}{k}W_{[ky],n}\right)-L\left(\frac{n}{k}U_{[kx],n},\frac{n}{k}W_{[ky],n}\right)\right\}$$
$$-W\left(\frac{n}{k}U_{[kx],n},\frac{n}{k}W_{[ky],n}\right)\to 0 \qquad\text{a.s.}$$

uniformly. We consider the three terms separately. First note that

$$V_{n,k}\left(\frac{n}{k}U_{[kx],n},\frac{n}{k}W_{[ky],n}\right)=\hat{L}(x,y)\ .$$

Further, (7.2.14) implies by the continuity of W that

$$\sup_{0\le x,y\le 1}\left|W\left(\frac{n}{k}U_{[kx],n},\frac{n}{k}W_{[ky],n}\right)-W(x,y)\right|\to 0 \qquad\text{a.s.}$$

Finally, relations (7.2.12) and (7.2.13) imply when combined with Cramér's delta method and the differentiability conditions for L that

$$\sup_{0\le x,y\le 1}\left|\sqrt{k}\left(L\left(\frac{n}{k}U_{[kx],n},\frac{n}{k}W_{[ky],n}\right)-L(x,y)\right)\right.$$
$$\left.+L_1(x,y)W(x,0)+L_2(x,y)W(0,y)\right|\to 0 \qquad\text{a.s.}$$

The proof is complete.   ∎

Next we turn to estimating the set Q_1 introduced in Section 6.1.5. The set determines the distribution G_0, as we have seen in this section. Since the convexity of the set Q_1 is characteristic for G_0 being a simple max-stable distribution, looking at how close the estimated Q-set is to a convex set, it can provide a heuristic way of checking whether the distribution F is in some domain of attraction, as we shall see later on. The estimator of the set Q_1 is

$$\hat{Q}_1:=\{(x,y)\in\mathbb{R}_+^2\ :\ \hat{L}(x,y)\le 1\}$$

with $\hat{L}$ from (7.2.4).

Clearly the set includes the points $(1,0)$ and $(0,1)$. If one draws the points $(n-R(X_i)+1,n-R(Y_i)+1)$, $i=1,\ldots,n$, in $\mathbb{R}_+^2$ and one draws horizontal and vertical lines through these points, then the boundary of $\hat{Q}_1$ follows these lines. One can start for example from the point $(1,0)$ and follow the vertical line until one hits the first horizontal line. Then one goes either up or to the left depending on whether the point on the horizontal line is to the left or to the right respectively. This way the number of points in the L-shaped area to the left and below the graph is kept constant. The estimation procedure is naturally extended to estimate any Q_c curve: $\hat{Q}_c:=\{(x,y):\hat{L}(x,y)\le c\}$.

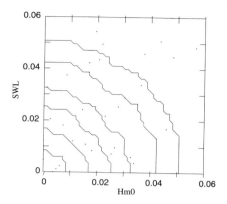

Fig. 7.1. Estimated Q-curves: $k = 7, 14, 21, 28, 35, 42$.

The resulting graph for a particular data set is shown in Figure 7.1. The estimated curves are for different levels of $c = k/n$, corresponding to $k = 7, 14, 21, 28, 35, 42$. The data set corresponds to 828 observations of wave height (HmO) and still water level (SWL) (these observations are illustrated in Figure 8.1). The characterizing properties of the Q-curve, convexity and equality of shape for different levels of c, seem to be true for this data set, giving some confidence in the extreme value model. Also, even for small values of c the curve seems to differ significantly from a straight line, giving indication of dependence between (high values of) the variables.

By estimating the Q-curve in this way, one does not make use of the extreme value conditions. An alternative way of estimating Q is via the homogeneity of L. Let $\{\rho(\theta)\}_{0 \leq \theta \leq \pi/2}$ be the polar representation of the boundary of the set Q_1 between the points $(1, 0)$ and $(0, 1)$,

$$L(\rho(\theta)\cos\theta, \rho(\theta)\sin\theta) = 1, \quad 0 \leq \theta \leq \pi/2,$$

and since L is homogeneous of degree 1,

$$\rho(\theta) = \frac{1}{L(\cos\theta, \sin\theta)}. \tag{7.2.15}$$

We can estimate the set Q_1 by estimating $\rho(\theta), 0 \leq \theta \leq \pi/2$. A natural estimator for $\rho(\theta)$ is

$$\hat{\rho}(\theta) := \frac{1}{\hat{L}(\cos\theta, \sin\theta)}.$$

In Figure 7.2 we find the estimation of Q_1 via $\hat{\rho}(\theta)$. Again the concavity of the Q-curve seems true and there is some indication of dependence between (high values of) the variables.

The asymptotic normality of $\hat{\rho}(\theta)$ follows straightforwardly from Theorem 7.2.2.

Corollary 7.2.5 *Let $(X_1, Y_1), (X_2, Y_2), \ldots$ be i.i.d. random vectors with distribution function F. Suppose F is in the domain of attraction of an extreme value distribution G with standard marginals. Suppose that for some $\alpha > 0$ and for all $x, y > 0$ the relation*

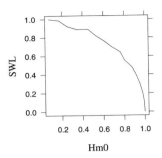

Fig. 7.2. Estimated Q_1-curve from $\hat{\rho}(\theta)$ with $k = 42$.

$$t \left\{ 1 - F\left(U_1\left(\frac{t}{x}\right), U_2\left(\frac{t}{y}\right) \right) \right\} = L(x, y) + O(t^{-\alpha}) \,,$$

$t \to \infty$, where $U_i := (1/(1 - F_i))^{\leftarrow}$ and F_i is the ith marginal distribution function, supposed continuous, $i = 1, 2$, holds uniformly on the set

$$\left\{ x^2 + y^2 = 1 \; : \; x \geq 0, y \geq 0 \right\} \,.$$

Suppose further that the function L has continuous first-order partial derivatives

$$L_1(x, y) := \frac{\partial}{\partial x} L(x, y) \quad and \quad L_2(x, y) := \frac{\partial}{\partial y} L(x, y)$$

for $x, y \geq 0$. Then for $k = k(n) \to \infty$, $k(n) = o\left(n^{2\alpha/(1+2\alpha)}\right)$ as $n \to \infty$,

$$\sqrt{k}\left(\frac{\hat{\rho}(\theta)}{\rho(\theta)} - 1 \right) \xrightarrow{d} -\rho(\theta) B(\cos\theta, \sin\theta)$$

in $D([0, \pi/2])$, with B the stochastic process defined in Theorem 7.2.2.

Finally, we remark that (for example) in the three-dimensional space one can estimate L by

$$\hat{L}(x, y, z) := \frac{1}{k} \sum_{i=1}^{n} 1_{\{R(X_i) \geq n - kx + 1 \text{ or } R(Y_i) \geq n - ky + 1 \text{ or } R(Z_i) \geq n - kz + 1\}}$$

and the Q-curve by

$$\hat{Q}_1 := \{(x, y, z) \in \mathbb{R}_+^3 \; : \; \hat{L}(x, y, z) \leq 1\} \,.$$

A sample graph of the Q-curve in $\mathbb{R}_+^3$ is in Figure 7.3. The variables involved are wave height (HmO) and still water level (SWL) as before, and wave period (Tpb) measured in seconds. The picture indicates no asymptotic independence since for asymptotic independence one expects a flat convex function.

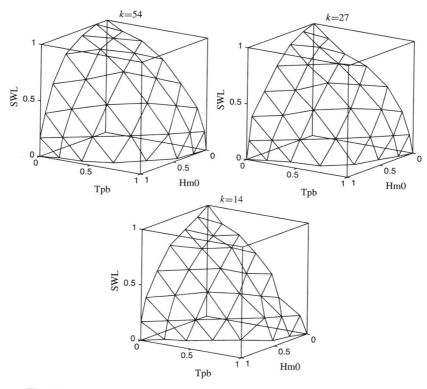

Fig. 7.3. Trivariate Q-surfaces: contours shown correspond to $k = 54, 27, 14$.

7.3 Estimation of the Spectral Measure (and L)

In Section 7.2 we were concerned with estimating the extreme value distribution G_0 via estimation of the function $L(x, y) := -\log G_0(1/x, 1/y), x, y > 0$. However, in general $\hat{G}_0 := \exp(-\hat{L}(1/x, 1/y))$ itself is *not* an extreme value distribution since it is not guaranteed that $\hat{L}$ satisfies the homogeneity property that is valid for the function L:

$$L(ax, ay) = aL(x, y)$$

for $a, x, y > 0$. Whether this is a problem depends on the application. However, it is useful to develop an estimator for G_0 that itself is an extreme value distribution. This can be done via Theorem 6.1.14 (e.g., via (6.1.31)), which states that *any* finite measure satisfying the side conditions, represented by the distribution function Φ (the spectral measure), gives rise to an extreme value distribution G_0 via (6.1.31). Hence now we focus on the estimation of the spectral measure and in order to do so we have to go back to the origin of this measure. We discuss only the spectral measure of Theorem 6.1.14 (3) and not the other two, since asymptotic normality has been proved so far only for the third form of the spectral measure.

Adapting the arguments in the beginning of Section 6.1.4 for this specific case, we consider the sets

$$D_{r,\theta} := \left\{ (x, y) \in \mathbb{R}_+^{2*} \; : \; x \vee y > r \text{ and } x/y \leq \tan\theta \right\}$$

for some $r > 0$ and $\theta \in [0, \pi/2]$. Then

$$\Phi(\theta) := r\nu(D_{r,\theta}) = \nu(D_{1,\theta}), \tag{7.3.1}$$

where ν is the exponent measure of Section 6.1.3. Since it is easier in this context to work with the uniform distribution as the basic distribution rather than with the distribution function $1 - 1/x$, $x \geq 1$, we reformulate (7.3.1) in terms of the measure μ, defined (as in Section 7.2) by

$$\mu\{(s, t) \in [0, \infty]^2 \setminus \{(\infty, \infty)\} \; : \; s < x \text{ or } t < y\}$$
$$:= \nu\{(s, t) \in [0, \infty]^2 \setminus \{(0, 0)\} \; : \; s > 1/x \text{ or } t > 1/y\} \, .$$

Clearly the two measures are the same modulo the marginal transformations $x \mapsto 1/x$, $y \mapsto 1/y$. Then

$$\Phi(\theta) = \mu(E_{1,\theta})$$

with

$$E_{q,\theta} := \left\{ (x, y) \in [0, \infty]^2 \setminus \{(\infty, \infty)\} \; : \; x \wedge y < q \text{ and } y/x \leq \tan\theta \right\} \tag{7.3.2}$$

for some $q > 0$ and $\theta \in [0, \pi/2]$.

Now assume for the moment just for the sake of simplifying this explanation that the marginal distributions of F, F_1, and F_2, are continuous. Then from the proof of Theorem 6.1.9,

$$\lim_{t \to \infty} t \, P\left(1 - F_1(X) < \frac{x}{t} \text{ or } 1 - F_2(Y) < \frac{y}{t}\right)$$
$$= \mu\{(s, t) \in [0, \infty]^2 \setminus \{(\infty, \infty)\} \; : \; s < x \text{ or } t < y\};$$

hence

$$\lim_{t \to \infty} t \, P\left((1 - F_1(X)) \wedge (1 - F_2(Y)) \leq \frac{1}{t} \text{ and } \frac{1 - F_2(Y)}{1 - F_1(X)} \leq \tan\theta\right)$$
$$= \mu\left(E_{1,\theta}\right) = \Phi(\theta) \tag{7.3.3}$$

for all continuity points θ of Φ, where (X, Y) is a random vector with distribution function F.

Now suppose that we have independent and identically distributed random vectors $(X_1, Y_1), (X_2, Y_2), \ldots, (X_n, Y_n)$ with distribution function F. In order to transform the left-hand side of (7.3.3) into an estimator of Φ we are going to replace the measure P by its empirical counterpart and we replace the measures symbolized by F_1 and F_2 by their empirical counterparts. But before doing so we need to choose t from (7.3.3) in relation to the sample size n. Since we want to deal with the tail of the distribution only, the choice $t = n/k$ imposes itself with $k = k(n)$, $k \to \infty$, and $k/n \to 0$.

Next we replace the measure P by its empirical counterpart. Then the left-hand side of (7.3.3) becomes

$$\frac{n}{k}\frac{1}{n}\sum_{i=1}^{n} 1_{\{(1-F_1(X_i))\wedge(1-F_2(Y_i))\leq k/n \text{ and } 1-F_2(Y)\leq(1-F_1(X))\tan\theta\}} \cdot$$

Further, we replace $1 - F_1(x)$ by its empirical counterpart, the left-continuous version of the empirical distribution function,

$$1 - F_1^{(n)}(x) := \frac{1}{n}\sum_{j=1}^{n} 1_{\{X_j \geq x\}} \cdot \tag{7.3.4}$$

Then $1 - F_1(X_i)$ should be replaced by

$$1 - F_1^{(n)}(X_i) := \frac{1}{n}\sum_{j=1}^{n} 1_{\{X_j \geq X_i\}} = \frac{n+1-R(X_i)}{n},$$

where $R(X_i)$ is the rank of the ith observation X_i, $i = 1, \ldots, n$, among $(X_1, X_2, \ldots, X_n)$. Similarly we replace $1 - F_2(Y_i)$ by $(n+1-R(Y_i))/n$ where $R(Y_i)$ is the rank of Y_i among $(Y_1, Y_2, \ldots, Y_n)$. Taking everything together we get the following estimator for Φ:

$$\hat{\Phi}(\theta) := \frac{1}{k}\sum_{i=1}^{n} 1_{\{R(X_i)\vee R(Y_i)\geq n+1-k \text{ and } n+1-R(Y_i)\leq(n+1-R(X_i))\tan\theta\}} \cdot \tag{7.3.5}$$

The estimator is nonparametric in that the statistic is invariant under monotone transformations of the marginals of the observations. So it does not depend on the marginal distributions.

In Figure 7.4 we find $\hat{\Phi}(\theta)$ for the 828 observations of wave height (HmO) and still water level (SWL). Note that there is some indication of dependence between the variables since the angular coordinates are not clustered in the neighborhood of 0 and $\pi/2$.

Another way of displaying the estimator of the spectral measure makes use of its discrete character: it gives equal weight to a limited number of points. Hence we can just display on the line segment $[0, \pi/2]$ the points

$$\left\{ \arctan \frac{n+1-R(Y_i)}{n+1-R(X_i)} \right\}$$

for those observations (X_i, Y_i) for which $R(X_i) \vee R(Y_i) \geq n+1-k$. This is done in Figure 7.4 too.

One can do this similarly in higher dimensions. For example, in $\mathbb{R}^3$ one can display the intersections of the lines through the points

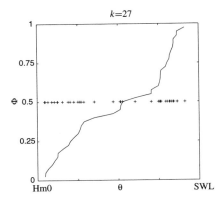

Fig. 7.4. Estimated spectral measure and angular coordinates θ_i (shown as "+" signs); the solid line represents the corresponding distribution function Φ scaled down from $39/27$ to 1.

$$(n + 1 - R(X_i), n + 1 - R(Y_i), n + 1 - R(Z_i))$$

and the origin with the plane $\{x, y, z \geq 0 : x + y + z = 1\}$. To display the intersection points on this triangle we do the following. Figure 7.5 shows the situation in which P is the point

$$\left(\frac{n + 1 - R(X_i)}{3n + 3 - R(X_i) - R(Y_i) - R(Z_i)}, \frac{n + 1 - R(X_i)}{3n + 3 - R(X_i) - R(Y_i) - R(Z_i)}, \frac{n + 1 - R(X_i)}{3n + 3 - R(X_i) - R(Y_i) - R(Z_i)}\right)$$

and O is the origin.

In order to find the distance DB consider the triangle OAB; see Figure 7.5. Note that

$$OF = (n + 1 - R(Y_i))/(3n + 3 - R(X_i) - R(Y_i) - R(Z_i)),$$
$$FG = EF = (n + 1 - R(X_i))/(3n + 3 - R(X_i) - R(Y_i) - R(Z_i)),$$

and hence

$$FG = (n + 1 - R(X_i))\sqrt{2}/(3n + 3 - R(X_i) - R(Y_i) - R(Z_i)) .$$

Now notice that

$$\frac{EF}{DB} = \frac{OF}{OB} .$$

It follows that

$$DB = \sqrt{2} \, \frac{n + 1 - R(X_i)}{2n + 2 - R(X_i) - R(Y_i)} .$$

Similar relations hold for the lines connecting P with the other edges, A and C.

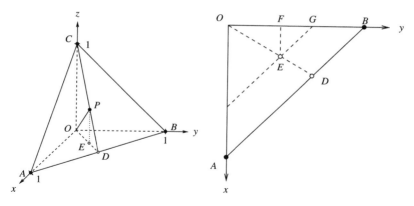

Fig. 7.5. The point P in the triangle and its projection in the plane.

Figure 7.6 displays the empirical spectral measure for the three-dimensional sea state data. Again the picture indicates no asymptotic independence since for asymptotic independence one expects the point to be concentrated near the three vertices.

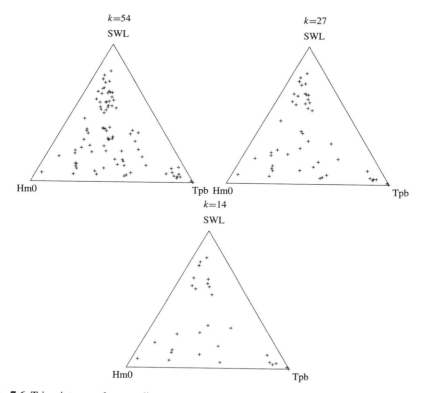

Fig. 7.6. Trivariate angular coordinates representing the spectral measure; scatter plots shown correspond to $k = 54, 27, 14$.

Now recall the connection between the function L of Section 6.1.5 and Φ:

$$L(x, y) = \int_0^{\pi/2} \{(x(1 \wedge \tan \theta)) \vee (y(1 \wedge \cot \theta))\} \; \Phi(d\theta) \qquad (7.3.6)$$

for $x, y > 0$ (see Theorem 6.1.14(3)). After splitting the integration interval into several parts and applying partial integration, we obtain (cf. proof of Theorem 7.3.1 below) the alternative expression

$$L(x, y) = x \; \Phi\left(\frac{\pi}{2}\right) + (x \vee y) \int_{\pi/4}^{\arctan(y/x)} \Phi(\theta) \left(\frac{1}{\sin^2 \theta} \wedge \frac{1}{\cos^2 \theta}\right) d\theta . \qquad (7.3.7)$$

This leads to an alternative estimator of the function L:

$$\hat{L}_\Phi(x, y) := x \; \hat{\Phi}\left(\frac{\pi}{2}\right) + (x \vee y) \int_{\pi/4}^{\arctan(y/x)} \hat{\Phi}(\theta) \left(\frac{1}{\sin^2 \theta} \wedge \frac{1}{\cos^2 \theta}\right) d\theta . \qquad (7.3.8)$$

This estimator is somewhat more complicated than the one in Section 7.2. On the other hand, the present estimator has the advantage that it is homogeneous, i.e.,

$$\hat{L}_\Phi(ax, ay) = a \; \hat{L}_\Phi(x, y)$$

for $a, x, y > 0$ and therefore the function

$$\hat{G}_0(x, y) := \exp\left(-\hat{L}_\Phi(1/x, 1/y)\right)$$

is an estimator of the max-stable distribution function G_0, which itself is a max-stable distribution function.

We now proceed to prove consistency and asymptotic normality of both estimators.

Theorem 7.3.1 *Let* $(X_1, Y_1), (X_2, Y_2), \ldots$ *be i.i.d. random vectors with continuous distribution function* F. *Let* $F_1(x) := F(x, \infty)$ *and* $F_2(x) := F(\infty, x)$, *and let* U_i *be the inverse of the function* $1/(1 - F_i)$, $i = 1, 2$. *Suppose for* $x, y \geq 0$,

$$\lim_{t \to \infty} t \left\{1 - F\left(U_1\left(\frac{t}{x}\right), U_2\left(\frac{t}{y}\right)\right)\right\} = L(x, y).$$

Let $k = k(n)$ *be a sequence of integers such that* $k \to \infty, k/n \to 0, n \to \infty$. *Then*

$$\hat{\Phi}(\theta) \overset{P}{\to} \Phi(\theta) \qquad (7.3.9)$$

for $\theta = \pi/2$ *and each* $\theta \in [0, \pi/2)$ *that is a continuity point of* Φ. *Moreover,*

$$\hat{L}_\Phi(x, y) \overset{P}{\to} L(x, y) \qquad (7.3.10)$$

for $x, y \geq 0$.

Corollary 7.3.2 *The statements of Theorem 7.3.1 imply the seemingly stronger statements*

$$\lim_{n\to\infty} P\left(\lambda\left(\hat{\Phi}, \Phi\right) > \varepsilon\right) = 0 \tag{7.3.11}$$

for each $\varepsilon > 0$, where λ is the Lévy distance:

$$\lambda\left(\hat{\Phi}, \Phi\right)$$

$$= \inf\left\{\delta \; : \; \hat{\Phi}(\theta - \delta) - \delta \le \Phi(\theta) \le \hat{\Phi}(\theta + \delta) + \delta \text{ for all } 0 \le \theta \le \pi/2\right\}$$

and for all $L > 0$,

$$\sup_{0 \le x, y \le L} \left|\hat{L}_\Phi(x, y) - L(x, y)\right| \xrightarrow{P} 0. \tag{7.3.12}$$

Proof (of Theorem 7.3.1). Define the measures μ and $\hat{\mu}$ as follows: for a Borel set A in $[0, \infty]^2 \setminus \{(\infty, \infty)\}$,

$$\mu(A) := \lim_{n\to\infty} \frac{n}{k} P\left((1 - F_1(X), 1 - F_2(Y)) \in \frac{k}{n} A\right), \tag{7.3.13}$$

where (X, Y) is a random vector with distribution function F and as before Theorem 7.3.1, we define

$$\hat{\mu}(A) := \frac{1}{k} \sum_{i=1}^{n} 1_{\{(n+1-R(X_i), n+1-R(Y_i)) \in kA\}}. \tag{7.3.14}$$

By Theorem 7.2.1 for all $T > 0$,

$$\sup_{0 \le x, y \le T} \left|\hat{\mu}\left(([x, \infty] \times [y, \infty])^c\right) - \mu\left(([x, \infty] \times [y, \infty])^c\right)\right| \xrightarrow{P} 0. \tag{7.3.15}$$

We invoke a Skorohod construction and proceed (without changing notation) as if this convergence held almost surely. By subtracting two sets as in (7.3.15) we then find that for $0 \le x_1 < x_2 < \infty, y \ge 0$,

$$\hat{\mu}\left([x_1, x_2] \times [y, \infty]\right) \xrightarrow{P} \mu\left([x_1, x_2] \times [y, \infty]\right). \tag{7.3.16}$$

By subtracting two sets as in (7.3.16) we get for $0 \le x_1 < x_2 < \infty, y \ge 0$,

$$\hat{\mu}\left([x_1, x_2] \times [0, y]\right) \xrightarrow{P} \mu\left([x_1, x_2] \times [0, y]\right). \tag{7.3.17}$$

This is also true with $x_2 = \infty, y = \infty$. Let θ be a continuity point of $\Phi(\theta)$. Clearly for $\varepsilon > 0$ we can find two finite unions of sets as in (7.3.17), L_ε and U_ε, such that

$$L_\varepsilon \subset E_{1,\theta} \subset U_\varepsilon$$

and

$$\mu\left(U_\varepsilon\right) - \varepsilon \leq \mu\left(E_{1,\theta}\right) \leq \mu\left(L_\varepsilon\right) + \varepsilon$$

with $E_{1,\theta}$ from (7.3.2). Also we have

$$\hat{\mu}(L_\varepsilon) \leq \hat{\Phi}(\theta) = \hat{\mu}\left(E_{1,\theta}\right) \leq \hat{\mu}(U_\varepsilon) \ .$$

Since $\hat{\mu}(L_\varepsilon)$ and $\hat{\mu}(U_\varepsilon)$ are clearly weakly consistent estimators of $\mu(L_\varepsilon)$ and $\mu(U_\varepsilon)$ respectively and since $\mu(U_\varepsilon) - \mu(L_\varepsilon) \to 0$ as $\varepsilon \downarrow 0$, we conclude that

$$\hat{\Phi}(\theta) = \hat{\mu}(E_{1,\theta}) \xrightarrow{P} \mu(E_{1,\theta}) = \Phi(\theta) \ .$$

Finally we are going to prove the statement about $\hat{L}_\Phi$. Before doing so we first prove the alternative representation of L in terms of Φ. First note that

$$L(x, y) = \int_0^{\pi/4} ((x \tan\theta) \vee y) \ \Phi(d\theta) + \int_{\pi/4}^{\pi/2} (x \vee (y \cot\theta)) \ \Phi(d\theta) \ .$$

Let $x < y$. Then the two integrals become

$$y \int_0^{\pi/4} \Phi(d\theta) + y \int_{\pi/4}^{\arctan(y/x)} \cot\theta \ \Phi(d\theta) + x \int_{\arctan(y/x)}^{\pi/2} \Phi(d\theta) \ .$$

Since $\cot\theta = 1 - \int_{\pi/4}^\theta \sin^{-2} q \ dq$, this equals

$$y \int_0^{\arctan(y/x)} \Phi(d\theta) + x \int_{\arctan(y/x)}^{\pi/2} \Phi(d\theta) - y \int_{\pi/4}^{\arctan(y/x)} \int_{\pi/4}^\theta \frac{dq}{\sin^2 q} \ \Phi(d\theta),$$

and by interchanging the integrals and some simplifications we get

$$x \ \Phi\left(\frac{\pi}{2}\right) + y \int_{\pi/4}^{\arctan(y/x)} \frac{\Phi(\theta)}{\sin^2\theta} \ d\theta \ .$$

Similarly for $x > y$ we get

$$x \ \Phi\left(\frac{\pi}{2}\right) - x \int_{\arctan(y/x)}^{\pi/4} \frac{\Phi(\theta)}{\cos^2\theta} \ d\theta \ .$$

This gives the stated form. Similarly we have

$$\hat{L}_\Phi(x, y) = x \ \hat{\Phi}\left(\frac{\pi}{2}\right) + (x \vee y) \int_{\pi/4}^{\arctan(y/x)} \hat{\Phi}(\theta) \left(\frac{1}{\sin^2\theta} \wedge \frac{1}{\cos^2\theta}\right) d\theta. \quad (7.3.18)$$

Since $\hat{\Phi}(\theta) \to \Phi(\theta)$ for $\theta = \pi/2$ and all θ in some set S where $[0, \pi/2] \cap S^c$ is a countable set,

$$\lim_{n\to\infty} \left(\hat{L}_\Phi(x, y) - L(x, y)\right) = x \lim_{n\to\infty} \left(\hat{\Phi}\left(\frac{\pi}{2}\right) - \Phi\left(\frac{\pi}{2}\right)\right)$$

$$+ (x \vee y) \int_{\substack{\pi/4 \leq \theta \leq \arctan(y/x) \\ \theta \in S}} \lim_{n\to\infty} \left(\hat{\Phi}(\theta) - \Phi(\theta)\right) \left(\frac{1}{\sin^2\theta} \wedge \frac{1}{\cos^2\theta}\right) d\theta = 0. \quad \blacksquare$$

Proof (of Corollary 7.3.2). We have already shown that (7.3.10) is sufficient for (7.3.12) (cf. proof of Theorem 7.2.1). Now we show that (7.3.9) is sufficient for (7.3.11). Fix $\varepsilon > 0$. Take $0 \leq \theta_0 < \theta_1 < \cdots < \theta_r < \theta_{r+1} = \pi/2$ such that θ_i is a continuity point of Φ for $i = 1, 2, \ldots, r$ and $\theta_{i+1} - \theta_i < \varepsilon$ for $i = 1, 2, \ldots, r$. Then as $n \to \infty$,

$$P\left(\sup_{1 \leq i \leq r} \left|\hat{\Phi}(\theta_i) - \Phi(\theta_i)\right| > \varepsilon\right) \to 0 .$$

For any $\theta \in \left[0, \frac{\pi}{2}\right]$ there exists θ_k such that $\theta \leq \theta_k \leq \theta + \varepsilon$. Then if $\hat{\Phi}(\theta_k) \geq \Phi(\theta_k) - \varepsilon$, we have

$$\hat{\Phi}(\theta + \varepsilon) \geq \hat{\Phi}(\theta_k) \geq \Phi(\theta_k) - \varepsilon \geq \Phi(\theta) - \varepsilon .$$

Similarly there exists θ_j such that $\theta - \varepsilon \leq \theta_j \leq \theta$. Then if $\hat{\Phi}(\theta_j) \leq \Phi(\theta_j) + \varepsilon$,

$$\hat{\Phi}(\theta - \varepsilon) \leq \hat{\Phi}(\theta_j) \leq \Phi(\theta_j) + \varepsilon \leq \Phi(\theta) + \varepsilon .$$

It follows that

$$P\left(\hat{\Phi}(\theta - \varepsilon) \leq \Phi(\theta) + \varepsilon \text{ and } \hat{\Phi}(\theta + \varepsilon) \geq \Phi(\theta) - \varepsilon \text{ for } 0 \leq \theta \leq \pi/2\right)$$
$$\geq P\left(\Phi(\theta_i) - \varepsilon \leq \hat{\Phi}(\theta_i) \leq \Phi(\theta_i) + \varepsilon \text{ for } i = 0, 1, \ldots, r + 1\right) \to 1$$

as $n \to \infty$. ∎

For the asymptotic normality of $\hat{\Phi}$ and $\hat{L}_\Phi$ we need two conditions, both of which strengthen the domain of attraction condition

$$\lim_{t \to \infty} t\, P\left(1 - F_1(X) < \frac{x}{t} \text{ or } 1 - F_2(Y) < \frac{y}{t}\right) = L(x, y)$$

considerably. Also we need to impose a further restriction on the growth of the sequence $k(n)$.

Let $\delta \in \left\{1, \frac{1}{2}, \frac{1}{3}, \ldots\right\}$, $p = 0, 1, 2, \ldots, 1/\delta - 1$, and define $I_\delta(p) := [p\delta/\tan\theta, (p+1)\delta/\tan\theta]$, $\theta \in [0, \pi/4]$. Let $\tilde{A}$ be the class containing all the following sets:

1. $\bigcup_{p=0}^{1/\delta - 1} \{(x, y) : x \in I_\delta(p), 0 \leq y \leq x \tan\theta + C_p(x \tan\theta)^{1/16}\}$, for some $\theta \in [0, \pi/4]$ and $C_0, C_1, \ldots, C_{1/\delta - 1} \in [-1, 1]$;
2. $\{(x, y) : y \leq b\}$, for some $b \leq 2$;
3. $\{(x, y) : x \leq a\}$, $\{(x, y) : x \leq M, y \leq 2\}$, for some $a \leq M$ (later on M will be taken large);
4. $\{(x, y) : x \geq 1/\tan\theta, y \leq b\}$, for some $\theta \in [0, \pi/4]$ and $b \leq 2$.

Next define $\tilde{A}_s = \left\{A_s : A \in \tilde{A}\right\}$, where for $A \in \tilde{A}$, $A_s = \{(x, y) : (y, x) \in A\}$. Finally define $\mathcal{A} = \tilde{A} \cup \tilde{A}_s$.

Condition 7.3.3 *For all* $\delta \in \left\{1, \frac{1}{2}, \frac{1}{3}, \ldots\right\}$ *and* $M > 1$,

$$\sup_{A \in \mathcal{A}'} \left| t^{-1} P\{(1 - F_1(X), 1 - F_2(X)) \in tA\} - \mu(A) \right| \to 0 ,$$

as $t \downarrow 0$, *with* $\mathcal{A}' := \{A_1 \cap A_2 : A_1, A_2 \in \mathcal{A}\}$.

We also need uniform convergence over a second class of sets. Consider the class of sets $\mathcal{C}_1 = \mathcal{C}_1(\beta)$ defined by

$$\mathcal{C}_1 := \Big\{ \{(x, y) \ : \ 0 \le y \le b(x) \text{for some nondecreasing function } b\}$$
$$\in \mathcal{B}([0, \infty]^2 \setminus \{(\infty, \infty)\}) :$$
$$\sup_{0 < x \le 2/\tan\theta} \frac{|b(x) - ((x\tan\theta) \wedge 1)|}{(x\tan\theta)^{1/16}} \le \beta \text{ for some } \theta \in [0, \pi/4],$$
$$\text{and } b(x) = b\left(\frac{2}{\tan\theta}\right) \text{ for } x > \frac{2}{\tan\theta}\Big\},$$

where $\mathcal{B}([0, \infty]^2 \setminus \{(\infty, \infty)\})$ denotes the class of Borel sets on $[0, \infty]^2 \setminus \{(\infty, \infty)\}$. The class of sets $\mathcal{C}_2 = \mathcal{C}_2(\beta)$ is like $\mathcal{C}_1$ but with x and y interchanged.

Condition 7.3.4 *For some* $\beta > 0$,

$$D(t) := \sup_{C \in \mathcal{C}_1 \cup \mathcal{C}_2} \left| t^{-1} P\{(1 - F_1(X), 1 - F_2(X)) \in tC\} - \mu(C) \right| \to 0 , \quad t \downarrow 0 .$$

Finally, we need a condition bounding the growth of the sequence $k(n)$.

Condition 7.3.5 *The sequence* $k = k(n)$ *should be such that (for the same* β*)*

$$\sqrt{k} D\left(\frac{k}{n}\right) \to 0 , \quad n \to \infty .$$

Theorem 7.3.6 *Let* $(X, Y), (X_1, Y_1), (X_2, Y_2), \ldots$ *be independent and identically distributed random vectors with continuous distribution function* F. *Let* $F_1(x) := F(x, \infty)$ *and* $F_2(x) := F(\infty, x)$. *Suppose that for* $x, y \ge 0$,

$$\lim_{t \to \infty} t \, P\left(1 - F_1(X) < \frac{x}{t} \text{ or } 1 - F_2(Y) < \frac{y}{t}\right) = L(x, y)$$

and moreover, the uniform extensions Conditions 7.3.3 and 7.3.4 hold. Suppose that μ *has a continuous density* λ *in* $[0, \infty)^2 \setminus \{(0, 0)\}$. *Let* $k = k(n) \to \infty, n \to \infty$ *and suppose Condition 7.3.5 holds for* k. *Then, as* $n \to \infty$,

$$\sqrt{k}\left(\hat{\Phi}(\theta) - \Phi(\theta)\right) \xrightarrow{d} W_\mu\left(E_{1,\theta}\right) + Z(\theta)$$

in $D([0, \pi/2])$, *where* W_μ *is a Wiener process indexed by sets, that is, a centered Gaussian process with* $E W_\mu(C_1) W_\mu(C_2) = \mu(C_1 \cap C_2)$. *Note that*

$$\left\{ W_\mu(E_{1,\theta}), \theta \in [0, \pi/2] \right\} \overset{d}{=} \left\{ W\left(\Phi(\theta) \right), \theta \in [0, \pi/2] \right\} ,$$

with W a standard Wiener process. The process $\{Z(\theta)\}$ is defined by

$$Z(\theta) := \int_0^{1 \vee (1/\tan \theta)} \lambda(x, x \tan \theta) \left\{ W_1(x) \tan \theta - W_2(x \tan \theta) \right\} dx$$

with $W_1(x) := W_\mu([0, x] \times [0, \infty])$ and $W_2(x) := W_\mu([0, \infty] \times [0, y])$. Note that W_1 and W_2 are also standard Wiener processes. Finally,

$$\sqrt{k} \left(\hat{L}_\Phi(x, y) - L(x, y) \right) \overset{d}{\to} Q(x, y)$$

in $D([0, T] \times [0, T])$, for all $T > 0$, where

$$Q(x, y) := x \left(W_\mu \left(E_{1, \pi/2} \right) + Z \left(\frac{\pi}{2} \right) \right)$$
$$+ (x \vee y) \int_{\pi/4}^{\arctan(y/x)} \left(\frac{1}{\sin^2 \theta} \wedge \frac{1}{\cos^2 \theta} \right) \left(W_\mu \left(E_{1,\theta} \right) + Z(\theta) \right) d\theta .$$

Proof. The proof is very intricate. We give here a sketch of the reasoning and refer to the paper Einmahl, de Haan, and Piterbarg (2001) for full details. We can write

$$\hat{\Phi}(\theta) = \frac{n}{k} \hat{P}_n \left(\frac{k}{n} C_\theta \right)$$

with $C_\theta = E_{1,\theta}$,

$$\hat{P}_n (C) := \frac{1}{n} \sum_{i=1}^n 1_{\left\{ (\hat{U}_i, \hat{V}_i) \in C \right\}},$$

and

$$\hat{U}_i := 1 - F_1^{(n)}(X_i) , \qquad \hat{V}_i := 1 - F_2^{(n)}(Y_i) ,$$

$i = 1, 2, \ldots, n$, where $F_1^{(n)}$ and $F_2^{(n)}$ are the marginal empirical distribution functions (cf. (7.3.4)). We also introduce

$$\hat{C}_\theta := \frac{n}{k} \left\{ (x, y) \in [0, \infty]^2 : \left(\hat{U}_i, \hat{V}_i \right) \in \frac{k}{n} C_\theta \right\} .$$

Now it is important to notice that

$$\frac{n}{k} \hat{P}_n \left(\frac{k}{n} C_\theta \right) = \frac{n}{k} P_n \left(\frac{k}{n} \hat{C}_\theta \right)$$

with

$$P_n (C) := \frac{1}{n} \sum_{i=1}^n 1_{\{ (U_i, V_i) \in C \}},$$

where
$$U_i := 1 - F_1(X_i) , \qquad V_i := 1 - F_2(Y_i) ,$$
$i = 1, 2, \ldots, n$. We now have

$$\sqrt{k} \left(\hat{\Phi}(\theta) - \Phi(\theta) \right)$$

$$= \sqrt{k} \left(\frac{n}{k} P_n \left(\frac{k}{n} \hat{C}_\theta \right) - \frac{n}{k} P \left(\frac{k}{n} \hat{C}_\theta \right) \right) \quad \text{(empirical measure term)}$$

$$+ \sqrt{k} \left(\frac{n}{k} P \left(\frac{k}{n} \hat{C}_\theta \right) - \mu \left(\hat{C}_\theta \right) \right) \quad \text{(bias term)}$$

$$+ \sqrt{k} \left(\mu \left(\hat{C}_\theta \right) - \mu \left(C_\theta \right) \right) \quad \text{(random set term)}$$

$$=: V_1(\theta) + r(\theta) + V_2(\theta) ,$$

$\theta \in [0, \pi/2]$. The part $\sup_\theta |r(\theta)|$ is negligible by Conditions 7.3.4 and 7.3.5 and the well-known behavior of weighted tail empirical and quantile processes.

The part $V_1(\theta)$ deals with the set $\hat{C}_\theta$, which is a random perturbation of the set C_θ. The set C_θ is the union of a rectangle and a triangle and hence it is a nice set, but $\hat{C}_\theta$ is not: it is not in a Vapnick–Červonenkis class or even in a Donsker class. But in each segment of length δ the set $\hat{C}_\theta$ can be majorized and minorized by two sets that are manageable using Conditions 7.3.3 and 7.3.4 and do not differ too much.

Finally, the part $V_2(\theta)$ also contributes to the limit result via the marginal empirical process. This part can be dealt with by writing it as an integral using the density of the measure μ. ∎

7.4 A Dependence Coefficient

Theorem 7.3.6 allows one to accurately estimate the asymptotic dependence structure, which can be very complicated. Nonetheless, it is sometimes useful to employ a simple dependence measure that summarizes the dependence information albeit in a rather crude way.

Consider the d-dimensional setting, i.e., consider a random vector $(X_1, \ldots, X_d)$ with distribution function F in the domain of attraction of some extreme value distribution. Let $K(t) := K_1(t) + \cdots + K_d(t)$ with $K_i(t) = 1_{\{X_i \geq U_i(t)\}}$, where $U_i = (1/(1 - F_i))^\leftarrow$, with $F_i = F(\infty, \ldots, x_i, \infty)$ the marginal distributions, $i = 1, \ldots, d$. Define

$$\kappa := \lim_{t \to \infty} E(K(t) | K(t) \geq 1)$$

$$= \lim_{t \to \infty} \frac{\sum_{j=1}^d P\left(X_j > U_j(t)\right)}{P\left(\cup_{j=1}^d X_j > U_j(t)\right)}$$

$$= \frac{L(1, 0, \ldots, 0) + L(0, 1, \ldots, 0) + \cdots + L(0, \ldots, 0, 1)}{L(1, 1, \ldots, 1)}$$

$$= \frac{d}{L(1, 1, \ldots, 1)} =: \frac{d}{L} .$$

One possible interpretation for this coefficient is that κ quantifies, on average, how many disasters will happen given that one disaster is sure to happen.

The case of asymptotic independence corresponds to $\kappa = 1$, and the case of full dependence corresponds to $\kappa = d$ (cf. Proposition 6.1.21). So in order to make things somewhat reminiscent of the correlation coefficient in that the case of asymptotic independence corresponds to 0 and the case of full dependence to 1, we define the following dependence coefficient (Embrechts, de Haan, and Huang (2000)):

$$H := \frac{\kappa - 1}{d - 1} = \frac{d - L}{(d - 1)L};$$

$H = 0$ is equivalent to asymptotic independence and $H = 1$ to full dependence.

In $\mathbb{R}^2$ it is somewhat usual to consider the dependence coefficient (Sibuya (1960))

$$\lambda := \lim_{t \to \infty} t P\left(X_1 > U_1(t), X_2 > U_2(t)\right)$$

$$= 2 - L(1, 1) = R(1, 1) = -\chi(1) = 2\left(1 - A\left(\frac{1}{2}\right)\right).$$

Hence $\lambda = HL(1, 1), 0 \leq \lambda \leq 1$ and $\lambda = 0$ corresponds to asymptotic independence and $\lambda = 1$ to full dependence in $\mathbb{R}^2$.

The straightforward generalization of λ to $\mathbb{R}^3$ is

$$\lambda = \lim_{t \to \infty} t \, P\left(X_1 > U_1(t), X_2 > U_2(t), X_3 > U_3(t)\right).$$

However, the extension of λ to higher dimensions does not share this property. Consider the random vector (Y_1, Y_1, Y_2) with Y_1, Y_2 independent and identically distributed with distribution function $\exp(-1/x), x > 0$. Since the first two components are the same, the exponent measure must be concentrated on the set $\{(x_1, x_2, x_3) \in \mathbb{R}_+^3 : x_1 = x_2\}$. On the other hand, the first and last components are independent hence the exponent measure must concentrate on the set $\{(x_1, x_2, x_3) \in \mathbb{R}_+^3 : x_1 = 0 \text{ or } x_2 = 0\}$. Since also the second and last components are independent, the exponent measure must be concentrated on $\{(x_1, x_2, x_3) \in \mathbb{R}_+^3 : x_2 = 0 \text{ or } x_3 = 0\}$. Putting everything together we find that the exponent measure is concentrated on the intersection of these sets, that is, the lines $\{(x_1, x_2, x_3) \in \mathbb{R}_+^3 : x_1 = x_2, \ x_3 = 0\}$ and $\{(x_1, x_2, x_3) \in \mathbb{R}_+^3 : x_1 = x_2 = 0\}$. Since clearly the exponent measure is not concentrated on the coordinate axis, there is no asymptotic independence. But

$$\lim_{t \to \infty} t \, P\left(X_1 > U_1(t), X_2 > U_2(t), X_3 > U_3(t)\right)$$

$$= \lim_{t \to \infty} t \, P(Y_1 > t) \, P(Y_2 > t) = 0 .$$

When dealing with observations from the domain of attraction of an extreme value distribution one can estimate H by

$$\hat{H} := \frac{d - \hat{L}}{(d - 1)\,\hat{L}} := \frac{d - \hat{L}(1, 1, \ldots, 1)}{(d - 1)\,\hat{L}(1, 1, \ldots, 1)}$$

with $\hat{L}$ similarly as in Section 7.2, i.e.,

$$\hat{L} := \hat{L}(1, 1, \ldots, 1) := \frac{1}{k} \sum_{i=1}^{n} 1_{\left\{ X_i^{(1)} \geq X_{n-k+1,n}^{(1)} \text{ or } \ldots \text{ or } X_i^{(d)} \geq X_{n-k+1,n}^{(d)} \right\}} \ ,$$

where $(X_1^{(1)}, \ldots, X_1^{(d)}), \ldots, (X_n^{(1)}, \ldots, X_n^{(d)})$ are independent and identically distributed observations from the distribution function F. Similarly as in Theorem 7.2.1 (cf. Exercise 7.1) we have that under the domain of attraction condition and $k = k(n) \to \infty, k/n \to 0, n \to \infty$,

$$\hat{H} \xrightarrow{P} H \ . \tag{7.4.1}$$

Let W be a d-dimensional continuous Gaussian process with mean zero and covariance structure given by the natural extension from the two-dimensional case considered in Theorem 7.2.2. For simplicity of notation define $W(1) := W(1, 1, \ldots, 1)$, $W^{(i)} := W(0, \ldots, 1, \ldots, 0)$, where the 1 is in the x_i coordinate and $L^{(i,j)} := L(0, \ldots, 1, \ldots, 1, \ldots, 0)$, where the 1's are in the x_i and x_j coordinates. Then, under the conditions of Theorem 7.2.2, with the obvious extensions to d dimensions, we have

$$\sqrt{k} \left(\hat{L} - L \right) \xrightarrow{d} W(1) - \sum_{i=1}^{d} L_i(1) W^{(i)}$$

with

$$\begin{aligned}
\mathrm{Var}(W(1)) &= L \ , \\
\mathrm{Var}(W^{(i)}) &= 1 \ , \\
E W(1) W^{(i)} &= 1 \ , \qquad i = 1, \ldots, d \ , \\
E W^{(i)} W^{(j)} &= d - L^{(i,j)}, \qquad i, j = 1, \ldots, d \ , \\
L_i(1) &= \frac{\partial}{\partial x_i} L(x_1, \ldots, x_d) \big|_{(x_1, \ldots, x_d) = (1, \ldots, 1)} \ , \qquad i = 1, \ldots, d
\end{aligned}$$

(cf. Exercise 7.3), i.e., $W(1) - \sum_{i=1}^{d} L_i(1) W^{(i)} =^d N(0, \sigma_L)$, where N is a standard normal random variable and

$$\sigma_L^2 := L + \sum_{i=1}^{d} \left(L_i^2(1) - 2L_i(1) \right) + 2 \sum_{i=1}^{d} \sum_{j=1, j \neq i}^{d} L_i(1) L_j(1) \left(2 - L^{(i,j)} \right) .$$

Hence, by Cramér's delta method,

$$\sqrt{k} \left(\hat{H} - H \right) \xrightarrow{d} N \left(0, \frac{d \, \sigma_L}{(d-1) \, L^2} \right) . \tag{7.4.2}$$

In order to be able to apply this limit result for testing, one needs to estimate $L_j(1)$ consistently, $j = 1, \ldots, d$. A consistent estimator is

$$\hat{L}_j(1) := k^{1/4} \left(\frac{1}{k} \sum_{i=1}^{n} 1_{\{X_i^{(1)} \geq X_{n-k+1,n}^{(1)}, \ldots, X_i^{(j-1)} \geq X_{n-k+1,n}^{(j-1)},}} \right.$$

$$\left. {}_{X_i^{(j)} \geq X_{n-[k(1+k^{-1/4})]+1,n}^{(j)}, X_i^{(j+1)} \geq X_{n-k+1,n}^{(j+1)}, \ldots, X_i^{(d)} \geq X_{n-k+1,n}^{(d)}\}} - \hat{L} \right) .$$

For the proof note that

$$k^{1/4} \left(\hat{L}(x, y) - L(x, y) \right) \xrightarrow{P} 0$$

locally uniformly and that for example,

$$k^{1/4} \left(L(1 + k^{-1/4}, 1, \ldots, 1) - L \right) \xrightarrow{P} L_1(1) .$$

As mentioned before, asymptotic independence is equivalent to $H = 0$. Hence one is tempted to use (7.4.2) to test for asymptotic independence. However, when $H = 0$ one has $L = d$, $L_i(1) = 1$, $L^{(i,j)} = 2$, $i, j = 1, \ldots, d$, $i \neq j$, so that the asymptotic variance of $\sqrt{k} \, (\hat{H} - H)$ is zero and hence the result cannot be used to construct an asymptotic confidence interval.

In fact, in order to test for asymptotic independence, it is better to work with a more refined model, which will be discussed in the next section.

7.5 Tail Probability Estimation and Asymptotic Independence: A Simple Case

In order to show the usefulness of a more refined model in case of asymptotic independence let us consider the following problem.

Suppose one has independent observations $(X_1, Y_1), (X_2, Y_2), \ldots, (X_n, Y_n)$ with distribution function F and suppose that we are interested in estimating the probability

$$1 - F(w, z),$$

where $w > \max_{1 \leq i \leq n} X_i$ and $z > \max_{1 \leq i \leq n} Y_i$.

One may think, for example, of an athlete who wants to compete in the Olympic Games in two disciplines. Her past records in the two disciplines are the observations above. The values w and z are the thresholds that one has to reach, at least one of them, in order to qualify. The athlete has never reached the thresholds.

This problem is a simple multivariate version of the problem of tail estimation (Section 4.4). We want to consider here the simplest situation just for the sake of exposition. We assume that both marginal distributions of F are $1 - 1/x$, $x > 1$. A much more general situation will be considered in Chapter 8. We want to look at the problem from an asymptotic point of view, hence with $n \to \infty$, and assuming that F is in the domain of attraction of an extreme value distribution. Since the condition $w > \max_{1 \leq i \leq n} X_i$ and $z > \max_{1 \leq i \leq n} Y_i$ is an essential feature of the problem, we want to preserve it in the asymptotic analysis. Hence we assume $w = w_n \to \infty$ and $z = z_n \to \infty$ and moreover that

$$n\,(1 - F(w_n, z_n)) \tag{7.5.1}$$

is bounded.

The aim is to estimate $p_n^\star := 1 - F(w_n, z_n)$. We further assume for simplicity that $w_n = cr_n$ and $z_n = dr_n$, for some positive sequence $r_n \to \infty$ and c, d positive constants. The domain of attraction condition is

$$\lim_{t \to \infty} t\,(1 - F(tx, ty)) = -\log G_0(x, y) = L\left(\frac{1}{x}, \frac{1}{y}\right). \tag{7.5.2}$$

Hence

$$p_n^\star = 1 - F(w_n, z_n) = 1 - F(cr_n, dr_n) \sim \frac{1}{r_n} L\left(\frac{1}{c}, \frac{1}{d}\right).$$

This limit relation suggests that we estimate $p_n^\star$ by

$$\hat{p}_n^\star := \frac{1}{r_n} V_{n,k}\left(\frac{1}{c}, \frac{1}{d}\right) = \frac{1}{r_n} \frac{1}{k} \sum_{i=1}^{n} 1_{\{X_i \geq nc/k \text{ or } Y_i \geq nd/k\}}$$

with V from (7.2.6). Indeed, we find by Proposition 7.2.3 that

$$\lim_{n \to \infty} \frac{\hat{p}_n^\star}{p_n^\star} = \lim_{n \to \infty} \frac{V_{n,k}\left(\frac{1}{c}, \frac{1}{d}\right)}{L\left(\frac{1}{c}, \frac{1}{d}\right)} = 1, \tag{7.5.3}$$

in probability.

This is straightforward. But let us now look at the problem of how to estimate

$$p_n := P\,(X > w_n, Y > z_n) = P\,(X > cr_n, Y > dr_n),$$

where (X, Y) has distribution function F with the same simplifications as before. Suppose moreover that the distribution function F is in the domain of attraction of an extreme value distribution with independent components. One can try to estimate p_n as before by

$$\frac{1}{r_n} \frac{1}{k} \sum_{i=1}^{n} 1_{\{X_i \geq nc/k \text{ and } Y_i \geq nd/k\}}$$

$$= \frac{1}{r_n} \frac{1}{k} \sum_{i=1}^{n} 1_{\{X_i \geq nc/k\}} + \frac{1}{r_n} \frac{1}{k} \sum_{i=1}^{n} 1_{\{Y_i \geq nd/k\}} - \frac{1}{r_n} \frac{1}{k} \sum_{i=1}^{n} 1_{\{X_i \geq nc/k \text{ or } Y_i \geq nd/k\}},$$

but this, multiplied by r_n, converges to $c^{-1} + d^{-1} - (c^{-1} + d^{-1}) = 0$. The problem is that in the case of asymptotic independence we know only that $P(X > tc \text{ and } Y > td)$ is of lower order than $P(X > tc \text{ or } Y > td)$, as $t \to \infty$, but the theory does not say anything about the asymptotic behavior of this probability itself. So it seems that in order to estimate p_n consistently we need a more refined model.

In fact, the condition of Theorem 7.2.2 on the asymptotic normality of $\hat{L}(x, y)$ gives a clue about where to look for further conditions. The condition is in our case

$$t\left(1 - F(tx, ty)\right) = L\left(\frac{1}{x}, \frac{1}{y}\right) + O\left(t^{-\alpha}\right)$$

as $t \to \infty$ for some $\alpha > 0$. A somewhat stronger form of this condition can serve as a second-order condition, quite similar to the one used in Section 2.3: there exists a function A, positive or negative, such that for all $0 < x, y \le \infty$,

$$\lim_{t \to \infty} \frac{t\left(1 - F(tx, ty)\right) - L\left(\frac{1}{x}, \frac{1}{y}\right)}{A(t)} = Q(x, y), \qquad (7.5.4)$$

where Q is not identically zero.

In case of asymptotic independence this second-order condition takes a simple form. Taking $x = \infty$ or $y = \infty$ in (7.5.4) we get

$$\frac{t\left(1 - F(tx, \infty)\right) - x^{-1}}{A(t)} \to Q(x, \infty), \qquad (7.5.5)$$

$$\frac{t\left(1 - F(\infty, ty)\right) - y^{-1}}{A(t)} \to Q(\infty, y). \qquad (7.5.6)$$

Now note that

$$P\left(X > tx, Y > ty\right) = P\left(X > tx\right) + P\left(Y > ty\right) - P\left(X > tx \text{ or } Y > ty\right)$$

and that in the present case $L\left(1/x, 1/y\right) = 1/x + 1/y$. Then (7.5.4)–(7.5.6) imply

$$\frac{t P\left(X > tx, Y > ty\right)}{A(t)} = -Q(x, y) + Q(x, \infty) + Q(\infty, y) =: S(x, y) \quad (7.5.7)$$

as $t \to \infty$ for $0 < x, y < \infty$. Comparing this relation with (7.5.2), we see that $P(X > t \text{ or } Y > t)$ is a regularly varying function of order -1 and $P(X > t \text{ and } Y > t)$ is of lower order in case of asymptotic independence. In fact, $P(X > t \text{ and } Y > t)$ is a regularly varying function of order $\rho - 1$, where $\rho \le 0$ is the index of the regularly varying function $|A|$.

We now show that condition (7.5.7) allows us to estimate p_n consistently. It is common to write (7.5.7) as

$$\lim_{t \to \infty} \frac{P\left(X > tx, Y > ty\right)}{P\left(X > t, Y > t\right)} = S(x, y) . \qquad (7.5.8)$$

In particular, $q(t) := P\left(X > t, Y > t\right)$ is a regularly varying function with index less than or equal to -1. In the original papers (Ledford and Tawn (1996,1997,1998)) the index is written as $-1/\eta$ with $\eta \le 1$. Clearly if there is no asymptotic independence

$A(t)$ can be taken constant and hence $\eta = 1$. Also S is the distribution function of a measure, say ρ, that is,

$$S(x, y) = \rho \left\{ (s, t) \in \mathbb{R}_+^2 \; : \; s > x, t > y \right\} .$$

We take

$$\hat{p}_n := \left(\frac{k}{n} r_n \right)^{-1/\hat{\eta}} \frac{k}{n} \frac{1}{k} \sum_{i=1}^{n} 1_{\{X_i \geq nc/k, Y_i \geq nd/k\}} ,$$

where $\hat{\eta}$ is an estimator of η to be discussed later.

Similarly as in the proof of the consistency of $\hat{L}$ (Theorem 7.2.1) one can prove that

$$\frac{\frac{1}{k} \sum_{i=1}^{n} 1_{\{X_i \geq nc/k, Y_i \geq nd/k\}}}{\frac{n}{k} q \left(\frac{n}{k} \right)} \xrightarrow{P} S(c, d) \tag{7.5.9}$$

provided that

$$\lim_{n \to \infty} nq \left(\frac{n}{k} \right) = \infty . \tag{7.5.10}$$

This condition sets a lower bound for the sequence $k = k(n)$. We now write

$$\frac{\hat{p}_n}{p_n} = \frac{\frac{1}{k} \sum_{i=1}^{n} 1_{\{X_i \geq nc/k, Y_i \geq nd/k\}}}{\frac{n}{k} q \left(\frac{n}{k} \right) S(c, d)} \frac{S(c, d) q \left(\frac{n}{k} \right) \left(\frac{k}{n} r_n \right)^{-1/\hat{\eta}}}{P(X > w_n, Y > z_n)} .$$

By (7.5.8)

$$P(X > w_n, Y > z_n) = P(X > r_n c, Y > r_n d) \sim q(r_n) S(c, d) .$$

Hence we get $\hat{p}_n / p_n \xrightarrow{P} 1$ if

$$\lim_{n \to \infty} \frac{q \left(\frac{n}{k} \right) \left(\frac{k}{n} r_n \right)^{-1/\eta}}{q(r_n)} = 1 \tag{7.5.11}$$

and

$$\left(\frac{k}{n} r_n \right)^{1/\eta - 1/\hat{\eta}} \xrightarrow{P} 1 . \tag{7.5.12}$$

Now (7.5.11) is implied by

$$\lim_{\substack{t \to \infty \\ x = x(t) \to \infty}} \frac{q(t) x^{-1/\eta}}{q(tx)} = 1,$$

which can be achieved by imposing a second-order condition on the regular variation condition for q as in Appendix B, Remark B.3.15.

In order for (7.5.12) to be true we need an estimator $\hat{\eta}$ that converges to η at a certain rate. So let us look now at the estimation of η.

7.6 Estimation of the Residual Dependence Index η

After the preparation in Section 7.5 we now define the residual dependence parameter η generally. Let F be a probability distribution function in the domain of attraction of an extreme value distribution; $F_1(x) := F(x, \infty)$ and $F_2(y) := F(\infty, y)$ the marginal distribution functions, which are supposed to be continuous; and (X, Y) a random vector with distribution function F.

Suppose that for $x, y > 0$,

$$\lim_{t \downarrow 0} \frac{P\left(1 - F_1(X) < tx, 1 - F_2(Y) < ty\right)}{P\left(1 - F_1(X) < t, 1 - F_2(Y) < t\right)} =: S(x, y) \qquad (7.6.1)$$

exists and is positive. Then, $P\left(1 - F_1(X) < t, 1 - F_2(Y) < t\right)$ is a regularly varying function with index $1/\eta$, say, and as in Theorem 6.1.9, for $a, x, y > 0$,

$$S(ax, ay) = a^{1/\eta} S(x, y) . \qquad (7.6.2)$$

The *residual dependence index* $\eta \in (0, 1]$ was introduced by Ledford and Tawn (1996, 1997, 1998).

Note that the domain of attraction condition implies

$$\lim_{t \downarrow 0} t^{-1} P\left(1 - F_1(X) < tx, 1 - F_2(Y) < ty\right)$$

$$= \lim_{t \downarrow 0} t^{-1} P\left(1 - F_1(X) < tx\right) + \lim_{t \downarrow 0} t^{-1} P\left(1 - F_2(Y) < ty\right)$$

$$- \lim_{t \downarrow 0} t^{-1} P\left(1 - F_1(X) < tx \text{ or } 1 - F_2(Y) < ty\right)$$

$$= x + y - L(x, y) .$$

This expression is zero for all $x, y > 0$ in the case of asymptotic independence and positive in all other cases (cf. Proposition 6.1.21). It follows that if there is no asymptotic independence, the index η in (7.6.2) has to be one. In other words, (7.6.1) and (7.6.2) with $\eta < 1$ imply asymptotic independence. On the other hand, $\eta = 1$ does not imply asymptotic independence.

This opens the possibility to devise a test for asymptotic independence in the framework of (7.6.1). The null hypothesis is $\eta = 1$, and the alternative one is $\eta < 1$. In order to carry out this test we need an estimator for η and that is what we discuss next.

Condition (7.6.1) implies

$$\lim_{t \to \infty} \frac{P\left(\frac{1}{1 - F_1(X)} \wedge \frac{1}{1 - F_2(Y)} > tx\right)}{P\left(\frac{1}{1 - F_1(X)} \wedge \frac{1}{1 - F_2(Y)} > t\right)} = S\left(\frac{1}{x}, \frac{1}{x}\right) = x^{-1/\eta} S(1, 1) = x^{-1/\eta}$$

for $x > 0$, i.e., the probability distribution of the random variable $((1 - F_1(X)) \vee (1 - F_2(Y)))^{-1}$ is regularly varying with index $-1/\eta$. This suggests that we use a Hill-type estimator as in Section 3.2.

If F_1 and F_2 were known, we could use

$$\frac{1}{k} \sum_{i=0}^{k-1} \log V_{n-i,n} - \log V_{n-k,n}$$

as an estimator, where $\{V_{j,n}\}$ are the the order statistics of the independent and identically distributed sequence $V_j := 1/\left((1 - F_1(X_j)) \vee (1 - F_2(Y_j))\right)$, $i = 1, 2, \ldots, n$.

Since F_1 and F_2 are not known, we replace them with their empirical counterparts $F_1^{(n)}$ and $F_2^{(n)}$ as defined in (7.3.4) (to prevent division by 0). This leads to the random variables

$$T_i^{(n)} := \frac{1}{\left((1 - F_1^{(n)}(X_i)) \vee (1 - F_2^{(n)}(Y_i))\right)}$$

$$= \frac{1}{\left(\left(\frac{n+1-R(X_i)}{n}\right) \vee \left(\frac{n+1-R(Y_i)}{n}\right)\right)}$$

$$= \frac{n}{n + 1 - (R(X_i) \wedge R(Y_i))},$$

where $R(X_i)$ is the rank of X_i among $X_1, X_2, \ldots, X_n$ and $R(Y_i)$ that of Y_i among $Y_1, Y_2, \ldots, Y_n$. The Hill-type estimator then becomes

$$\hat{\eta} := \frac{1}{k} \sum_{i=0}^{k-1} \log T_{n-i,n}^{(n)} - \log T_{n-k,n}^{(n)},$$

where $\{T_{j,n}\}$ are the order statistics of the non–independent and identically distributed sequence $T_i^{(n)}$, $i = 1, 2, \ldots, n$.

Asymptotic normality can be proved under a refinement of condition (7.6.1).

Theorem 7.6.1 *Let $(X_1, Y_1), (X_2, Y_2), \ldots$ be i.i.d. random vectors with distribution function F. Suppose (7.6.1) and (7.6.2) hold for some $\eta \in (0, 1]$. We also assume that the following second-order refinement of (7.6.1) holds:*

$$\lim_{t \downarrow 0} \frac{\frac{P(1 - F_1(X) < tx \text{ and } 1 - F_2(Y) < ty)}{P(1 - F_1(X) < t \text{ and } 1 - F_2(Y) < t)} - S(x, y)}{q_1(t)} =: Q(x, y)$$

exists for all $x, y \geq 0$ with $x + y > 0$, where q_1 is some positive function and Q is neither a constant nor a multiple of S. Moreover, we assume that the convergence is uniform on $\{(x, y) \in \mathbb{R}_+^2 : x^2 + y^2 = 1\}$ and that the function S has first-order partial derivatives $S_x := \partial S(x, y)/\partial x$ and $S_y := \partial S(x, y)/\partial y$. Finally, we assume that

$$\lim_{t \downarrow 0} t^{-1} P\left(1 - F_1(X) < t \text{ and } 1 - F_2(Y) < t\right) =: l$$

exists. Let $q^{\leftarrow}$ be the inverse of the function $q(t) = P(1 - F_1(X) < t \text{ and } 1 - F_2(Y) < t)$. For a sequence $k = k(n)$ of integers with $k \to \infty$, $k/n \to 0$, and $\sqrt{k}q_1(q^{\leftarrow}(k/n)) \to 0$, $n \to \infty$,

$$\sqrt{k}\,(\hat{\eta} - \eta)$$

is asymptotic normal with mean zero and variance

$$\eta^2(1 - l)(1 - 2l\,S_x(1, 1)\,S_y(1, 1)) .$$

Proof. We provide a sketch of the proof. The original elaborate proof (Draisma, Drees, Ferreira, and de Haan (2004)) is beyond the scope of this book.

Similarly to Section 7.2 one obtains, with $m := nq(k/n)$ and $m \to \infty$,

$$\sqrt{m}\left(\frac{1}{m}\sum_{i=1}^{n}1_{\{X_i \geq X_{n-[kx]+1,n} \text{ and } Y_i \geq Y_{n-[ky]+1,n}\}} - S(x, y)\right) \xrightarrow{d} W(x, y)$$

in $D([0, T] \times [0, T])$, for every $T > 0$, where W is a zero-mean Gaussian process with in case $l = 0$,

$$EW(x_1, y_1)W(x_2, y_2) = S(x_1 \wedge x_2, y_1 \wedge y_2),$$

and in case $l > 0$,

$$W(x, y) = \frac{1}{\sqrt{l}}\,(W_1(x, 0) + W_1(0, y) - W_1(x, y))$$
$$- \sqrt{l}S_x(x, y)W_1(x, 0) - \sqrt{l}S_y(x, y)W_1(0, y)$$

and

$$EW_1(x_1, y_1)W_1(x_2, y_2) = x_1 \wedge x_2 + y_1 \wedge y_2 - lS(x_1, y_1)$$
$$- lS(x_2, y_2) + lS(x_1 \vee x_2, y_1 \vee y_2) .$$

Next note that

$$\sum_{i=1}^{n}1_{\{X_i \geq X_{n-[kx]+1,n} \text{ and } Y_i \geq Y_{n-[kx]+1,n}\}}$$

$$= \sum_{i=1}^{n}1_{\{1-F_1^{(n)}(X_i) \leq 1-F_1^{(n)}(X_{n-[kx]+1,n})=[kx]/n \text{ and }}$$
$$1-F_2^{(n)}(Y_i) \leq 1-F_2^{(n)}(X_{n-[kx]+1,n})=[kx]/n\}}$$

$$= \sum_{i=1}^{n}1_{\{T_i^{(n)} \geq n/[kx]\}} .$$

Hence with

$$1 - F_T^{(n)}(x) := \frac{1}{n}\sum_{i=1}^{n}1_{\{kT_i^{(n)}/n \geq x\}}$$

we obtain

$$\sqrt{m}\left(\frac{1 - F_T^{(n)}(x)}{q\left(\frac{k}{n}\right)} - x^{1/\eta}\right) \xrightarrow{d} W(x, x)$$

in $D([0, T])$, for every $T > 0$. This relation is somewhat similar to Theorem 5.1.4, which led to the asymptotic normality of the "usual" Hill estimator. For further details we refer to the mentioned paper. ∎

Exercises

7.1. Prove Theorem 7.2.1 for the d-dimensional case, i.e., if $(X_1^{(1)}, \ldots, X_1^{(d)}), \ldots,$ $(X_n^{(1)}, \ldots, X_n^{(d)})$ are independent and identically distributed random vectors with distribution function F in the domain of attraction of an extreme value distribution G, with

$$L(x_1, x_2, \ldots, x_n) := -\log G\left(\frac{x_1^{-\gamma_1} - 1}{\gamma_1}, \frac{x_2^{-\gamma_2} - 1}{\gamma_2}, \ldots, \frac{x_d^{-\gamma_d} - 1}{\gamma_d}\right)$$

for $(x_1, x_2, \ldots, x_d) \in \mathbb{R}_+^d$, where $\gamma_1, \gamma_2, \ldots, \gamma_d$ are the marginal extreme value indices and $L(x, 0, \ldots, 0) = L(0, x, \ldots, 0) = L(0, 0, \ldots, x) = x$, then for $T > 0$, as $n \to \infty$, $k = k(n) \to \infty$, $k/n \to 0$,

$$\sup_{0 \leq x_1, x_2, \ldots, x_d \leq T} \left|\hat{L}(x_1, x_2, \ldots, x_d) - L(x_1, x_2, \ldots, x_d)\right| \xrightarrow{P} 0.$$

7.2. Consider Example 5.5.3. Determine the dependence coefficient H of the random vector (X_n, X_{n+1}).

7.3. Prove Theorem 7.2.2 with the natural extension to d dimensions, i.e., that under the given conditions and for $k = k(n) \to \infty$, $k(n) = o\left(n^{2\alpha/(1+2\alpha)}\right)$, $\alpha > 0$, as $n \to \infty$,

$$\sqrt{k}\left(\hat{L}(x_1, x_2, \ldots, x_d) - L(x_1, x_2, \ldots, x_d)\right) \xrightarrow{d} B(x_1, x_2, \ldots, x_d)$$

in $D([0, T]^d)$, for every $T > 0$, and for $(x_1, x_2, \ldots, x_d) \in \mathbb{R}_+^d$, where

$$B(x_1, x_2, \ldots, x_d) = W(x_1, x_2, \ldots, x_d) - L_1(x_1, x_2, \ldots, x_d)W(x_1, 0, \ldots, 0)$$
$$-L_2(x_1, x_2, \ldots, x_d)W(0, x_2, 0, \ldots, 0) - \ldots - L_d(x_1, x_2, \ldots, x_d)W(0, 0, \ldots, x_d)$$

and W is a continuous Gaussian process with mean zero and covariance structure $EW(x_1, \ldots, x_d)W(x_1, \ldots, x_d) = \mu\left(R(x_1, \ldots, x_d) \cap R(x_1, \ldots, x_d)\right)$ with $R(x_1, \ldots, x_d) := \{(u_1, \ldots, u_d) \in \mathbb{R}_+^d : 0 \leq u_1 \leq x_1 \text{ or } \ldots \text{ or } 0 \leq u_d \leq x_d\}$.

7.4. Show that under the conditions of Theorem 7.2.2 the proposed estimator of Sibuya's dependence coefficient $\hat{\lambda}$ (Section 7.4) satisfies

$$\sqrt{k}\left(\hat{\lambda} - \lambda\right) \overset{d}{\to} N\left(0, L(1 - 2L_1 L_2) + (L_1 + L_2)^2 + 2(1 - L_1)(1 - L_2) - 2\right),$$

where $L := L(1, 1)$, $L_1 := L_1(1, 1)$, $L_2 := L_2(1, 1)$, and N is a standard normal random variable.

7.5. Let $F(x_1, x_2)$ be a probability distribution function in the domain of attraction of an extreme value distribution $G(x_1, x_2)$, i.e., there are functions $a_1, a_2 > 0, b_1$, and b_2 such that for all x_1, x_2 for which $0 < G(x_1, x_2) < 1$, $\lim_{t \to \infty} t\,\{1 - F\,(b_1(t) + x_1 a_1(t),\ b_2(t) + x_2 a_2(t))\} = -\log G(x_1, x_2) =: \Phi(x_1, x_2)$. Suppose that the following second-order condition holds: there exists a positive or negative function A with $\lim_{t \to \infty} A(t) = 0$ and a function Ψ not a multiple of Φ such that for each (x_1, x_2) for which $0 < G(x_1, x_2) < 1$,

$$\lim_{t \to \infty} \frac{t\,(1 - F\,(b_1(t) + x_1 a_1(t), b_2(t) + x_2 a_2(t))) - \Phi(x_1, x_2)}{A(t)} = \Psi(x_1, x_2)$$

locally uniformly for $(x_1, x_2) \in (0, \infty] \times (0, \infty]$. Show that this second-order condition implies the second-order condition of Section 2.3 for the two marginal distributions. Show that the function A is regularly varying. Show that if the index of A is smaller than zero, condition (7.2.8) of Theorem 7.2.2 holds pointwise.

7.6. Show that if X and Y are independent, the residual dependence parameter η from Section 7.6 for (X, Y) is $\frac{1}{2}$.

7.7. Let X, Y, U be independent random variables, where X and Y have distribution function $1 - 1/x$, $x > 1$, and U has distribution function $1 - 1/(x^\alpha \log x)$, $x > 1$, for some $\alpha \in [1, 2)$. Show that the distribution function of the random vector $(X \vee U, Y \vee U)$ is in the domain of attraction of an extreme value distribution with independent components and residual dependence index $1/\alpha$.

8

Estimation of the Probability of a Failure Set

8.1 Introduction

In this chapter we are going to deal with methods to solve the problem posed in a graphical way in Chapter 6. The wave height (HmO) and still water level (SWL) have been recorded during 828 storm events that are relevant for the Pettemer Zeewering. Engineers of RIKZ (Institute for Coastal and Marine Management) have determined *failure conditions*, that is, those combinations of HmO and SWL that result in overtopping the seawall, thus creating a dangerous situation. The set of those combinations forms a failure set C.

Figure 8.1 displays the failure set

$$C = \{(\text{HmO, SWL}) \ : \ 0.3\text{HmO} + \text{SWL} > 7.6\} ,$$

as well as 828 independent and identically distributed observations of HmO and SWL. This set is such that if an independent observation should fall into C, it would (could) lead to a disaster. The problem is how to determine the probability that an independent observation falls into this set.

In order to develop statistical methods to deal with this problem, we use the theory developed in Chapters 6 and 7. We start by assuming that there exist normalizing functions $a_1, a_2 > 0$ and b_1, b_2 real, and a distribution function G with nondegenerate marginals, such that for all continuity points (x, y) of G,

$$\lim_{t \to \infty} F^t(a_1(t)x + b_1(t), a_2(t)y + b_2(t)) = G(x, y) . \tag{8.1.1}$$

Moreover, we choose the functions a_1, a_2, b_1, b_2 such that

$$G(x, \infty) = \exp\left(-(1 + \gamma_1 x)^{-1/\gamma_1}\right), \quad 1 + \gamma_1 x > 0, \tag{8.1.2}$$

and

$$G(\infty, y) = \exp\left(-(1 + \gamma_2 y)^{-1/\gamma_2}\right), \quad 1 + \gamma_2 x > 0, \tag{8.1.3}$$

where γ_1 and γ_2 are the marginal extreme value indices. Then from Section 6.1 (e.g., Theorem 6.1.11),

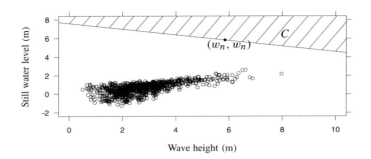

Fig. 8.1. Failure set C, boundary point (w_n, w_n) and observations.

$$\lim_{t \to \infty} t \, P\left(\left(1 + \gamma_1 \frac{X - b_1(t)}{a_1(t)}\right)^{1/\gamma_1} > x \text{ or } \left(1 + \gamma_2 \frac{Y - b_2(t)}{a_2(t)}\right)^{1/\gamma_2} > y\right)$$

$$= \lim_{t \to \infty} t \, P\left(\frac{X - b_1(t)}{a_1(t)} > \frac{x^{\gamma_1} - 1}{\gamma_1} \text{ or } \frac{Y - b_2(t)}{a_2(t)} > \frac{y^{\gamma_2} - 1}{\gamma_2}\right)$$

$$= -\log G\left(\frac{x^{\gamma_1} - 1}{\gamma_1}, \frac{y^{\gamma_2} - 1}{\gamma_2}\right).$$

Or more generally, with ν the exponent measure defined in Section 6.1.3,

$$\lim_{t \to \infty} t \, P\left\{\left(\left(1 + \gamma_1 \frac{X - b_1(t)}{a_1(t)}\right)^{1/\gamma_1}, \left(1 + \gamma_2 \frac{Y - b_2(t)}{a_2(t)}\right)^{1/\gamma_2}\right) \in Q\right\} = \nu(Q)$$

(8.1.4)

for all Borel sets $Q \subset \mathbb{R}_+^2$ with $\inf_{(x,y) \in Q} \max(x, y) > 0$ and $\nu(\partial Q) = 0$. Then, for any $a > 0$ we know that

$$\nu(aQ) = a^{-1}\nu(Q),$$

(8.1.5)

where

$$aQ := \{(ax, ay) : (x, y) \in Q\}$$

(cf. Theorem 6.1.9), and this property will be our main tool. Note that ν is the approximate mean measure of the point process formed by the observations (Theorem 6.1.11). Hence in principle $\nu(Q)$ can be estimated by just counting the number of observations in the set Q.

Now recall that we want to estimate $P((X, Y) \in C_n)$. Clearly there is no observation in the failure set. In fact, the observations are all some distance away from the failure set. There has been no dangerous situation around the dike during the observation period. This suggests that in a first approximation, $P(C) < 1/n$. This particular feature is essential for extreme value problems and we want to capture this in our approach, based on a limit situation in which the number of observations grows to infinity.

We have n independent observations $(X_1, Y_1), (X_2, Y_2), \ldots, (X_n, Y_n)$ with common distribution function F and we have a failure set C with $P(C) < 1/n$. This means that if we assume $n \to \infty$, in order to preserve the extreme value situation, we have to assume that the failure set is not fixed but depends on n: $C = C_n$ with $P(C_n) \to 0, n \to \infty$.

Now we write the probability we want to estimate in terms of the transformed variables:

$$p_n := P((X, Y) \in C_n)$$
$$= P\left\{\left(\left(1 + \gamma_1 \frac{X - b_1(t)}{a_1(t)}\right)^{1/\gamma_1}, \left(1 + \gamma_2 \frac{Y - b_2(t)}{a_2(t)}\right)^{1/\gamma_2}\right) \in Q_n\right\} \quad (8.1.6)$$

with

$$Q_n := \left\{\left(\left(1 + \gamma_1 \frac{x - b_1(t)}{a_1(t)}\right)^{1/\gamma_1}, \left(1 + \gamma_2 \frac{x - b_2(t)}{a_2(t)}\right)^{1/\gamma_2}\right) : (x, y) \in C_n\right\}. \quad (8.1.7)$$

Since the set Q_n, like the set C_n, does not contain any observations, we divide the set Q_n by a large positive constant c_n such that Q_n/c_n contains a small portion of the observations. This way we can estimate $\nu(Q_n/c_n)$ and hence $\nu(Q_n) = \nu(Q_n/c_n)/c_n$.

Summing up, the procedure involves the following steps:

1. Marginal transformations

$$X_i \to \left(1 + \gamma_1 \frac{X_i - b_1(t)}{a_1(t)}\right)^{1/\gamma_1},$$
$$Y_i \to \left(1 + \gamma_2 \frac{Y_i - b_2(t)}{a_2(t)}\right)^{1/\gamma_2}, \quad (8.1.8)$$

$i = 1, \ldots, n$, in order to transform the marginal distributions approximately to a Pareto distribution with distribution function $1 - 1/x, x \geq 1$. Note that after this transformation the probabilities can be approximated by the measure ν as in (8.1.4).

2. Use the homogeneity property of the measure ν, (8.1.5), in order to pull the transformed failure set to the observations. Next estimate ν by its empirical measure.

The two steps for the given data set are indicated in Figures 8.2 and 8.3.

Next we write analytically (although not yet in a formal way) the reasoning developed above. Let k be an intermediate sequence, i.e., $k = k(n) \to \infty, k/n \to 0$, $n \to \infty$. If the failure set C_n can be written as

$$C_n = \left\{\left(a_1\left(\frac{n}{k}\right) \frac{(c_n x)^{\gamma_1} - 1}{\gamma_1} + b_1\left(\frac{n}{k}\right), \right.\right.$$
$$\left.\left. a_2\left(\frac{n}{k}\right) \frac{(c_n y)^{\gamma_2} - 1}{\gamma_2} + b_2\left(\frac{n}{k}\right)\right) : (x, y) \in S\right\}, \quad (8.1.9)$$

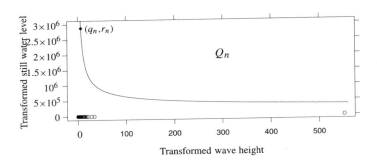

Fig. 8.2. Transformed: failure set (8.1.7) (area above the curved line), boundary point (q_n, r_n) and data set (8.1.8).

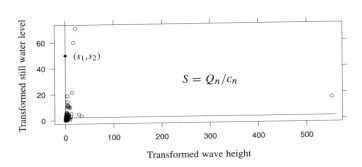

Fig. 8.3. Transformed data set (8.1.8), boundary point $(s_1, s_2) := (q_n/c_n, r_n/c_n)$ and pulled set S (area above the curved line) from (8.1.9).

where c_n is a positive sequence (generally $c_n \to \infty, n \to \infty$) and S is a fixed open set of $\mathbb{R}^2$, and the marginal transformations (8.1.8) applied to C_n give the set $c_n S$ (called Q_n above). Note that n/k is playing the role of the running variable t considered before. Then, for some fixed open Borel set $S \subset \mathbb{R}^2_+$ with $\inf_{(x,y) \in S} \max(x, y) > 0$ and $\nu(\partial S) = 0$ we can write (8.1.6) as

$$P\left\{\left(\left(1 + \gamma_1 \frac{X - b_1\left(\frac{n}{k}\right)}{a_1\left(\frac{n}{k}\right)}\right)^{1/\gamma_1}, \left(1 + \gamma_2 \frac{Y - b_2\left(\frac{n}{k}\right)}{a_2\left(\frac{n}{k}\right)}\right)^{1/\gamma_2}\right) \in c_n S\right\}. \quad (8.1.10)$$

This, by (8.1.4), is approximately equal to

$$\frac{k}{n} \nu\left(c_n S\right) = \frac{k}{nc_n} \nu(S), \quad (8.1.11)$$

where the last equality follows from (8.1.5). This leads to the estimator (defined in more detail below; cf. Theorems 8.2.1 and 8.3.1)

$$\hat{p}_n := \frac{k}{nc_n}\,\hat{v}(\hat{S})\ .$$

Note that S is not known since γ_1, γ_2, a_1, a_2, b_1, b_2 are not known.

Up to this point we have dealt with c_n as if it were known. That is, it has played a similar role to that of the intermediate sequence $k = k(n)$ in the univariate estimation. This way it is to be chosen (under certain bounds) by the statistician. An alternative way to deal with c_n is to incorporate it in the problem itself, and consequently to estimate it along with the other unknown quantities. We shall discuss these two approaches in two separate subsections.

We add some comments at this point. In the above discussion we assumed $v(S)$ positive, and this will be the case considered in the next section. In fact this is the case if the random variables X and Y are not asymptotically independent or S contains (at least part of) the axis

$$\{(x, y)\ :\ x > 0 \text{ and } y = 0\}\ \cup\ \{(x, y)\ :\ x = 0 \text{ and } y > 0\}\ . \tag{8.1.12}$$

The notion of asymptotic independence was first introduced in Section 6.2. Recall that a random vector (X, Y) is said to be asymptotically independent if its distribution function is in the domain of attraction of some extreme value distribution with independent components, i.e., the limiting distribution is the product of its marginals.

In this case we know that the exponent measure from Section 6.1.3 is concentrated in the positive axes given in (8.1.12). In terms of the spectral representation discussed in Section 6.1.4, recall that for $0 \leq \theta_1 \leq \theta_2 \leq \pi/2$ either for all $0 < r_1 < r_2 < \infty$ the set

$$\left\{(x, y) : r_1 < \sqrt{x^2 + y^2} < r_2, \theta_1 \leq \arctan \frac{x}{y} \leq \theta_2\right\}$$

has positive μ-mass or for no choice of $0 < r_1 < r_2 < \infty$ it has positive μ-mass, depending on whether the spectral measure has positive or zero mass in $[\theta_1, \theta_2]$.

This means that $v(S)$ is always positive as long as we do not have asymptotic independence. But $v(S)$ can be positive even under asymptotic independence, e.g., in case $S \supset (x_1, x_2) \times [0, y)$ for some $0 < x_1 < x_2, y > 0$.

Note that the proposed transformations of the failure set C_n that lead to the set S are such that certain features of the original set C_n are preserved after the transformation to S. For instance, if $C_n \subset [x, \infty) \times [y, \infty)$ then S will also satisfy this, for possibly some other x, y, or if $C_n \supset (x_1, x_2) \times (-\infty, y)$ then $S \supset (x_1, x_2) \times [0, y)$, for possibly some other x_1, x_2, y.

The case $v(S) = 0$ is discussed in Section 8.3. Clearly $v(S) = 0$ under asymptotic independence and if S is contained in a set of the form $(x, \infty) \times (y, \infty)$, for some $x, y > 0$. The procedure is quite similar to that with $v(S) > 0$, and additionally it involves the residual independence index η introduced in Section 7.6. For testing asymptotic independence we refer to Section 7.6.

For simplicity we consider only the two-dimensional case. The generalization to higher dimensions is straightforward. For a similar result in function space see Section 10.5.

Though the procedure is developed with the same choice of the intermediate sequence k for both marginals, it can be adapted for different intermediate sequences. Hence in practice one can use different k's in the estimation of the marginals.

8.2 Failure Set with Positive Exponent Measure

Recall that on the basis of independent and identically distributed random vectors $(X_1, Y_1), \ldots, (X_n, Y_n)$ from F we want to estimate $p_n = P((X, Y) \in C_n)$, for some given failure set. In this section we assume that there exists some boundary point of C_n, (v_n, w_n) such that

$$C_n \subset \{(x, y) : x \geq v_n \text{ or } y \geq w_n\} \tag{8.2.1}$$

for all n. Note that this is a rather weak assumption. For instance in Figure 8.1 we took the diagonal point (w_n, w_n) for (v_n, w_n).

Now let us see what happens to this point after the marginal transformations (8.1.8), with t replaced by n/k, where $k = k(n) \to \infty, k/n \to 0, n \to \infty$. They are ilustrated in Figure 8.2. Define

$$q_n := \left(1 + \gamma_1 \frac{v_n - b_1\left(\frac{n}{k}\right)}{a_1\left(\frac{n}{k}\right)}\right)^{1/\gamma_1}, \quad r_n := \left(1 + \gamma_2 \frac{w_n - b_2\left(\frac{n}{k}\right)}{a_2\left(\frac{n}{k}\right)}\right)^{1/\gamma_2} \tag{8.2.2}$$

and assume that $\lim_{n\to\infty} q_n/r_n$ exists and is finite; this avoids the predominance of one marginal over the other so that the problem does not become a univariate one in the limit.

8.2.1 First Approach: c_n Known

In the next theorem we state the necessary conditions for the consistency of $\hat{p}_n$. We opted for a long theorem, which in turn is mostly self-contained in all its conditions and definitions.

Theorem 8.2.1 *Let* $(X_1, Y_1), \ldots, (X_n, Y_n)$ *be an i.i.d. sample from F. Suppose F is in the domain of attraction of an extreme value distribution with normalizing functions $a_i > 0$, b_i real, marginal extreme value indices γ_i, $i = 1, 2$, and exponent measure ν (cf. (8.1.1)–(8.1.4)).*

Consider some estimators of γ_i, $a_i(n/k) > 0$, $b_i(n/k)$ such that for some sequence $k = k(n) \to \infty, k/n \to 0, n \to \infty$,

$$\sqrt{k}\left(\hat{\gamma}_i - \gamma_i, \frac{\hat{a}_i\left(\frac{n}{k}\right)}{a_i\left(\frac{n}{k}\right)} - 1, \frac{\hat{b}_i\left(\frac{n}{k}\right) - b_i\left(\frac{n}{k}\right)}{a_i\left(\frac{n}{k}\right)}\right) = (O_P(1), O_P(1), O_P(1)), \tag{8.2.3}$$

$i = 1, 2$.

Suppose the failure set C_n is an open set for which (8.2.1) holds. Suppose further that C_n can be written as

$$C_n = \left\{ \left(a_1 \left(\frac{n}{k} \right) \frac{(c_n x)^{\gamma_1} - 1}{\gamma_1} + b_1 \left(\frac{n}{k} \right), \right. \right.$$
$$\left. \left. a_2 \left(\frac{n}{k} \right) \frac{(c_n y)^{\gamma_2} - 1}{\gamma_2} + b_2 \left(\frac{n}{k} \right) \right) : (x, y) \in S \right\}, \qquad (8.2.4)$$

where S is an open Borel set in $\mathbb{R}_+^2$ with $\nu(\partial S) = 0$ and $\nu(S) > 0$, and c_n a sequence of positive numbers with $c_n \to \infty$, $n \to \infty$.

Finally, suppose $0 < q_n/r_n < \infty$ (our conditions imply that q_n/r_n does not depend on n),

$$\lim_{n \to \infty} \frac{w_{\gamma_1 \wedge \gamma_2}(c_n)}{\sqrt{k}} = 0, \qquad (8.2.5)$$

where

$$w_\gamma(t) = t^{-\gamma} \int_1^t s^{\gamma-1} \log s \, ds, \quad t > 1$$

(the function $w_\gamma(t)$ has been defined in Theorem 4.4.1), and that (8.1.4) holds with Q replaced by $c_n S$, i.e.,

$$\lim_{n \to \infty} \frac{n}{k} P \left\{ \left(\left(1 + \gamma_1 \frac{X - b_1 \left(\frac{n}{k} \right)}{a_1 \left(\frac{n}{k} \right)} \right)^{1/\gamma_1}, \right. \right.$$
$$\left. \left. \left(1 + \gamma_2 \frac{Y - b_2 \left(\frac{n}{k} \right)}{a_2 \left(\frac{n}{k} \right)} \right)^{1/\gamma_2} \right) \in c_n S \right\} \Big/ \nu(c_n S) = 1 . \quad (8.2.6)$$

Then, with

$$\hat{p}_n := \frac{1}{nc_n} \sum_{i=1}^n 1 \left\{ \left(\left(1 + \hat{\gamma}_1 \frac{X_i - \hat{b}_1 \left(\frac{n}{k} \right)}{\hat{a}_1 \left(\frac{n}{k} \right)} \right)^{1/\hat{\gamma}_1}, \left(1 + \hat{\gamma}_2 \frac{Y_i - \hat{b}_2 \left(\frac{n}{k} \right)}{\hat{a}_2 \left(\frac{n}{k} \right)} \right)^{1/\hat{\gamma}_2} \right) \in \hat{S} \right\}, \qquad (8.2.7)$$

where

$$\hat{S} := \left\{ \left(\frac{1}{c_n} \left(1 + \hat{\gamma}_1 \frac{x - \hat{b}_1 \left(\frac{n}{k} \right)}{\hat{a}_1 \left(\frac{n}{k} \right)} \right)^{1/\hat{\gamma}_1}, \right. \right.$$
$$\left. \left. \frac{1}{c_n} \left(1 + \hat{\gamma}_2 \frac{y - \hat{b}_2 \left(\frac{n}{k} \right)}{\hat{a}_2 \left(\frac{n}{k} \right)} \right)^{1/\hat{\gamma}_2} \right) : (x, y) \in C_n \right\}, \qquad (8.2.8)$$

we have

$$\frac{\hat{p}_n}{p_n} \xrightarrow{P} 1 .$$

We postpone the proof to Section 8.2.3. We want to add some comments at this point.

The estimation of γ_i, $a_i(n/k)$, and $b_i(n/k)$, $i = 1, 2$, is known from univariate extreme value statistics (cf. Chapters 3 and 4). For instance one can use the moment-type estimators of Sections 3.5 and 4.2:

$$M_{n,1}^{(j)} = \frac{1}{k} \sum_{i=1}^{k} (\log X_{n,n-i+1} - \log X_{n,n-k})^j, \quad j = 1, 2,$$

$$\hat{\gamma}_1 = M_{n,1}^{(1)} + 1 - \frac{1}{2} \left(1 - \frac{\left(M_{n,1}^{(1)} \right)^2}{M_{n,1}^{(2)}} \right)^{-1},$$

$$\hat{a}_1 \left(\frac{n}{k} \right) = X_{n-k,n} M_{n,1}^{(1)} \left(1 - \hat{\gamma}_1 + M_{n,1}^{(1)} \right),$$

and

$$\hat{b}_1 \left(\frac{n}{k} \right) = X_{n,n-k} ,$$

where for $M_{n,2}^{(j)}$, $\hat{\gamma}_2$, $\hat{a}_2(n/k)$, and $\hat{b}_2(n/k)$ replace X by Y in the previous formulas. Then under the second-order regular variation condition (cf. Definition 2.3.1) for both marginals with auxiliary functions A_i, $i = 1, 2$, and provided $k = k(n) \to \infty$, $k/n \to 0$, $\sqrt{k} A_i(n/k) = O(1)$, $i = 1, 2$, as $n \to \infty$, the O_P-property for the individual terms in (8.2.3) follows from Sections 2.2, 3.5, and 4.2. Then they are also jointly $O_P(1)$.

Remark 8.2.2 Note that $\hat{p}_n$ and $\hat{S}$ may not be defined if $1 + \hat{\gamma}_1 (X_i - \hat{b}_1(n/k)/\hat{a}_1(n/k)$ ≤ 0 for some X_i and similarly with the second component. However, when checking the proofs one sees that when $n \to \infty$, the probability that this happens tends to zero.

Remark 8.2.3 Note that the relation between $k = k(n)$ and c_n may restrict the range of possible values of the marginal extreme value indices. For $\gamma_1 \wedge \gamma_2 < 0$, condition (8.2.5) implies

$$\lim_{n \to \infty} \frac{c_n^{-(\gamma_1 \wedge \gamma_2)}}{\sqrt{k}} = \lim_{n \to \infty} k^{-1/2 - (\gamma_1 \wedge \gamma_2)} \left(\frac{k}{c_n} \right)^{\gamma_1 \wedge \gamma_2} = 0 .$$

For instance, if we want to allow $k/c_n = O(1)$, we must have $k^{-1/2 - (\gamma_1 \wedge \gamma_2)} \to 0$, which is true only if $\gamma_1 \wedge \gamma_2 > -\frac{1}{2}$.

8.2.2 Alternative Approach: Estimate c_n

Define, for some $r > 0$,

$$c_n := \frac{\sqrt{q_n^2 + r_n^2}}{r} , \tag{8.2.9}$$

where q_n and r_n are as in (8.2.2). According to (8.1.9) the point $(s_1, s_2) :=$ $(q_n/c_n, r_n/c_n)$ is on the boundary of S. Moreover, from (8.2.9) we have $s_1^2 + s_2^2 =$

$(q_n/c_n)^2 + (r_n/c_n)^2 = r^2$, that is, (s_1, s_2) is on a circle of radius r and hence close enough to the observations (cf. Figure 8.3).

Let x_1^* and x_2^* be the right endpoints of the marginal distributions. Note that $v_n \uparrow x_1^*$ and $w_n \uparrow x_2^*$ imply $q_n \to \infty$ and $r_n \to \infty$ respectively, under the domain of attraction condition.

Corollary 8.2.4 *Under the conditions of Theorem 8.2.1, with $\gamma_1 \wedge \gamma_2 > -\frac{1}{2}$, define the estimators*

$$\hat{q}_n := \left(1 + \hat{\gamma}_1 \frac{v_n - \hat{b}_1\left(\frac{n}{k}\right)}{\hat{a}_1\left(\frac{n}{k}\right)}\right)^{1/\hat{\gamma}_1}, \tag{8.2.10}$$

$$\hat{r}_n := \left(1 + \hat{\gamma}_2 \frac{w_n - \hat{b}_2\left(\frac{n}{k}\right)}{\hat{a}_2\left(\frac{n}{k}\right)}\right)^{1/\hat{\gamma}_2}, \tag{8.2.11}$$

$$\hat{c}_n := \frac{\sqrt{\hat{q}_n^2 + \hat{r}_n^2}}{r}, \tag{8.2.12}$$

for some $r > 0$ (to be chosen by the statistician), and

$$\hat{p}_n^\star := \frac{1}{n\hat{c}_n} \sum_{i=1}^n 1\left\{\left(\left(1 + \hat{\gamma}_1 \frac{x_i - \hat{b}_1\left(\frac{n}{k}\right)}{\hat{a}_1\left(\frac{n}{k}\right)}\right)^{1/\hat{\gamma}_1}, \left(1 + \hat{\gamma}_2 \frac{y_i - \hat{b}_2\left(\frac{n}{k}\right)}{\hat{a}_2\left(\frac{n}{k}\right)}\right)^{1/\hat{\gamma}_2}\right) \in \hat{S}^\star\right\} \tag{8.2.13}$$

with

$$\hat{S}^\star := \left\{\left(\frac{1}{\hat{c}_n}\left(1 + \hat{\gamma}_1 \frac{x - \hat{b}_1\left(\frac{n}{k}\right)}{\hat{a}_1\left(\frac{n}{k}\right)}\right)^{1/\hat{\gamma}_1},\right.\right.$$

$$\left.\left.\frac{1}{\hat{c}_n}\left(1 + \hat{\gamma}_2 \frac{y - \hat{b}_2\left(\frac{n}{k}\right)}{\hat{a}_2\left(\frac{n}{k}\right)}\right)^{1/\hat{\gamma}_2}\right) : (x, y) \in C_n\right\}. \tag{8.2.14}$$

Then

$$\frac{\hat{p}_n^\star}{p_n} \overset{P}{\to} 1.$$

Remark 8.2.5 Under much more stringent conditions it can be proved that in case $\nu(S) > 0$,

$$\frac{\sqrt{k}}{q_{\gamma_1 \wedge \gamma_2}(c_n)}\left(\frac{\hat{p}_n^\star}{p_n} - 1\right)$$

is asymptotically normal (de Haan and Sinha (1999)).

8.2.3 Proofs

We give some intermediate results from which the consistency of $\hat{p}_n$ and $\hat{p}_n^\star$ will follow.

Lemma 8.2.6 *Let $f_n(x)$ and $g_n(x)$ be strictly increasing continuous functions for all n, $\lim_{n\to\infty} f_n(x) = x$, and $\lim_{n\to\infty} g_n(x) = x$ for $x > 0$. For an open set O, let*

$$O_n := \{(f_n(x), g_n(y)) : (x, y) \in O\} \, .$$

Then

$$1_{O_n}(x, y) := 1_{\{(x,y)\in O_n\}} \to 1_O(x, y) := 1_{\{(x,y)\in O\}}$$

for $(x, y) \in O$.

Proof. Take $(x, y) \in O$ and $\varepsilon > 0$ such that $(x - \varepsilon, y - \varepsilon) \in O$. For $n \geq n_0$ we have $f_n^{\leftarrow}(x) > x - \varepsilon$ and $g_n^{\leftarrow}(y) > y - \varepsilon$. Hence $\left(f_n^{\leftarrow}(x), g_n^{\leftarrow}(y)\right) \in O$ for $n \geq n_0$. It follows that

$$1_O\left(f_n^{\leftarrow}(x), g_n^{\leftarrow}(y)\right) \to 1$$

for $(x, y) \in O$. Now

$$(x, y) \in O_n \Leftrightarrow \left(f_n^{\leftarrow}(x), g_n^{\leftarrow}(y)\right) \in O \Leftrightarrow 1_O\left(f_n^{\leftarrow}(x), g_n^{\leftarrow}(y)\right) = 1 \, .$$

Hence the conclusion. ∎

Proposition 8.2.7 *Let $(X_1, Y_1), (X_2, Y_2), \ldots$ be an i.i.d. sample from F. Suppose F is in the domain of attraction of an extreme value distribution with normalizing functions $a_i > 0$, b_i real, marginal extreme value indices γ_i, $i = 1, 2$, and exponent measure ν. Let S be an open Borel set in $\mathbb{R}_+^2$ with $\inf_{(x,y)\in S} \max(x, y) > 0$, $\nu(\partial S) = 0$, and $\nu(S) > 0$. For the random variable*

$$\tilde{\nu}_n(S) := \frac{1}{k} \sum_{i=1}^{n} 1\left\{\left(\left(1 + \gamma_1 \frac{X_i - b_1\left(\frac{n}{k}\right)}{a_1\left(\frac{n}{k}\right)}\right)^{1/\gamma_1}, \left(1 + \gamma_2 \frac{Y_i - b_2\left(\frac{n}{k}\right)}{a_2\left(\frac{n}{k}\right)}\right)^{1/\gamma_2}\right) \in S\right\}$$

with k satisfying $k = k(n) \to \infty$, $k/n \to 0$, $n \to \infty$, we have

$$\tilde{\nu}_n(S) \xrightarrow{P} \nu(S) \, .$$

Proof. By the domain of attraction condition (8.1.4) we easily obtain

$$\lim_{n\to\infty} E e^{it\tilde{\nu}_n(S)} = e^{it\nu(S)}$$

for all t. ∎

Next we define

$$\hat{\nu}(S) := \frac{1}{k} \sum_{i=1}^{n} 1\left\{\left(\left(1 + \hat{\gamma}_1 \frac{X_i - \hat{b}_1\left(\frac{n}{k}\right)}{\hat{a}_1\left(\frac{n}{k}\right)}\right)^{1/\hat{\gamma}_1}, \left(1 + \hat{\gamma}_2 \frac{Y_i - \hat{b}_2\left(\frac{n}{k}\right)}{\hat{a}_2\left(\frac{n}{k}\right)}\right)^{1/\hat{\gamma}_2}\right) \in S\right\} \, . \tag{8.2.15}$$

Note that

$$\hat{\nu}(S) = \tilde{\nu}_n(S_n),$$

where

$$S_n := \{(f_n(x), g_n(y)) : (x, y) \in S\},\tag{8.2.16}$$

$$f_n(x) := \left(1 + \gamma_1 \left(\frac{\hat{a}_1\left(\frac{n}{k}\right) x^{\hat{\gamma}_1} - 1}{a_1\left(\frac{n}{k}\right)} \cdot \frac{1}{\hat{\gamma}_1} + \frac{\hat{b}_1\left(\frac{n}{k}\right) - b_1\left(\frac{n}{k}\right)}{a_1\left(\frac{n}{k}\right)}\right)\right)^{1/\gamma_1},$$

$$g_n(x) := \left(1 + \gamma_2 \left(\frac{\hat{a}_2\left(\frac{n}{k}\right) x^{\hat{\gamma}_2} - 1}{a_2\left(\frac{n}{k}\right)} \cdot \frac{1}{\hat{\gamma}_2} + \frac{\hat{b}_2\left(\frac{n}{k}\right) - b_2\left(\frac{n}{k}\right)}{a_2\left(\frac{n}{k}\right)}\right)\right)^{1/\gamma_2}.$$

Proposition 8.2.8 *Assume the conditions of Proposition 8.2.7 with k satisfying $k = k(n) \to \infty$, $k/n \to 0$, $n \to \infty$, and take $\hat{v}(S)$ as in (8.2.15). Then*

$$\hat{v}(S) \xrightarrow{P} v(S).$$

Proof. Under the domain of attraction condition,

$$\left(\tilde{v}_n(S), \hat{\gamma}_1, \hat{\gamma}_2, \frac{\hat{a}_1\left(\frac{n}{k}\right)}{a_1\left(\frac{n}{k}\right)}, \frac{\hat{a}_2\left(\frac{n}{k}\right)}{a_2\left(\frac{n}{k}\right)}, \frac{\hat{b}_1\left(\frac{n}{k}\right) - b_1\left(\frac{n}{k}\right)}{a_1\left(\frac{n}{k}\right)}, \frac{\hat{b}_2\left(\frac{n}{k}\right) - b_2\left(\frac{n}{k}\right)}{a_2\left(\frac{n}{k}\right)}\right)$$

converges, in probability, to $(v(S), \gamma_1, \gamma_2, 1, 1, 0, 0)$. Next invoke a Skorohod construction so that we may pretend that this relation holds almost surely. Let S_n be as in (8.2.16). By Lemma 8.2.6 we have

$$1_{S_n}(x, y) \to 1_S(x, y)\tag{8.2.17}$$

for $(x, y) \in S$. Note that the given conditions for C_n imply that there exist $s_1, s_2 > 0$, $(s_1, s_2) \in \partial S$ such that $x > s_1$ or $y > s_2$ for all $(x, y) \in S$. It follows that

$$S \subset \{(x, y) : x > s_1 \text{ or } y > s_2\} =: D.$$

Define D_n as

$$D_n := \{(f_n(x), g_n(y)) : (x, y) \in D\}.$$

Since $f_n(s_1) \to s_1$ and $g_n(s_2) \to s_2$, for $n \geq n_0$,

$$D_n \subset D_\varepsilon := \{(x - \varepsilon, y - \varepsilon) : (x, y) \in D\}.$$

Hence for $n \geq n_0$,

$$1_{S_n}(x, y) \leq 1_{D_n}(x, y) \leq 1_{D_\varepsilon}(x, y)\tag{8.2.18}$$

for all $(x, y) \in D$.

Define the measure v^* by

$$v^* := \sum_{n=0}^{\infty} 2^{-n} \tilde{v}_n$$

with the convention $\tilde{v}_0 := v$. Let h_n be the density of $\tilde{v}_n$ with respect to v^*. We know that

$$\tilde{v}_n(S) = \int_S h_n \, dv^* \to \int_S h_0 \, dv^* = v(S) \qquad \text{a.s.}$$

By Lemma 8.2.6, Proposition 8.2.7, (8.2.17), (8.2.18), and Pratt's (1960) lemma (summarizing: if $g_n \to g$ pointwise, $|g_n| \le f_n$ for all n, and $\int f_n \to \int f$ for some functions f_n, g_n, f, and g, then $\int g_n \to \int g$) we have

$$\hat{v}(S) = \tilde{v}_n(S_n) = \int 1_{S_n} h_n \, dv^* \to \int 1_S h_0 \, dv^* = v(S) \, .$$

It follows that

$$\hat{v}(S) \xrightarrow{P} v(S) \, . \qquad \blacksquare$$

Proposition 8.2.9 *Let* $(X_1, Y_1), (X_2, Y_2), \ldots$ *be an i.i.d. sample from F. Suppose F is in the domain of attraction of an extreme value distribution with normalizing functions $a_i > 0$, b_i real, and marginal extreme value indices γ_i, $i = 1, 2$. Suppose (8.2.3) for some sequence $k = k(n) \to \infty$, $k/n \to 0$, $n \to \infty$. Redefine $f_n(x)$ and $g_n(x)$ as*

$$f_n(x) := \frac{1}{c_n}\left(1 + \hat{\gamma}_1 \left(\frac{a_1\left(\frac{n}{k}\right)}{\hat{a}_1\left(\frac{n}{k}\right)} \frac{(c_n x)^{\gamma_1} - 1}{\gamma_1} + \frac{b_1\left(\frac{n}{k}\right) - \hat{b}_1\left(\frac{n}{k}\right)}{\hat{a}_1\left(\frac{n}{k}\right)} \right)\right)^{1/\hat{\gamma}_1},$$

$$g_n(x) := \frac{1}{c_n}\left(1 + \hat{\gamma}_2 \left(\frac{a_2\left(\frac{n}{k}\right)}{\hat{a}_2\left(\frac{n}{k}\right)} \frac{(c_n x)^{\gamma_2} - 1}{\gamma_2} + \frac{b_2\left(\frac{n}{k}\right) - \hat{b}_2\left(\frac{n}{k}\right)}{\hat{a}_2\left(\frac{n}{k}\right)} \right)\right)^{1/\hat{\gamma}_2},$$

with $c_n \to \infty$, *as* $n \to \infty$, *and*

$$\lim_{n\to\infty} \frac{w_{\gamma_1 \wedge \gamma_2}(c_n)}{\sqrt{k}} = 0 \, . \tag{8.2.19}$$

Then

$$f_n(x) \xrightarrow{P} x \quad \text{and} \quad g_n(x) \xrightarrow{P} x \, ,$$

for all $x > 0$.

Proof. We prove the first of the two statements. Invoke the Skorohod construction, so that (8.2.3) holds almost surely. Next we consider the cases $\gamma_1 \ne 0$ and $\gamma_1 = 0$ separately.

Start with $\gamma_1 \ne 0$. Then from (8.2.3),

$$f_n(x) = \frac{1}{c_n}\left\{ 1 + \left(1 + O\left(\frac{1}{\sqrt{k}}\right)\right) ((c_n x)^{\gamma_1} - 1) + O\left(\frac{1}{\sqrt{k}}\right)\right\}^{1/\hat{\gamma}_1}$$

$$= \frac{1}{c_n}\left\{ (c_n x)^{\gamma_1} + (c_n x)^{\gamma_1} O\left(\frac{1}{\sqrt{k}}\right) + O\left(\frac{1}{\sqrt{k}}\right)\right\}^{1/\hat{\gamma}_1}$$

$$= \frac{(c_n x)^{\gamma_1/\hat{\gamma}_1}}{c_n}\left\{ 1 + O\left(\frac{1}{\sqrt{k}}\right) + (c_n x)^{-\gamma_1} O\left(\frac{1}{\sqrt{k}}\right)\right\}^{1/\gamma_1 + o(1/\sqrt{k})} \, .$$

Now we deal with both factors separately and for that we use condition (8.2.19). Note that (8.2.19) implies $w_{\gamma_i}(c_n)/\sqrt{k} \to 0$, $i = 1, 2$. Now

$$\frac{w_{\gamma_1}(c_n)}{\sqrt{k}} \sim \begin{cases} \frac{\log c_n}{\gamma_1 \sqrt{k}} \to 0, & \gamma_1 > 0, \\[2mm] \frac{c_n^{-\gamma_1}}{\gamma_1^2 \sqrt{k}} \to 0, & \gamma_1 < 0; \end{cases}$$

hence regardless of whether $\gamma_1 > 0$ or $\gamma_1 < 0$, we have $c_n^{-\gamma_1}/\sqrt{k} \to 0$ and $(\log c_n)/\sqrt{k} \to 0$, $n \to \infty$. The result follows for $\gamma_1 \neq 0$.

Next consider $\gamma_1 = 0$. Then

$$f_n(x) = \frac{1}{c_n}\left(1 + \hat{\gamma}_1\left(\frac{a_1\left(\frac{n}{k}\right)}{\hat{a}_1\left(\frac{n}{k}\right)}\log(c_n x) + \frac{b_1\left(\frac{n}{k}\right) - \hat{b}_1\left(\frac{n}{k}\right)}{\hat{a}_1\left(\frac{n}{k}\right)}\right)\right)^{1/\hat{\gamma}_1}.$$

We first prove that $\lim_{n\to\infty} f_n^{\leftarrow}(x) = x$ and we write

$$f_n^{\leftarrow}(x) = \frac{1}{c_n}\exp\left\{\frac{\hat{a}_1\left(\frac{n}{k}\right)}{a_1\left(\frac{n}{k}\right)}\frac{(c_n x)^{\hat{\gamma}_1} - 1}{\hat{\gamma}_1} + \frac{\hat{b}_1\left(\frac{n}{k}\right) - b_1\left(\frac{n}{k}\right)}{a_1\left(\frac{n}{k}\right)}\right\}$$

$$= x\exp\left\{\frac{\hat{a}_1\left(\frac{n}{k}\right)}{a_1\left(\frac{n}{k}\right)}\left(\frac{(c_n x)^{\hat{\gamma}_1} - 1}{\hat{\gamma}_1} - \log(c_n x)\right)\right.$$

$$\left. + \left(\frac{\hat{a}_1\left(\frac{n}{k}\right)}{a_1\left(\frac{n}{k}\right)} - 1\right)\log(c_n x) + \frac{\hat{b}_1\left(\frac{n}{k}\right) - b_1\left(\frac{n}{k}\right)}{a_1\left(\frac{n}{k}\right)}\right\}.$$

Now note that

$$\left|\frac{(c_n x)^{\hat{\gamma}_1} - 1}{\hat{\gamma}_1} - \log(c_n x)\right|$$

$$= \left|\int_1^{c_n x}\left(s^{\hat{\gamma}_1} - 1\right)\frac{ds}{s}\right| = \left|\hat{\gamma}_1\int_1^{c_n x}\int_1^s u^{\hat{\gamma}_1}\frac{du}{u}\frac{ds}{s}\right|$$

$$\leq |\hat{\gamma}_1|(c_n x)^{|\hat{\gamma}_1|}\frac{1}{2}\log^2(c_n x)$$

$$= \sqrt{k}|\hat{\gamma}_1|x^{|\hat{\gamma}_1|}e^{\sqrt{k}|\hat{\gamma}_1|(\log c_n)/\sqrt{k}}\frac{\log^2(c_n x)}{2}.$$

Then use (8.2.3) with the Skorohod construction and assumption (8.2.19), which for $\gamma = 0$ implies $\log^2(c_n)/\sqrt{k} \to 0$, $n \to \infty$. Hence $\lim_{n\to\infty} f_n^{\leftarrow}(x) = x$, and hence also $\lim_{n\to\infty} f_n(x) = x$. ∎

Proposition 8.2.10 *Assume the conditions of Theorem 8.2.1. Let $\hat{v}(\hat{S})$ be as in* (8.2.15) *with S replaced by $\hat{S}$. Then*

$$\hat{v}(\hat{S}) \xrightarrow{P} v(S).$$

Proof. From Proposition 8.2.8 we know that $\hat{v}(S) \to^P v(S)$. Invoke a Skorohod construction, so that

$$\hat{v}(S) - v(S) = o(1) \quad \text{a.s.}$$

and

$$\sqrt{k}\left(\hat{\gamma}_i - \gamma_i, \frac{\hat{a}_i\left(\frac{n}{k}\right)}{a_i\left(\frac{n}{k}\right)} - 1, \frac{\hat{b}_i\left(\frac{n}{k}\right) - b_i\left(\frac{n}{k}\right)}{a_i\left(\frac{n}{k}\right)}\right) = (O(1), O(1), O(1)) \quad \text{a.s.}$$

for $i = 1, 2$. Then from Lemma 8.2.6 and Proposition 8.2.9 we have

$$1_{\hat{S}}(x, y) \to 1_S(x, y)$$

for $(x, y) \in S$. The rest of the proof is like that of Proposition 8.2.8. ∎

Proof (of Theorem 8.2.1). By (8.2.6) and (8.1.11),

$$p_n = \frac{k}{nc_n} v(S)(1 + o_p(1)) ,$$

as $n \to \infty$. Therefore from Proposition 8.2.10,

$$\lim_{n\to\infty} \frac{\hat{p}_n}{p_n} = \lim_{n\to\infty} \frac{\frac{k}{nc_n}\hat{v}(\hat{S})}{p_n} = \lim_{n\to\infty} \frac{\hat{v}(\hat{S})}{v(S)} \xrightarrow{P} 1 .$$ ∎

Proof (of Corollary 8.2.4). It is enough to prove that $\hat{c}_n/c_n \to^P 1, n \to \infty$. For this, note that

$$\hat{q}_n = \left(1 + \hat{\gamma}_1 \frac{v_n - \hat{b}_1\left(\frac{n}{k}\right)}{\hat{a}_1\left(\frac{n}{k}\right)}\right)^{1/\hat{\gamma}_1}$$

$$= \left(1 + \frac{\hat{\gamma}_1}{\hat{a}_1\left(\frac{n}{k}\right)}\left(a_1\left(\frac{n}{k}\right)\frac{q_n^{\gamma_1} - 1}{\gamma_1} + b_1\left(\frac{n}{k}\right)\right) - \frac{\hat{b}_1\left(\frac{n}{k}\right)}{\hat{a}_1\left(\frac{n}{k}\right)}\right)^{1/\hat{\gamma}_1}$$

$$= \left(1 + \hat{\gamma}_1\left(\frac{a_1\left(\frac{n}{k}\right)}{\hat{a}_1\left(\frac{n}{k}\right)}\frac{q_n^{\gamma_1} - 1}{\gamma_1} + \frac{b_1\left(\frac{n}{k}\right) - \hat{b}_1\left(\frac{n}{k}\right)}{\hat{a}_1\left(\frac{n}{k}\right)}\right)\right)^{1/\hat{\gamma}_1} .$$

Then from Proposition 8.2.9,

$$\frac{\hat{q}_n}{q_n} \xrightarrow{P} 1 .$$

Similarly $\hat{r}_n/r_n \to^P 1$. The result follows. ∎

Remark 8.2.11 The estimation of $\hat{q}_n$ and $\hat{r}_n$ is practically the same as tail estimation as discussed in Section 4.4. Since the conditions of Corollary 4.4.5 are satisfied (note that $q_n \to \infty$ corresponds to $d_n \to \infty$ there), one could alternatively invoke this result to prove the consistency of $\hat{c}_n$.

8.3 Failure Set Contained in an Upper Quadrant; Asymptotically Independent Components

Let (X, Y) be a random vector with distribution function F and suppose that F satisfies the domain of attraction condition (8.1.1). Under this condition, a particular case is that in which the limiting distribution is the product of its marginal distributions. Recall from Section 6.2 that a random vector (X, Y) whose distribution is in the domain of attraction of such a max-stable distribution was defined to be asymptotically independent.

Let us start with the failure set as an upper quadrant. From (8.1.1) one gets

$$\lim_{t \to \infty} t \, P \left(\frac{X - b_1(t)}{a_1(t)} > x \text{ or } \frac{Y - b_2(t)}{a_2(t)} > y \right) = - \log G(x, y) \,,$$

and hence

$$\lim_{t \to \infty} t \, P \left(\frac{X - b_1(t)}{a_1(t)} > x \text{ and } \frac{Y - b_2(t)}{a_2(t)} > y \right)$$
$$= \log G(x, y) - \log G(x, \infty) - \log G(\infty, y),$$

and in case of asymptotic independence the right-hand side is identically zero. More generally if Q is any Borel set contained in $[u, \infty) \times [v, \infty)$, with $u, v > 0$ and $v(\partial Q) = 0$, under asymptotic independence of (X, Y),

$$\lim_{t \to \infty} t \, P \left\{ \left(\left(1 + \gamma_1 \frac{X - b_1(t)}{a_1(t)} \right)^{1/\gamma_1}, \left(1 + \gamma_2 \frac{Y - b_2(t)}{a_2(t)} \right)^{1/\gamma_2} \right) \in Q \right\} = 0 \,.$$

This gives too little information on the probability of the set Q.

To estimate

$$p_n = P \left((X, Y) \in C_n \right) \quad \text{when} \quad C_n \subset [u_n, \infty) \times [v_n, \infty) \,,$$

we propose the following refinement of (8.1.4), which will lead to a new limit measure v: for $x, y > 0$ and some functions a_1, a_2, r positive and b_1, b_2 real, $r \to \infty$,

$$\lim_{t \to \infty} r(t) \, P \left\{ \left(1 + \gamma_1 \frac{X - b_1(t)}{a_1(t)} \right)^{1/\gamma_1} > x \text{ and } \left(1 + \gamma_2 \frac{Y - b_2(t)}{a_2(t)} \right)^{1/\gamma_2} > y \right\}$$
$$\tag{8.3.1}$$

exists, and it is positive and finite.

Then, similarly as in Section 6.1.3, one can define the measure v as follows: for any Borel set Q in $\mathbb{R}_+^2$ with $\inf_{(x,y) \in Q} \max(x, y) > 0$ and $v(\partial Q) = 0$ let

$$v(Q)$$
$$:= \lim_{t \to \infty} r(t) \, P \left\{ \left(\left(1 + \gamma_1 \frac{X - b_1(t)}{a_1(t)} \right)^{1/\gamma_1}, \left(1 + \gamma_2 \frac{Y - b_2(t)}{a_2(t)} \right)^{1/\gamma_2} \right) \in Q \right\} \,.$$
$$\tag{8.3.2}$$

Moreover, it follows that the function r is regularly varying with index greater than or equal to 1. Using the notation introduced in Section 7.6, the index of the regularly varying function r is $1/\eta$, where $\eta \in (0, 1]$ is the residual independence index. Also as in the proof of Theorem 6.1.9, it follows that v is homogeneous of order $-1/\eta$, i.e.,

$$v(aQ) = a^{-1/\eta}v(Q) , \tag{8.3.3}$$

for any $a > 0$, where aQ is the set obtained by multiplying all elements of Q by a.

Note that (8.3.2) is valid if X and Y are not asymptotically independent. In this case $r(t) = t$, $\eta = 1$, and $v = \mu$, the exponent measure of Section 6.1.3.

We are now ready to proceed with the estimation of p_n, which closely follows the reasoning developed in the previous section. Using again (8.1.8),

$$p_n = P((X, Y) \in C_n)$$
$$= P\left\{ \left(\left(1 + \gamma_1 \frac{X - b_1\left(\frac{n}{k}\right)}{a_1\left(\frac{n}{k}\right)}\right)^{1/\gamma_1} , \left(1 + \gamma_2 \frac{Y - b_2\left(\frac{n}{k}\right)}{a_2\left(\frac{n}{k}\right)}\right)^{1/\gamma_2} \right) \in c_n S \right\},$$

which, now by (8.3.2), is approximately equal to (cf. (8.1.9))

$$\frac{v(c_n S)}{r\left(\frac{n}{k}\right)} = \frac{v(S)}{c_n^{1/\eta} r\left(\frac{n}{k}\right)},$$

where the last equality follows from (8.3.3). Comparing with the previous section, apart from estimating S and v, we now have to deal with the parameter η. But this was the subject of Section 7.6, from where we know how to estimate η.

In the next theorem we state the necessary conditions for the consistency of $\hat{p}_n$. As in the previous section we opted for a long theorem that is mostly self-contained in all its conditions and definitions. The proof is left to the reader (Exercises 8.1–8.3).

Theorem 8.3.1 *Let $(X_1, Y_1), \ldots, (X_n, Y_n)$ be i.i.d. random vectors with distribution function F, satisfying (8.3.1) for some positive function r, with marginal extreme value indices γ_i and normalizing functions $a_i > 0$, b_i real, $i = 1, 2$.*

Consider some estimators of γ_i, $a_i(n/k) > 0$, $b_i(n/k)$, $i = 1, 2$, and η such that for some sequence $k/n \to 0$, $r(n/k)/n \to 0$ (this implies $k \to \infty$), $n \to \infty$,

$$\sqrt{k}\left(\hat{\gamma}_i - \gamma_i, \frac{\hat{a}_i\left(\frac{n}{k}\right)}{a_i\left(\frac{n}{k}\right)} - 1, \frac{\hat{b}_i\left(\frac{n}{k}\right) - b_i\left(\frac{n}{k}\right)}{a_i\left(\frac{n}{k}\right)}\right) = (O_p(1), O_p(1), O_p(1)) , \tag{8.3.4}$$

$i = 1, 2$, *and*

$$\sqrt{\frac{n}{r\left(\frac{n}{k}\right)}}\left(\hat{\eta} - \eta\right) = O_p(1) . \tag{8.3.5}$$

Suppose C_n is an open set and that there exists some boundary point of C_n, (v_n, w_n) such that

$$C_n \subset [v_n, \infty) \times [w_n, \infty)$$

for all n, and that

$$C_n = \left\{ \left(a_1 \left(\frac{n}{k} \right) \frac{(c_n x)^{\gamma_1} - 1}{\gamma_1} + b_1 \left(\frac{n}{k} \right), \right. \right.$$
$$\left. \left. a_2 \left(\frac{n}{k} \right) \frac{(c_n y)^{\gamma_2} - 1}{\gamma_2} + b_2 \left(\frac{n}{k} \right) \right) : (x, y) \in S \right\}, \quad (8.3.6)$$

where S is an open Borel set in $[0, \infty)^2$ with $\upsilon(\partial S) = 0$ and $\upsilon(S) > 0$, and c_n a sequence of positive numbers with $c_n \to \infty$, $n \to \infty$.

Finally, suppose $0 < q_n/r_n < \infty$ with q_n and r_n as in (8.2.2) (our conditions imply that q_n/r_n does not depend on n),

$$\lim_{n \to \infty} \frac{w_{\gamma_1 \wedge \gamma_2}(c_n)}{\sqrt{k}} = 0 , \quad (8.3.7)$$

$$\lim_{n \to \infty} \sqrt{\frac{r\left(\frac{n}{k} \right)}{n}} \log(q_n) = 0 , \quad (8.3.8)$$

and that (8.3.2) holds with Q replaced by $c_n S$, i.e.,

$$\lim_{n \to \infty} r \left(\frac{n}{k} \right) P \left\{ \left(\left(1 + \gamma_1 \frac{X - b_1 \left(\frac{n}{k} \right)}{a_1 \left(\frac{n}{k} \right)} \right)^{1/\gamma_1} , \right. \right.$$
$$\left. \left. \left(1 + \gamma_2 \frac{Y - b_2 \left(\frac{n}{k} \right)}{a_2 \left(\frac{n}{k} \right)} \right)^{1/\gamma_2} \right) \in c_n S \right\} \Big/ \upsilon(c_n S) = 1 .$$

Then with

$$\hat{p}_n := \frac{1}{n c_n^{1/\hat{\eta}}} \sum_{i=1}^{n} 1 \left\{ \left(\left(1 + \hat{\gamma}_1 \frac{X_i - \hat{b}_1 \left(\frac{n}{k} \right)}{\hat{a}_1 \left(\frac{n}{k} \right)} \right)^{1/\hat{\gamma}_1} , \left(1 + \hat{\gamma}_2 \frac{Y_i - \hat{b}_2 \left(\frac{n}{k} \right)}{\hat{a}_2 \left(\frac{n}{k} \right)} \right)^{1/\hat{\gamma}_2} \right) \in \hat{S} \right\}, \quad (8.3.9)$$

where

$$\hat{S} := \left\{ \left(\frac{1}{c_n} \left(1 + \hat{\gamma}_1 \frac{x - \hat{b}_1 \left(\frac{n}{k} \right)}{\hat{a}_1 \left(\frac{n}{k} \right)} \right)^{1/\hat{\gamma}_1} , \right. \right.$$
$$\left. \left. \frac{1}{c_n} \left(1 + \hat{\gamma}_2 \frac{y - \hat{b}_2 \left(\frac{n}{k} \right)}{\hat{a}_2 \left(\frac{n}{k} \right)} \right)^{1/\hat{\gamma}_2} \right) : (x, y) \in C_n \right\}, \quad (8.3.10)$$

we have

$$\frac{\hat{p}_n}{p_n} \xrightarrow{P} 1 .$$

Remark 8.3.2 The sequence c_n can be estimated as in Section 8.2.2.

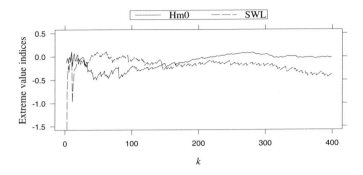

Fig. 8.4. Diagram of estimates of γ with moment estimators.

8.4 Sea Level Case Study

As described at the beginning of this chapter, we have 828 independent and identically distributed observations of the wave height (HmO) and the still water level (SWL). These are illustrated in Figure 8.1, as well as the failure set

$$C_n = \{(\text{HmO, SWL}) \ : \ 0.3\text{HmO} + \text{SWL} > 7.6\} \ .$$

The first step for the estimation of $p_n = P((X, Y) \in C_n)$ is the estimation of the marginals. For that see Sections 3.7.3 and 4.6.2. In Figure 8.4 we show the diagram of estimates of γ, i.e., the estimates against the number of upper order statistics k, for both samples. We use the moment-type estimators given after Theorem 8.2.1. According to earlier results (Section 3.7.3) we take zero for the extreme value index of SWL, $\gamma_{\text{SWL}} = 0$. We show results for the window $40 \leq k \leq 110$, which seems a quite reasonable one. In Table 8.1 we illustrate point estimates for $k = 100$, which were the ones used for Figures 8.2 and 8.3.

As discussed at the end of Section 8.1, the shape of the failure set may determine the method to estimate p_n. Since in our case the set S contains part of both axes, we are in the conditions of Section 8.2. In Figure 8.5 we show transformed sets S

Table 8.1. Point estimates for $k = 100$.

	HmO	SWL
$\hat{\gamma}$	-0.22	0.00
$\hat{a}(n/k)$	1.14	0.30
$\hat{b}(n/k)$	4.10	1.34
$\hat{q}_n(v_n = 5.85)$	6.85	$-$
$\hat{r}_n(w_n = 5.85)$	$-$	$2879640.$
$\hat{c}_n(r = 50)$	57592.80	

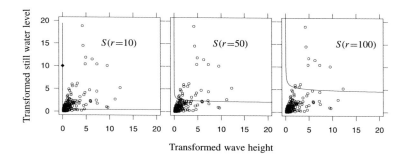

Fig. 8.5. Transformed data set and transformed failure sets (area above the curved line).

for $r = 10, 50, 100$ (we use the approach of Section 8.2.2, where, recall, r denotes the radius of the circle to which the boundary point (s_1, s_2) belongs, in the picture illustrated with the diamond point). One sees that there is quite a difference in the number of points belonging to each set S, but the effect of this in the estimates of p_n is quite negligible. In Figure 8.6 one finds the diagram of estimates of $\hat{p}_n$ over the window $40 \leq k \leq 110$ and for $r = 10, 50, 100$.

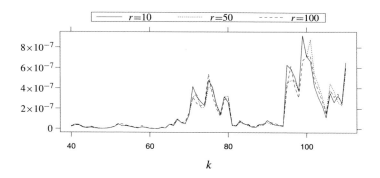

Fig. 8.6. Diagram of estimates of p_n.

Exercises

In the next three exercises one gradually proves Theorem 8.3.1.

8.1. Let $(X_1, Y_1), \ldots, (X_n, Y_n)$ be i.i.d. random vectors with distribution function F, satisfying (8.3.1) for some positive function r and limit measure υ, with normalizing functions $a_i > 0$, b_i real, and marginal extreme value indices γ_i, $i = 1, 2$. Let S be an open Borel set in $\mathbb{R}_+^2$ with $\inf_{(x,y)\in S} \max(x, y) > 0$, $\upsilon(\partial S) = 0$, and $\upsilon(S) > 0$. Introduce the random variable

$$\tilde{\upsilon}_n(S) := \frac{r\left(\frac{n}{k}\right)}{k} \sum_{i=1}^{n} 1\left\{ \left(\left(1 + \gamma_1 \frac{X_i - b_1\left(\frac{n}{k}\right)}{a_1\left(\frac{n}{k}\right)}\right)^{1/\gamma_1}, \left(1 + \gamma_2 \frac{Y_i - b_2\left(\frac{n}{k}\right)}{a_2\left(\frac{n}{k}\right)}\right)^{1/\gamma_2}\right) \in S \right\}$$

with k satisfying $k = k(n)$, $k/n \to 0$, and $r(n/k)/n \to 0$ (this implies $k \to \infty$), as $n \to \infty$. Under the stated conditions prove that

$$\tilde{\upsilon}_n(S) \overset{P}{\to} \upsilon(S) .$$

Hint: See Proposition 8.2.7.

8.2. Let $(X_1, Y_1), \ldots, (X_n, Y_n)$ be i.i.d. random vectors with distribution function F, satisfying (8.3.1) for some positive function r and limit measure υ, with normalizing functions $a_i > 0$, b_i real, and marginal extreme value indices γ_i, $i = 1, 2$. Suppose (8.3.4) for some sequence $k = k(n)$, $k/n \to 0$, $r(n/k)/n \to 0$ (this implies $k \to \infty$), $n \to \infty$. Let S be an open Borel set in $\mathbb{R}_+^2$ with $\inf_{(x,y)\in S} \max(x, y) > 0$, $\upsilon(\partial S) = 0$, and $\upsilon(S) > 0$. Define

$$\hat{\upsilon}(S) := \frac{r\left(\frac{n}{k}\right)}{n} \sum_{i=1}^{n} 1\left\{ \left(\left(1 + \hat{\gamma}_1 \frac{X_i - \hat{b}_1\left(\frac{n}{k}\right)}{\hat{a}_1\left(\frac{n}{k}\right)}\right)^{1/\hat{\gamma}_1}, \left(1 + \hat{\gamma}_2 \frac{Y_i - \hat{b}_2\left(\frac{n}{k}\right)}{\hat{a}_2\left(\frac{n}{k}\right)}\right)^{1/\hat{\gamma}_2}\right) \in S \right\} .$$

Prove that

$$\hat{\upsilon}(S) \overset{P}{\to} \upsilon(S) .$$

Hint: See Proposition 8.2.8.

8.3. Prove Theorem 8.3.1.

8.4. Under the conditions of Theorem 8.3.1 and with $\hat{c}_n$ as in (8.2.10)–(8.2.12) prove that $\hat{c}_n^{1/\hat{\eta}}/c_n^{1/\eta} \to^P 1$.

Part III

Observations That Are Stochastic Processes

9

Basic Theory in $C[0, 1]$

9.1 Introduction: An Example

Infinite-dimensional extreme value theory is not just a theoretical extension of the theory to a more abstract context. It serves to solve concrete problems. We start with a motivating example.

The two northern provinces of the Netherlands, Friesland and Groningen, lie almost completely below sea level. Since there are no natural coast defenses like sand dunes, the area is protected against inundations by a long dike. Since there is no subdivision of the area by dikes, a breach in the dike at any place could lead to flooding of the entire area. This leads to the following mathematical problem.

Suppose we have a deterministic function f defined on $[0, 1]$ (representing the top of the dike). Suppose we have independent and identically distributed random functions $X_1, X_2, \ldots$ defined on $[0, 1]$ (representing observations of high-tide water levels monitored along the coast). The question is, how can we estimate

$$P(X(s) \leq f(s) \text{ for all } s \in [0, 1])$$

on the basis of n observed independent realizations of the process X (n large)?

Now, a typical feature of this kind of problem is that none of the observed processes X comes even close to the boundary f, that is, during the observation period there has not been any flooding. This means that we have to extrapolate the distribution of X far into the tail. Since nonparametric methods cannot be used, we resort to limit theory; that is, we imagine that $n \to \infty$, but in doing so we wish to keep the essential feature that the observations are far from the boundary. This leads to the assumption that f is not a fixed function when $n \to \infty$ but that in fact f depends on n and moves to the upper boundary of the distribution of X when $n \to \infty$. So, as in the finite-dimensional case, in order to answer this question, we need a limit theory for the pointwise maximum of independent and identically distributed random functions, and this is the subject of the present chapter. In fact, this theory of infinite-dimensional extremes is quite analogous to the corresponding theory in finite-dimensional space explained in Chapter 6.

Let $X, X_1, X_2, \ldots$ be independent and identically distributed stochastic processes on $[0, 1]$ with continuous sample paths, i.e., belonging to $C[0, 1]$, the space of continuous functions f on $[0, 1]$ equipped with the supremum norm $|f|_\infty = \sup_{s \in [0,1]} |f(s)|$.

We consider extreme value theory in $C[0, 1]$. That is, we assume that there exist continuous functions $a_s(n)$ positive and $b_s(n)$ real, $n = 1, 2, \ldots$, such that the sequence of stochastic processes

$$\left\{ \max_{i \le n} \frac{X_i(s) - b_s(n)}{a_s(n)} \right\}_{s \in [0,1]}$$

converges weakly (or in distribution) in $C[0, 1]$ to a stochastic process $\{Y(s)\}_{s \in [0,1]}$ with non-degenerate marginals, i.e., $Y(s)$ is non-degenerate for all $s \in [0, 1]$. Hence we take the maximum pointwise and consider convergence of the resulting stochastic process.

The parameter set $[0, 1]$ has been chosen for convenience. Most results hold for any compact subset of a Euclidean space.

For convenience we shall sometimes refer to the parameter i as "time" and to the parameter s as "space."

We want to investigate the structure of the possible limit processes Y and for each of them we want to characterize the domain of attraction.

9.2 The Limit Distribution; Standardization

As in the finite-dimensional case the problem becomes more tractable if we first transform to processes with standard marginal distributions. For that assume that the marginal distribution functions $F_s(x) := P(X(s) \le x)$ are continuous in x. Then in order to transform to standard marginals note that the convergence in $C[0, 1]$ entails

$$F_s^n (a_s(n)x + b_s(n)) \to P(Y(s) \le x) \tag{9.2.1}$$

as $n \to \infty$, uniformly for $s \in [0, 1]$ and locally uniformly for x. In particular, note that it means that the distribution functions of $X(s)$ and $Y(s)$ are continuous in s. We can choose the functions $a_s(n)$ and $b_s(n)$ in such a way that $P(Y(s) \le x)$ is an extreme value distribution in the von Mises form (cf. Section 1.1.3). Then

$$P(Y(s) \le x) = \exp\left(-(1 + \gamma(s)x)^{-1/\gamma(s)} \right) \tag{9.2.2}$$

for $s \in [0, 1]$ and all x with $1 + \gamma(s)x > 0$, where γ is a continuous function.

From (9.2.1) and (9.2.2) we get

$$\lim_{n \to \infty} n \{1 - F_s (a_s(n)x + b_s(n))\} = (1 + \gamma(s)x)^{-1/\gamma(s)} \tag{9.2.3}$$

uniformly for $s \in [0, 1]$ and locally uniformly for x with $1 + \gamma(s)x > 0$. Since convergence of a sequence of monotone functions is equivalent to convergence of their inverses (Lemma 1.1.1), (9.2.3) is equivalent to

$$\lim_{n \to \infty} \frac{U_s(nu) - b_s(n)}{a_s(n)} = \frac{u^{\gamma(s)} - 1}{\gamma(s)}$$

uniformly for $s \in [0, 1]$ and locally uniformly for $u \in (0, \infty)$, where U_s is the left-continuous inverse of $1/(1 - F_s)$ for $s \in [0, 1]$.

We have

$$\left\{ \max_{i \leq n} \frac{X_i(s) - b_s(n)}{a_s(n)} \right\}_{s \in [0,1]} \xrightarrow{d} \{Y(s)\}_{s \in [0,1]} \tag{9.2.4}$$

in $C[0, 1]$ and (9.2.3). By combining the two and using the uniformity in both statements we get

$$\left\{ \max_{i \leq n} \frac{1}{n\{1 - F_s(X_i(s))\}} \right\}_{s \in [0,1]} \xrightarrow{d} \left\{ (1 + \gamma(s)Y(s))^{1/\gamma(s)} \right\}_{s \in [0,1]}$$

in $C[0, 1]$. We have proved the following result.

Theorem 9.2.1 *Let* $X, X_1, X_2, \ldots$ *be i.i.d. stochastic processes in* $C[0, 1]$. *Let* $a_s(n)$ *positive and* $b_s(n)$ *real be continuous functions,* $\{Y(s)\}_{s \in [0,1]}$ *a stochastic process in* $C[0, 1]$,

$$F_s(x) := P(X(s) \leq x)$$

continuous in x, *and* U_s *the left-continuous inverse of* $1/(1 - F_s)$. *The following statements are equivalent:*

1.

$$\left\{ \max_{i \leq n} \frac{X_i(s) - b_s(n)}{a_s(n)} \right\}_{s \in [0,1]} \xrightarrow{d} \{Y(s)\}_{s \in [0,1]}$$

in $C[0, 1]$, *where* $a_s(n)$ *and* $b_s(n)$ *are chosen such that* $-\log P(Y(s) \leq x) = (1 + \gamma(s)x)^{-1/\gamma(s)}$ *for all* x *with* $1 + \gamma(s)x > 0$.

2.

$$\left\{ \max_{i \leq n} \frac{1}{n(1 - F_s(X_i(s)))} \right\}_{s \in [0,1]} \xrightarrow{d} \{Z(s)\}_{s \in [0,1]} \tag{9.2.5}$$

in $C[0, 1]$, *and for all* $u \in (0, \infty)$,

$$\lim_{n \to \infty} \frac{U_s(nu) - b_s(n)}{a_s(n)} = \frac{u^{\gamma(s)} - 1}{\gamma(s)} \tag{9.2.6}$$

uniformly for $s \in [0, 1]$.

The relation between Y *and* Z *is* $\{Z(s)\}_{s \in [0,1]} =^d \{(1 + \gamma(s)Y(s))^{1/\gamma(s)}\}_{s \in [0,1]}$.

Remark 9.2.2 Relation (9.2.6) means that the function $U_s(t)$ is extended regularly varying with an extra parameter (see Section B.4). We call the continuous function $\gamma = \gamma(s)$ the *index function*.

Since relation (9.2.6) is not difficult to handle (cf. Section B.4 in Appendix B), this theorem reduces our problem to studying the limit relation

$$\frac{1}{n} \bigvee_{i=1}^{n} \xi_i \xrightarrow{d} \eta \tag{9.2.7}$$

in $C[0, 1]$ for independent and identically distributed stochastic processes $\xi_1, \xi_2, \ldots$ in

$$C^+[0, 1] := \{f \in C[0, 1] : f > 0\} ,$$

where for $s \in [0, 1]$,

$$P(\eta(s) \leq 1) = e^{-1} . \tag{9.2.8}$$

Theorem 9.2.3 *Suppose $\eta_1, \eta_2, \ldots$ are i.i.d. copies of the process η from (9.2.7). Then for all positive integers k,*

$$\frac{1}{k} \bigvee_{i=1}^{k} \eta_i \overset{d}{=} \eta . \tag{9.2.9}$$

Proof. Let k and n be positive integers. We write

$$\frac{1}{nk} \bigvee_{i=1}^{nk} \xi_i = \frac{1}{k} \bigvee_{j=1}^{k} \frac{1}{n} \bigvee_{r=1}^{n} \xi_{r,k},$$

where the $\xi_{r,k}$ are independent and they all have the distribution of the ξ_i. Now keep k fixed and let n tend to infinity. Then the left-hand side tends to η in distribution and the right-hand side tends to $k^{-1} \bigvee_{j=1}^{k} \eta_j$ in distribution where the η_j are independent and have the same distribution as η. ∎

Definition 9.2.4 A stochastic process on $C^+[0, 1]$ with nondegenerate marginals is called *simple max-stable* if (9.2.9) holds and $P(\eta(s) \leq 1) = e^{-1}$ for all $s \in [0, 1]$ (i.e. it has *standard* Fréchet distribution).

Hence the class of limit processes of Theorem 9.2.1(2) is the same as the class of simple max-stable processes.

More generally we have the following:

Definition 9.2.5 A stochastic process Y on $C[0, 1]$ with nondegenerate marginals is *max-stable* if there are continuous functions $a_s(n)$ positive and $b_s(n)$ real such that if $Y_1, Y_2, \ldots$, are i.i.d. copies of Y, then

$$\bigvee_{i=1}^{n} \frac{Y_i - b_n}{a_n} \overset{d}{=} Y$$

for $n = 1, 2, \ldots$.

Hence the class of max-stable processes coincides with the class of limit processes in (9.2.4).

9.3 The Exponent Measure

Next we take (9.2.7) as our point of departure. So let $\xi, \xi_1, \xi_2, \ldots$ be independent and identically distributed stochastic processes in $C^+[0, 1]$ for which (9.2.7) and (9.2.8) hold. It is useful at this point to introduce a sequence of measures v_n, $n = 1, 2, 3, \ldots$. For Borel sets $A \subset C^+[0, 1]$ define the measures v_n by

$$v_n(A) := n\, P\left(n^{-1}\xi \in A\right). \tag{9.3.1}$$

We shall prove that as in the finite-dimensional case, the sequence of measures v_n converges in a certain sense. In order to do so, we want to show first that the sequence v_n is relatively compact in a suitable *complete separable metric space* (CSMS, cf. Daley and Vere-Jones (1988)). Note that $C^+[0, 1]$ is not a CSMS. Hence we need to extend the space $C^+[0, 1]$ and we also need to change the metric.

Since the transformation

$$f \leftrightarrow (|f|_\infty, f/|f|_\infty)$$

is one-to-one on $C^+[0, 1]$, we can write

$$C^+[0, 1] = (0, \infty) \times C_1^+[0, 1],$$

where

$$C_1^+[0, 1] := \{f \in C[0, 1] : f > 0, |f|_\infty = 1\}.$$

Next we enlarge the space $(0, \infty) \times C_1^+[0, 1]$ to $(0, \infty] \times \overline{C}_1^+[0, 1]$, where

$$\overline{C}_1^+[0, 1] := \{f \in C[0, 1] : f \geq 0, |f|_\infty = 1\}. \tag{9.3.2}$$

The space $\overline{C}_1^+[0, 1]$ equipped with the supremum norm is CSMS, and we turn $(0, \infty]$ into a CSMS by introducing the metric $\varrho(x, y) := |1/x - 1/y|$. Hence finally we introduce

$$\overline{C}_\varrho^+[0, 1] := (0, \infty] \times \overline{C}_1^+[0, 1], \tag{9.3.3}$$

with the lower index ϱ meaning that the space $(0, \infty]$ is equipped with the metric ϱ, and we consider v_n as a measure on $\overline{C}_\varrho^+[0, 1]$ for each n. Despite the introduction of the new normed space, for convenience we still use the notation $|f|_\infty$ for $\sup_{0 \leq s \leq 1} f(s)$.

Theorem 9.3.1 *Let $\xi, \xi_1, \xi_2, \ldots$ be i.i.d. stochastic processes in $C^+[0, 1]$. If*

$$\frac{1}{n}\bigvee_{i=1}^{n} \xi_i \xrightarrow{d} \eta$$

in $C^+[0, 1]$, then

$$v_n \xrightarrow{d} v$$

in $\overline{C}_\varrho^+[0, 1]$, where $v_n(A) := nP(n^{-1}\xi \in A)$ for $n = 1, 2, \ldots$.

Equivalently, for every Borel set A in $\{f \in C[0, 1] : f \geq 0\}$ such that $\inf\{|f|_\infty : f \in A\} > 0$ and $v(\partial A) = 0$, we have

$$\lim_{n \to \infty} v_n(A) = v(A).$$

The relation between the probability distribution of η and the measure ν is that for $m = 1, 2, \ldots,$

$$P(\eta \in A_{\mathbf{K},\mathbf{x}}) = \exp\left(-\nu\left(A_{\mathbf{K},\mathbf{x}}^c\right)\right)$$

with, for $\mathbf{K} = (K_1, \ldots, K_m)$ compact sets in $[0, 1]$ and $\mathbf{x} = (x_1, \ldots, x_m)$ positive,

$$A_{\mathbf{K},\mathbf{x}} := \left\{ f \in \overline{C}_\varrho^+[0, 1] : f(s) \le x_j \text{ for } s \in K_j, j = 1, 2, \ldots, m \right\} .$$

Later on we shall need a refinement of the result.

Corollary 9.3.2 *The conditions of the theorem imply*

$$\lim_{t \to \infty} t P(t^{-1}\xi \in A) = \nu(A) < \infty,$$

where t runs through the reals, for each Borel set A in $\{f \in C[0, 1] : f \ge 0\}$ such that $\inf\{|f|_\infty : f \in A\} > 0$ and $\nu(\partial A) = 0$.

Proof. Let $m = 1, 2, 3, \ldots$, let $K_1, K_2, \ldots, K_m$ be compact sets, and $x_1, x_2, \ldots, x_m$ positive numbers. Define

$$B := \left\{ f : f(s) \le x_j \text{ for } s \in K_j, j = 1, 2, \ldots, m \right\}^c . \tag{9.3.4}$$

For $t \ge 1$,

$$[t] P\left([t]^{-1}\xi \in B\right) \ge \frac{[t]}{t} t P\left(t^{-1}\xi \in B\right) \ge \frac{[t]}{1+[t]}(1+[t]) P\left((1+[t])^{-1}\xi \in B\right),$$

and both the right- and left-hand sides converge to $\nu(B)$, for all sets B of the form (9.3.4), when $t \to \infty$. Since the measure ν is determined by its value on sets of that form, the proof is complete. ∎

Remark 9.3.3 Note (Daley and Vere-Jones (1988), A.2.6) that the conclusion of the theorem amounts to $\nu_n \to^d \nu$ in "weak hash" topology ($w^\#$) or equivalently to weak convergence in any subspace of the form $\{f : |f|_\infty > a\}, a > 0$.

For the proof of the theorem we need two lemmas.

Lemma 9.3.4 *Let η be a simple max-stable process on $[0, 1]$. Then*

$$P(|\eta|_\infty \le x) = \exp\left(-\frac{c}{x}\right) , \quad x > 0 ,$$

with c a positive constant.

Proof. With $\eta_1, \eta_2, \ldots$ independent and identically distributed copies of η we have

$$\eta \overset{d}{=} \frac{1}{n} \bigvee_{i=1}^{n} \eta_i$$

for all n. Hence

$$P(|\eta|_\infty \le x) = P^n(|\eta|_\infty \le nx) .$$

The result follows. ∎

The most important step in the proof of Theorem 9.3.1 is the following result.

Lemma 9.3.5 *Under the conditions of Theorem 9.3.1, for each $\varepsilon > 0$ the sequence of measures $\{v_{n,\varepsilon}\}$ defined by*

$$v_{n,\varepsilon}(A) := v_n\{f \in A \ : \ |f|_\infty > \varepsilon\}$$

for each $A \in \overline{C}_\varrho^+[0, 1]$ is relatively compact.

Proof. We need to prove two things: First, that the sequence

$$v_{n,\varepsilon}\left(\overline{C}_\varrho^+[0, 1]\right) = v_n\left\{f \in \overline{C}_\varrho^+[0, 1] \ : \ |f|_\infty > \varepsilon\right\}$$

is bounded. This follows from Lemma 9.3.4 since

$$v_{n,\varepsilon}\left(\overline{C}_\varrho^+[0, 1]\right) = nP\left(n^{-1}|\xi|_\infty > \varepsilon\right) \to -\log P(|\eta|_\infty \leq \varepsilon) = \frac{c}{\varepsilon}\,. \qquad (9.3.5)$$

Second, that the sequence $\{v_{n,\varepsilon}\}$ is tight. Note that since $v_{n,\varepsilon}(\overline{C}_\varrho^+[0, 1])$ has a finite limit as $n \to \infty$, we can check tightness for the sequence $\{v_{n,\varepsilon}\}$ as if it were a sequence of probability measures. According to Billingsley (1968), Theorem 15.3, this is equivalent to the following:

1. For each positive β there exists an $\alpha > 0$ such that

$$v_{n,\varepsilon}(S_\alpha) \leq \beta$$

   for all n, where

$$S_\alpha := \{f \in \overline{C}_\varrho^+[0, 1] \ : \ |f|_\infty \geq \alpha\}$$

   for each $\alpha > 0$.

2. For each positive α and β, there exist a δ, $0 < \delta < 1$, and an integer n_0 such that
   (a)

$$v_{n,\varepsilon}\left\{f : w_f''(\delta) \geq \alpha\right\} \leq \beta$$

   for $n \geq n_0$ with

$$w_f''(\delta) := \sup_{\substack{s_1 \leq s \leq s_2 \\ s_2 - s_1 \leq \delta}} \min(|f(s) - f(s_1)|, |f(s_2) - f(s)|)\,;$$

   (b)

$$v_{n,\varepsilon}\left\{f : \sup_{0 \leq s,t < \delta} |f(s) - f(t)| \geq \alpha\right\} \leq \beta$$

   for $n \geq n_0$;
   (c)

$$v_{n,\varepsilon}\left\{f : \sup_{1 - \delta \leq s,t < 1} |f(s) - f(t)| \geq \alpha\right\} \leq \beta$$

   for $n \geq n_0$.

Now (1) follows from the first part of the proof. Next we prove (2a); the other parts are similar. Relation (9.2.7) implies convergence in distribution, hence tightness, of $\{M_n \vee (\alpha/2)\}_{n=1}^{\infty}$ with $M_n := n^{-1} \bigvee_{i=1}^{n} \xi_i$. Consequently, for any $\beta^\star > 0$,

$$P\left(\omega''_{M_n \vee (\alpha/2)}(\delta) \geq (\alpha/2)\right) \leq \beta^\star$$

for $n \geq n_0^\star$. Define

$$Q_{n,\alpha} := (M_n \vee (\alpha/2)) 1_{\{|\xi_i|_\infty \geq n(\alpha/2) \text{ for some } i, \ |\xi_j|_\infty < n(\alpha/2) \text{ for } j \neq i\}}.$$

Since $Q_{n,\alpha}$ is either 0 or $M_n \vee (\alpha/2)$, we have

$$P\left(\omega''_{Q_{n,a}}(\delta) \geq (\alpha/2)\right)$$
$$= P\left(\omega''_{Q_{n,\alpha}}(\delta) \geq (\alpha/2), \ 1_{\{|\xi_i|_\infty \geq n(\alpha/2) \text{ for some } i, \ |\xi_j|_\infty < n(\alpha/2) \text{ for } j \neq i\}} = 0\right)$$
$$+ P\left(\omega''_{Q_{n,\alpha}}(\delta) \geq (\alpha/2), \ 1_{\{|\xi_i|_\infty \geq n(\alpha/2) \text{ for some } i, \ |\xi_j|_\infty < n(\alpha/2) \text{ for } j \neq i\}} = 1\right)$$
$$\leq P\left(\omega''_{M_n \vee (\alpha/2)}(\delta) \geq (\alpha/2)\right) \leq \beta^\star$$

for $n \geq n_0^\star$. Hence by the definition of $Q_{n,\alpha}$,

$$n \, P^{n-1}\left(|\xi|_\infty < \frac{n\alpha}{2}\right) P\left(\omega''_{\xi/n \vee (\alpha/2)}(\delta) \geq \frac{\alpha}{2}\right) = P\left(\omega''_{Q_{n,\alpha}}(\delta) \geq \frac{\alpha}{2}\right) \leq \beta^\star$$

$$(9.3.6)$$

for $n \geq n_0^\star$.

Now

$$P^{n-1}\left(|\xi|_\infty < \frac{n\alpha}{2}\right) = P^{(n-1)/n}\left(|M_n|_\infty < \frac{\alpha}{2}\right) \rightarrow P\left(|\eta|_\infty < \frac{\alpha}{2}\right) =: d > 0 .$$

$$(9.3.7)$$

Hence by (9.3.6) and (9.3.7),

$$n \, P\left(\omega''_{\xi/n \vee (\alpha/2)}(\delta) \geq \frac{\alpha}{2}\right) \leq \frac{2\beta^\star}{d} =: \beta$$

for $n \geq n_0$. Since

$$\omega''_f(\delta) \leq \omega''_{f \vee (\alpha/2)}(\delta) + \omega''_{f \wedge (\alpha/2)}(\delta) \leq \omega''_{f \vee (\alpha/2)}(\delta) + \frac{\alpha}{2} \, ,$$

we obtain

$$n \, P\left(\omega''_{\xi/n}(\delta) \geq \alpha\right) \leq \beta \, ,$$

i.e.,

$$\nu_{n,\varepsilon}\left\{f : \omega''_f(\delta) \geq \alpha\right\} \leq \beta \, .$$

∎

Proof (of Theorem 9.3.1). Note that since we have convergence in $C^+[0, 1]$, for $m = 1, 2, \ldots, K_1, K_2, \ldots, K_m$ compact sets in $[0, 1]$, and positive $x_1, x_2, \ldots, x_m$,

$$\lim_{n \to \infty} \nu_n \{ f : f(s) \le x_j, \text{ for } s \in K_j, j = 1, 2, \ldots, m \}^c$$

$$= \lim_{n \to \infty} n \left(1 - P(n^{-1}\xi(s) \le x_j, \text{ for } s \in K_j, j = 1, 2, \ldots, m) \right)$$

$$= \lim_{n \to \infty} -n \log P(n^{-1}\xi(s) \le x_j, \text{ for } s \in K_j, j = 1, 2, \ldots, m)$$

$$= \lim_{n \to \infty} -\log P \left(\frac{1}{n} \bigvee_{i=1}^{n} \xi_i(s) \le x_j, \text{ for } s \in K_j, j = 1, 2, \ldots, m \right)$$

$$= -\log P(\eta(s) \le x_j, \text{ for } s \in K_j, j = 1, 2, \ldots, m) .$$

Now there is exactly one measure, say ν, satisfying, for each choice of m, $K_1, \ldots, K_m, x_1, \ldots, x_m$,

$$\nu \{ f : f(s) \le x_j, \text{ for } s \in K_j, j = 1, 2, \ldots, m \}^c$$
$$= -\log P(\eta(s) \le x_j, \text{ for } s \in K_j, j = 1, 2, \ldots, m) .$$

Since for any $\varepsilon > 0$ the sequence $\{\nu_{n,\varepsilon}\}$ is relatively compact by Lemma 9.3.5, every convergent subsequence has this same limit. ∎

Definition 9.3.6 We call the measure ν the *exponent measure* of the simple max-stable process. This is analogous to the exponent measure in finite-dimensional space (Section 6.1.3).

The characterizing property of the exponent measure is the following homogeneity relation.

Theorem 9.3.7 *For any Borel set A in $\{ f \in C[0, 1] : f \ge 0 \}$ such that $\inf \{ \|f\|_\infty : f \in A \} > 0$ and $\nu(\partial A) = 0$, and any $a > 0$,*

$$\nu(aA) = a^{-1}\nu(A), \tag{9.3.8}$$

where the set aA is obtained by multiplying all elements of A by a.

Proof. On the one hand, from Corollary 9.3.2, for any $a > 0$,

$$\lim_{t \to \infty} t \, a \, P \left(t^{-1}\xi \in aA \right) = \nu(A) .$$

But the left-hand side also converges to $a \, \nu(aA)$. ∎

Remark 9.3.8 Hence by Theorems 9.3.1 and 9.3.7, for any K some compact subset of $[0, 1]$, and each $x > 0$

$$P \left(\sup_{s \in K} \eta(s) \le x \right) = \exp \left(-\nu\{ f : f(s) > x \text{ for some } s \in K \} \right)$$

$$= \exp \left(-\frac{1}{x} \nu\{ f : f(s) > 1 \text{ for some } s \in K \} \right) ,$$

i.e., $\sup_{s \in K} \eta(s)$ has an extreme value distribution.

As in the finite-dimensional case, a nice intuitive background for the role of the exponent measure is provided by the following theorem.

Theorem 9.3.9 *Assume the conditions of Theorem 9.3.1. Define the random measures* N_n *on* $\overline{C}_\varrho^+[0, 1]$ *as follows: for any Borel set* A *with* $\nu(\partial A) = 0$,

$$N_n(A) := \sum_{i=1}^{n} 1_{\{n^{-1}\xi_i \in A\}} \ .$$

Let N *be a Poisson process on* $\overline{C}_\varrho^+[0, 1]$ *with mean measure* ν. *Then* N_n *converges in distribution to* N, *i.e., for* $m = 1, 2, \ldots$ *and Borel sets* A_j *with* $\nu(\partial A_j) = 0$ *for* $j = 1, 2, \ldots, m$,

$$(N_n(A_1), N_n(A_2), \ldots, N_n(A_m)) \overset{d}{\to} (N(A_1), N(A_2), \ldots, N(A_m)) \ .$$

Proof. Without loss of generality we consider the sets A_j, $j = 1, 2, \ldots, m$, disjoint. Let $\lambda_1, \lambda_2, \ldots, \lambda_m > 0$. It is sufficient to prove that the Laplace transform of the left-hand side,

$$E \exp\left(-\sum_{j=1}^{m} \lambda_j N_n(A_j)\right) = \left(E \exp\left(-\sum_{j=1}^{m} \lambda_j 1_{\{n^{-1}\xi \in A_j\}}\right)\right)^n$$

$$= \left(1 + \sum_{j=1}^{m} P\left(n^{-1}\xi \in A_j\right)\left(e^{-\lambda_j} - 1\right)\right)^n , \quad (9.3.9)$$

converges to that of the right-hand side,

$$E \exp\left(-\sum_{j=1}^{m} \lambda_j N(A_j)\right) = \exp\left(\sum_{j=1}^{m} \nu\left(A_j\right)\left(e^{-\lambda_j} - 1\right)\right) \ .$$

The convergence follows from the conclusion of Theorem 9.3.1. ∎

9.4 The Spectral Measure

Recall that
$$\overline{C}_1^+[0, 1] := \{f \in C[0, 1] \ : \ f \geq 0 , |f|_\infty = 1\} \ .$$

In Section 9.3 we proved that the exponent measure ν satisfies a homogeneity property: for $a > 0$ and any Borel set A in $\overline{C}_\varrho^+[0, 1] = (0, \infty] \times \overline{C}_1^+[0, 1]$,

$$\nu(aA) = a^{-1}\nu(A) \ . \quad (9.4.1)$$

As we did in Proposition 6.1.12 for the finite-dimensional case, we now apply a polar coordinate–type transformation $f \to (|f|_\infty, f/|f|_\infty)$ that leads to a spectral measure.

Let A be a Borel set in $\overline{C}_1^+[0, 1]$. For $r > 0$ define the Borel set $B_{r,A} \subset \overline{C}_\varrho^+[0, 1]$ by

$$B_{r,A} := (r, \infty] \times A .$$

Clearly

$$B_{r,A} := r B_{1,A} ;$$

hence by (9.4.1)

$$\nu(B_{r,A}) = r^{-1} \nu(B_{1,A}) .$$

This relation means that after the transformation $f \to (|f|_\infty, f/|f|_\infty)$ the measure ν becomes a product measure. Define the measure ρ on $\overline{C}_1^+[0, 1]$ by

$$\rho(A) := \nu(B_{1,A}) \tag{9.4.2}$$

for each Borel set A in $\overline{C}_1^+[0, 1]$. This finite measure is called the *spectral measure* of the limiting process η in the relation

$$\frac{1}{n} \bigvee_{i=1}^{n} \xi_i \xrightarrow{d} \eta .$$

Theorem 9.4.1 (Giné, Hahn, and Vatan (1990)) *Suppose* $\xi_1, \xi_2, \ldots$ *are i.i.d. stochastic processes in* $C^+[0, 1]$,

$$\frac{1}{n} \bigvee_{i=1}^{n} \xi_i \xrightarrow{d} \eta \tag{9.4.3}$$

in $C^+[0, 1]$, *and* $P(\eta(s) \leq 1) = e^{-1}$ *for* $s \in [0, 1]$, *i.e.,* η *is simple max-stable in* $C^+[0, 1]$. *Then there exists a finite measure* ρ *on* $\overline{C}_1^+[0, 1]$ *with*

$$\int_{\overline{C}_1^+[0,1]} f(s) \, d\rho(f) = 1 \tag{9.4.4}$$

for all $s \in [0, 1]$ *such that for* $m = 1, 2, \ldots$, $K_1, K_2, \ldots, K_m$ *compact sets in* $[0, 1]$, *and* $x_1, x_2, \ldots, x_m > 0$,

$$- \log P(\eta(s) \leq x_j, \text{ for } s \in K_j, j = 1, 2, \ldots, m)$$

$$= \int_{\overline{C}_1^+[0,1]} \max_{1 \leq j \leq m} \left(x_j^{-1} \sup_{s \in K_j} g(s) \right) d\rho(g) . \tag{9.4.5}$$

Conversely, any finite measure ρ *on* $\overline{C}_1^+[0, 1]$ *satisfying (9.4.4) gives rise to a simple max-stable stochastic process in* $C^+[0, 1]$. *The connection is given by (9.4.5) (note that even the finite-dimensional distributions determine the distribution of a process in* $C^+[0, 1]$; *cf. Billingsley (1968), p. 20).*

Proof. Let η be simple max-stable in $C^+[0, 1]$. We have already obtained the measure ρ in (9.4.2). Next we prove (9.4.5) for this measure ρ. We proceed as in the proof of Theorem 9.3.1. On the one hand, as in the mentioned proof,

$$\lim_{n\to\infty} \nu_n \left\{ f : f(s) \le x_j, \text{ for } s \in K_j, j = 1, 2, \ldots, m \right\}^c$$

$$= \lim_{n\to\infty} n P \left(\left\{ \xi(s) \le n x_j, \text{ for } s \in K_j, j = 1, 2, \ldots, m \right\}^c \right)$$

$$= -\log P(\eta(s) \le x_j, \text{ for } s \in K_j, j = 1, 2, \ldots, m)$$

and on the other hand,

$$\lim_{n\to\infty} \nu_n \left\{ f : f(s) \le x_j, \text{ for } s \in K_j, j = 1, 2, \ldots, m \right\}^c$$

$$= \nu \left\{ f : f(s) \le x_j, \text{ for } s \in K_j, j = 1, 2, \ldots, m \right\}^c$$

$$= \nu \left\{ f : |f|_\infty f(s)/|f|_\infty \le x_j, \text{ for } s \in K_j, j = 1, 2, \ldots, m \right\}^c$$

$$= \nu \left\{ f : |f|_\infty > \min_{1\le j\le m} x_j \Big/ \sup_{s\in K_j} \frac{f(s)}{|f|_\infty} \right\}$$

$$= \int_{g\in \overline{C}_1^+[0,1]} \int_{r > \min_{1\le j\le m} x_j / \sup_{s\in K_j} g(s)} \frac{dr}{r^2} \, d\rho(g)$$

$$= \int_{\overline{C}_1^+[0,1]} \max_{1\le j\le m} \frac{\sup_{s\in K_j} g(s)}{x_j} \, d\rho(g) .$$

For (9.4.4) note that $P(\eta(s) \le 1) = e^{-1}$, $s \in [0, 1]$. Hence for each $s \in [0, 1]$,

$$1 = -\log P(\eta(s) \le 1) = \nu\{f : f(s) > 1\}$$

$$= \nu \left\{ f : |f|_\infty > (f(s)/|f|_\infty)^{-1} \right\}$$

$$= \int_{\overline{C}_1^+[0,1]} \int_{r > 1/g(s)} \frac{dr}{r^2} \, d\rho(g) = \int_{\overline{C}_1^+[0,1]} g(s) \, d\rho(g) .$$

For the converse statement of the theorem assume that ρ is a finite measure on $\overline{C}_1^+[0, 1]$ satisfying (9.4.4). The measure ν on $\overline{C}_\varrho^+[0, 1]$ is defined by

$$\nu\{f : |f|_\infty > r \text{ and } f/|f|_\infty \in A\} = r^{-1}\rho(A)$$

for $r > 0$ and A a Borel set in $\overline{C}_1^+[0, 1]$. Let N be a Poisson point process on $\overline{C}_\varrho^+[0, 1]$ with mean measure ν (cf. Theorem 9.3.9). Let

$$\zeta_1, \zeta_2, \zeta_3, \ldots$$

be a realization of the point process. Define

$$\eta := \bigvee_{i=1}^{\infty} \zeta_i . \tag{9.4.6}$$

We claim that η is a simple max-stable process in $C^+[0, 1]$.

First we show that the process η is finite:

$$P(|\eta|_\infty \leq x) = P(N \text{ has no points in the set } \{f : |f|_\infty > x\})$$
$$= \exp(-\nu\{f : |f|_\infty > x\}) = \exp\left(-x^{-1}\rho\left(\overline{C_1^+}[0, 1]\right)\right).$$

Next we check the distribution of η: for $m = 1, 2, \ldots, K_1, K_2, \ldots, K_m$ compact sets in $[0, 1]$, $x_1, x_2, \ldots, x_m > 0$,

$$P(\eta(s) \leq x_j, \text{ for } s \in K_j, j = 1, 2, \ldots, m)$$
$$= P(\text{the graph of every } f \in N \text{ avoids the set } K_j \times (x_j, \infty], j = 1, 2, \ldots, m)$$
$$= \exp\left(-\int_{\overline{C_1^+}[0,1]} \max_{1 \leq j \leq m}\left(x_j^{-1} \sup_{s \in K_j} g(s)\right) d\rho(g)\right).$$

Then we check that the process η is max-stable. Take k independent copies of the process as defined by (9.4.6). Then

$$\bigvee_{j=1}^{k} \eta_j = \bigvee_{j=1}^{k} \bigvee_{i=1}^{\infty} \zeta_i^{(j)}$$

and it is clear that this process has the same structure as the process η except that the measure ν is replaced by $k\nu$. On the other hand we write

$$k\eta = \bigvee_{i=1}^{\infty} k\zeta_i$$

and again the process has the same structure as the process η except that for a Borel set $A \subset \overline{C_\varrho^+}[0, 1]$ the mean measure is now

$$\nu\{f : kf \in A\} = \nu\left\{k^{-1}A\right\} = k\nu\{A\}$$

by Theorem 9.3.7. Since the two processes $\bigvee_{j=1}^{k} \eta_j$ and $k\eta$ have the same distribution, the process η is max-stable.

Finally, we prove that η is in $C^+[0, 1]$, that is, we prove that η has continuous sample paths and that $P(\eta > 0) = 1$. In order to prove continuity we show that (1) $\liminf_{s \to s_0} \eta(s) \geq \eta(s_0)$; (2) $\limsup_{s \to s_0} \eta(s) \leq \eta(s_0)$ for each $s_0 \in [0, 1]$ with probability one.

1. Take any realization $\zeta_1, \zeta_2, \zeta_3, \ldots \in \overline{C_\varrho^+}[0, 1]$. Since $\eta := \bigvee_{i=1}^{\infty} \zeta_i$, for each $\varepsilon > 0$ there is a ζ_i such that $\zeta_i(s_0) > \eta(s_0) - \varepsilon$. Since ζ_i is continuous, $\lim_{s \to s_0} \zeta_i(s) = \zeta_i(s_0)$. Hence

$$\liminf_{s \to s_0} \eta(s) \geq \lim_{s \to s_0} \zeta_i(s) > \eta(s_0) - \varepsilon.$$

2. Define

$$A_I^x := \left\{ f \in C^+[0, 1] : f(s) \le x \text{ for } s \in I \right\}^c,$$

where $x > 0$ and I is a closed interval in $[0, 1]$. Then

$$P\left(N\left(A_I^x \right) < \infty \right) = 1$$

since $v\left(A_I^x \right) < \infty$. Then also

$$P\left(N\left(A_I^x \right) < \infty \text{ for all } x > 0 \text{ rational and } I \in \mathcal{I}_Q \right) = 1, \tag{9.4.7}$$

where $\mathcal{I}_Q$ is the set of closed intervals in $[0, 1]$ with rational endpoints.

Now let us take a realization of the point process satisfying the statement in (9.4.7). Suppose that for some $s_0 \in [0, 1]$ and real $y > 0$,

$$\eta(s_0) := \bigvee_{i=1}^{\infty} \zeta_i(s_0) \le y .$$

It is sufficient to prove that $\limsup_{s \to s_0} \bigvee_{i=1}^{\infty} \zeta_i(s) \le y$.

First we note that this implies

$$N\left(A_{\{s_0\}}^y \right) = 0 . \tag{9.4.8}$$

Next take a monotone sequence of intervals $I_n \in \mathcal{I}_Q$ such that $\cap_{n=1}^{\infty} I_n = \{s_0\}$. Then

$$\cap_{n=1}^{\infty} A_{I_n}^y = A_{\{s_0\}}^y .$$

Hence, since N is a measure,

$$\lim_{n \to \infty} N\left(A_{I_n}^y \right) = N\left(A_{\{s_0\}}^y \right) = 0 .$$

It follows that $N\left(A_{I_n}^y \right) = 0$ for $n \ge n_0$ and hence $\sup_{s \in I_n} \bigvee_{i=1}^{\infty} \zeta_i(s) \le y$ for $n \ge n_0$. In particular, $\limsup_{s \to s_0} \bigvee_{i=1}^{\infty} \zeta_i(s) \le y$. This proves the continuity.

The last statement we need to prove is

$$P(\eta > 0) = 1 .$$

We have

$$\eta \overset{d}{=} \frac{1}{n} \bigvee_{i=1}^{n} \eta_i$$

with $\eta, \eta_1, \eta_2, \ldots, \eta_n$ independent and identically distributed.

Note that for $s \in [0, 1]$ we have $n^{-1} \bigvee_{i=1}^{n} \eta_i(s) = 0$ if and only if $\eta_i(s) = 0$ for $i = 1, 2, \ldots, n$.

Define $A := \{s \in [0, 1] : \eta(s) = 0\}$ and $A_i := \{s \in [0, 1] : \eta_i(s) = 0\}$, $i = 1, 2, \ldots, n$. We have

$$P(A \ne \emptyset) = P\left(\cap_{i=1}^{n} A_i \ne \emptyset \right) \le P\left(\cap_{i=1}^{n} \{A_i \ne \emptyset\} \right) = P^n(A \ne \emptyset)$$

for all n. Hence either $P(A \ne \emptyset) = 1$, i.e., there is some s with $\eta(s) = 0$, which is impossible since $P(\eta(s) \le x) = \exp(-1/x)$ for $x > 0$, or $P(A \ne \emptyset) = 0$, and that is what we want to prove. ∎

In the course of the proof we established the following representation.

Corollary 9.4.2 *Let η be simple max-stable in $C^+[0, 1]$. Then*

$$\eta \stackrel{d}{=} \bigvee_{i=1}^{\infty} \zeta_i,$$

where $\zeta_i = Z_i \pi_i$ and the (Z_i, π_i) form a realization of a Poisson point process on $(0, \infty] \times \overline{C}_1^+[0, 1]$ with mean measure ν satisfying $d\nu = (dr/r^2) \times d\rho$.

Conversely every stochastic process with the given representation is simple max-stable in $C^+[0, 1]$.

Example 9.4.3 Consider a Poisson point process on $\mathbb{R}^2 \setminus \{(0, 0)\}$ with mean measure $(x^2 + y^2)^{-3/2} \, dx \, dy$. Let $\{(X_i, Y_i)\}$ be an enumeration of the points of the point process. Note that there are only finitely many points outside the unit circle. We show that the simple max-stable process $\{2^{-1} \vee_{i=1}^{\infty} X_i \cos \theta + Y_i \sin \theta\}_{0 \le \theta \le 2\pi}$ has the representation of Corollary 9.4.2. With $x = r \cos \phi$ and $y = r \sin \phi$ we have $(x^2 + y^2)^{-3/2} \, dx \, dy = r^{-2} \, dr \, d\phi$. Write $X_i = R_i \cos \Phi_i$ and $Y_i = R_i \sin \Phi_i$. Note that for each θ the half-plane $\{(x, y) : x \cos \theta + y \sin \theta > 0\}$ contains infinitely many points of the point process. Hence for $0 \le \theta \le 2\pi$,

$$\frac{1}{2} \bigvee_{i=1}^{\infty} X_i \cos \theta + Y_i \sin \theta = \bigvee_{i=1}^{\infty} \frac{1}{2} R_i \left((\cos \Phi_i \cos \theta + \sin \Phi_i \sin \theta) \vee 0 \right) .$$

Corollary 9.4.4 *With probability one there exists a finite collection $\zeta_1, \ldots, \zeta_k$ (hence k is random) from Corollary 9.4.2 such that*

$$\eta(s) \stackrel{d}{=} \bigvee_{i=1}^{k} \zeta_i(s)$$

for all $s \in [0, 1]$.

Proof. Excluding the null set we can assume that $\eta(s) > 0$ for $s \in [0, 1]$ and that for each $\varepsilon > 0$ only finitely many Z_i (of Corollary 9.4.2) are larger than ε. The result follows. ∎

Corollary 9.4.2 leads to the following simple representation.

Corollary 9.4.5 *All simple max-stable processes in $C^+[0, 1]$ can be generated in the following way. Consider a Poisson point process on $(0, \infty]$ with mean measure $r^{-2} \, dr$. Let $\{Z_i\}_{i=1}^{\infty}$ be a realization of this point process. Further consider i.i.d. stochastic processes $V, V_1, V_2, \ldots$ in $C^+[0, 1]$ with $EV(s) = 1$ for all $s \in [0, 1]$ and $E \sup_{0 \le s \le 1} V(s) < \infty$. Let the point process and the sequence $V, V_1, V_2, \ldots$ be independent. Then*

$$\eta \stackrel{d}{=} \bigvee_{i=1}^{\infty} Z_i \, V_i.$$

Conversely, each process with this representation is simple max-stable.
One can take the stochastic process V such that

$$\sup_{0 \le s \le 1} V(s) = c \quad a.s. \tag{9.4.9}$$

with c some positive constant.

Example 9.4.6 A nice example of a simple max-stable process has already been given by Brown and Resnick (1977): for the independent and identically distributed processes $\{V_i\}_{i=1}^{\infty}$ of Corollary 9.4.5 take

$$\{V_i(s)\}_{s \in \mathbb{R}} := \left\{ e^{W_i(s) - |s|/2} \right\}_{s \in \mathbb{R}},$$

where the W_i are independent Brownian motions, i.e.,

$$\{\eta(s)\}_{s \in \mathbb{R}} := \left\{ \bigvee_{i=1}^{\infty} Z_i \, e^{W_i(s) - |s|/2} \right\}_{s \in \mathbb{R}},$$

where $\{Z_i\}_{i=1}^{\infty}$ is a realization of a Poisson point process on $(0, \infty]$ with mean measure dr/r^2 and independent of $\{W_i\}_{i=1}^{\infty}$. The process η is stationary (cf. Section 9.8 below).

For the proof of Corollary 9.4.5 we use the following result:

Lemma 9.4.7 *Suppose P is a Poisson point process on the product space $S_1 \times S_2$ with S_1 and S_2 metric spaces and the intensity measure is $\nu = \nu_1 \times \nu_2$, where ν_1 is not bounded and ν_2 is a probability measure. The process can be generated in the following way: let $\{U_i\}$ be an enumeration of the points of a Poisson point process on S_1 with intensity measure ν_1 and let $V_1, V_2, \ldots$ be independent and identically distributed random elements of S_2 with probability distribution ν_2. Then the counting measure N defined by*

$$N(A_1 \times A_2) := \sum_{i=1}^{\infty} 1_{\{(U_i, V_i) \in A_1 \times A_2\}}$$

for Borel sets $A_1 \subset S_1$, $A_2 \subset S_2$, has the same distribution as the point process P.

Proof. We need to prove that the number of points of the set $\{(U_i, V_i)\}_{i=1}^{\infty}$ in two disjoint Borel sets are independent (which is trivial) and that the number of points $N(A_1 \times A_2)$ in a Borel set $A_1 \times A_2$, with $A_1 \subset S_1$ and $A_2 \subset S_2$, has a Poisson distribution with mean measure $\nu_1(A_1)\nu_2(A_2)$. Now

$$P\left(N(A_1 \times A_2) = k\right)$$

$$= \sum_{m=k}^{\infty} P\left(N(A_1 \times A_2) = k \mid \text{the number of points in } A_1 = m\right)$$

$$\times \frac{(\nu_1 (A_1))^m}{m!} e^{-\nu_1 (A_1)}$$

$$= \sum_{m=k}^{\infty} \frac{m!}{(m-k)!k!} (\nu_2 (A_2))^k (1 - \nu_2 (A_2))^{m-k} \frac{(\nu_1 (A_1))^m}{m!} e^{-\nu_1 (A_1)}$$

$$= \frac{(\nu_1 (A_1) \nu_2 (A_2))^k}{k!} e^{-\nu_1 (A_1)} \sum_{m=k}^{\infty} \frac{(1 - \nu_2 (A_2))^{m-k}}{(m-k)!} (\nu_1 (A_1))^{m-k}$$

$$= \frac{(\nu_1 (A_1) \nu_2 (A_2))^k}{k!} e^{-\nu_1 (A_1)\nu_2(A_2)} \; . \qquad \blacksquare$$

Proof (of Corollary 9.4.5). In order to establish the representation we start from the result of Corollary 9.4.2. Let $\{(\tilde{Z}_i, \tilde{\pi}_i)\}_{i=1}^{\infty}$ be an enumeration of the points of a Poisson point process on $(0, \infty] \times \overline{C}_1^+ [0, 1]$ with mean measure

$$\rho(\overline{C}_1^+ [0, 1]) \frac{dr}{r^2} \times \frac{d\rho}{\rho(\overline{C}_1^+ [0, 1])} \; .$$

Then the Poisson point process represented by $\{\tilde{Z}_i \tilde{\pi}_i\}_{i=1}^{\infty}$ has the same distribution as that represented by $\{Z_i \pi_i\}_{i=1}^{\infty}$.

Next define for $i = 1, 2, \ldots,$

$$\tilde{\tilde{Z}}_i := \frac{\tilde{Z}_i}{\rho(\overline{C}_1^+ [0, 1])},$$

$$\tilde{\tilde{\pi}}_i := \tilde{\pi}_i \, \rho(\overline{C}_1^+ [0, 1]) \; .$$

Then $\{(\tilde{\tilde{Z}}_i, \tilde{\tilde{\pi}}_i)\}_{i=1}^{\infty}$ represents a Poisson point process on

$$(0, \infty] \times \left\{ f \in C^+[0, 1] \; : \; |f|_{\infty} = \rho \left(\overline{C}_1^+ [0, 1] \right) \right\} \; .$$

We now argue that its intensity measure is $r^{-2} dr \times dQ$ with Q a probability measure. The intensity measure of the first component is

$$\int_{\{z:z/\rho(\overline{C}_1^+[0,1]) \in A\}} \rho \left(\overline{C}_1^+ [0, 1] \right) \frac{dz}{z^2} = \int_A \frac{dz}{z^2}$$

for a Borel set A of $(0, \infty]$. The intensity measure of the second component is

$$Q(\cdot) := \frac{\rho \left\{ f \; : \; f \geq 0 \, , \; |f|_{\infty} = 1 \, , \, f\rho \left(\overline{C}_1^+ [0, 1] \right) \in \cdot \right\}}{\rho \left(\overline{C}_1^+ [0, 1] \right)},$$

which is a probability measure.

Moreover, for a random element V with probability measure Q we have $E V(s) = 1$ for $s \in [0, 1]$ by (9.4.4). Hence we have the stated representation with V satisfying (9.4.9) by Lemma 9.4.7.

In order to prove that conversely the stated construction represents a simple max-stable process, just follow the steps back of this proof.

It remains to prove that for the converse the requirement $\sup_{0 \leq s \leq 1} V(s) = c$ a.s. can be relaxed to $E \sup_{0 \leq s \leq 1} V(s) < \infty$. Note that the former is used to ensure the finiteness of the process η. But this also follows from the following weaker assumption: we consider now a probability measure Q on the space

$$C^\star := \{ f \in C[0, 1] : f \geq 0, |f|_\infty > 0 \}$$

with the property

$$\int_{C^\star} |f|_\infty \, dQ(f) < \infty .$$

Then

$$P \left\{ \sup_{0 \leq s \leq 1} \eta(s) \leq x \right\} = P \left\{ \tilde{Z}_i(s) \tilde{\tilde{\pi}}_i \leq x \text{ for } 0 \leq s \leq 1, i = 1, 2, \ldots \right\}$$

$$= \exp \left(- \iint_{z|f|_\infty > x} \frac{dz}{z^2} \, dQ(f) \right)$$

$$= \exp \left(- \frac{1}{x} \int_{C^\star} |f|_\infty \, dQ(f) \right) > 0 .$$

Hence the process η is bounded. ∎

Remark 9.4.8 Note that in the finite-dimensional situation of Part II an analogous result holds.

Corollary 9.4.9 *Under the conditions of Theorem 9.4.1, for any positive continuous function f,*

$$- \log P(\eta(s) < f(s) \text{ for } 0 \leq s \leq 1) = \int_{\overline{C}_1^+[0,1]} |g/f|_\infty \, d\rho(g) .$$

The proof of Corollary 9.4.9 is left to the reader (cf. Giné, Hahn, and Vatan (1990)).

Combining the results of Theorems 9.2.1 and 9.4.1, we get the following characterization of max-stable processes in $C[0, 1]$.

Theorem 9.4.10 *For each limit process $\{Y(s)\}_{s \in [0,1]}$ in (9.2.4) that satisfies*

$$P(Y(s) \leq x) = \exp \left(-(1 + \gamma(s)x)^{-1/\gamma(s)} \right)$$

for $s \in [0, 1]$ there exist a continuous function γ and a finite measure ρ on $\overline{C}_1^+[0, 1]$, satisfying (9.4.4) of Theorem 9.4.1, such that with η from Theorem 9.4.1,

$$\{Y(s)\}_{s \in [0,1]} \stackrel{d}{=} \left\{ \frac{(\eta(s))^{\gamma(s)} - 1}{\gamma(s)} \right\}_{s \in [0,1]} . \tag{9.4.10}$$

Conversely, any pair (γ, ρ), *with* γ *a continuous function and* ρ *a finite measure on* $\overline{C}_1^+[0, 1]$ *satisfying* (9.4.4) *of Theorem 9.4.1, gives rise to a max-stable process via* (9.4.10).

9.5 Domain of Attraction

Once again we consider the limit relation

$$\left\{\max_{i \leq n} \frac{X_i(s) - b_s(n)}{a_s(n)}\right\}_{s \in [0,1]} \xrightarrow{d} \{Y(s)\}_{s \in [0,1]} \tag{9.5.1}$$

in $C[0, 1]$. Define

$$U_s(x) := F_s^{\leftarrow}\left(1 - \frac{1}{x}\right)$$

for $x > 1$, $s \in [0, 1]$. Theorem 9.2.1 states that if (9.5.1) holds with proper choices of $a_s(n)$ positive and $b_s(n)$ real, then

$$\lim_{n \to \infty} \frac{U_s(nu) - b_s(n)}{a_s(n)} = \frac{u^{\gamma(s)} - 1}{\gamma(s)}$$

uniformly for $s \in [0, 1]$ and locally uniformly for $u \in (0, \infty)$. Moreover, the processes

$$\xi_i(s) := \frac{1}{1 - F_s(X_i(s))},$$

$s \in [0, 1]$, satisfy

$$\left\{\frac{1}{n} \bigvee_{i=1}^{n} \xi_i(s)\right\}_{s \in [0,1]} \xrightarrow{d} \left\{(1 + \gamma(s)Y(s))^{1/\gamma(s)}\right\}_{s \in [0,1]} =: \{\eta(s)\}_{s \in [0,1]}$$

in $C^+[0, 1]$, where according to Theorem 9.4.1 the probability distribution of η is characterized by a spectral measure ρ. This means that any limit process Y in (9.5.1) is characterized by a continuous function γ and a finite spectral measure ρ. We call the function γ the index function. This situation is quite similar to the finite-dimensional case (Chapter 6).

We shall now establish domain of attraction conditions, that is, for each choice of γ and ρ we shall find necessary and sufficient conditions on the distribution of X such that (9.5.1) holds with a limit process Y characterized by these γ and ρ.

Theorem 9.5.1 *Suppose* $X_1, X_2, \ldots$ *are i.i.d. random elements of* $C[0, 1]$. *Let* $\{Y(s)\}_{s \in [0,1]}$ *be a max-stable stochastic process in* $C[0, 1]$ *with index function* γ *and spectral measure* ρ *on* $\overline{C}_1^+[0, 1]$. *Define*

$$U_s(x) := F_s^{\leftarrow}\left(1 - \frac{1}{x}\right)$$

for $x > 1$, $s \in [0, 1]$. *The following statements are equivalent.*

1.

$$\left\{\max_{i \leq n} \frac{X_i(s) - b_s(n)}{a_s(n)}\right\}_{s \in [0,1]} \xrightarrow{d} \{Y(s)\}_{s \in [0,1]}$$

in $C[0, 1]$, where $a_s(n)$ positive and $b_s(n)$ real are chosen such that $-\log P(Y(s) \leq x) = (1 + \gamma(s)x)^{-1/\gamma(s)}$ for all x with $1 + \gamma(s)x > 0$;

2.

$$\lim_{n \to \infty} \frac{U_s(nu) - b_s(n)}{a_s(n)} = \frac{u^{\gamma(s)} - 1}{\gamma(s)} \tag{9.5.2}$$

uniformly for $s \in [0, 1]$ and one of the following equivalent conditions holds (and then all of them are true) with $\xi_i(s) := 1/(1 - F_s(X_i(s)))$, $i = 1, 2, \ldots$, $s \in [0, 1]$:

(a) $n^{-1} \bigvee_{i=1}^n \xi_i \to^d \eta$ in $C^+[0, 1]$ with $\eta(s) := (1 + \gamma(s)Y(s))^{1/\gamma(s)}$ for $s \in [0, 1]$.

(b) For each Borel set A in $\{f \in C[0, 1] : f \geq 0\}$ such that $\inf\{|f|_\infty : f \in A\} > 0$ and $\nu(\partial A) = 0$,

$$\lim_{t \to \infty} tP\left(t^{-1}\xi \in A\right) = \nu(A) \tag{9.5.3}$$

(with t running through the reals), where

$$\nu(A) = \int\int_{rg \in A} \frac{dr}{r^2} d\rho(g) . \tag{9.5.4}$$

(c) For each $r > 0$ and each Borel set $B \subset \overline{C}_1^+[0, 1]$ (defined in (9.3.2)) with $\rho(\partial B) = 0$,

$$\lim_{t \to \infty} tP(|\xi|_\infty > tr) = r^{-1}\rho(\overline{C}_1^+[0, 1]) \tag{9.5.5}$$

and

$$\lim_{t \to \infty} P\left(\frac{\xi}{|\xi|_\infty} \in B \mid |\xi|_\infty > t\right) = \frac{\rho(B)}{\rho(\overline{C}_1^+[0, 1])} . \tag{9.5.6}$$

Proof. We have already proved in Theorem 9.2.1 that (1) is equivalent to (9.5.2), and (2a). It remains to prove that (2a), (2b), and (2c) are equivalent.

We start with the equivalence of (2a) and (2b). The direct statement has been proved in Theorem 9.3.1 and Corollary 9.3.2. The proof that (2b) implies (2a) consists in rearrangement of the equalities in the proof of Theorem 9.3.1. For (9.5.4) note that with $r := |f|_\infty$ and $g := f/|f|_\infty$,

$$\nu(A) = \nu\left\{f : |f|_\infty \frac{f}{|f|_\infty} \in A\right\} = \nu\{f : rg \in A\},$$

where

$$\nu\{f : r > r_0 \text{ and } g \in B\} = r_0^{-1}\rho(B)$$

for $B \in \overline{C}_1^+[0, 1]$.

Next we prove that (2b) is equivalent to (2c). Take A in (9.5.3) to be

$$\{f \ : \ |f|_\infty > r \, , f/|f|_\infty \in B\}, \tag{9.5.7}$$

where $r > 0$ and B is a Borel set in $\overline{C}_1^+[0, 1]$. By taking $B = \overline{C}_1^+[0, 1]$ we get for $r > 0$,

$$\lim_{t \to \infty} t P(|\xi|_\infty > tr) = r^{-1} \rho \left(\overline{C}_1^+[0, 1] \right), \tag{9.5.8}$$

which is (9.5.5). For general B and $r = 1$ we have

$$\lim_{t \to \infty} t P(|\xi|_\infty > t \text{ and } \xi/|\xi|_\infty \in B) = \nu\{f \ : \ |f|_\infty > 1 \, , f/|f|_\infty \in B\} = \rho(B) \, .$$

Combining with (9.5.8) gives (9.5.6).

Next assume (9.5.5) and (9.5.6). It suffices to prove convergence for a family of sets that is a convergence-determining class for convergence in C. According to Theorem 2.2, p. 14, of Billingsley (1968) this is the case if the family is closed under the formation of finite intersections and if each open set is a finite or countable union of elements of the family. Clearly it is sufficient to prove that any open sphere in C,

$$S_{f_0,c} := \{f \ : \ |f - f_0|_\infty < c\} \, ,$$

with $f_0 \in \overline{C}_\varrho^+[0, 1]$ and $c > 0$, is a countable union of elements of the family.

We use the family

$$\left\{ f \ : \ p < |f|_\infty < q \text{ and } a_i < \frac{f(s)}{|f|_\infty} < b_i \leq 1 \, , \right.$$

$$\left. \text{for } s \text{ in } s_i \leq s \leq s_{i+1} \, , \ i = 1, 2, \ldots, m \right\}, \tag{9.5.9}$$

where $s_1, s_2 \ldots, s_{m+1}$ are rationals such that $0 = s_1 \leq s_2 \leq \cdots \leq s_{m+1} = 1$, and p, $q, a_1, \ldots, a_m, b_1, \ldots, b_m$ are positive rational numbers.

Take $s_1, s_2 \ldots, s_{m+1}$ such that for all $i = 1, 2, \ldots, m$,

$$\sup_{s_i \leq s \leq s_{i+1}} f_0(s) - \inf_{s_i \leq s \leq s_{i+1}} f_0(s) < 2\varepsilon \, .$$

Now for any function f with $|f - f_0| < \varepsilon$ take $a_i < b_i$ such that for $i = 1, 2, \ldots, m$,

$$\frac{\sup_{s_i \leq s \leq s_{i+1}} f_0(s) - 3\varepsilon}{|f|_\infty} < a_i < b_i < \frac{\inf_{s_i \leq s \leq s_{i+1}} f_0(s) + 3\varepsilon}{|f|_\infty},$$

and on $s_i \leq s \leq s_{i+1}$,

$$a_i < \frac{f(s)}{|f|_\infty} < b_i \, ,$$

and p, q with $|f_0|_\infty - \varepsilon < p < q < |f_0|_\infty + \varepsilon$ such that $p < |f|_\infty < q$. ∎

Remark 9.5.2 It is easy to see that (9.5.2) can be extended to

$$\lim_{t \to \infty} \frac{U_s(tu) - b_s([t])}{a_s([t])} = \frac{u^{\gamma(s)} - 1}{\gamma(s)}$$

uniformly, where t runs through the reals.

9.6 Spectral Representation and Stationarity

This section is a continuation of Section 9.4, i.e., we study max-stable processes, not their domains of attraction. Our point of departure is Corollary 9.4.5: a simple max-stable process η in $C^+[0, 1]$ can be written

$$\eta \stackrel{d}{=} \bigvee_{i=1}^{\infty} Z_i V_i, \tag{9.6.1}$$

where $\{Z_i\}$ is an enumeration of the points of a Poisson point process on $(0, \infty]$ with mean measure dr/r^2 and $V, V_1, V_2, \ldots$ are independent and identically distributed nonnegative stochastic processes in $C^+[0, 1]$ with $EV(s) = 1$ for all $s \in [0, 1]$ and $\sup_{0 \le s \le 1} V(s) = c$ a.s., where c is a positive constant. The point process and $V, V_1, V_2, \ldots$ are independent.

Let Q be the probability distribution of the process V.

9.6.1 Spectral Representation

In order to make this representation more analytical, we use Theorem 3.2 of Billingsley (1971), which says that for each probability measure on a metric space S with its Borel sets, there is a random element of S, defined on the unit interval (that is the unit interval with its Borel sets and Lebesgue measure λ as the probability measure) with the same probability distribution. Let

$$\overline{C}_c^+[0, 1] := \{f \in C[0, 1] \ : \ f \ge 0 \, , \ |f|_\infty = c\}$$

for some $c > 0$. It follows that there is a measurable mapping $h : [0, 1] \to \overline{C}_c^+[0, 1]$ such that for each Borel set A of $\overline{C}_c^+[0, 1]$,

$$Q(A) = \lambda \left(\{t \in [0, 1] : h(t) \in A\}\right) . \tag{9.6.2}$$

We are going to use the mapping h to build an alternative version of (9.6.1). Note that with $d\nu_1 := (dr/r^2) \times d\lambda$ (λ Lebesgue measure on $[0,1]$), A_1 a Borel set of $(0, \infty]$, and A a Borel set of $\overline{C}_c^+[0, 1]$,

$$\nu_1 \left(\{(z, t) \in (0, \infty] \times [0, 1] \ : \ (z, h(t)) \in A_1 \times A\}\right) = Q(A) \int_{A_1} \frac{dr}{r^2} \ .$$

Hence if $\{(Z_i, T_i)\}_{i=1}^{\infty}$ is a realization of a Poisson point process on $(0, \infty] \times [0, 1]$ with mean measure $d\nu_1 := (dr/r^2) \times d\lambda$, then

$$\{(Z_i, h(T_i))\}_{i=1}^{\infty} \tag{9.6.3}$$

is a realization of a Poisson point process on $(0, \infty] \times \overline{C}_c^+[0, 1]$ with mean measure $d\nu := (dr/r^2) \times dQ$, where Q is the probability measure of $h(T)$ on $\overline{C}_c^+[0, 1]$. It follows that

$$\eta \stackrel{d}{=} \bigvee_{i=1}^{\infty} Z_i \, h(T_i) . \tag{9.6.4}$$

Now note that h is a mapping from $[0, 1]$ into $\overline{C}_c^+[0, 1]$. Hence for each $t \in [0, 1]$ the mapping provides us with a continuous function, $f_s(t)$ say, with $f_s(t) \in [0, \infty)$, $\int_0^1 f_s(t)\, dt = 1$ for $0 \leq s \leq 1$, and $\sup_{0 \leq s \leq 1} f_s(u) = c$ for all $t \in [0, 1]$.

This leads to the following result.

Theorem 9.6.1 (Resnick and Roy (1991)) *Let $\{(Z_i, T_i)\}_{i=1}^{\infty}$ be a realization of a Poisson point process on $(0, \infty] \times [0, 1]$ with mean measure $(dr/r^2) \times d\lambda$ (λ Lebesgue measure). If the process η is simple max-stable in $C^+[0, 1]$, then there is a family of functions $f_s(t)$ with*

1. *for each $t \in [0, 1]$ we have a nonnegative continuous function $f_s(t) : [0, 1] \to [0, \infty)$,*
2. *for each $s \in [0, 1]$,*

$$\int_0^1 f_s(t)\, dt = 1 , \tag{9.6.5}$$

3.

$$\int_0^1 \sup_{0 \leq s \leq 1} f_s(t)\, dt < \infty ,$$

such that

$$\{\eta(s)\}_{s \in [0,1]} \overset{d}{=} \left\{ \bigvee_{i=1}^{\infty} Z_i\, f_s(T_i) \right\}_{s \in [0,1]} . \tag{9.6.6}$$

Conversely, every process of the form exhibited at the right-hand side of (9.6.6) with the stated conditions is a simple max-stable process in $C^+[0, 1]$.

Remark 9.6.2 The family of functions $\{f_s\}$ is called a family of *spectral functions* of the simple max-stable process. Note that the spectral functions are by no means unique.

Remark 9.6.3 By defining $f_s^{\star}(u) := H'(u) f_s(H(u))$ for $u \in \mathbb{R}$, where H is a probability distribution function and H' its density, one can take the spectral functions in $L_1(\mathbb{R})$ rather than $L_1([0, 1])$.

Remark 9.6.4 There is also a weaker form of this theorem, where for the process η a.s. continuity is replaced by continuity in probability and for the functions $f_s(t)$ continuity is replaced by continuity in measure: $\lambda\{t : |f_{s_n}(t) - f_s(t)| > \varepsilon\} \to 0$ as $n \to \infty$ for each $\varepsilon > 0$ when $s_n \to s$ (de Haan (1984)).

Remark 9.6.5 It is not difficult to see that it is not essential that the max-stable process be defined on $[0, 1]$. One can take any compact set in a Euclidean space.

9.6.2 Stationarity

In this subsection we consider stochastic processes defined on the whole real line rather than on the unit interval as in the previous sections. We do this mainly in view of applications and of some examples to be considered in Sections 9.7 and 9.8. Since the proofs of the results in this section are quite lengthly, we refer to the original papers for some key points.

Definition 9.6.6 *A stochastic process η on $C^+(\mathbb{R})$ with non-degenerate marginals is called* simple max-stable *if for $\eta_1, \eta_2, \ldots$, i.i.d. copies of the process η,*

$$\frac{1}{k} \bigvee_{i=1}^{k} \eta_i \stackrel{d}{=} \eta$$

and $P(\eta(s) \leq 1) = e^{-1}$ for all $s \in \mathbb{R}$.

We start by reproving Theorem 9.6.1 in the present setting.

Theorem 9.6.7 *Let $\{(Z_i, T_i)\}_{i=1}^{\infty}$ be a realization of a Poisson point process on $(0, \infty] \times [0, 1]$ with mean measure $(dr/r^2) \times d\lambda$ (λ Lebesgue measure). If η is a simple max-stable process in $C^+(\mathbb{R})$, then there exists a family of functions $f_s(t)$ ($s \in \mathbb{R}$, $t \in [0, 1]$) with*

1. for each $t \in [0, 1]$ we have a non-negative continuous function $f_s(t) : \mathbb{R} \to [0, \infty)$,
2. for each $s \in \mathbb{R}$

$$\int_0^1 f_s(t)\, dt = 1 , \tag{9.6.7}$$

3. for each compact interval $I \in \mathbb{R}$

$$\int_0^1 \sup_{s \in I} f_s(t)\, dt < \infty ,$$

such that

$$\{\eta(s)\}_{s \in \mathbb{R}} \stackrel{d}{=} \left\{ \bigvee_{i=1}^{\infty} Z_i\, f_s(T_i) \right\}_{s \in \mathbb{R}} . \tag{9.6.8}$$

Conversely every process of the form exhibited at the right-hand side of (9.6.8) *with the stated conditions, is a simple max-stable process in $C^+(\mathbb{R})$.*

Remark 9.6.8 The family of functions $\{f_s\}$ is called a family of *spectral functions* of the simple max-stable process.

Proof (of Theorem 9.6.7). The proof is semi-constructive. First consider an infinite sequence of positive random variables $W := (Y_1, Y_2, \ldots)$. We assume that this sequence is simple max-stable, i.e. for $W_1, W_2, \ldots$ independent and identically distributed copies of the sequence W and all k

$$\frac{1}{k} \bigvee_{i=1}^{k} W_i =^d W .$$

Moreover we assume that $P(Y_i \leq 1) = e^{-1}, i \geq 1$.

We extend the line of reasoning of Chapter 6 (finite-dimensional extremes) to this situation. The process W introduces a probability measure on the infinite product $S := \mathbb{R}_+ \times \mathbb{R}_+ \times \cdots$. Since for any $n \geq 1$, $Y_1, Y_2, \ldots, Y_n > 0$

$$P^k\{Y_1 \le ky_1, \ldots, Y_n \le ky_n\} = P\{Y_1 \le y_1, \ldots, Y_n \le y_n\} \qquad (9.6.9)$$

for all $k = 1, 2, \ldots$, we find (similar to the reasoning in Section 6.1.3)

$$-\log P\{Y_1 \le y_1, \ldots, Y_n \le y_n\} = \lim_{k \to \infty} -k \log P\{Y_1 \le ky_1, \ldots, Y_n \le ky_n\}$$
$$= \lim_{k \to \infty} kP\left\{(Y_1 \le ky_1, \ldots, Y_n \le ky_n)^c\right\}.$$

Take $a_1, a_2, \ldots$ positive such that $\sum_{i=1}^{\infty} a_i^{-1/2} < \infty$. Then

$$E \sup_{i \ge 1} \left(\frac{Y_i}{a_i}\right)^{1/2} \le E \sum_{i=1}^{\infty} \left(\frac{Y_i}{a_i}\right)^{1/2} = EY_1^{1/2} \sum_{i=1}^{\infty} a_i^{-1/2} < \infty.$$

It follows that $\sup_{i \ge 1} Y_i/a_i < \infty$ almost surely. For any $n = 1, 2, \ldots$ the random variable $\sup_{1 \le i \le n} Y_i/a_i$ has a Fréchet distribution by the results of Chapter 6, i.e. there exists positive constants $b_1, b_2, \ldots$ such that $P(\sup_{1 \le i \le n} Y_i/a_i \le x) = \exp(-b_n/x)$, for $x > 0$. Hence $b := \lim_{n \to \infty} b_n$ exists in $(0, \infty)$ and

$$P\left(\sup_{i \ge 1} \frac{Y_i}{a_i} \le x\right) = e^{-b/x}, \qquad \text{for } x > 0.$$

Next we introduce a set function v on S. For any $Y_1, Y_2, \ldots, Y_n > 0$

$$v\left\{((x_1, x_2, \ldots) : x_i \le y_i \text{ for } i = 1, \ldots, n)^c\right\}$$
$$:= -\log P\{Y_1 \le y_1, \ldots, Y_n \le y_n\}.$$

Next for any $\varepsilon > 0$ we determine a consistent family of finite-dimensional probability distributions v_ε on the set

$$S_{\varepsilon,\mathbf{a}} := \left\{((x_1, x_2, \ldots) : x_i \le \varepsilon a_i \text{ for } i = 1, 2, \ldots)^c\right\},$$

with $\mathbf{a} := (a_1, \ldots, a_n)$, as follows: for $y_i > \varepsilon a_i$ for $i = 1, \ldots, n$

$$v_\varepsilon\left\{((x_1, x_2, \ldots) : x_i \le y_i \text{ for } i = 1, \ldots, n)^c\right\}$$
$$:= \frac{-\log P\{Y_1 \le y_1, \ldots, Y_n \le y_n\}}{-\log P\{Y_1 \le \varepsilon a_1, Y_2 \le \varepsilon a_2, \ldots\}}.$$

By Kolmogorov's existence theorem this defines a probability measure v_ε on $S_{\varepsilon,\mathbf{a}}$ consistent with the finite-dimensional distributions. Hence the set function v_ε can be extended in a unique way to a measure on the set $S_{\varepsilon,\mathbf{a}}$. Now ε is arbitrary, hence in fact the measure v is defined on the whole of S. Since for $Y_1, Y_2, \ldots, Y_n > 0$ and $k = 1, 2, \ldots$ we have

$$kv\left\{k\left([0, y_1] \times \cdots \times [0, y_n]\right)^c\right\} = v\left\{\left([0, y_1] \times \cdots \times [0, y_n]\right)^c\right\},$$

we have for any Borel set B in S

$$kv(kB) = v(B) \qquad \text{for } k = 1, 2, \ldots . \tag{9.6.10}$$

As in the finite-dimensional case k may in fact be any positive number. Moreover for $\varepsilon > 0$

$$v\left\{((x_1, x_2, \ldots) : x_i \le \varepsilon a_i \text{ for } i = 1, 2, \ldots)^c\right\} < \infty \tag{9.6.11}$$

(i.e. the measure v is finite outside a neighbourhood of the origin).

Next, as in Section 6.1.4, we move towards a spectral measure. Using the transformation L, with the a_i's as before:

$$w := \sup_{i \ge 1} (x_i / a_i)$$

$$z_k := \begin{cases} x_k / w, & w > 0 \\ 0, & w = 0 \end{cases} \qquad k = 1, 2, \ldots$$

(mapping from S into S) we get for $c > 0$, $n = 1, 2, \ldots$, $u_i \ge 0$ $(i = 1, 2, \ldots, n)$ using (9.6.10) that

$$cv\{(x_1, x_2, \ldots) : w > c, z_1 \le u_1, \ldots, z_n \le u_n\}$$
$$= v\left\{(c^{-1}x_1, c^{-1}x_2, \ldots) : w > c, z_1 \le u_1, \ldots, z_n \le u_n\right\}$$
$$= v\{(x_1, x_2, \ldots) : w > 1, z_1 \le u_1, \ldots, z_n \le u_n\}$$
$$\le v\left\{(x_1, x_2, \ldots) : \sup_{i \ge 1}\left(\frac{x_i}{a_i}\right) > 1\right\} < \infty$$

by (9.6.11). This gives

$$v\{(x_1, x_2, \ldots) : w > c, z_1 \le u_1, \ldots, z_n \le u_n\}$$
$$= c^{-1}v\{(x_1, x_2, \ldots) : w > 1, z_1 \le u_1, \ldots, z_n \le u_n\}$$

i.e. the transformed measure $d(v \circ L^{\leftarrow}) = (dr/r^2) \times d\mu$ on $[0, \infty) \times S$ for some measure μ on S. Note that

$$\mu(S) = v\left\{(x_1, x_2, \ldots) : \sup_{i \ge 1}\left(\frac{x_i}{a_i}\right) > 1\right\}$$
$$= v\left\{((x_1, x_2, \ldots) : x_i \le a_i \text{ for } i = 1, 2, \ldots)^c\right\} < \infty$$

by (9.6.11). Hence μ is a finite measure.

From the definition of v we can write, for $n \ge 1$, $Y_1, Y_2, \ldots, Y_n > 0$,

$$P\{Y_1 \le y_1, \ldots, Y_n \le y_n\}$$
$$= \exp -v\left\{((x_1, x_2, \ldots) : x_i \le y_i \text{ for } i = 1, \ldots, n)^c\right\}$$
$$= \exp -v\left\{((w, z_1, z_2, \ldots) : z_i w \le y_i \text{ for } i = 1, \ldots, n)^c\right\}$$
$$= \exp -v\left\{(w, z_1, z_2, \ldots) : \min_{1 \le i \le n} \frac{y_i}{z_i} < w\right\}$$

$$= \exp - \iint \cdots \int_{\min_{1 \le i \le n}(y_i/z_i) < w} \frac{dr}{r^2} \, \mu(d(z_1, z_2, \ldots))$$

$$= \exp - \int \cdots \int_S \max_{1 \le i \le n} \left(\frac{z_i}{y_i} \right) \mu(d(z_1, z_2 \ldots)) \, .$$

Note that since ν is homogeneous (cf. (9.6.10)), if we multiply all $a_i's$ in (9.6.11) by a constant, we can transform μ into a probability measure.

Next we apply Theorem 3.2 of Billingsley (1971): for each probability measure on a metric space S with its Borel sets there is a random element of S defined on the unit interval ([0, 1] with its Borel sets and Lebesgue measure λ as the probability measure) with the same probability distribution. It follows that there are non-negative functions f_n defined on [0, 1] such that

$$P\{Y_1 \le y_1, \ldots, Y_n \le y_n\}$$

$$= \exp - \int \cdots \int_S \max_{1 \le i \le n} \left(\frac{z_i}{y_i} \right) \mu(d(z_1, z_2 \ldots)) = \exp - \int_0^1 \max_{1 \le i \le n} \frac{f_i(t)}{y_i} dt \, .$$

Next we give a representation of the process $(Y_1, Y_2, \ldots)$ using the Poisson point process of the statement of the theorem: we have

$$(Y_1, Y_2, \ldots) \overset{d}{=} \left(\bigvee_{i=1}^{\infty} Z_i f_1(T_i), \bigvee_{i=1}^{\infty} Z_i f_2(T_i), \ldots \right)$$

since

$$P \left\{ \bigvee_{i=1}^{\infty} Z_i f_1(T_i) \le x_1, \ldots, \bigvee_{i=1}^{\infty} Z_i f_n(T_i) \le x_n \right\}$$

$$= P \left\{ Z_i f_j(T_i) \le x_j \text{ for } j = 1, \ldots, n; \; i = 1, 2, \ldots \right\}$$

$$= P \left\{ Z_i \le \inf_{1 \le j \le n} \frac{x_j}{f_j(T_i)} \text{ for } i = 1, 2, \ldots \right\}$$

$$= \exp - \iint_{r > \inf_{1 \le j \le n} \frac{x_j}{f_j(t)}} \frac{dr}{r^2} dt$$

$$= \exp - \int_0^1 \max_{1 \le i \le n} \frac{f_i(t)}{x_i} dt \, .$$

Now let us consider the process η. By the results obtained so far we have a spectral representation for the process $\{\eta(r_n)\}_{n=1}^{\infty}$ where $r_1, r_2, \ldots$ is an enumeration of the rationals of $\mathbb{R}$: with some abuse of notation we can write

$$\{\eta(r_n)\}_{n=1}^{\infty} \overset{d}{=} \left\{ \bigvee_{i=1}^{\infty} Z_i f_{r_n}(T_i) \right\}_{n=1}^{\infty} \, .$$

The next step - finding a similar representation of the process $\eta(s)$ for real s - is done by using continuity. The process η has continuous sample paths hence in particular it is continuous in probability.

We use without proof the auxiliary result: any sequence of random variables $\{\bigvee_{i=1}^{\infty} Z_i f_n^*(Y_i)\}_{n=1}^{\infty}$ with f_n^* spectral functions converges in probability as $n \to \infty$ if and only if the sequence f_n^* converges in Lebesgue measure. This gives representation (9.6.8) of the process $\eta(s)$ for real s.

The final step, proving the continuity of $f_s(t)$ for almost each t and the convergence of the integrals in 3., is provided by Theorem 3.2 of Resnick and Roy (1991): for a compact interval I in $\mathbb{R}$, the process $\{\eta(s)\}_{s \in I}$ has continuous sample paths if and only if the family $f_s(t)$ of spectral functions is continuous in s for almost all t and if moreover

$$\int_0^1 \sup_{s \in I} f_s(t) \, dt < \infty .$$

The proof of the converse statement of the theorem is easy.

Next we turn to the issue of stationarity.

Definition 9.6.9 *A mapping Φ from L_1^+ (the non-negative integrable functions on $[0, 1]$) to L_1^+ is called a* piston *if for $h \in L_1^+$*

$$\Phi(h(t)) = r(t)h(H(t))$$

with H a one-to-one measurable mapping from $[0, 1]$ to $[0, 1]$ and r a positive measurable function, such that for every $h \in L_1^+$

$$\int_0^1 \Phi(h(t)) \, dt = \int_0^1 h(t) \, dt .$$

Theorem 9.6.10 *Let $\{(Z_i, T_i)\}_{i=1}^{\infty}$ be a realization of a Poisson process on $(0, \infty] \times [0, 1]$ with mean measure $(dr/r^2) \times d\lambda$ (λ Lebesgue measure).*

If the stochastic process $\{\eta(s)\}_{s \in \mathbb{R}}$ is simple max-stable, strictly stationary and continuous a.s., then there is a function h in L_1^+ with $\int_0^1 h(t)dt = 1$ and a continuous group of pistons $\{\Phi_s\}_{s \in \mathbb{R}}$ (continuous, i.e., $\Phi_{s_n}(h(t)) \to \Phi_s(h(t))$ as $s_n \to s$ for almost all $t \in [0, 1]$) with

$$\int_0^1 \sup_{s \in I} \Phi_s(h(t)) \, dt < \infty$$

for each compact interval I, such that

$$\{\eta(s)\}_{s \in \mathbb{R}} \overset{d}{=} \left\{ \bigvee_{i=1}^{\infty} Z_i \, \Phi_s(h(T_i)) \right\}_{s \in \mathbb{R}} . \tag{9.6.12}$$

Conversely every stochastic process of the form exhibited at the right-hand side of (9.6.12) with the stated conditions, is simple max-stable, strictly stationary and a.s. continuous.

Proof. We apply de Haan and Pickands (1986). It says that if η is a simple max-stable process on $\mathbb{R}$ the representation of this theorem holds with "η has continuous sample paths" replaced by "η is continuous in probability" and with the statement "$\Phi_{s_n}(f(u)) \rightarrow \Phi_s(f(u))$ for almost all $u \in \mathbb{R}$" replaced by "$\int_0^1 |\Phi_{s_n}(f(u)) - \Phi_s(f(u))| du \rightarrow 0$."

Next, replacing convergence in probability with a.s. convergence on the one hand and replacing convergence in L_1-norm by a.s. convergence on the other hand can be done locally, i.e., for each compact interval. Theorem 3.2 of Resnick and Roy (1991) again justifies such replacement (cf. the proof of Theorem 9.6.7).

Conversely consider a process with the given representation.

For $s_1, s_2 \in \mathbb{R}$ we have with $f_s(t) := \Phi_s(h(t))$

$$- \log P\left(\eta(s_1) \le x_1, \eta(s_2) \le x_2\right)$$
$$= - \log P\left(Z_i f_{s_1}(T_i) \le x_1 \text{ and } Z_i f_{s_2}(T_i) \le x_2 \text{ for all } i\right)$$
$$= - \log P\left(\max_i \max \left(\frac{Z_i f_{s_1}(T_i)}{x_1}, \frac{Z_i f_{s_2}(T_i)}{x_2}\right) \le 1\right)$$
$$= \int \int_{\max(r f_{s_1}(t)/x_1, r f_{s_2}(t)/x_2) > 1} \frac{dr}{r^2} \times dt$$
$$= \int_0^1 \max \left(\frac{f_{s_1}(t)}{x_1}, \frac{f_{s_2}(t)}{x_2}\right) dt .$$

Hence it is sufficient to prove that for $s \in \mathbb{R}$

$$\int_0^1 \max \left(\frac{f_{s_1+s}(t)}{x_1}, \frac{f_{s_2+s}(t)}{x_2}\right) dt = \int_0^1 \max \left(\frac{f_{s_1}(t)}{x_1}, \frac{f_{s_2}(t)}{x_2}\right) dt . \qquad (9.6.13)$$

Now $f_s(t) = \Phi_s(h(t))$ for $s \in \mathbb{R}$. Hence, since the Φ_s form a group, the left-hand side of (9.6.13) is

$$\int_0^1 \max \left(\frac{\Phi_s(f_{s_1}(t))}{x_1}, \frac{\Phi_s(f_{s_2}(t))}{x_2}\right) dt$$
$$= \int_0^1 \max \left(\frac{f_{s_1}(H_s(t))}{x_1}, \frac{f_{s_2}(H_s(t))}{x_2}\right) r_s(t) dt$$
$$= \int_0^1 \Phi_s \left(\max \left(\frac{f_{s_1}(t)}{x_1}, \frac{f_{s_2}(t)}{x_2}\right)\right) dt$$
$$= \int_0^1 \max \left(\frac{f_{s_1}(t)}{x_1}, \frac{f_{s_2}(t)}{x_2}\right) dt$$

by assumption. In a similar way one deals with the higher-dimensional marginal distributions.

9.7 Special Cases

An interesting special case of Theorem 9.6.10 occurs when Φ is a shift: $\Phi_s(h(u)) := h(u - s)$. Here h is a probability density function. Hence examples of stationary simple max-stable processes can be constructed using some well-known probability densities:

1. The exponential model: take for $\beta > 0$,

$$h(u) := \frac{\beta}{2} \exp(-\beta|u|) .$$

2. The normal model: take for $\beta > 0$,

$$h(u) := \beta(2\pi)^{-1/2} \exp(-\beta^2 u^2/2) .$$

3. The Student-t model: take for $\beta > 0$ and ν a positive integer,

$$h(u) := \frac{\beta\Gamma((\nu + 1)/2)}{\Gamma(\nu/2)\sqrt{\pi}}(1 + \beta^2 u^2)^{-(\nu+1)/2} .$$

Note that all conditions, in particular condition (3) of Theorem 9.6.1, are fulfilled since the densities are continuous and unimodal.

In all three cases the parameter β has been introduced in order to control the amount of spatial dependence: note that when β increases the amount of dependence between values at two fixed sites decreases.

The two-dimensional marginal distributions can be calculated explicitly in all three cases (and also for their two-dimensional analogues where s is a vector in $\mathbb{R}^2$; see de Haan and Pereira (2006)).

1. For the exponential model:

$$-\log P(\eta(0) \le x, \eta(s) \le y) = \begin{cases} \frac{1}{y} , & 0 < y \le xe^{-\beta|s|}, \\ \frac{1}{x} + \frac{1}{y} - \frac{e^{-\beta|s|/2}}{\sqrt{xy}} , & xe^{-\beta|s|} < y \le xe^{\beta|s|}, \\ \frac{1}{x} , & y > xe^{\beta|s|} ; \end{cases}$$

hence the spectral measure is concentrated on a proper subinterval of $[0, \pi/2]$ and has two atoms.

2. For the normal model:

$$-\log P(\eta(0) \le x, \eta(s) \le y)$$
$$= \frac{1}{x}\Phi\left(\frac{|s|\beta}{2} + \frac{1}{|s|\beta}\log\frac{y}{x}\right) + \frac{1}{y}\Phi\left(\frac{|s|\beta}{2} + \frac{1}{|s|\beta}\log\frac{x}{y}\right) .$$

Compare with Example 9.4.6.

3. For the Student-t model:

$$-\log P(\eta(0) \le x, \eta(s) \le y)$$

$$
= \begin{cases}
\frac{1}{y}, & 0 < y \le xL_2^{-(v+1)/2}, \\
\frac{1}{x} P_1(\beta, s, z) + \frac{1}{y}(1 - P_2(\beta, s, z)), & xL_2^{-(v+1)/2} < y < x, \\
\frac{2}{x} P\left(T_{v,1} \le \frac{\beta|s|}{2}\right), & x = y, \\
\frac{1}{x}(1 - P_1(\beta, s, z)) + \frac{1}{y} P_2(\beta, s, z), & x < y \le xL_1^{-(v+1)/2}, \\
\frac{1}{x}, & y > xL_1^{-(v+1)/2},
\end{cases}
$$

where

$$
L_1 = 1 + \frac{\beta^2 s^2}{2} - \beta|s|\sqrt{1 + \frac{\beta^2 s^2}{4}},
$$

$$
L_2 = 1 + \frac{\beta^2 s^2}{2} + \beta|s|\sqrt{1 + \frac{\beta^2 s^2}{4}},
$$

$$
P_1(\beta, s, z) = P\left(\left|T_{v,1} - \frac{\beta s}{1 - z}\right| \le \beta\sqrt{\frac{s^2 z}{(1 - z)^2} - \frac{1}{\beta^2}}\right),
$$

$$
P_2(\beta, s, z) = P\left(\left|T_{v,1} - \frac{\beta s z}{1 - z}\right| \le \beta\sqrt{\frac{s^2 z}{(1 - z)^2} - \frac{1}{\beta^2}}\right),
$$

$T_{v,1}$ is a random variable with a Student-t distribution with v degrees of freedom and scale parameter one, and $z = (x/y)^{2/(v+1)}$. These results can be used for constructing an estimator for the dependence parameter β.

9.8 Two Examples

Let us go back to Example 9.4.6 of Section 9.4:

$$
\{\eta(s)\}_{s \in \mathbb{R}} := \left\{\bigvee_{i=1}^{\infty} Z_i e^{W_i(s) - |s|/2}\right\}_{s \in \mathbb{R}}, \tag{9.8.1}
$$

where $\{Z_i\}_{i=1}^{\infty}$ is a realization of a Poisson point process on $(0, \infty]$ with mean measure dr/r^2 and independently, $\{W_i\}_{i=1}^{\infty}$ is a sequence of independent Brownian motions.

The process η is stationary on $\mathbb{R}$. For the proof it is sufficient to prove that all marginal distributions are stationary. We shall show this for the two-dimensional distributions. Let $0 < s_1 < s_2$ and write $u := s_2 - s_1$. Then for $x, y \in \mathbb{R}$,

$$
- \log P(\eta(s_1) \le e^x, \eta(s_2) \le e^y)
$$
$$
= E \max\left(e^{W(s_1) - s_1/2 - x}, e^{W(s_2) - s_2/2 - y}\right)
$$

$$= E \; e^{W(s_1)-s_1/2} \max \left(e^{-x}, e^{W(s_2)-W(s_1)-(s_2-s_1)/2-y} \right)$$

$$= E \; \max \left(e^{-x}, e^{W(s_2)-W(s_1)-(s_2-s_1)/2-y} \right)$$

$$= e^{-x} P \left(W(s_2) - W(s_1) - \frac{s_2 - s_1}{2} \le y - x \right)$$

$$+ e^{-y} \frac{1}{\sqrt{2\pi}} \int_{t\sqrt{u}-u/2 > y - x} e^{t\sqrt{u}-u/2} e^{-t^2/2} \, dt$$

$$= e^{-x} \Phi \left(\frac{\sqrt{u}}{2} - \frac{y-x}{\sqrt{u}} \right) + e^{-y} \frac{1}{\sqrt{2\pi}} \int_{t-\sqrt{u} > -\sqrt{u}/2 + (y-x)/\sqrt{u}} e^{-(t-\sqrt{u})^2/u} \, dt$$

$$= e^{-x} \Phi \left(\frac{\sqrt{u}}{2} - \frac{y-x}{\sqrt{u}} \right) + e^{-y} \left(1 - \Phi \left(-\frac{\sqrt{u}}{2} + \frac{y-x}{\sqrt{u}} \right) \right)$$

$$= e^{-x} \Phi \left(\frac{\sqrt{u}}{2} + \frac{y-x}{\sqrt{u}} \right) + e^{-y} \Phi \left(\frac{\sqrt{u}}{2} + \frac{x-y}{\sqrt{u}} \right)$$

with Φ the standard normal distribution function. Clearly the distribution depends on s_1 and s_2 only through $u = s_2 - s_1$. The reasoning is similar when $s_1 < s_2 < 0$.

Finally consider the case $s_1 < 0 < s_2$:

$$-\log P(\eta(s_1) \le e^x, \eta(s_2) \le e^y) = E \; \max \left(e^{W(s_1)+s_1/2-x}, e^{W(s_2)-s_2/2-y} \right).$$

Denote the distribution function of $e^{W(s_1)+s_1/2-x}$ by F_1 and the distribution function of $e^{W(s_2)-s_2/2-y}$ by F_2. Then the expectation is

$$\int_0^\infty t \, dF_1(t) F_2(t) = \int_0^\infty t \left(F_1'(t) \int_0^t F_2'(u) \, du + F_2'(t) \int_0^t F_1'(u) \, du \right) dt$$

$$= \int_0^\infty \int_u^\infty t F_1'(t) \, dt \, F_2'(u) \, du + \int_0^\infty \int_u^\infty t F_2'(t) \, dt \, F_1'(u) \, du.$$

Note that

$$F_1'(t) = \frac{1}{t\sqrt{|s_1|}} \phi \left(\frac{\sqrt{|s_1|}}{2} + \frac{x}{\sqrt{|s_1|}} + \frac{\log t}{\sqrt{|s_1|}} \right)$$

and

$$F_2'(t) = \frac{1}{t\sqrt{s_2}} \phi \left(\frac{\sqrt{s_2}}{2} + \frac{x}{\sqrt{s_2}} + \frac{\log t}{\sqrt{s_2}} \right)$$

with $\phi = \Phi'$. Hence

$$\int_u^\infty t F_1'(t) \, dt = \int_u^\infty \frac{1}{\sqrt{|s_1|}} \phi \left(\frac{\sqrt{|s_1|}}{2} + \frac{x}{\sqrt{|s_1|}} + \frac{\log t}{\sqrt{|s_1|}} \right) dt$$

$$= \int_{\log u}^\infty \frac{1}{\sqrt{|s_1|}} \phi \left(\frac{\sqrt{|s_1|}}{2} + \frac{x}{\sqrt{|s_1|}} + \frac{v}{\sqrt{|s_1|}} \right) e^v \, dv.$$

Now

$$e^v \phi \left(\frac{\sqrt{|s_1|}}{2} + \frac{x}{\sqrt{|s_1|}} + \frac{v}{\sqrt{|s_1|}} \right)$$

$$= \frac{e^v}{\sqrt{2\pi}} \exp \left(-\frac{1}{2} \left(\frac{|s_1|}{4} + \frac{x^2}{|s_1|} + \frac{v^2}{|s_1|} + x + v + \frac{2vx}{|s_1|} \right) \right)$$

$$= \frac{e^{-x}}{\sqrt{2\pi}} \exp \left(-\frac{1}{2} \left(\frac{|s_1|}{4} + \frac{x^2}{|s_1|} + \frac{v^2}{|s_1|} - x - v + \frac{2vx}{|s_1|} \right) \right)$$

$$= e^{-x} \phi \left(\frac{\sqrt{|s_1|}}{2} - \frac{x}{\sqrt{|s_1|}} - \frac{v}{\sqrt{|s_1|}} \right).$$

Hence

$$\int_u^\infty t F_1'(t) \, dt = e^{-x} \int_{\log u}^\infty \phi \left(\frac{v}{\sqrt{|s_1|}} + \frac{x}{\sqrt{|s_1|}} - \frac{\sqrt{|s_1|}}{2} \right) \frac{dv}{|s_1|}$$

$$= e^{-x} \left(1 - \Phi \left(\frac{\log u}{\sqrt{|s_1|}} + \frac{x}{\sqrt{|s_1|}} - \frac{\sqrt{|s_1|}}{2} \right) \right) = e^{-x} P \left(e^{W(s_1) - s_1/2 - x} > u \right)$$

and

$$\int_0^\infty \int_u^\infty t F_1'(t) \, dt \, F_2'(u) \, du$$

$$= e^{-x} \int_0^\infty P \left(e^{W(s_1) - s_1/2 - x} > u \right) \, dP \left(e^{W(s_2) - s_2/2 - y} \le u \right)$$

$$= e^{-x} P \left(e^{W(s_1) - s_1/2 - x} > e^{W(s_2) - s_2/2 - y} \right)$$

$$= e^{-x} P \left(W(s_1) - W(s_2) > x - y - \frac{s_2 - s_1}{2} \right)$$

$$= e^{-x} \Phi \left(\frac{-x + y}{\sqrt{s_2 - s_1}} + \frac{\sqrt{s_2 - s_1}}{2} \right).$$

Similarly

$$\int_0^\infty \int_u^\infty t F_2'(t) \, dt \, F_1'(u) \, du = e^{-y} \Phi \left(\frac{x - y}{\sqrt{s_2 - s_1}} + \frac{\sqrt{s_2 - s_1}}{2} \right).$$

Hence for $s_1 < 0 < s_2$

$$- \log P(\eta(s_1) \le e^x, \eta(s_2) \le e^y)$$

$$= e^{-x} \Phi \left(\frac{-x + y}{\sqrt{s_2 - s_1}} + \frac{\sqrt{s_2 - s_1}}{2} \right) + e^{-y} \Phi \left(\frac{x - y}{\sqrt{s_2 - s_1}} + \frac{\sqrt{s_2 - s_1}}{2} \right),$$

which depends on s_1 and s_2 only through $s_2 - s_1$.

We now exhibit a stochastic process in the domain of attraction of this simple max-stable process.

Example 9.8.1 Let Y be a random variable with distribution function $1 - 1/x, x \geq 1$. Let W be Brownian motion independent of Y. Consider the process

$$\{\xi(s)\}_{s \in \mathbb{R}} := \left\{ Y e^{W(s) - |s|/2} \right\}_{s \in \mathbb{R}} . \tag{9.8.2}$$

We claim that this process is in the domain of attraction of the process in (9.8.1). Consider independent and identically distributed copies of the process: $\left\{ Y_i e^{W_i(s) - |s|/2} \right\}_{s \in \mathbb{R}}$ for $i = 1, 2 \dots$. Now consider the point process consisting of the points

$$\left\{ \left(\frac{Y_i}{n}, e^{W_i(s) - |s|/2} \right) \right\}_{i=1}^{n} . \tag{9.8.3}$$

These are elements of $(0, \infty] \times C^+(\mathbb{R})$. We already know from Theorem 2.1.2 that the point process constructed from the points $\{Y_i / n\}_{i=1}^{n}$ converges in distribution to the point process constructed from the points $\{Z_i\}_{i=1}^{\infty}$ of (9.8.1). Since the second component is independent and does not change, the point process (9.8.3) converges in distribution to the point process constructed from the points $\left\{ \left(Z_i, e^{W_i(s) - |s|/2} \right) \right\}_{i=1}^{\infty}$. Then the point process constructed from the points $\left\{ \left(n^{-1} Y_i e^{W_i(s) - |s|/2} \right) \right\}_{i=1}^{n}$ converges to the Poisson point process constructed from the points $\left\{ \left(Z_i e^{W_i(s) - |s|/2} \right) \right\}_{i=1}^{\infty}$. The points are continuous functions. Since $\left\{ \sup_{i \leq n} n^{-1} Y_i e^{W_i(s) - |s|/2} \right\}_{s \in \mathbb{R}}$ is a continuous functional of the point process, we have indeed

$$\left\{ \sup_{i \leq n} n^{-1} Y_i e^{W_i(s) - |s|/2} \right\}_{s \in \mathbb{R}} \xrightarrow{d} \{\eta(s)\}_{s \in \mathbb{R}}$$

in $C^+(\mathbb{R})$.

The process (9.8.2) has an interesting property. For $a > 0$ the distribution of $\{\xi(s)/a\}_{s \in \mathbb{R}}$ given $\xi(0) > a$ is the same as the distribution of $\{\xi(s)\}_{s \in \mathbb{R}}$.

This property—which we call *excursion stability*—is analogous to the defining property for the (generalized) Pareto distribution; see Exercise 3.1.

Next we consider maxima of independent and identically distributed Ornstein–Uhlenbeck processes. We show that if we apply a suitable time transformation the limit process is max-stable.

Example 9.8.2 Let $\{X(s)\}_{s \in \mathbb{R}}$ be a Ornstein–Uhlenbeck process, i.e.,

$$X(s) = \int_{-\infty}^{s} e^{-(s-u)/2} dW(u)$$

for all $s \in \mathbb{R}$ with W Brownian motion on $(-\infty, \infty)$, i.e., two independent Brownian motions starting at 0 and going off in two directions of time. Since for $s \neq t$ the random vector $(X(s), X(t))$ is multivariate normal with correlation coefficient less than one, Example 6.2.6 tells us that, relation (9.5.1) can not hold for any max-stable process in $C[0, 1]$: since Y has continuous sample paths, $Y(s)$ and $Y(t)$ can not be independent. Hence we compress time in order to create more dependence, i.e., we consider the convergence of

$$\left\{ \bigvee_{i=1}^{n} b_n \left(X_i \left(\frac{s}{b_n^2} \right) - b_n \right) \right\}_{s \in \mathbb{R}} \tag{9.8.4}$$

in $C[-s_0, s_0]$ for arbitrary $s_0 > 0$, where $X_1, X_2, \ldots$ are independent and identically distributed copies of X and the b_n are the correct normality constants for the standard one-dimensional normal distribution, e.g., $b_n = (2 \log n - \log \log n - \log(4\pi))^{1/2}$ (cf. Example 1.1.7). In order to show convergence we write

$$X(s) = e^{-s/2} \left(X(0) + \int_0^s e^{u/2} dW(u) \right) ;$$

hence

$$b_n \left(X \left(\frac{s}{b_n^2} \right) - b_n \right)$$

$$= e^{-s/(2b_n^2)} \left(b_n (X(0) - b_n) + b_n \int_0^{s/b_n^2} e^{u/2} dW(u) + \left(1 - e^{s/(2b_n^2)} \right) b_n^2 \right) .$$

Note that uniformly for $|s| \leq s_0$,

$$e^{-s/(2b_n^2)} = 1 + O \left(\frac{1}{b_n^2} \right) .$$

Further, since $e^{u/2} = 1 + O \left(1/b_n^2 \right)$ for $|u| < s_0/b_n^2$,

$$b_n \int_0^{s/b_n^2} e^{u/2} dW(u) = \left(1 + O \left(\frac{1}{b_n^2} \right) \right) b_n W \left(\frac{s}{b_n^2} \right) .$$

Finally, for $|s| \leq s_0$,

$$\left(1 - e^{s/(2b_n^2)} \right) b_n^2 = -\frac{s}{2} + O \left(\frac{1}{b_n^2} \right) .$$

It follows that

$$b_n \left(X \left(\frac{s}{b_n^2} \right) - b_n \right)$$

$$= \left(1 + O \left(\frac{1}{b_n^2} \right) \right) \left(b_n (X(0) - b_n) + b_n W \left(\frac{s}{b_n^2} \right) - \frac{s}{2} \right) + O \left(\frac{1}{b_n^2} \right) .$$

We write $W^\star(s) := b_n W \left(s/b_n^2 \right)$. Then $W^\star$ is also Brownian motion. We have

$$\bigvee_{i=1}^{n} b_n \left(X_i \left(\frac{s}{b_n^2} \right) - b_n \right)$$

$$= \left(1 + O\left(\frac{1}{b_n^2} \right) \right) \left\{ \bigvee_{i=1}^{n} \left(b_n \left(X_i(0) - b_n \right) + W_i^{\star}(s) \right) - \frac{s}{2} \right\} + O\left(\frac{1}{b_n^2} \right).$$

Hence the limit of (9.8.4) is the same as that of

$$\left\{ \bigvee_{i=1}^{n} \left(b_n \left(X_i(0) - b_n \right) + W_i^{\star}(s) \right) - \frac{s}{2} \right\}_{s \in \mathbb{R}}. \tag{9.8.5}$$

The rest of the proof runs as in the previous example.

One finds that the sequence of processes (9.8.5) converges weakly in $C[-s_0, s_0]$, hence in $C(\mathbb{R})$, to

$$\left\{ \bigvee_{i=1}^{\infty} \left(\log Z_i + W_i^{\star}(s) \right) - \frac{s}{2} \right\}_{s \in \mathbb{R}}.$$

Exercises

9.1. Show that the constant c in Lemma 9.3.4 is $\rho(\overline{C}_1^{+}[0, 1])$, where ρ is the spectral measure of Section 9.4. Argue that this constant is an analogue of $L(1, \ldots, 1)$, where L is the dependence function defined in Section 6.1.5.

9.2. In Section 7.4 (finite-dimensional extremes) a quantity κ has been introduced that quantifies the strength of dependence. It was shown that for d-dimensions κ is $d/(- \log P(Z_1 \leq 1, \ldots, Z_d \leq 1)) = d/L(1, \ldots, 1)$, where $(Z_1, \ldots, Z_d)$ is a random vector with distribution function G_0, where G_0 is from Theorem 6.1.1. Argue that $1/c$ could serve as an infinite-dimensional analogue of this coefficient, where c is the constant from Lemma 9.3.4. How would one estimate this quantity c?

9.3. Show that all marginal distributions of the process of Example 9.4.3 are indeed $\exp(-1/x)$, $x > 0$.

9.4. Check that the regular variation condition of Theorem 9.5.1(2c) implies the regular variation condition of Theorem 6.2.1(1) for all marginal distributions.

9.5. Consider the stochastic process ξ defined by $\xi(s) := Y V(s)$ for $s \in \mathbb{R}$, where Y has distribution function $1 - 1/x$, $x \geq 1$, and V is a continuous stochastic process independent of V satisfying $EV(s) = 1$, for all s and $E \sup_{a \leq s \leq b} V(s) < \infty$ for $a < b$. Show that ξ is in the domain of attraction of a simple max-stable process that has the representation of Corollary 9.4.5 with the same auxiliary process V. Moreover, for $a > 1$ and $V(0) = 1$, $\{\xi(s)/a\}_{s>0}$ given $\xi(0) > a$ has the same distribution as ξ. This property resembles a corresponding property for a generalized Pareto distribution in finite-dimensional space. Find that the one-dimensional marginal distribution equals, for each $s > 0$ and $V(s) > 0$ a.s.,

$$P(\xi(s) > x) = \frac{1}{x} \int_0^x P(V(s) > u) \, du \,, \qquad u > 0 \,.$$

Note that they depend on s and do not follow a generalized Pareto distribution (cf. Section 3.1).

9.6. Consider independent and identically distributed random vectors $(R, \Phi), (R_1, \Phi_1),$ $(R_2, \Phi_2), \ldots$, where R and Φ are independent, $P(R > r) = \exp(-r^2/2)$, and Φ has a uniform distribution over $[0, 2\pi]$. This means that $(R \cos \Phi, R \sin \Phi)$ has a standard normal distribution. Prove that $\{\vee_{i=1}^n b_n (R_i \cos(\theta/b_n - \Phi_i) - b_n)\}_\theta$ converges to $\{\vee_{i=1}^n T_i + \theta Z_i - \theta^2/2\}_\theta$, where $\{T_i\}$ is an enumeration of the points of a point process on $\mathbb{R}$ with mean measure $e^{-x} \, dx$ and Z_i independent and identically distributed random variables (Eddy and Gale (1981)).
Hint: Expand $\cos(\theta/b_n - \Phi_i)$ and proceed as in Example 9.8.2. Note that by Corollary 5.4.2, $\vee_{i=1}^\infty R_i/b_n \to^P 1$.

10

Estimation in $C[0, 1]$

10.1 Introduction: An Example

In Section 9.1 we considered the following mathematical problem: given n independent and identically distributed random functions $X, X_1, X_2, \ldots, X_n$ in $C[0, 1]$ whose distribution is in the domain of attraction of an extreme value distribution in $C[0, 1]$, estimate the probability

$$P\left(X(s) > f(s), \text{ for some } s \in [0, 1]\right),$$

where f is a given continuous function.

An essential feature of the problem is that all the observed processes X_i, $i = 1, 2, \ldots, n$, are well below f. This means that in an asymptotic setting, i.e., with $n \to \infty$, we are forced to assume that f is not constant but depends on n and moves to the tail of the distribution as $n \to \infty$. In fact, analogous to the finite-dimensional case, we assume

$$f_n(s) = U_s\left(\frac{nc_n}{k}h(s)\right), \tag{10.1.1}$$

where U_s is the inverse of $1/(1 - F_s)$ with $F_s(x) := P(X(s) \le x)$, $k = k(n) \to \infty$, $k/n \to 0$, $n \to \infty$, c_n is a sequence of positive constants, and h is a fixed (but unknown) function.

Let

$$p_n := P\left(X(s) > f_n(s), \text{ for some } s \in [0, 1]\right).$$

We show informally how to approximate p_n. As in the beginning of Section 9.5 define

$$\zeta(s) := \frac{1}{1 - F_s(X(s))}. \tag{10.1.2}$$

We write

$$p_n = P\left(X(s) > f_n(s), \text{ for some } s \in [0, 1]\right)$$
$$= P\left(\frac{1}{1 - F_s(X(s))} > \frac{1}{1 - F_s(f_n(s))} \text{ for some } s \in [0, 1]\right)$$

$$= P\left(\frac{k}{n}\zeta(s) > c_n h(s) \text{ for some } s \in [0, 1]\right),$$

and by Corollary 9.3.2, this is approximately equal to

$$\frac{k}{n}\nu\left\{g \in C^+[0, 1] : g(s) > c_n h(s) \text{ for some } s \in [0, 1]\right\},$$

which equals, by Theorem 9.3.7,

$$\frac{k}{nc_n}\nu\left\{g \in C^+[0, 1] : g(s) > h(s) \text{ for some } s \in [0, 1]\right\}.$$

For the estimation of p_n we thus need an estimator for ν as well as for c_n and the function h from (10.1.1). This involves estimation of F_s in the tail and for this we also need estimators for the index function $\gamma(s)$, the scale $a_s(n/k)$, and location $b_s(n/k)$. Our aim in Sections 10.2–10.4 is to develop estimators for those four quantities. We come back to the estimation of p_n in Section 10.5.

10.2 Estimation of the Exponent Measure: A Simple Case

For ease of exposition let us consider now the "simple" case. Suppose $\zeta_1, \zeta_2, \zeta_3, \ldots$ are independent and identically distributed stochastic processes in $C^+[0, 1]$, i.e., the processes are continuous and positive, and assume that

$$\left\{\frac{1}{n}\bigvee_{1 \le i \le n} \zeta_i(s)\right\}_{s \in [0, 1]} \xrightarrow{d} \{\eta(s)\}_{s \in [0, 1]}$$

in $C^+[0, 1]$ with $P(\eta(s) \le 1) = e^{-1}$ for $0 \le s \le 1$ (i.e., standard Fréchet). Then η is a simple max-stable process. Obviously in this case the exponent measure ν is the only unknown feature characterizing the process. We are going to develop an estimator for ν. As in the finite-dimensional case the estimator is based on a small fraction of higher observations only. Define for $k < n$ the estimator $\bar{\nu}_{n,k}$ as

$$\bar{\nu}_{n,k}(\cdot) := \frac{1}{k}\sum_{i=1}^{n} 1_{\{\zeta_i k/n \in \cdot\}}.$$

We claim that $\bar{\nu}_{n,k}$ is a consistent estimator for ν if $k = k(n) \to \infty$, $k/n \to 0$, $n \to \infty$.

Theorem 10.2.1 *Let $\zeta, \zeta_1, \zeta_2, \zeta_3, \ldots$ be i.i.d. stochastic processes in $C^+[0, 1]$. If*

$$\frac{1}{n}\bigvee_{i=1}^{n}\zeta_i \xrightarrow{d} \eta \tag{10.2.1}$$

in $C^+[0, 1]$, then for any $c > 0$ as $k = k(n) \to \infty$, $k(n)/n \to 0$, $n \to \infty$,

$$\overline{\nu}_{n,k|S_c} \overset{P}{\to} \nu_{|S_c}, \tag{10.2.2}$$

where at both sides we consider the restrictions of the measures to the set

$$S_c := \left\{ f \in \overline{C}^+[0, 1] \; : \; |f|_\infty \geq c \right\}$$

and convergence is in the space of finite measures on $\overline{C}^+[0, 1]$. The measure ν is the exponent measure of the process η (cf. Section 9.3).

Proof. According to Daley and Vere-Jones (1988), Theorem 9.1.VI, we need only to prove that the finite-dimensional marginal distributions converge, i.e., for any Borel ν-continuous sets $E_1, E_2, \ldots, E_m \subset S_c$,

$$(\overline{\nu}_{n,k}(E_1), \overline{\nu}_{n,k}(E_2), \ldots, \overline{\nu}_{n,k}(E_m)) \overset{P}{\to} (\nu(E_1), \nu(E_2), \ldots, \nu(E_m)) \,.$$

Since the limit is not random, this is equivalent to the following: for any Borel ν-continuous set $E \subset S_c$,

$$\overline{\nu}_{n,k}(E) \overset{P}{\to} \nu(E) \,.$$

Using characteristic functions, we see that this is equivalent to

$$\frac{n}{k} P\left(\frac{k}{n}\zeta \in E\right) \to \nu(E) \,,$$

which has been proved in Corollary 9.3.2. ∎

A corollary of this theorem is the uniform convergence of the marginal tail empirical distribution functions as well as the tail quantile functions. This will be useful later on.

Corollary 10.2.2 *For each s let* $\zeta_{1,n}(s) \leq \zeta_{2,n}(s) \leq \cdots \leq \zeta_{n,n}(s)$ *be the order statistics of* $\zeta_1(s), \zeta_2(s), \ldots, \zeta_n(s)$ *and define*

$$1 - G_{n,s}(x) := \frac{1}{k} \sum_{i=1}^{n} 1_{\{k\zeta_i(s)/n > x\}} \,.$$

Suppose the domain of attraction condition (10.2.1) holds. Then for any $c > 0$,

$$\sup_{0 \leq s \leq 1, x \geq c} \left| 1 - G_{n,s}(x) - \frac{1}{x} \right| \overset{P}{\to} 0 \tag{10.2.3}$$

and

$$\sup_{0 \leq s \leq 1, \, x \geq c} \left| \frac{k}{n}\zeta_{n-[kx],n}(s) - \frac{1}{x} \right| \overset{P}{\to} 0 \,. \tag{10.2.4}$$

Later on we shall also need

$$\sup_{0 \le s \le 1, \, x \le c} \left| \frac{1}{1 - G_{n,s}(x)} - x \right| \overset{P}{\to} 0 \tag{10.2.5}$$

and

$$\sup_{0 \le s \le 1, x \le c} \left| \frac{k}{n} \zeta_{n - [k/x], n}(s) - x \right| \overset{P}{\to} 0 . \tag{10.2.6}$$

Proof. Fix $c > 0$. By changing the probability space and using a Skorohod construction, we can pretend that the result of Theorem 10.2.1 holds a.s., i.e.,

$$\bar{\nu}_{n,k | S_c} \to \nu_{| S_c} \qquad \text{a.s.} \tag{10.2.7}$$

This means convergence of finite random measures. A metric characterizing this type of convergence is given in Daley and Vere-Jones (1988), A.2.5:

$$d(\nu, \mu) := \inf \left\{ \varepsilon > 0 : \nu(F) \le \mu(F^\varepsilon) + \varepsilon \text{ and } \mu(F) \le \nu(F^\varepsilon) + \varepsilon \right.$$
$$\left. \text{for all closed sets } F \in C^+[0, 1] \right\}, \tag{10.2.8}$$

where $F^\varepsilon := \left\{ f \in C^+[0, 1] : |f - g|_\infty < \varepsilon \text{ for some } g \in F \right\}$.

Take $0 < \varepsilon < c/2$ and take n so large that (from (10.2.7))

$$d \left(\bar{\nu}_{n,k | S_c}, \nu_{| S_c} \right) < \varepsilon \qquad \text{a.s.} \tag{10.2.9}$$

For $x > 0, 0 \le s \le 1$, define the closed set

$$E_{x,s} := \left\{ f \in C^+[0, 1] : f(s) \ge x \right\} .$$

Note that $E_{x,s}^\varepsilon$ is in fact the same as $E_{x - \varepsilon, s}$ and that $\nu\left(E_{x,s} \right) = 1/x$. It follows from (10.2.9) that for $x > c, 0 \le s \le 1$,

$$1 - G_{n,s}(x) = \bar{\nu}_{n,k} \left\{ f \in C^+[0, 1] : f(s) > x \right\}$$
$$\le \bar{\nu}_{n,k} \left(E_{x,s} \right) \le \nu \left(E_{x - \varepsilon, s} \right) + \varepsilon$$
$$= \frac{1}{x - \varepsilon} + \varepsilon$$

and

$$1 - G_{n,s}(x) \ge \bar{\nu}_{n,k} \left(E_{x + \varepsilon, s} \right) \ge \nu \left(E_{x + 2\varepsilon, s} \right) - \varepsilon = \frac{1}{x + 2\varepsilon} - \varepsilon .$$

This proves that

$$\sup_{0 \le s \le 1, x \ge c} \left| 1 - G_{n,s}(x) - \frac{1}{x} \right| \to 0 \qquad \text{a.s.}$$

as $n \to \infty$, hence we have convergence in probability, i.e., (10.2.3). ∎

Now clearly from

$$\sup_{0\le s\le 1,\, c\le x\le b}\left|1 - G_{n,s}(x) - \frac{1}{x}\right| \to 0 \qquad \text{a.s.} \qquad (10.2.10)$$

it follows, since $k/n\zeta_{n-[kx],n}(s)$ is the inverse function of $1 - G_{n,s}(x)$, that for $0 < c < b < \infty$,

$$\sup_{0\le s\le 1,\, c\le x\le b}\left|\frac{k}{n}\zeta_{n-[kx],n}(s) - \frac{1}{x}\right| \to 0 \qquad \text{a.s.} \qquad (10.2.11)$$

and hence by monotonicity

$$\sup_{0\le s\le 1,\, x\ge c}\left|\frac{k}{n}\zeta_{n-[kx],n}(s) - \frac{1}{x}\right| \to 0 \qquad \text{a.s.}$$

This proves (10.2.4) and (10.2.6). Next from (10.2.10) we get

$$\sup_{0\le s\le 1,\, c\le x\le b}\left|\frac{1}{1 - G_{n,s}(x)} - x\right| \to 0 \qquad \text{a.s.}$$

and by monotonicity (in x) we obtain (10.2.5).

10.3 Estimation of the Exponent Measure

Next we look at the general case. Let $X, X_1, X_2, \ldots$ be independent and identically distributed stochastic processes in $C[0, 1]$ with continuous marginal distribution functions and suppose that there are continuous functions $a_s(n) > 0$ and $b_s(n)$ such that

$$\left\{\max_{i\le n}\frac{X_i(s) - b_s(n)}{a_s(n)}\right\}_{s\in[0,1]} \overset{d}{\to} \{Y(s)\}_{s\in[0,1]} \qquad (10.3.1)$$

in $C[0, 1]$. From (10.3.1) we get that (cf. Theorem 9.2.1)

$$\left\{\max_{i\le n}\frac{1}{n\{1 - F_s(X_i(s))\}}\right\}_{s\in[0,1]} \overset{d}{\to} \left\{(1 + \gamma(s)Y(s))^{1/\gamma(s)}\right\}_{s\in[0,1]}, \qquad (10.3.2)$$

i.e., (10.2.1) holds with

$$\zeta_i(s) := \frac{1}{1 - F_s(X_i(s))}$$

and $\eta(s) := (1 + \gamma(s)Y(s))^{1/\gamma(s)}, 0 \le s \le 1$. Hence, according to the results of the previous section, we would be inclined to define as an estimator for ν the quantity

$$\hat{\nu}'_{n,k}(A) := \frac{1}{k}\sum_{i=1}^{n} 1_{\{k/(n\{1-F_\cdot(X_i(\cdot))\})\in A\}},$$

where $A \subset \overline{C}^+[0, 1]$. This is consistent for v. However, this is not a statistic since $1 - F_s$ is unknown. Hence we replace $1 - F_s$ by its empirical counterpart

$$1 - \hat{F}_{n,s}(x) := \frac{1}{n} \sum_{j=1}^{n} 1_{\{X_i(s) > x\}} .$$

This leads to the estimator

$$\hat{v}_{n,k}(\cdot) := \frac{1}{k} \sum_{i=1}^{n} 1_{\{k\hat{\zeta}_i/n \in \cdot\}} \tag{10.3.3}$$

with

$$\hat{\zeta}_i(s) := \frac{1}{1 - \hat{F}_{n,s}(X_i(s))} = \frac{1}{n^{-1} \sum_{j=1}^{n} 1_{\{X_j(s) > X_i(s)\}}} . \tag{10.3.4}$$

We know by Theorem 9.2.1 that for these processes (10.2.1) holds with $\eta(s) := (1 + \gamma(s)Y(s))^{1/\gamma(s)}$, $s \in [0, 1]$. This leads to a simpler way of writing (10.3.4):

$$\frac{k}{n}\hat{\zeta}_i(s) := \frac{1}{k^{-1} \sum_{j=1}^{n} 1_{\{\zeta_j(s) > \zeta_i(s)\}}} = \frac{1}{1 - G_{n,s}(k\zeta_i(s)/n)} \tag{10.3.5}$$

with $G_{n,s}$ as in Corollary 10.2.2. Hence we can analyze this estimator using the results of Section 10.1.

Theorem 10.3.1 *Let $X, X_1, X_2, \ldots$ be i.i.d. stochastic processes in $C[0, 1]$ and assume that their distribution is in the domain of attraction of a max-stable process in $C[0, 1]$, i.e., (10.3.1) holds.*
    Let

$$1 - \hat{F}_{n,s}(x) := \frac{1}{n} \sum_{j=1}^{n} 1_{\{X_j(s) > x\}}$$

for $n = 1, 2, \ldots, 0 \leq s \leq 1$. Define

$$\hat{\zeta}_i(s) := \frac{1}{1 - \hat{F}_{n,s}(X_i(s))}$$

and

$$\hat{v}_{n,k}(\cdot) := \frac{1}{k} \sum_{i=1}^{n} 1_{\{k\hat{\zeta}_i/n \in \cdot\}} . \tag{10.3.6}$$

Then, as $k \to \infty$, $k/n \to 0$, $n \to \infty$, for all $c > 0$,

$$\hat{v}_{n,k|S_c} \xrightarrow{P} v_{|S_c} \tag{10.3.7}$$

in the space of finite measures on $\overline{C}^+[0, 1]$, where on both sides we consider the restriction of the measure to the set

$$S_c := \left\{ f \in \overline{C}^+[0, 1] : |f|_\infty \geq c \right\} .$$

Proof. We have only to prove that for a ν-continuous Borel set $E \subset S_c$,

$$\hat{\nu}_{n,k}(E) \to \nu(E)$$

in probability (cf. proof of Theorem 10.2.1). Write

$$E = \left(E \cap S_{[c,b]}\right)\bigcup (E \cap S_b) =: E_1 \cup E_2$$

with

$$S_{[c,b]} := \left\{f \in \overline{C}^{+}[0, 1] \ : \ c \leq |f|_\infty \leq b\right\} .$$

Let $k\zeta/n \in E_1$. Then $k\zeta(s)/n \leq b$ for $0 \leq s \leq 1$, hence by (10.2.5) of Corollary 10.2.2 for sufficiently large n,

$$\frac{k}{n}\zeta(s) - \varepsilon < \frac{1}{1 - G_{n,s}\left(k\zeta(s)/n\right)} < \frac{k}{n}\zeta(s) + \varepsilon ,$$

$0 \leq s \leq 1$; hence the function $\left(1 - G_{n,s}\left(k\zeta(s)/n\right)\right)^{-1}$ is in

$$E_1^\varepsilon := \left\{f \in \overline{C}^{+}[0, 1] \ : \ |f - g|_\infty < \varepsilon \text{ for some } g \in E_1\right\} .$$

It follows that

$$\overline{\nu}_{n,k}(E_1) \leq \hat{\nu}_{n,k}(E_1^\varepsilon) .$$

Similarly one can prove, now using (10.2.6), that

$$\hat{\nu}_{n,k}(E_1) \leq \overline{\nu}_{n,k}(E_1^\varepsilon) .$$

We already know by Theorem 10.2.1 that $\overline{\nu}_{n,k}(E_1) \to^P \nu(E_1)$ and $\overline{\nu}_{n,k}(E_1^\varepsilon) \to^P \nu(E_1^\varepsilon)$. Also by the ν-continuous property of E_1 we have $\nu(E_1^\varepsilon) \to \nu(E_1)$ as $\varepsilon \downarrow 0$. This proves

$$\hat{\nu}_{n,k}\left(E \cap S_{[c,b]}\right) \xrightarrow{P} \nu\left(E \cap S_{[c,b]}\right) \tag{10.3.8}$$

for $0 < c < b < \infty$. Next we consider $\hat{\nu}_{n,k}(E_2)$. According to (10.3.5),

$$\hat{\nu}_{n,k}(E_1) = \frac{1}{k}\sum_{i=1}^{n} 1_{\left\{k\hat{\zeta}_i/n \in E_2\right\}}$$

$$= \frac{1}{k}\sum_{i=1}^{n} 1_{\left\{\left(1 - G_{n,\cdot}(k\zeta_i(\cdot)/n)\right)^{-1} \in E_2\right\}}$$

$$= \frac{1}{k}\sum_{i=1}^{n} 1_{\left\{k\zeta_i(\cdot)/n = (1/(1 - G_{n,\cdot}))^{\leftarrow} g(\cdot) \text{ with } g \in E_2\right\}}$$

$$= \overline{\nu}_{n,k}\left\{f : f(\cdot) = \left(\frac{1}{1 - G_{n,\cdot}}\right)^{\leftarrow} g(\cdot) \text{ with } g \in E_2\right\} .$$

Hence

$$\hat{v}_{n,k}(E_2) \le \hat{v}(S_b) = \overline{v}_{n,k} \left\{ f : f(\cdot) = \left(\frac{1}{1 - G_{n,\cdot}} \right)^{\leftarrow} g(\cdot), \text{ with } |g|_\infty > b \right\} .$$

Since $g \in S_b$ we have $g(s_0) \ge b$ for some $s_0 \in [0, 1]$. Hence

$$\left(\frac{1}{1 - G_{n,s_0}} \right)^{\leftarrow} (g(s_0)) \ge \left(\frac{1}{1 - G_{n,s_0}} \right)^{\leftarrow} (b) = \zeta_{n-[k/b],n}(s_0) \overset{P}{\to} b$$

by (10.2.6). Hence

$$P \left\{ \sup_{s \in [0,1]} \left(\frac{1}{1 - G_{n,s}} \right)^{\leftarrow} (g(s)) > b - \varepsilon \right\} \to 1, \quad n \to \infty ,$$

i.e., with probability tending to 1,

$$\hat{v}(E_2) \le \overline{v}_{n,k} (S_{b-\varepsilon}) ,$$

and by (10.2.2) the right-hand side tends to $v (S_{b-\varepsilon})$, which by Theorem 9.3.7 equals $C/(b - \varepsilon)$ for some positive constant C. By choosing b large enough this can be made smaller then ε. Then as $n \to \infty$,

$$P(\hat{v}_{n,k} (E \cap S_b) > \varepsilon) \to 0 . \tag{10.3.9}$$

The proof is completed by combining (10.3.8) and (10.3.9).   ∎

10.4 Estimation of the Index Function, Scale and Location

Recall the domain of attraction condition. Let $X_1, X_2, \ldots$ be independent and identically distributed stochastic processes in $C[0, 1]$. Suppose that there are continuous functions $a_s (n) > 0$ and $b_s (n)$ such that

$$\left\{ \max_{i \le n} \frac{X_i(t) - b_s(n)}{a_s(n)} \right\}_{s \in [0,1]} \overset{d}{\to} \{Y(s)\}_{s \in [0,1]} \tag{10.4.1}$$

in $C[0, 1]$. Then Y is a random element in $C[0, 1]$ satisfying (for a judicious choice of scale function a and location function b)

$$P(Y(s) \le x) = \exp \left\{ -(1 + \gamma(s)x)^{-1/\gamma(s)} \right\} \tag{10.4.2}$$

for each $s \in [0, 1]$, where $\gamma \in C[0, 1]$. The function γ is called the *index function*. In this section we develop estimators for γ, the scale and the location. The estimators will be based on the moment estimator of Section 3.5, but similar results should hold, for example, for the maximum likelihood estimator.

Now, since in applications one does not need estimators of scale and location for extreme order statistics, but rather for intermediate ones such as those of Section 2.4, we shall specify what we want to estimate.

Define for $t > 0, 0 \leq s \leq 1$,

$$
\begin{aligned}
a_s(t) &:= a_s([t]), \\
b_s(t) &:= b_s([t]) .
\end{aligned}
\tag{10.4.3}
$$

In (10.4.1) we used those with $t = n$. But as in the finite-dimensional situation (cf. Section 4.2 and Chapter 8) we need estimators of the quantities in (10.4.3) with $t = n/k$, where $k = k(n) \to \infty$, $k/n \to 0$, as $n \to \infty$, i.e., we need estimators for $a_s(n/k)$ and $b_s(n/k)$. In fact, by Theorem 1.1.2 we can replace $b_s(n/k)$ by $U_s(n/k)$ with, for $x > 1$,

$$
U_s(x) := F_s^{\leftarrow}\left(1 - \frac{1}{x}\right),
\tag{10.4.4}
$$

where $F_s(x) := P(X(s) \leq x)$.

Finally, in order to construct the moment estimator we need positive random variables; hence we assume

$$
\inf_{0 \leq s \leq 1} U_s(\infty) > 0,
\tag{10.4.5}
$$

which can be achieved by a shift. Next we introduce the estimators. They are simple extensions of the ones used in the finite-dimensional case. Define the sample functions

$$
M_n^{(j)}(s) := \frac{1}{k} \sum_{i=0}^{k-1} \left(\log X_{n-i,n}(s) - \log X_{n-k,n}(s)\right)^j ,
\tag{10.4.6}
$$

$j = 1, 2$, where $X_{1,n}(s) \leq X_{2,n}(s) \leq \cdots \leq X_{n,n}(s)$ are the order statistics of $X_1(s), X_2(s), \ldots, X_n(s)$.

Next define

$$
\hat{\gamma}_+(s) := M_n^{(1)}(s),
\tag{10.4.7}
$$

$$
\hat{\gamma}_-(s) := 1 - \frac{1}{2}\left\{1 - \frac{\left(M_n^{(1)}(s)\right)^2}{M_n^{(2)}(s)}\right\}^{-1},
\tag{10.4.8}
$$

$$
\hat{\gamma}(s) := \hat{\gamma}_+(s) + \hat{\gamma}_-(s),
\tag{10.4.9}
$$

$$
\hat{a}_s(n/k) := X_{n-k,n}(s)\hat{\gamma}_+(s)\left(1 - \hat{\gamma}_-(s)\right),
\tag{10.4.10}
$$

$$
\hat{b}_s(n/k) := X_{n-k,n}(s).
\tag{10.4.11}
$$

10.4.1 Consistency

We have the following consistency result.

Theorem 10.4.1 *Let $X_1, X_2, \ldots$ be i.i.d. stochastic processes in $C[0, 1]$ and assume that their distribution is in the domain of attraction of a max-stable process in $C[0, 1]$, i.e., (10.4.1) holds. If $k = k(n) \to \infty$, $k/n \to 0$, $n \to \infty$, then*

$$\sup_{0 \le s \le 1} |\hat{\gamma}_+(s) - \gamma_+(s)| \overset{P}{\to} 0 \; \text{with} \; \gamma_+(s) := \gamma(s) \vee 0 , \qquad (10.4.12)$$

$$\sup_{0 \le s \le 1} |\hat{\gamma}_-(s) - \gamma_-(s)| \overset{P}{\to} 0 \; \text{with} \; \gamma_-(s) := \gamma(s) \wedge 0 , \qquad (10.4.13)$$

$$\sup_{0 \le s \le 1} |\hat{\gamma}(s) - \gamma(s)| \overset{P}{\to} 0 , \qquad (10.4.14)$$

$$\sup_{0 \le s \le 1} \left| \frac{\hat{a}_s(n/k)}{a_s(n/k)} - 1 \right| \overset{P}{\to} 0 , \qquad (10.4.15)$$

$$\sup_{0 \le s \le 1} \left| \frac{\hat{b}_s(n/k) - U_s(n/k)}{a_s(n/k)} \right| \overset{P}{\to} 0 . \qquad (10.4.16)$$

For the proof of Theorem 10.4.1 we need two technical lemmas. The first one has been taken from Appendix B.

Lemma 10.4.2 *Suppose that the functions* $\log a_s(t)$ *and* $g_s(t) > 0$ *are locally bounded in* $0 \le s \le 1$, $0 < t < \infty$, *and for some* $\gamma \in C[0, 1]$ *and all* $x > 0$,

$$\lim_{t \to \infty} \frac{g_s(tx) - g_s(t)}{a_s(t)} = \frac{x^{\gamma(s)} - 1}{\gamma(s)} \qquad (10.4.17)$$

uniformly for $0 \le s \le 1$. *Then for* $x > 0$,

$$\lim_{t \to \infty} \frac{a_s(tx)}{a_s(t)} = x^{\gamma(s)} \qquad (10.4.18)$$

uniformly for $0 \le s \le 1$, *and for any* $\varepsilon > 0$ *there exists* $t_0 > 0$ *such that for* $t \ge t_0$, $tx \ge t_0$,

$$\left| \frac{a_s(tx)}{a_s(t)} - x^{\gamma(s)} \right| < \varepsilon x^{\gamma(s)} \max \left(x^\varepsilon, x^{-\varepsilon} \right) , \qquad (10.4.19)$$

or alternatively,

$$(1 - \varepsilon)x^{\gamma(s)} \min \left(x^\varepsilon, x^{-\varepsilon} \right) < \frac{a_s(tx)}{a_s(t)} < (1 + \varepsilon)x^{\gamma(s)} \max \left(x^\varepsilon, x^{-\varepsilon} \right) \qquad (10.4.20)$$

and

$$\left| \frac{g_s(tx) - g_s(t)}{g_s(t)} - \frac{x^{\gamma(s)} - 1}{\gamma(s)} \right| < \varepsilon x^{\gamma(s)} \max \left(x^\varepsilon, x^{-\varepsilon} \right) . \qquad (10.4.21)$$

Further,

$$\lim_{t \to \infty} \frac{a_s(t)}{g_s(t)} = \gamma_+(s) \qquad (10.4.22)$$

uniformly for $0 \le s \le 1$ *with* $\gamma_+(s) := \max(\gamma(s), 0)$, *and for any* $\varepsilon > 0$ *there exists* t_0 *such that for* $t, tx \ge t_0$,

$$\left| \frac{\log g_s(tx) - \log g_s(t)}{a_s(t)/g_s(t)} - \frac{x^{\gamma_-(s)} - 1}{\gamma_-(s)} \right| < \varepsilon x^{\gamma_-(s)} \max\left(x^\varepsilon, x^{-\varepsilon}\right) \tag{10.4.23}$$

with $\gamma_-(s) := \min(\gamma(s), 0)$.

The second lemma is probabilistic in nature.

Lemma 10.4.3 *Let* $\zeta_1, \zeta_2, \ldots$ *be i.i.d. stochastic processes in* $C^+[0, 1]$ *and suppose*

$$\left\{ \frac{1}{n} \bigvee_{i=1}^{n} \zeta_i(s) \right\}_{s \in [0,1]} \xrightarrow{d} \{\eta(s)\}_{s \in [0,1]} \tag{10.4.24}$$

in $C^+[0, 1]$. *Let* $\zeta_{1,n}(s) \leq \zeta_{2,n}(s) \leq \cdots \leq \zeta_{n,n}(s)$ *be the order statistics of* $\zeta_1(s), \zeta_2(s), \ldots, \zeta_n(s)$. *Also let* μ *and* λ *be continuous functions defined on* $[0, 1]$ *with* $\mu < 1$, $\lambda < 1$, $\mu + \lambda < 1$. *Then*

$$\sup_{0 \leq s \leq 1} \left| \frac{1}{k} \sum_{i=0}^{k-1} \frac{\left(\zeta_{n-i,n}(s)/\zeta_{n-k,n}(s)\right)^{\mu(s)} - 1}{\mu(s)} - \frac{1}{1 - \mu(s)} \right| \xrightarrow{P} 0 \tag{10.4.25}$$

and

$$\sup_{0 \leq s \leq 1} \left| \frac{1}{k} \sum_{i=0}^{k-1} \frac{\left(\zeta_{n-i,n}(s)/\zeta_{n-k,n}(s)\right)^{\mu(s)} - 1}{\mu(s)} \frac{\left(\zeta_{n-i,n}(s)/\zeta_{n-k,n}(s)\right)^{\lambda(s)} - 1}{\lambda(s)} \right.$$
$$\left. - \frac{2 - \mu(s) - \lambda(s)}{(1 - \mu(s) - \lambda(s))(1 - \mu(s))(1 - \lambda(s))} \right| \xrightarrow{P} 0. \tag{10.4.26}$$

Proof. We shall prove (10.4.25). The proof of (10.4.26) is similar. Observe that

$$\frac{1}{k} \sum_{i=0}^{k-1} \frac{\left(\zeta_{n-i,n}(s)/\zeta_{n-k,n}(s)\right)^{\mu(s)} - 1}{\mu(s)}$$

$$= \left(\frac{n}{k} \frac{1}{\zeta_{n-k,n}(s)}\right)^{\mu(s)} \frac{1}{k} \sum_{i=0}^{k-1} \frac{\left(k\zeta_{n-i,n}(s)/n\right)^{\mu(s)} - \left(k\zeta_{n-k,n}(s)/n\right)^{\mu(s)}}{\mu(s)}$$

$$= \left(\frac{n}{k} \frac{1}{\zeta_{n-k,n}(s)}\right)^{\mu(s)} \int_{(k/n)\zeta_{n-k,n}(s)}^{\infty} \left(\int_{(k/n)\zeta_{n-k,n}(s)}^{x} y^{\mu(s)-1} dy\right) dG_{n,s}(x)$$

$$= \left(\frac{n}{k} \frac{1}{\zeta_{n-k,n}(s)}\right)^{\mu(s)} \int_{(k/n)\zeta_{n-k+1,n}(s)}^{\infty} \left(1 - G_{n,s}(x)\right) x^{\mu(s)-1} dx$$

$$= \left(\frac{n}{k} \frac{1}{\zeta_{n-k,n}(s)}\right)^{\mu(s)} \int_{1}^{\infty} \left(1 - G_{n,s}(x)\right) x^{\mu(s)-1} dx$$

$$+ \left(\frac{n}{k} \frac{1}{\zeta_{n-k,n}(s)}\right)^{\mu(s)} \int_{(k/n)\zeta_{n-k,n}(s)}^{1} \left(1 - G_{n,s}(x)\right) x^{\mu(s)-1} dx.$$

By Corollary 10.2.2 ((10.2.3) and (10.2.4)) the second part converges to 0. So we need only to prove

$$\sup_{0 \le s \le 1} \left| \int_1^\infty \left(1 - G_{n,s}(x)\right) x^{\mu(s)-1} dx - \frac{1}{1 - \mu(s)} \right| \xrightarrow{P} 0 . \tag{10.4.27}$$

Let $Y_i := \sup_{s \in [0,1]} \zeta_i(s)$, $i = 1, 2, \ldots, n$. These are independent and identically distributed random variables. By the continuous mapping theorem,

$$\bigvee_{i=1}^n \frac{1}{n} Y_i = \bigvee_{i=1}^n \frac{1}{n} \sup_{s \in [0,1]} \zeta_i(s) = \sup_{s \in [0,1]} \bigvee_{i=1}^n \frac{1}{n} \zeta_i(s) \to \sup_{s \in [0,1]} \eta(s) =: Y$$

in distribution, where by Lemma 9.3.4,

$$P(Y \le x) = \exp\left(-\left(\frac{c}{x}\right)\right)$$

for some $c > 0$.

Let $Y_{n-i,n}$ be the order statistics of Y_i, $i = 1, 2, \ldots, n$, and

$$1 - F_n(x) := \frac{1}{k} \sum_{i=1}^n 1_{\{Y_{i,n} > xn/k\}} .$$

We have for $0 \le s \le 1$,

$$1 - G_{n,s}(x) \le 1 - F_n(x).$$

Hence by one-dimensional results,

$$\int_1^\infty \left(1 - G_{n,s}(x)\right) x^{\mu(s)-1} dx \le \int_1^\infty \left(1 - F_n(x)\right) x^{\mu(s)-1} dx$$

$$\to \int_1^\infty \frac{c}{x} x^{\mu(s)-1} dx = \frac{c}{1 - \mu(s)} .$$

Hence by Pratt's lemma (summarizing, if $g_n \to g$ pointwise, $|g_n| \le f_n$ for all n and $\int f_n \to \int f$, for some functions f_n, g_n, f, g, then $\int g_n \to \int g$; Pratt (1960))

$$\int_1^\infty \left(1 - G_{n,s}(x)\right) x^{\mu(s)-1} dx \to \int_1^\infty x^{\mu(s)-2} dx = \frac{1}{1 - \mu(s)}$$

uniformly in s. Hence we have proved (10.4.25).   ∎

Proof (of Theorem 10.4.1). Define for $0 \le s \le 1$ and $i = 1, 2, \ldots$,

$$\zeta_i(s) := \frac{1}{1 - F_s(X_i(s))} \tag{10.4.28}$$

as in Section 10.3. Then, according to Theorem 9.2.1, for $\zeta_1(s), \zeta_2(s), \ldots$ the results of Section 10.2 hold (the "simple" case).

We first prove (10.4.16). Note that by (10.4.28)

$$X_i(s) = U_s(\zeta_i(s)) \; ;$$

hence

$$\frac{\hat{b}_s\left(\frac{n}{k}\right) - U_s\left(\frac{n}{k}\right)}{a_s\left(\frac{n}{k}\right)} = \frac{X_{n-k,n}(s) - U_s\left(\frac{n}{k}\right)}{a_s\left(\frac{n}{k}\right)} = \frac{U_s\left(\frac{n}{k}\left\{\frac{k}{n}\zeta_{n-k,n}(s)\right\}\right) - U_s\left(\frac{n}{k}\right)}{a_s\left(\frac{n}{k}\right)},$$

which by Corollary 10.2.2 (10.2.4) and Theorem 9.2.1 (note that as in Theorem 1.1.2 one sees that (9.2.6) holds with n replaced by t running through the reals) converges to zero, in probability and uniformly in s. For the proof of the other statements of the theorem we start with the following:

$$\frac{M_n^{(1)}}{a_s(\zeta_{n-k,n}(s))/U_s(\zeta_{n-k,n}(s))}$$

$$= \frac{1}{k}\sum_{i=0}^{k-1}\frac{\log X_{n-i,n}(s) - \log X_{n-k,n}(s)}{a_s(\zeta_{n-k,n}(s))/U_s(\zeta_{n-k,n}(s))}$$

$$= \frac{1}{k}\sum_{i=0}^{k-1}\frac{\log U_s\left(\zeta_{n-k,n}(s)\left\{\zeta_{n-i,n}(s)/\zeta_{n-k,n}(s)\right\}\right) - \log U_s\left(\zeta_{n-k,n}(s)\right)}{a_s(\zeta_{n-k,n}(s))/U_s(\zeta_{n-k,n}(s))} \; .$$

Since $\zeta_{n-k,n}(s) \to \infty$ in probability, as $n \to \infty$, uniformly in s by Corollary 10.2.2, we have with high probability by Lemma 10.4.2,

$$\left| \frac{M_n^{(1)}(s)}{a_s(\zeta_{n-k,n}(s))/U_s(\zeta_{n-k,n}(s))} - \frac{1}{k}\sum_{i=0}^{k-1}\frac{(\zeta_{n-i,n}(s)/\zeta_{n-k,n}(s))^{\gamma_-(s)} - 1}{\gamma_-(s)} \right|$$

$$< \varepsilon\frac{1}{k}\sum_{i=0}^{k-1}\left(\zeta_{n-i,n}(s)/\zeta_{n-k,n}(s)\right)^{\gamma_-(s)+\varepsilon} \; .$$

Upon applying Lemma 10.4.3 we then get

$$\sup_{0\le s\le 1}\left| \frac{\hat{\gamma}_+(s)}{a_s(\zeta_{n-k,n}(s))/U_s(\zeta_{n-k,n}(s))} - \frac{1}{1-\gamma_-(s)} \right| \overset{P}{\to} 0 \; . \tag{10.4.29}$$

Similarly we get

$$\sup_{0\le s\le 1}\left| \frac{M_n^{(2)}(s)}{\left(a_s(\zeta_{n-k,n}(s))/U_s(\zeta_{n-k,n}(s))\right)^2} - \frac{2}{(1-\gamma_-(s))(1-2\gamma_-(s))} \right| \overset{P}{\to} 0$$

$$\tag{10.4.30}$$

and, using (10.4.22) of Lemma 10.4.2,

$$\sup_{0\le s\le 1}\left| \frac{a_s(\zeta_{n-k,n}(s))}{U_s(\zeta_{n-k,n}(s))} - \gamma_+(s) \right| \to 0 \; . \tag{10.4.31}$$

Combining (10.4.29) and (10.4.31) gives (10.4.12). Combining (10.4.29), (10.4.30), and (10.4.31) gives (10.4.13). Then (10.4.14) follows. For (10.4.15) note that

$$\frac{\hat{a}_s\left(\frac{n}{k}\right)}{a_s\left(\frac{n}{k}\right)} = \frac{a_s\left(\frac{n}{k}\left\{\frac{k}{n}\zeta_{n-k,n}(s)\right\}\right)}{a_s\left(\frac{n}{k}\right)}\frac{M_n^{(1)}(s)(1-\hat{\gamma}_-(s))}{a_s(\zeta_{n-k,n}(s))/U_s(\zeta_{n-k,n}(s))}$$

and $a_s(tx)/a_s(t) \to x^{\gamma(s)}$ uniformly by Theorem 9.2.1 and Lemma 10.4.2, relation (10.4.18). ∎

10.4.2 Asymptotic Normality

Next we discuss the asymptotic normality of the estimators (10.4.7)–(10.4.11) in the appropriate function space. Recall that for the consistency, the uniform convergence result for the "simple" tail empirical distribution function was essential (cf. Corollary 10.2.2). Now we start with the corresponding asymptotic normality in the "simple" case, which is basic for the later results. The proof of the result is rather technical and follows from an entropy with bracketing central limit theorem for empirical processes in van der Vaart and Wellner (1996) (Einmahl and Lin (2006)).

Theorem 10.4.4 *Let* $\zeta, \zeta_1, \zeta_2, \zeta_3, \ldots$ *be i.i.d. stochastic processes in* $C^+[0, 1]$. *Suppose*

$$\frac{1}{n}\bigvee_{i=1}^{n} \zeta_i \xrightarrow{d} \eta$$

in $C^+[0, 1]$. *Assume also that the following smoothness condition holds: for all* $0 \leq \beta < \frac{1}{2}$ *and* $c > 0$ *there exists* $K > 0$ *and for large enough* v *there exists* $\delta_0 > 0$ *such that for all* $\delta \in [0, \delta_0]$,

$$\sup_{0 \leq s \leq 1} P\left\{\zeta \notin E_{s,\delta} \middle| \sup_{s \leq u \leq s+\delta} \zeta(u) \geq v\right\} \leq c(-\log\delta)^{-(2+2\beta)/(1-2\beta)}, \quad (10.4.32)$$

where

$$E_{s,\delta} = \left\{h \in C[0, 1] : h \geq 0, \frac{|h(u) - h(s)|}{h(s)} \leq K(-\log\delta)^{-3}, \right.$$

$$\left. \text{for all } u \in [s, s+\delta]\right\}. \quad (10.4.33)$$

Then for a special construction,

$$\sup_{0 \leq s \leq 1, x \geq c} x^{\beta}\left|\sqrt{k}\left((1 - G_{n,s}(x)) - \frac{1}{x}\right) - W(C_{s,x})\right| \xrightarrow{P} 0$$

as $n \to \infty$, *where* W *is a zero-mean Gaussian process indexed by sets* $C_{s,x}$ *defined for* $0 \leq s \leq 1$ *and* $x > 0$ *by*

$$C_{s,x} := \{h \in C[0, 1] : h(s) \geq x\}$$

and, with $0 \leq u \leq 1$,

$$EW(C_{u,x})W(C_{s,y}) = v(C_{u,x} \cap C_{s,y}),$$

where v *is the exponent measure of Section 10.3.*

Remark 10.4.5 Note that $v\left(C_{s,1/y_1} \cap C_{s,1/y_2}\right) = v\left(C_{s_1,1/y_1 \wedge y_2}\right) = y_1 \wedge y_2$. Hence for fixed s the process $W(C_{s,1/x})$ is a standard Wiener process.

We also need the following result on the inverse empirical distribution function.

Corollary 10.4.6 *Under the conditions of Theorem 10.4.4, for any function* $\alpha \in C[0, 1]$, *with a special construction,*

$$\sup_{0 \leq s \leq 1} \left| \sqrt{k} \left(\left(\frac{k}{n} \zeta_{n-k,n}(s) \right)^{\alpha(s)} - 1 \right) - \alpha(s) W(C_{s,1}) \right| \xrightarrow{P} 0$$

as $n \to \infty$.

The asymptotic normality for our estimators is as follows.

Theorem 10.4.7 *Let $X_1, X_2, X_3, \ldots$ be i.i.d. stochastic processes in $C[0, 1]$. Assume (10.4.1)–(10.4.11). Moreover, adopt assumption (10.4.32) of Theorem 10.4.4 with $\zeta(s) := \{1 - F_s(X(s))\}^{-1}$. Finally, we need a uniform second-order condition: for some positive or negative function $A_s(t)$ defined for $0 \leq s \leq 1$ and $t > 0$ and satisfying $\sup_{0 \leq s \leq 1} A_s(t) \to 0$ as $t \to \infty$,*

$$\frac{\frac{\log U_s(xt) - \log U_s(t)}{a_s(t)/U_s(t)} - \frac{x^{\gamma_-(s)} - 1}{\gamma_-(s)}}{A_s(t)} \to H_{\gamma_-(s),\rho(s)}(x) \tag{10.4.34}$$

uniformly for $s \in [0, 1]$ with $\rho(s) \in C[0, 1]$, $\rho(s) \leq 0$, $0 \leq s \leq 1$, and

$$H_{\gamma_-(s),\rho(s)}(x) = \int_1^x y^{\gamma_-(s)-1} \int_1^y u^{\rho(s)-1} du \, dy .$$

If, as $n \to \infty$,

$$\sqrt{k} \sup_{0 \leq s \leq 1} \left| A_s\left(\frac{n}{k}\right) \right| \to 0 \tag{10.4.35}$$

and

$$\sqrt{k} \sup_{0 \leq s \leq 1} \left| \frac{a_s\left(\frac{n}{k}\right)}{U_s\left(\frac{n}{k}\right)} - \gamma_+(s) \right| \to 0 , \tag{10.4.36}$$

then we have

$$\sup_{0 \leq s \leq 1} \left| \sqrt{k}\left(\hat{\gamma}_+(s) - \gamma_+(s)\right) - \gamma_+(s)\mathcal{P}(s) \right| \to 0 , \tag{10.4.37}$$

$$\sup_{0 \leq s \leq 1} \left| \sqrt{k}\left(\hat{\gamma}(s) - \gamma(s)\right) - \mathcal{G}(s) \right| \xrightarrow{P} 0 , \tag{10.4.38}$$

$$\sup_{0 \leq s \leq 1} \left| \sqrt{k}\left(\frac{\hat{b}_s\left(\frac{n}{k}\right) - U_s\left(\frac{n}{k}\right)}{a_s\left(\frac{n}{k}\right)}\right) - \mathcal{U}(s) \right| \xrightarrow{P} 0 , \tag{10.4.39}$$

$$\sup_{0 \leq s \leq 1} \left| \sqrt{k}\left(\frac{\hat{a}_s\left(\frac{n}{k}\right)}{a_s\left(\frac{n}{k}\right)} - 1\right) - \mathcal{A}(s) \right| \xrightarrow{P} 0 , \tag{10.4.40}$$

where $\mathcal{P}, \mathcal{G}, \mathcal{U},$ and $\mathcal{A}$ are defined in terms of the process W as follows:

$$\mathcal{P}(s) = \int_1^\infty W(C_{s,x}) \frac{dx}{x^{1-\gamma_-(s)}} - \frac{1}{1-\gamma_-(s)} W(C_{s,1}) ,$$

$$\mathcal{Q}(s) = 2 \int_1^\infty W(C_{s,x}) \frac{x^{\gamma_-(s)} - 1}{\gamma_-(s)} \frac{dx}{x^{1-\gamma_-(s)}}$$
$$-2((1-\gamma_-(s))(1-2\gamma_-(s)))^{-1} W(C_{s,1}) ,$$

$$\mathcal{G}(s) = \left\{ \gamma_+(s) - 2(1-\gamma_-(s))^2((1-2\gamma_-(s)) \right\} \mathcal{P}(s)$$
$$+ \frac{1}{2}(1-\gamma_-(s))^2(1-2\gamma_-(s))^2 \mathcal{Q}(s) ,$$

$$\mathcal{U}(s) = W(C_{s,1}),$$

$$\mathcal{A}(s) = \gamma(s)W(C_{s,1}) + (3 - 4\gamma_-(s))(1-\gamma_-(s))\mathcal{P}(s)$$
$$- \frac{1}{2}(1-\gamma_-(s))(1-2\gamma_-(s))^2 \mathcal{Q}(s) ,$$

$s \in [0, 1]$.

Proof. The proof is somewhat similar to that of Theorem 10.4.1. We sketch the line of reasoning. Condition (10.4.34) implies that for any $\varepsilon > 0$ there exists t_0 such that for $t \geq t_0, x \geq 1,$ and $0 \leq s \leq 1,$

$$\left| \frac{\frac{\log U_s(tx) - \log U_s(t)}{a_s(t)/U_s(t)} - \frac{x^{\gamma_-(s)} - 1}{\gamma_-(s)}}{A_s(t)} - H_{\gamma_-(s),\rho(s)}(x) \right| \leq \varepsilon \left(1 + x^{\gamma_-(s)+\varepsilon} \right) . \quad (10.4.41)$$

We start with (10.4.39) and write

$$\sqrt{k} \, \frac{\hat{b}_s\left(\frac{n}{k}\right) - U_s\left(\frac{n}{k}\right)}{a_s\left(\frac{n}{k}\right)}$$

$$= \sqrt{k} \left(\frac{\hat{b}_s\left(\frac{n}{k}\right)}{U_s\left(\frac{n}{k}\right)} - 1 \right) \frac{\log \hat{b}_s\left(\frac{n}{k}\right) - \log U_s\left(\frac{n}{k}\right)}{a_s\left(\frac{n}{k}\right)/U_s\left(\frac{n}{k}\right)} \left\{ \log\left(\frac{\hat{b}_s\left(\frac{n}{k}\right)}{U_s\left(\frac{n}{k}\right)} \right) \right\}^{-1}$$

$$= \left(\sqrt{k} \, \frac{\log U_s\left(\frac{n}{k}\left(\frac{k}{n}\zeta_{n-k,n}(s)\right)\right) - \log U_s\left(\frac{n}{k}\right)}{a_s\left(\frac{n}{k}\right)/U_s\left(\frac{n}{k}\right)} \right)$$

$$\times \left(\frac{\log\left(1 - \left\{ \frac{U_s\left(\frac{n}{k}\left(\frac{k}{n}\zeta_{n-k,n}(s)\right)\right)}{U_s\left(\frac{n}{k}\right)} - 1 \right\}\right)}{\frac{U_s\left(\frac{n}{k}\left(\frac{k}{n}\zeta_{n-k,n}(s)\right)\right)}{U_s\left(\frac{n}{k}\right)} - 1} \right)^{-1} . \quad (10.4.42)$$

We consider the second factor of (10.4.42) first. Note that Corollary 10.4.6 implies

$$\sup_{0 \leq s \leq 1} \left| \frac{k}{n} \zeta_{n-k,n}(s) - 1 \right| \overset{P}{\to} 0 . \quad (10.4.43)$$

Since (10.4.17) implies

$$\lim_{t\to\infty} \frac{U_s(tx)}{U_s(t)} = x^{\max(0,\gamma(s))}$$

locally uniformly for $x > 0$, we have

$$\sup_{0\le s\le 1} \frac{U_s\left(\frac{n}{k}\left\{\frac{k}{n}\zeta_{n-k,n}(s)\right\}\right)}{U_s\left(\frac{n}{k}\right)} \xrightarrow{P} 1 .$$

Hence the second factor of (10.4.42) tends to one in probability, uniformly in s.
For the first factor of (10.4.42) note that by (10.4.41) and (10.4.43),

$$\sqrt{k}\, \frac{\log U_s\left(\frac{n}{k}\left\{\frac{k}{n}\zeta_{n-k,n}(s)\right\}\right) - \log U_s\left(\frac{n}{k}\right)}{a_s\left(\frac{n}{k}\right)/U_s\left(\frac{n}{k}\right)}$$

$$= \sqrt{k}\, \frac{\left\{\frac{k}{n}\zeta_{n-k,n}(s)\right\}^{\gamma_-(s)} - 1}{\gamma_-(s)}$$

$$+ \sqrt{k}A_s\left(\frac{n}{k}\right) H_{\gamma_-(s),\rho(t)}\left(\frac{k}{n}\zeta_{n-k,n}(s)\right)$$

$$+ o_p(1)\sqrt{k}A_s\left(\frac{n}{k}\right)\left\{1 + \left(\frac{k}{n}\zeta_{n-k,n}(s)\right)^{\gamma_-(s)+\varepsilon}\right\} .$$

For the first term apply Corollary 10.4.6 with $\alpha(s) := \gamma_-(s)$. Further, note that by the boundedness of $H_{\gamma_-(s),\rho(s)}((k/n)\,\zeta_{n-k,n}(s))$ and (10.4.40) the second and third terms converge to zero in probability uniformly in s. Relation (10.4.39) follows.
For the other relations we use again (10.4.41). Applying this with

$$x := \zeta_{n-i,n}(s)/\zeta_{n-k,n}(s) \quad \text{and} \quad t := \zeta_{n-k,n}(s),$$

we get with

$$M_n^{(1)}(s) := \frac{1}{k}\sum_{i=0}^{k-1} \log X_{n-i,n}(s) - \log X_{n-k,n}(s)$$

as before that

$$\frac{M_n^{(1)}(s)}{a_s(X_{n-k,n}(s))/U_s(X_{n-k,n}(s))}$$

has the same limit behavior as

$$\frac{1}{k}\sum_{i=0}^{k-1} \frac{(\zeta_{n-i,n}(s)/\zeta_{n-k,n}(s))^{\gamma_-(s)} - 1}{\gamma_-(s)} .$$

The latter equals (cf. proof of Lemma 10.4.3)

$$\frac{1}{\gamma_-(s)}\left\{\left(\frac{k}{n}\zeta_{n-k,n}(s)\right)^{-\gamma_-(s)}\int_{(k/n)\zeta_{n-k,n}(s)}^{\infty}(1 - \mathcal{G}_{n,s}(x))\, x^{\gamma_-(s)-1}\, dx - 1\right\},$$

with

$$1 - \mathcal{G}_{n,s}(x) := \frac{1}{k}\sum_{i=1}^{n} 1_{\{(k/n)\zeta_i(s)>x\}}$$

as before. Now we have expressed everything in terms of $(k/n)\zeta_{n-k,n}(s)$ and $1 - \mathcal{G}_{n,s}(x)$, so that we can find the asymptotic behavior by Theorem 10.4.4 and Corollary 10.4.6. We still need some rearrangement:

$$\sqrt{k}\left(\frac{M_n^{(1)}(s)}{a_s(X_{n-k,n}(s))/U_s(X_{n-k,n}(s))} - \frac{1}{1 - \gamma_-(s)}\right)$$

$$\approx \sqrt{k}\left\{\left(\frac{k}{n}\zeta_{n-k,n}(s)\right)^{-\gamma_-(s)}\int_{(k/n)\zeta_{n-k,n}(s)}^{\infty}(1 - \mathcal{G}_{n,s}(x))\,x^{\gamma_-(s)-1}\,dx\right.$$

$$\left. - \int_1^{\infty}x^{\gamma_-(s)-2}\,dx\right\}$$

$$= \left(\frac{k}{n}\zeta_{n-k,n}(s)\right)^{-\gamma_-(s)}\int_{(k/n)\zeta_{n-k,n}(s)}^{\infty}\sqrt{k}\left\{(1 - \mathcal{G}_{n,s}(x)) - x^{-1}\right\}x^{\gamma_-(s)-1}dx$$

$$+ \sqrt{k}\left\{\left(\frac{k}{n}\zeta_{n-k,n}(s)\right)^{-\gamma_-(s)} - 1\right\}\int_{(k/n)\zeta_{n-k,n}(s)}^{\infty}x^{\gamma_-(s)-2}\,dx$$

$$+ \sqrt{k}\int_{(k/n)\zeta_{n-k,n}(s)}^{\infty}x^{\gamma_-(s)-2}\,dx\;.$$

Hence

$$\sqrt{k}\left(\frac{M_n^{(1)}(s)}{a_s(X_{n-k,n}(s))/U_s(X_{n-k,n}(s))} - \frac{1}{1 - \gamma_-(s)}\right) - \mathcal{P}(s)$$

$$\approx \left(\frac{k}{n}\zeta_{n-k,n}(s)\right)^{-\gamma_-(s)}\int_{(k/n)\zeta_{n-k,n}(s)}^{\infty}\left(\sqrt{k}\left\{1 - \mathcal{G}_{n,s}(x) - x^{-1}\right\}\right.$$

$$\left. - W(C_{s,x})\right)x^{\gamma_-(s)-1}\,dx$$

$$+ \left(\left(\frac{k}{n}\zeta_{n-k,n}(s)\right)^{-\gamma_-(s)} - 1\right)\int_{(k/n)\zeta_{n-k,n}(s)}^{\infty}W(C_{s,x})x^{\gamma_-(s)-1}\,dx$$

$$+ \left\{\sqrt{k}\left(\left(\frac{k}{n}\zeta_{n-k,n}(s)\right)^{-\gamma_-(s)} - 1\right)\right.$$

$$\left. + \gamma_-(s)W(C_{s,1})\right\}\int_{(k/n)\zeta_{n-k,n}(s)}^{\infty}x^{\gamma_-(s)-2}\,dx$$

$$+ \sqrt{k}\int_{(k/n)\zeta_{n-k,n}(s)}^{1}x^{\gamma_-(s)-2}\,dx + W(C_{s,1})$$

$$- \int_1^{(k/n)\zeta_{n-k,n}(s)}W(C_{s,x})\,x^{\gamma_-(s)-1}\,dx$$

$$+ \gamma_-(s) W(C_{s,1}) \int_1^{(k/n)\zeta_{n-k,n}(s)} x^{\gamma_-(s)-2}\, dx,$$

which by Theorem 10.4.4 and Corollary 10.4.6 converges to zero in probability, uniformly in s. The proofs of the other parts of the theorem are similar. ∎

10.5 Estimation of the Probability of a Failure Set

We return to the problem sketched in Section 1.1.4. The coastal provinces Friesland and Groningen of the Netherlands are protected against inundation by a long dike and a breach in the dike at one place could mean inundation of the whole area. Suppose that we have n independent observations of the water level along the dike during a wind storm over a certain period. We want to estimate the probability

$$P\{X(s) > f(s) \text{ for some } 0 \le s \le 1\},$$

where $\{X(s)\}_{s\in[0,1]}$ is the stochastic process representing the water level along the dike during a wind storm, and the continuous function f represents the top of the dike. Since in modern times no flood has been recorded, all our observations $X_1, X_2, \dots, X_n$ are well below f. So we have to estimate the probability of an event that might never have taken place. Since this is an essential feature of the problem, we want to keep it in the asymptotic analysis, where we imagine that the number of observations n tends to infinity.

More generally, consider a failure set C_n with $P(C_n) = O(1/n)$ and a sample $X_1, X_2, \dots$, of elements of $C[0, 1]$. We want to estimate $P(C_n)$.

Our conditions are quite similar to those in Chapter 8:

1. The domain of attraction condition (cf. Theorem 9.3.1): for each Borel set $A \subset \{f \in C[0, 1] : f \ge 0\}$ with $\nu(\partial A) = 0$ and $\inf\{\|f\|_\infty : f \in A\} > 0$,

$$\lim_{n\to\infty} \frac{n}{k} P(R_n X \in A) = \nu(A)$$

for some $k = k(n) \to \infty$, $k/n \to 0$, where for $0 \le s \le 1$,

$$R_n X(s) := \left(1 + \gamma(s)\frac{X(s) - b_s\left(\frac{n}{k}\right)}{a_s\left(\frac{n}{k}\right)}\right)^{1/\gamma(s)}.$$

The functions $a_s(n/k) > 0$ and $b_s(n/k)$ are suitable continuous normalizing functions.

2. Estimators $\hat\gamma(s), \hat a_s(n/k), \hat b_s(n/k)$ such that with some sequence $k = k(n) \to \infty$, $k(n) = o(n)$, $n \to \infty$,

$$\sup_{0\le s\le 1}\left(\sqrt{k}\left|\hat\gamma(s) - \gamma(s)\right| \vee \sqrt{k}\left|\frac{\hat a_s\left(\frac{n}{k}\right)}{a_s\left(\frac{n}{k}\right)} - 1\right| \vee \sqrt{k}\left|\frac{\hat b_s\left(\frac{n}{k}\right) - b_s\left(\frac{n}{k}\right)}{a_s\left(\frac{n}{k}\right)}\right|\right)$$
$$= O_P(1).$$

3. C_n is open in $C[0, 1]$ and there exists $h_n \in \partial C_n$ such that

$$f \le h_n \Rightarrow f \notin C_n .$$

4. (Stability property) We require

$$C_n = \left\{ \left(a_s \left(\frac{n}{k} \right) \frac{(c_n f(s))^{\gamma(s)} - 1}{\gamma(s)} + b_s \left(\frac{n}{k} \right) \right)_{s \in [0,1]} : f \in S \right\},$$

where S is a fixed set in $C[0, 1]$ with
   (a) $f \ge 0$ for $f \in S$;
   (b) $v(\partial S) = 0$ and $\inf\{|f|_\infty : f \in S\} > 0$;
   (c)

$$c_n := \sup_{0 \le s \le 1} \left(1 + \gamma(s) \frac{h_n(s) - b_s \left(\frac{n}{k} \right)}{a_s \left(\frac{n}{k} \right)} \right)^{1/\gamma(s)}.$$

5. Sharpening of (1):

$$\frac{n \, P\left(R_n(X) \in c_n S\right)}{k \, v\left(c_n S\right)} \xrightarrow{P} 1 , \quad n \to \infty .$$

6. With $\underline{\gamma} := \inf_{0 \le s \le 1} \gamma(s)$:

$$\underline{\gamma} > -\frac{1}{2}$$

and

$$\lim_{n \to \infty} \frac{w_{\underline{\gamma}}(c_n)}{\sqrt{k}} = 0 .$$

Finally, we define the estimator $\hat{p}_n$ for $p_n := P(C_n)$:

$$\hat{p}_n := \frac{1}{n \, \hat{c}_n} \sum_{i=1}^{n} 1_{\left\{ \hat{R}_n(X) \in \hat{S}_n \right\}},$$

where

$$\hat{c}_n := \sup_{0 \le s \le 1} \left(1 + \hat{\gamma}(s) \frac{h_n(s) - \hat{b}_s \left(\frac{n}{k} \right)}{\hat{a}_s \left(\frac{n}{k} \right)} \right)^{1/\hat{\gamma}(s)},$$

$$\hat{R}_n f(s) := \left(1 + \hat{\gamma}(s) \frac{f(s) - \hat{b}_s \left(\frac{n}{k} \right)}{\hat{a}_s \left(\frac{n}{k} \right)} \right)^{1/\hat{\gamma}(s)} , \quad \text{for } 0 \le s \le 1 ,$$

and

$$\hat{S}_n := \frac{1}{\hat{c}_n} \hat{R}_n(C_n) .$$

Remark 10.5.1 Note that $\hat{c}_n$ is not defined if

$$1 + \hat{\gamma}(s) \frac{h_n(s) - \hat{b}_s\left(\frac{n}{k}\right)}{\hat{a}_s\left(\frac{n}{k}\right)} \leq 0$$

for some $s \in [0, 1]$. However, when checking the proof, one sees that as $n \to \infty$, the probability that this happens tends to zero.

Theorem 10.5.2 *Under our conditions,*

$$\frac{\hat{p}_n}{p_n} \xrightarrow{P} 1, \quad n \to \infty,$$

provided $v(S) > 0$.

The proof of Theorem 10.5.2 follows from three lemmas and four propositions. The proofs are very similar to those in Chapter 8 and will be mostly omitted.

Lemma 10.5.3 *Let G_n be an increasing and invertible mapping: $C[0, 1] \to C[0, 1]$. Suppose that $\lim_{n \to \infty} G_n f = f$ in $C[0, 1]$ for all $f \in C[0, 1]$. For an open set O let*

$$O_n := \{G_n f : f \in O\}.$$

Then for all $f \in O$,

$$1_{O_n}(f) := 1_{\{f \in O_n\}} \to 1_O(f) := 1_{\{f \in O\}}.$$

Lemma 10.5.4 *For all $x > 0$,*

$$\lim_{n \to \infty} \left(1 + \gamma(s) \left\{(1 + o_1(1)) \frac{x^{\gamma(s) + o_2(1)} - 1}{\gamma(s) + o_2(1)} + o_3(1)\right\}\right)^{1/\gamma(s)} = x,$$

uniformly for $0 \leq s \leq 1$, provided γ is a continuous function on $[0, 1]$ and the o-terms tend to zero uniformly in s.

Lemma 10.5.5 *For all $x > 0$ and $c_n \to \infty$,*

$$\lim_{n \to \infty} \frac{1}{c_n} \left(1 + \gamma_n(s) \left\{(1 + O_1(\gamma_n(s) - \gamma(s))) \frac{(c_n x)^{\gamma(s)} - 1}{\gamma(s)}\right.\right.$$

$$\left.\left. + O_2(\gamma_n(s) - \gamma(s))\right\}\right)^{1/\gamma_n(s)} = x,$$

uniformly for $0 \leq s \leq 1$, provided γ_n and γ are continuous functions,

$$\sup_{0 \leq s \leq 1} |\gamma_n(s) - \gamma(s)| = 0,$$

and

$$\lim_{n \to \infty} \sup_{0 \leq s \leq 1} |\gamma_n(s) - \gamma(s)| c_n^{-\gamma(s)} \int_1^{c_n} s^{\gamma(s) - 1} \log s \, ds = 0.$$

Proposition 10.5.6 *Let S be an open Borel set in $\{f \in C[0, 1] : f \geq 0\}$ with $v(S) > 0$, $v(\partial S) = 0$ and such that $\inf\{\|f\|_\infty : f \in S\} > 0$. Suppose condition (1) holds. Define*

$$\tilde{v}_n(S) := \frac{1}{k} \sum_{i=1}^n 1_{\{R_n X \in S\}} .$$

Then as $n \to \infty$,

$$\tilde{v}_n(S) \xrightarrow{P} v_n(S) .$$

Proposition 10.5.7 *Assume the conditions of Proposition 10.5.6. Let $\hat{\gamma}(s)$, $\hat{a}_s(n/k)$, $\hat{b}_s(n/k)$ be estimators such that*

$$\sup_{0 \leq s \leq 1} \left| \hat{\gamma}(s) - \gamma(s) \right| \xrightarrow{P} 0$$

$$\sup_{0 \leq s \leq 1} \left| \frac{\hat{a}_s\left(\frac{n}{k}\right)}{a_s\left(\frac{n}{k}\right)} - 1 \right| \xrightarrow{P} 0$$

$$\sup_{0 \leq s \leq 1} \left| \frac{\hat{b}_s\left(\frac{n}{k}\right) - b_s\left(\frac{n}{k}\right)}{a_s\left(\frac{n}{k}\right)} \right| \xrightarrow{P} 0 .$$

Define

$$\hat{v}_n(S) := \frac{1}{k} \sum_{i=1}^n 1_{\{\hat{R}_n X \in S\}} .$$

Then, as $n \to \infty$,

$$\hat{v}_n(S) \xrightarrow{P} v(S) .$$

Proof. Invoke a Skorohod construction so that we may assume that by virtue of Lemma 10.5.4,

$$R_n \hat{R}_n^{\leftarrow} \to \text{Identity} \quad \text{a.s.}$$

Write

$$c := \inf\{\|f\|_\infty : f \in S\} > 0 .$$

For all $0 < c_0 < c$, $f \in S \Rightarrow$ there exist $s \in [0, 1]$ such that $f(s) > c_0$.
    Take n_0 such that for $n \geq n_0$,

$$R_n \hat{R}_n^{\leftarrow} c_0 > \frac{c_0}{2} .$$

Then for each $f \in S$ there exists $s \in [0, 1]$ such that

$$R_n \hat{R}_n^{\leftarrow} f(s) > R_n \hat{R}_n^{\leftarrow} c_0 > \frac{c_0}{2} .$$

Hence

$$\left\{ R_n \hat{R}_n^{\leftarrow} f : f \in S \right\} \subset \left\{ f : f \leq R_n \hat{R}_n^{\leftarrow} c_0 \right\}^c \subset \left\{ f : f \leq \frac{c_0}{2} \right\}^c =: D_\varepsilon \; ,$$

i.e.,

$$1_{R_n \hat{R}_n^{\leftarrow}(S)} \leq 1_{D_\varepsilon} \; .$$

Now

$$\tilde{v}_n(D_\varepsilon) \to v(D_\varepsilon)$$

by Proposition 10.5.6. Hence as in the proof of Proposition 8.2.8,

$$\tilde{v}_n \left(R_n \hat{R}_n^{\leftarrow}(S) \right) \to v(S)$$

almost surely, and hence in probability. ∎

Proposition 10.5.8 *Under the conditions of the theorem,*

$$\frac{\hat{c}_n}{c_n} \xrightarrow{P} 1 \; , \qquad n \to \infty \; .$$

Proof. For $0 \leq s \leq 1$, write

$$r_n(s) := \left(1 + \gamma(s) \frac{h_n(s) - b_s \left(\frac{n}{k} \right)}{a_s \left(\frac{n}{k} \right)} \right)^{1/\gamma(s)} .$$

Then

$$\hat{c}_n = \sup_{0 \leq s \leq 1} \left(1 + \hat{\gamma}(s) \frac{h_n(s) - \hat{b}_s \left(\frac{n}{k} \right)}{\hat{a}_s \left(\frac{n}{k} \right)} \right)^{1/\hat{\gamma}(s)}$$

$$= \sup_{0 \leq s \leq 1} r_n(s) \left\{ \frac{1}{r_n(s)} \left(1 + \hat{\gamma}(s) \left[\frac{a_s \left(\frac{n}{k} \right)}{\hat{a}_s \left(\frac{n}{k} \right)} \frac{(r_n(s))^{\gamma(s)} - 1}{\gamma(s)} \right. \right. \right.$$

$$\left. \left. \left. + \frac{b_s \left(\frac{n}{k} \right) - \hat{b}_s \left(\frac{n}{k} \right)}{\hat{a}_s \left(\frac{n}{k} \right)} \right] \right)^{1/\hat{\gamma}(s)} \right\} .$$

Hence, since the expression inside the curly brackets tends to one in probability uniformly in s by Lemma 10.5.5, we have

$$\frac{\hat{c}_n}{c_n} = \frac{\hat{c}_n}{\sup_{0 \leq s \leq 1} r_n(s)} \xrightarrow{P} 1 \; . \qquad ∎$$

Proposition 10.5.9 *Under the conditions of the theorem*

$$\hat{v}_n \left(\frac{1}{\hat{c}_n} \hat{R}_n \, R_n^{\leftarrow}(c_n S) \right) \xrightarrow{P} v_n(S) \; .$$

Part IV

Appendix

A

Skorohod Theorem and Vervaat's Lemma

One form of the Skorohod representation theorem is as follows.

Theorem A.0.1 *Let S be a complete and separable metric space. Suppose that the sequence of probability measures P_n on S converges to the probability measure P_0 in distribution. Then one can find Borel measurable functions (random variables) $X_0, X_1, \ldots$ defined on the Lebesgue interval $([0, 1], \mathcal{B}, \lambda)$, with $\mathcal{B}$ the Borel sets and λ Lebesgue measure, such that X_n has probability distribution P_n for $n = 0, 1, 2, \ldots$ and*

$$\lim_{n \to \infty} X_n(\omega) = X(\omega)$$

for all $\omega \in [0, 1]$. For a proof we refer to Billingsley (1971).

We combine this theorem with Vervaat's lemma and give an example of the application of both.

Lemma A.0.2 (Vervaat (1972)) *Suppose y is a continuous function and $x_1, x_2, \ldots$ are nondecreasing functions on some interval $[a, b]$. Further, we have a function g on $[a, b]$ with a positive derivative g'. Let δ_n be a sequence of positive numbers such that $\delta_n \to 0, n \to \infty$, and*

$$\lim_{n \to \infty} \frac{x_n(s) - g(s)}{\delta_n} = y(s) \tag{A.0.1}$$

uniformly on $[a, b]$. Then

$$\lim_{n \to \infty} \frac{x_n^{\leftarrow}(s) - g^{\leftarrow}(s)}{\delta_n} = -(g^{\leftarrow})'(s)\, y\, (g^{\leftarrow}(s)) \tag{A.0.2}$$

uniformly on $[g(a), g(b)]$, where $g^{\leftarrow}, x_1^{\leftarrow}, x_2^{\leftarrow}, \ldots$ are the inverse functions (right- or left-continuous or defined in any way consistent with monotonicity).

Proof. We first prove the result for $g(s) = s$ for all s. Note that (A.0.1) implies that $x_n(s)$ converges uniformly to s and hence $x_n^{\leftarrow}(s)$ converges uniformly to s. Take any sequence $s_n \to s_0 \in (g^{\leftarrow}(a), g^{\leftarrow}(b))$. The local uniformity in (A.0.1) gives

$$\lim_{n\to\infty} \frac{x_n\left(x_n^{\leftarrow}(s_n)\pm\varepsilon_n\right)-x_n^{\leftarrow}(s_n)\mp\varepsilon_n}{\delta_n} = \lim_{n\to\infty} y\left(x_n^{\leftarrow}(s_n)\pm\varepsilon_n\right) = y(s_0)$$

with ε_n positive and $0 < \delta_n \vee (g(b)-x_n^{\leftarrow}(s_n)) \vee (x_n^{\leftarrow}(s_n)-g(a))$. Now, $x_n\left(x_n^{\leftarrow}(s_n)-\varepsilon_n\right)$
$\leq s_n \leq x_n\left(x_n^{\leftarrow}(s_n)+\varepsilon_n\right)$ and hence

$$\left|\frac{s_n-x_n^{\leftarrow}(s_n)}{\delta_n}-y(s_0)\right|$$

$$\leq \max\left(\left|\frac{x_n\left(x_n^{\leftarrow}(s_n)+\varepsilon_n\right)-x_n^{\leftarrow}(s)}{\delta_n}-y(s_0)\right|,\right.$$

$$\left.\left|\frac{x_n\left(x_n^{\leftarrow}(s_n)-\varepsilon_n\right)-x_n^{\leftarrow}(s)}{\delta_n}-y(s_0)\right|\right)$$

$$\to 0.$$

Next note that from (A.0.1),

$$\frac{x_n(g^{\leftarrow}(s))-s}{\delta_n} \to y(g^{\leftarrow}(s))$$

uniformly for $g(a) \leq s \leq g(b)$, which implies, as we just proved,

$$\frac{g^{\leftarrow}\left(x_n^{\leftarrow}(s)\right)-s}{\delta_n} \to -y(g^{\leftarrow}(s)).$$

Now

$$g^{\leftarrow}\left(x_n^{\leftarrow}(s)\right)-s = g^{\leftarrow}\left(x_n^{\leftarrow}(s)\right)-g^{\leftarrow}(g(s))$$
$$= \left(x_n^{\leftarrow}(s)-g(s)\right)(g^{\leftarrow})'\left(g(s)+\theta\left(x_n^{\leftarrow}(s)-g(s)\right)\right)$$

for some $\theta \in [0,1]$. The result follows. ∎

An application of the combination of both results is the following.

Example A.0.3 Let $U_1, U_2, \ldots$ be independent and identically uniformly distributed random variables on $[0,1]$. The empirical distribution function U_n is defined as

$$U_n(x) := \frac{1}{n}\sum_{i=1}^{n} 1_{\{U_i\leq x\}}.$$

From Billingsley (1968), Theorem 16.4, we have

$$\left\{\sqrt{n}\left(U_n(x)-x\right)\right\}_{0\leq x\leq 1} \overset{d}{\to} \{B_0(x)\}_{0\leq x\leq 1}$$

in $D[0,1]$-space, where B_0 is a Brownian bridge. Skorohod's theorem implies that there are random processes $\{U_n^\star\}$ and $\{B_0^\star\}$, Brownian bridges, such that $U_n^\star =^d U_n$ for all n and

$$\lim_{n\to\infty} \sup_{0\le x\le 1} \left| \sqrt{n} \left(U_n^\star(x) - x \right) - B_0^\star(x) \right| = 0 \qquad \text{(A.0.3)}$$

with probability one.

Next consider the nth order statistics of $U_1, U_2, \ldots, U_n$, indicated by $U_{1,n}, U_{2,n}, \ldots, U_{n,n}$. The random function

$$\{U_{[nx],n}\}_{0\le x\le 1}$$

is easily seen to satisfy

$$\sup_{0\le x\le 1} \left| U_n \left(U_{[nx],n} \right) - x \right| = O\left(\frac{1}{n} \right);$$

hence

$$\sup_{0\le x\le 1} \left| \sqrt{n} \left(U_{[nx],n} - U_n^\leftarrow(x) \right) \right| \to 0 . \qquad \text{(A.0.4)}$$

Since $\left(U_n^\star \right)^\leftarrow =^d U_n^\leftarrow$ for all n, we obtain from (A.0.3) via Vervaat's lemma

$$\lim_{n\to\infty} \sup_{0\le x\le 1} \left| \sqrt{n} \left(\left(U_n^\star \right)^\leftarrow (x) - x \right) + B_0^\star(x) \right| = 0$$

with probability one, which implies

$$\left\{ \sqrt{n} \left(U_n^\leftarrow(x) - x \right) \right\}_{0\le x\le 1} \xrightarrow{d} \{-B_0(x)\}_{0\le x\le 1}$$

in $D[0, 1]$-space and finally via (A.0.4),

$$\left\{ \sqrt{n} \left(U_{[nx],n} - x \right) \right\}_{0\le x\le 1} \xrightarrow{d} \{-B_0(x)\}_{0\le x\le 1} \stackrel{d}{=} \{B_0(x)\}_{0\le x\le 1}$$

in $D[0, 1]$-space.

B

Regular Variation and Extensions

B.1 Regularly Varying (RV) Functions

Regular variation is an important tool in this book, mainly the more recent developments. Since these cannot be found in any book, we need to discuss them in detail. But then it seems more natural to start from the beginning and develop the main results on regular variation from scratch. This, along with the newer developments, is the topic of this appendix. The basic material has been taken from Geluk and de Haan (1987) with kind permission of J.L. Geluk.

One way to think about regular variation is as a derivative at infinity. For a real measurable function g write the differential quotient

$$\frac{g(y+h) - g(y)}{h}, \tag{B.1.1}$$

where $h \neq 0$. Now, we do not take the limit $h \to 0$ for fixed y as usual, but we take the limit $y \to \infty$ for fixed h. If this limit exists for all $h \neq 0$, then it follows (Theorem B.1.3 below) that the limit does not depend on h and we can write (see Proposition B.1.9(3)) $g(y) = g_0(y) + o(1)$, as $y \to \infty$, where g_0 is differentiable and

$$\lim_{y \to \infty} g_0'(y) = \lim_{y \to \infty} \frac{g(y+h) - g(y)}{h}. \tag{B.1.2}$$

If the limit in (B.1.2) as $y \to \infty$ exists, the function $f : \mathbb{R}^+ \to \mathbb{R}^+$ defined by $f(t) = \exp g(\log t)$ satisfies

$$\lim_{t \to \infty} \frac{f(tx)}{f(t)} = x^\alpha \tag{B.1.3}$$

for all $x \in \mathbb{R}^+$ for some $\alpha \in \mathbb{R}$. Then f is called a regularly varying function.

In this appendix these functions are studied thoroughly. Moreover, we study the more general class of functions $f : \mathbb{R}^+ \to \mathbb{R}$ for which

$$\lim_{t \to \infty} \frac{f(tx) - b(t)}{a(t)} \tag{B.1.4}$$

exists for all $x \in \mathbb{R}^+$, where $a > 0$ and b are suitably chosen auxiliary functions. The results for functions satisfying (B.1.4) are surprisingly similar to those for functions satisfying (B.1.3).

Definition B.1.1 A Lebesgue measurable function $f : \mathbb{R}^+ \to \mathbb{R}$ that is eventually positive is *regularly varying* (at infinity) if for some $\alpha \in \mathbb{R}$,

$$\lim_{t \to \infty} \frac{f(tx)}{f(t)} = x^\alpha , \qquad x > 0 . \tag{B.1.5}$$

Notation: $f \in \mathrm{RV}_\alpha$.

The number α in the above definition is called the *index* of regular variation. A function satisfying (B.1.5) with $\alpha = 0$ is called *slowly varying*.

Example B.1.2 For $\alpha, \beta \in \mathbb{R}$ the functions x^α, $x^\alpha (\log x)^\beta$, $x^\alpha (\log \log x)^\beta$ are RV_α. The functions $2 + \sin(\log \log x)$, $\exp((\log x)^\alpha)$, with $0 < \alpha < 1$, $x^{-1} \log \Gamma(x)$, $\sum_{k \le x} 1/k$, $(\log t)^{\sin(\log \log t)}$ are slowly varying. The functions $2 + \sin x$, $\exp[\log x]$, $2 + \sin \log x$, $x \exp \sin \log x$ are not regularly varying.

Our next result shows that it is possible to weaken the conditions in Definition B.1.1.

Theorem B.1.3 *Suppose* $f : \mathbb{R}^+ \to \mathbb{R}$ *is measurable, eventually positive, and*

$$\lim_{t \to \infty} \frac{f(tx)}{f(t)} \tag{B.1.6}$$

exists, and is finite and positive for all x *in a set of positive Lebesgue measure. Then* $f \in \mathrm{RV}_\alpha$ *for some* $\alpha \in \mathbb{R}$.

Proof. Define $F(t) := \log f(e^t)$. Then $\lim_{t \to \infty} (F(t+x) - F(t))$ exists for all x in a set K of positive Lebesgue measure. Define $\phi : K \to \mathbb{R}$ by $\phi(x) := \lim_{t \to \infty} \{F(t+x) - F(t)\}$. By Steinhaus's theorem (cf. Hewitt and Stromberg (1969) p. 143) the set $K - K := \{x - y : x, y \in K\}$ contains a neighborhood of zero. Since K is an additive subgroup of $\mathbb{R}$, we have $K = \mathbb{R}$ and thus $\phi(x)$ is defined for all $x \in \mathbb{R}$ and

$$\phi(x + y) = \phi(x) + \phi(y) \tag{B.1.7}$$

for all $x, y \in \mathbb{R}$.

It remains to solve equation (B.1.7) for measurable ϕ: Consider the restriction of ϕ to an interval $L \subset \mathbb{R}$. By Lusin's theorem (cf. Halmos (1950) p. 242) there exists a compact set $M \subset L$ with positive Lebesgue measure λM such that the restriction of ϕ to M is continuous. Let $\varepsilon > 0$ be arbitrary. Then there exists $\delta > 0$ such that $\phi(y) - \phi(x) \in (-\varepsilon, \varepsilon)$ whenever $x, y \in M$ and $|x - y| < \delta$ (since the restriction of ϕ to M is uniformly continuous) and also such that $M - M$ contains the interval $(-\delta, \delta)$ (by Steinhaus's theorem). For each $s \in (-\delta, \delta) \subset M - M$ there exists $x_0 \in M$ such that also $x_0 + s \in M$. Then $\phi(x + s) - \phi(x) = \phi(s) = \phi(x_0 + s) - \phi(x_0) \in (-\varepsilon, \varepsilon)$ for all $x \in \mathbb{R}$; hence ϕ is uniformly continuous on $\mathbb{R}$. Since $\phi(n/m) = n\phi(1/m) = n\phi(1)/m$ for $n, m \in \mathbb{Z}$, $m \ne 0$, we have by the continuity of ϕ that $\phi(x) = \phi(1)x$ for $x \in \mathbb{R}$. Now (B.1.5) follows. ∎

Theorem B.1.4 (Uniform convergence theorem) *If $f \in RV_\alpha$, then relation* (B.1.5) *holds uniformly for $x \in [a, b]$ with $0 < a < b < \infty$.*

Proof. Without loss of generality we may suppose $\alpha = 0$ (if not, replace $f(t)$ by $f(t)/t^\alpha$).

We define the function F by $F(x) := \log f(e^x)$. It is sufficient to deduce a contradiction from the following assumption: Suppose there exist $\delta > 0$ and sequences $t_n \to \infty$, $x_n \to 0$ as $n \to \infty$ such that

$$|F(t_n + x_n) - F(t_n)| > \delta$$

for $n = 1, 2, \ldots$. For an arbitrary finite interval $J \subset \mathbb{R}$ we consider the sets

$$Y_{1,n} = \left\{ y \in J \; : \; |F(t_n + y) - F(t_n)| > \frac{\delta}{2} \right\}$$

and

$$Y_{2,n} = \left\{ y \in J \; : \; |F(t_n + x_n) - F(t_n + y)| > \frac{\delta}{2} \right\} .$$

The above sets are measurable for each n and $Y_{1,n} \cup Y_{2,n} = J$; hence either $\lambda(Y_{1,n}) \geq \lambda(J)/2$ or $\lambda(Y_{2,n}) \geq \lambda(J)/2$ (or both), where λ denotes Lebesgue measure.

Now we define

$$Z_n = \left\{ z \; : \; |F(t_n + x_n) - F(t_n + x_n - z)| > \frac{\delta}{2} \; , \; x_n - z \in J \right\}$$

$$= \left\{ z \; : \; x_n - z \in Y_{2,n} \right\} .$$

Then $\lambda(Z_n) = \lambda(Y_{2,n})$ and thus we have either $\lambda(Y_{1,n}) \geq \lambda(J)/2$ infinitely often or $\lambda(Z_n) \geq \lambda(J)/2$ infinitely often (or both).

Since all the $Y_{1,n}$'s are subsets of a fixed finite interval we have $\lambda(\lim_{n\to\infty} \sup Y_{1,n}) = \lim_{k\to\infty} \lambda(\cup_{n=k} Y_{1,n}) \geq \lambda(J)/2$ or a similar statement for the Z_n's (or both). This implies the existence of a real number x_0 contained in infinitely many $Y_{1,n}$ or infinitely many Z_n, which contradicts the assumption $\lim_{t\to\infty} F(t + x_0) - F(t) = 0$. ∎

Theorem B.1.5 (Karamata's theorem) *Suppose $f \in RV_\alpha$. There exists $t_0 > 0$ such that $f(t)$ is positive and locally bounded for $t \geq t_0$. If $\alpha \geq -1$ then*

$$\lim_{t\to\infty} \frac{tf(t)}{\int_{t_0}^t f(s)ds} = \alpha + 1. \tag{B.1.8}$$

If $\alpha < -1$, or $\alpha = -1$ and $\int_0^\infty f(s)\,ds < \infty$, then

$$\lim_{t\to\infty} \frac{tf(t)}{\int_t^\infty f(s)\,ds} = -\alpha - 1 . \tag{B.1.9}$$

Conversely, if (B.1.8) *holds with $-1 < \alpha < \infty$, then $f \in RV_\alpha$; if* (B.1.9) *holds with $-\infty < \alpha < -1$, then $f \in RV_\alpha$.*

Proof. Suppose $f \in RV_\alpha$. By Theorem B.1.4, there exist t_0, c such that $f(tx)/f(t) < c$ for $t \geq t_0$, $x \in [1, 2]$. Then for $t \in [2^n t_0, 2^{n+1} t_0]$ we have

$$\frac{f(t)}{f(t_0)} = \frac{f(t)}{f(2^{-1}t)} \frac{f(2^{-1}t)}{f(2^{-2}t)} \cdots \frac{f(2^{-n}t)}{f(t_0)} < c^{n+1} .$$

Hence $f(t)$ is locally bounded for $t \geq t_0$ and $\int_{t_0}^t f(s)\, ds < \infty$ for $t \geq t_0$.

In order to prove (B.1.8), we first show that $\int_{t_0}^\infty f(s)\, ds = \infty$ for $\alpha > -1$. Since $f(2s) \geq 2^{-1} f(s)$ for s sufficiently large, we have for $n \geq n_0$,

$$\int_{2^n}^{2^{n+1}} f(s)\, ds = 2 \int_{2^{n-1}}^{2^n} f(2s)\, ds \geq \int_{2^{n-1}}^{2^n} f(s)\, ds .$$

Hence

$$\int_{2^{n_0}}^\infty f(s)\, ds = \sum_{n=n_0}^\infty \int_{2^n}^{2^{n+1}} f(s)\, ds \geq \sum_{n=n_0}^\infty \int_{2^{n_0}}^{2^{n_0+1}} f(s)\, ds = \infty .$$

Next we prove $F(t) := \int_{t_0}^t f(s)\, ds \in RV_{\alpha+1}$ for $\alpha > -1$. Fix $x > 0$. For arbitrary $\varepsilon > 0$ there exists $t_1 = t_1(\varepsilon)$ such that $f(xt) < (1 + \varepsilon)x^\alpha f(t)$ for $t > t_1$. Since $\lim_{t \to \infty} F(t) = \infty$,

$$\frac{F(tx)}{F(t)} = \frac{\int_{t_0}^{tx} f(s)\, ds}{\int_{t_0}^t f(s)\, ds} \sim \frac{\int_{t_1 x}^{tx} f(s)\, ds}{\int_{t_1}^t f(s)\, ds} = \frac{x \int_{t_1}^t f(xs)\, ds}{\int_{t_1}^t f(s)\, ds} ,$$

as $t \to \infty$, and hence

$$\frac{F(tx)}{F(t)} < (1 + 2\varepsilon)x^{\alpha+1} \tag{B.1.10}$$

for t sufficiently large. A similar lower inequality is easily derived and we obtain $F \in RV_{\alpha+1}$ for $\alpha > -1$.

In case $\alpha = -1$ and $F(t) \to \infty$ the same proof applies. If $\alpha = -1$ and $F(t)$ has a finite limit, obviously $F \in RV_0$.

Now for all α,

$$\frac{F(tx) - F(t)}{tf(t)} = \int_1^x \frac{f(tu)}{f(t)}\, du \to \frac{x^{\alpha+1} - 1}{\alpha + 1} , \qquad t \to \infty , \tag{B.1.11}$$

by the uniform convergence theorem (Theorem B.1.4). Since $F \in RV_{\alpha+1}$, (B.1.8) follows. For the proof of (B.1.9) we first show the finiteness of the function G defined by

$$G(t) := \int_t^\infty f(s)\, ds .$$

Since in the case $\alpha < -1$ there exists $\delta > 0$ such that $f(2s) \leq 2^{-1-\delta} f(s)$ for s sufficiently large, we have, for n_1 sufficiently large,

$$\int_{2^{n_1}}^{\infty} f(s)\, ds = \sum_{n=n_1}^{\infty} \int_{2^n}^{2^{n+1}} f(s)\, ds \le \sum_{n=n_1}^{\infty} 2^{-\delta(n-n_1)} \int_{2^{n_1}}^{2^{n_1+1}} f(s)\, ds < \infty.$$

The rest of the proof is analogous.

Conversely, suppose (B.1.8) holds. Define

$$b(t) := t\, \frac{f(t)}{F(t)}\,. \tag{B.1.12}$$

Without loss of generality we suppose $f(t) > 0$, $t > 0$. Integrating both sides of $b(t)/t = f(t)/F(t)$ we obtain for some real c_1 and all $x > 0$ (note that $\log F$ is indeed an absolutely continuous function)

$$\int_1^x \frac{b(t)}{t}\, dt = \log F(x) + c_1 \tag{B.1.13}$$

(since the derivatives of the two parts exist and are equal almost everywhere). Using the definition of b again we obtain from (B.1.13)

$$f(x) = c\, b(x) \exp\left(\int_1^x \frac{b(t) - 1}{t}\, dt \right), \tag{B.1.14}$$

for all $x > 0$, with $c = e^{-c_1} > 0$; hence for all $x, t > 0$,

$$\frac{f(tx)}{f(t)} = \frac{b(tx)}{b(t)} \exp\left(\int_1^x \frac{b(ts) - 1}{s}\, ds \right).$$

Now for arbitrary $\varepsilon > 0$ there is a t_0 such that $|b(ts) - \alpha - 1| < \varepsilon$ for $t \ge t_0$ and $s \ge \min(1, x)$. Hence the function f satisfies (B.1.5).

The last statement of the theorem ((B.1.9) implies that $f \in RV_\alpha$) can be proved in a similar way. ∎

Theorem B.1.6 (Representation theorem) *If $f \in RV_\alpha$, there exist measurable functions $a : \mathbb{R}^+ \to \mathbb{R}$ and $c : \mathbb{R}^+ \to \mathbb{R}$ with*

$$\lim_{t \to \infty} c(t) = c_0\ (0 < c_0 < \infty) \quad \text{and} \quad \lim_{t \to \infty} a(t) = \alpha \tag{B.1.15}$$

and $t_0 \in \mathbb{R}^+$ such that for $t > t_0$,

$$f(t) = c(t) \exp\left(\int_{t_0}^t \frac{a(s)}{s}\, ds \right). \tag{B.1.16}$$

Conversely, if (B.1.6) holds with a and c satisfying (B.1.15), then $f \in RV_\alpha$.

Proof. Suppose $f \in RV_\alpha$. The function $t^{-\alpha} f(t)$ is slowly varying and hence has a representation as in (B.1.6) by (B.1.14). Then f has such a representation with $a(s)$ replaced by $a(s) + \alpha$ and $c(t)$ replaced by $t_0^\alpha c(t)$. Now the result follows. Conversely, one verifies directly that (B.1.5) follows from (B.1.6). ∎

Remark B.1.7   1. In formula (B.1.6) we may take $t_0 \in [0, \infty)$ arbitrarily by changing the functions $c(t)$ and $a(t)$ suitably on the interval $[0, t_0]$.
2. The functions $a(t)$ and $c(t)$ (given in (B.1.6)) are not uniquely determined. It can
easily be seen that it is possible to choose $a(t)$ continuous: define

$$f_0(t) := \exp\left(\int_{t_0}^{t} a(v)\, \frac{dv}{v}\right) \quad \text{and} \quad b_0(t) := t\, \frac{f_0(t)}{\int_{t_0}^{t} f_0(s)\, ds}\,.$$

Since $f_0 \in RV_\alpha$ we get (B.1.14) with f and b replaced by f_0 and b_0 respectively,
i.e.,

$$f(x) = c(x)\, c\, b_0(x) \exp\left(\int_{1}^{x} (b_0(t) - 1)\, \frac{dt}{t}\right)$$

for all $x > 0$ with $b_0(t) - 1$ continuous. It is possible to put all the undesirable
behavior of the function f into the function $c(t)$.

We are going to list a number of consequences of the above theorems. For that
we need the following definition.

Definition B.1.8 Suppose $f : (t_0, \infty) \to \mathbb{R}$ for some $t_0 \geq -\infty$ is bounded
on intervals of the form (t_0, a) with $a < \infty$ and $\lim_{t \to \infty} f(t) = \infty$. Since
$\lim_{t \to \infty} f(t) = \infty$, the set $\{y : f(y) \geq x\}$ is nonempty for all $x \in \mathbb{R}$. Hence
$-\infty \leq \inf\{y : f(y) \geq x\} < \infty$ for $x \in \mathbb{R}$. Note that this infimum is nondecreasing
in x. Since f is bounded on intervals of the form (t_0, a), $\lim_{x \to \infty} \inf\{y : f(y) \geq
x\} = \infty$. Hence there exists $x_0 \in \mathbb{R}$ such that $\inf\{y : f(y) \geq x\} > -\infty$ for all
$x \geq x_0$. The *generalized inverse function* $\bar{f} : (x_0, \infty) \to \mathbb{R}$ is defined by

$$\bar{f}(x) := \inf\{y : f(y) \geq x\}\,.$$

In case f is a nondecreasing function, $\bar{f}$ is its inverse function. In that case we write
$f^{\leftarrow}$ instead of $\bar{f}$.

Proposition B.1.9 (Properties of RV functions)   *1. If $f \in RV_\alpha$, then $\log f(t)/\log t$
$\to \alpha,\ t \to \infty$. This implies*

$$\lim_{t \to \infty} f(t) = \begin{cases} 0\,, & \alpha < 0, \\ \infty\,, & \alpha > 0. \end{cases}$$

2. *If $f_1 \in RV_{\alpha_1}$, $f_2 \in RV_{\alpha_2}$, then $f_1 + f_2 \in RV_{\max(\alpha_1, \alpha_2)}$. If moreover
$\lim_{t \to \infty} f_2(t) = \infty$, then the composition $f_1 \circ f_2 \in RV_{\alpha_1 \alpha_2}$.*
3. *If $f \in RV_\alpha$ with $\alpha > 0$ ($\alpha < 0$) then f is asymptotically equivalent to a strictly
increasing (decreasing) differentiable function g with derivative $g' \in RV_{\alpha-1}$ if
$\alpha > 0$ and $-g' \in RV_{\alpha-1}$ if $\alpha < 0$.
As a consequence of this; if $f \in RV_\alpha (\alpha > 0)$ is bounded on finite intervals of
$\mathbb{R}^+$, then*

$$\sup_{0 < x \leq t} f(x) \sim f(t) \qquad (t \to \infty)\,. \tag{B.1.17}$$

If $f \in RV_\alpha\ (\alpha < 0)$, then

$$\inf_{x \geq t} f(x) \sim f(t) \qquad (t \to \infty)\,. \tag{B.1.18}$$

4. If $f \in RV_\alpha$ is integrable on finite intervals of $\mathbb{R}^+$ and $\alpha \geq -1$, then $\int_0^t f(s)\, ds$ is regularly varying with exponent $\alpha + 1$.
   If $f \in RV_\alpha$ and $\alpha < -1$, then $\int_t^\infty f(s)\, ds$ exists for t sufficiently large and is regularly varying with exponent $\alpha + 1$. The same is true for $\alpha = -1$ provided $\int_1^\infty f(s)\, ds < \infty$.

5. (Potter, 1942) Suppose $f \in RV_\alpha$. If $\delta_1, \delta_2 > 0$ are arbitrary, there exists $t_0 = t_0(\delta_1, \delta_2)$ such that for $t \geq t_0$, $tx \geq t_0$,

$$(1 - \delta_1)x^\alpha \min(x^{\delta_2}, x^{-\delta_2}) < \frac{f(tx)}{f(t)} < (1 + \delta_1)x^\alpha \max(x^{\delta_2}, x^{-\delta_2}). \quad (B.1.19)$$

   Note that conversely, if f satisfies the above property, then $f \in RV_\alpha$.

6. Suppose $f \in RV_\alpha$ is bounded on finite intervals of $\mathbb{R}^+$ and $\alpha > 0$. For $\xi > 0$ arbitrary there exist $c > 0$ and t_0 such that for $t \geq t_0$ and $0 < x \leq \xi$,

$$\frac{f(tx)}{f(t)} \leq c. \quad (B.1.20)$$

7. If $f \in RV_\alpha, \alpha \leq 0$, is bounded on finite intervals of $\mathbb{R}^+$ and $\delta, \xi > 0$ are arbitrary, there exist $c > 0$ and t_0 such that for $t \geq t_0$ and $0 < x \leq \xi$,

$$\frac{f(tx)}{f(t)} < cx^{\alpha - \delta}. \quad (B.1.21)$$

8. If

$$f(t) = \exp\left(\int_0^t a(s)\frac{ds}{s}\right) \quad (B.1.22)$$

   with a continuous function $a(s) \to \alpha > 0$, $s \to \infty$, then $f^\leftarrow \in RV_{1/\alpha}$, where $f^\leftarrow$ is the inverse function of f.

9. Suppose $f \in RV_\alpha, \alpha > 0$, is bounded on finite intervals of $\mathbb{R}^+$. Then $\bar{f} \in RV_{1/\alpha}$. (Formally, $\bar{f}$ is defined only on a neighborhood of infinity; we can extend its domain of definition by taking $\bar{f}$ zero elsewhere). In particular, if $f \in RV_\alpha, \alpha > 0$, and f is increasing, the inverse function $f^\leftarrow$ is in $RV_{1/\alpha}$.

10. If $f \in RV_\alpha$, $\alpha > 0$, there exists an asymptotically unique function h such that $f(h(x)) \sim h(f(x)) \sim x$, $x \to \infty$. Moreover, $h \sim \bar{f}$ if f is bounded on finite intervals of $\mathbb{R}^+$.

11. If $f \in RV_\alpha$, $\alpha \geq 0$, and $f(t) = f(t_0) + \int_{t_0}^t \psi(s)ds$ for $t \geq t_0$ with ψ monotone, then

$$\lim_{t \to \infty} \frac{t\psi(t)}{f(t)} = \alpha.$$

   Hence in case $\alpha > 0$ we have $\psi \in RV_{\alpha - 1}$.
   Moreover, if $f \in RV_\alpha$, $\alpha \leq 0$, and $f(t) = \int_t^\infty \psi(s)\, ds < \infty$ with ψ nonincreasing, then $t\psi(t)/f(t) \to -\alpha$, as $t \to \infty$. Hence in case $\alpha < 0$ we have $\psi \in RV_{\alpha - 1}$.

Proof. (1)–(5) Properties 1, 3, and 5 follow immediately from the representation theorem (Theorem B.1.6). In order to prove regular variation of $|f'|$ in property 3 one also needs Remark B.1.7 following Theorem B.1.6. Properties 2 and 4 are easy consequences of the uniform convergence theorem (Theorem B.1.4) and Theorem B.1.5 respectively.

(6) Take $\xi > 0$. By property 5 there exists t_0' such that if $t \geq t_0'$,

$$\frac{f(tx)}{f(t)} < 2x^{\alpha+1} \quad \text{for} \quad x \geq 1 .$$

Also, by property 3, if $t \geq t_0''$,

$$\frac{f(tx)}{f(t)} \leq \frac{\sup_{u \leq t} f(u)}{f(t)} < 2 \quad \text{for} \quad 0 < x < 1 .$$

Hence if $t > t_0 := \max(t_0', t_0'')$,

$$\frac{f(tx)}{f(t)} < \max(2, 2\,\xi^{\alpha+1}) \quad \text{for} \quad 0 < x \leq \xi .$$

(7) Apply property 6 above to the function $t^{-\alpha+\delta} f(t)$.

(8) Since $f(t) \to \infty, t \to \infty$, and f is eventually strictly increasing and differentiable, there exists, for x sufficiently large, a unique differentiable inverse function $g(x) = f^{\leftarrow}(x)$ for all x and

$$f(g(x)) = g(f(x)) = x \quad \text{for} \quad x > x_0 . \tag{B.1.23}$$

Differentiating the second equality in (B.1.23) we get, using (B.1.22),

$$\frac{g'(f(x))f(x)}{g(f(x))} = \frac{1}{a(x)} . \tag{B.1.24}$$

Since f is continuous and $f(x) \to \infty, x \to \infty$, (B.1.24) implies

$$\frac{tg'(t)}{g(t)} \to \frac{1}{\alpha} , \quad t \to \infty .$$

Application of Theorem B.1.5 gives $g' \in RV_{-1+1/\alpha}$; hence $g = f^{\leftarrow} \in RV_{1/\alpha}$ by property 4 above.

(9) Suppose $f \in RV_\alpha, \alpha > 0$. By Theorem B.1.6 and the remarks thereafter f has the representation (B.1.6) with $t_0 = 1$ and a continuous. For arbitrary $\varepsilon > 0$ there exists $x_0 = x_0(\varepsilon)$ such that for $x > x_0$,

$$(c_0 - \varepsilon)g(x) \leq f(x) \leq (c_0 + \varepsilon)g(x) , \tag{B.1.25}$$

where $g(x) = \exp\left(\int_1^x a(s)/s\,ds\right)$.

The inequality (B.1.25) implies

$$g^\leftarrow \left(\frac{x}{c_0 - \varepsilon}\right) \geq \bar{f}(x) \geq g^\leftarrow \left(\frac{x}{c_0 + \varepsilon}\right) \tag{B.1.26}$$

for x sufficiently large. By property 8 above we have $g^\leftarrow \in RV_{1/\alpha}$. Hence $g^\leftarrow(x/(c_0 \pm \varepsilon)) \sim (c_0 \pm \varepsilon)^{-1/\alpha} g^\leftarrow(x)$. Since $\varepsilon > 0$ is arbitrary, (B.1.26) implies $\bar{f} \sim c_0^{-1/\alpha} g^\leftarrow \in RV_{1/\alpha}$.

(10) Without loss of generality we may and do suppose f bounded on finite intervals of $\mathbb{R}^+$. Then the proof of property 9 gives the existence of functions g and $g^\leftarrow$ such that $f(x) \sim g(x)$, $\bar{f}(x) \sim g^\leftarrow(x)$, as $x \to \infty$, $g(g^\leftarrow(x)) = g^\leftarrow(g(x)) = x$ for x sufficiently large. This implies $x = g^\leftarrow(g(x)) \sim g^\leftarrow(f(x)) \sim \bar{f}(f(x))$, as $x \to \infty$, where the first asymptotic equivalence follows from $f(x) \sim g(x)$, $x \to \infty$, $g^\leftarrow \in RV_{1/\alpha}$, and the uniform convergence theorem. The statement $f(\bar{f}(x)) \sim x$, $x \to \infty$ follows similarly.

Suppose now

$$f(h_i(x)) \sim h_i(f(x)) \sim x \, (x \to \infty) \quad \text{for} \quad i = 1, 2 \, .$$

Now $\lim_{n\to\infty} f(h_1(x_n))/f(h_2(x_n)) = \lim_{n\to\infty}(h_1(x_n)/h_2(x_n))^\alpha$ for any sequence $x_n \to \infty$ by the uniform convergence theorem; hence $h_1(x) \sim h_2(x)$, $x \to \infty$.

(11) Suppose first that ψ is nondecreasing and $f(t) = f(t_0) + \int_{t_0}^t \psi(s) \, ds$ for $t \geq t_0$. Then for $a > 1$ and $t \geq t_0$ we have

$$\frac{t(a-1)\psi(t)}{f(t)} \leq \int_1^a \frac{t\psi(tv)dv}{f(t)} = \frac{f(ta) - f(t)}{f(t)} \, .$$

Since $f \in RV_\alpha$ we find that $\limsup_{t\to\infty} t\psi(t)/f(t) \leq (a^\alpha - 1)/(a - 1)$ for all $a > 1$. Letting $a \to 1$ we get

$$\limsup_{t\to\infty} \frac{t\psi(t)}{f(t)} \leq \alpha \, .$$

Similar inequalities for $0 < a < 1$ lead to $\liminf_{t\to\infty} t\psi(t)/f(t) \geq \alpha$. The cases ψ nonincreasing and $\alpha < 0$ can be proved similarly. ∎

A stronger version of Proposition B.1.9(5) is as follows.

Proposition B.1.10 (Drees (1998)) *If $f \in RV_\alpha$, for each $\varepsilon, \delta > 0$ there is a $t_0 = t_0(\varepsilon, \delta)$ such that for $t, tx \geq t_0$,*

$$\left| \frac{f(tx)}{f(t)} - x^\alpha \right| \leq \varepsilon \max(x^{\alpha+\delta}, x^{\alpha-\delta}) \, .$$

Proof. Clearly it is sufficient to prove the statement for $\alpha = 0$. We apply Proposition B.1.9(5): for $t, tx \geq t_0$, and $\delta > \delta_2$,

$$e^{-\delta|\log x|}\left((1 - \delta_1)e^{-\delta_2|\log x|} - 1\right) \leq e^{-\delta|\log x|}\left(\frac{f(tx)}{f(t)} - 1\right)$$

$$\leq e^{-(\delta-\delta_2)|\log x|}\left((1 + \delta_1) - e^{-\delta_2|\log x|}\right) \, .$$

We prove that the left-hand side of the inequality tends to zero when $\delta_1, \delta_2 \to 0$ uniformly for $x > 0$. The proof for the right-hand side is similar. Let, as the δ's go to zero, x be such that $\delta_2 | \log x | \to 0$. Then the second factor goes to zero (note that the first factor is bounded). If $\delta_2 | \log x | \to c \in (0, \infty]$, then $| \log x | \to \infty$ and hence the first factor goes to zero (and the second is bounded). ∎

Remark B.1.11 1. There is no analogue of property (3) in case $\alpha = 0$; even if $\lim_{t \to \infty} f(t) = \infty$ with $f \in RV_0$, then f is not necessarily asymptotic to a nondecreasing function, as the following example (due to Karamata) shows. Define $f(x) := \exp \left(\int_0^x \varepsilon(s)/s \, ds \right)$, where

$$
\varepsilon(s) = \begin{cases}
0, & 0 \le s \le 1, \\
a_n, & (2n - 1)! < s \le (2n)!, \ n = 1, 2, 3, \ldots, \\
-a_n/2, & (2n)! < s \le (2n + 1)!, \ n = 1, 2, 3, \ldots,
\end{cases}
$$

where the sequence a_n is such that $a_n \to 0, n \to \infty$, and $a_n \log n \to \infty, n \to \infty$. Then

$$
\sup_{0 < x \le (2n+1)!} \frac{f(x)}{f((2n+1)!)} = \frac{f((2n)!)}{f((2n+1)!)}
$$

$$
= \exp \left(- \int_{(2n)!}^{(2n+1)!} \frac{\varepsilon(s)}{s} \, ds \right) = \exp \left(\frac{a_n}{2} \log(2n + 1) \right) \to \infty, \quad n \to \infty.
$$

Hence (B.1.13) does not hold.
2. Using the representation theorem for regularly varying functions, it is possible to show that if f is locally bounded and $f \in RV_0$, then the function $\sup_{0 < x < t} f(x)$ is slowly varying.
3. Note that $\int_0^t f(s) \, ds \in RV_{\alpha+1}$ with $\alpha > -1$ does not imply $f \in RV_\alpha$. Example: $f(t) = \exp[\log t]$.

The following result is a generalization of Theorem B.1.5 (the kernel function k below is constant in Theorem B.1.5).

Theorem B.1.12 *Let* $f \in RV_\alpha$ *and suppose* f *is (Lebesgue) integrable on finite intervals of* $\mathbb{R}^+$.

1. *If* $\alpha > -1$ *and the function* $k : \mathbb{R}^+ \to \mathbb{R}$ *is bounded on* $(0, 1)$, *then*

$$
\lim_{t \to \infty} \int_0^1 k(s) \frac{f(ts)}{f(t)} \, ds = \int_0^1 k(s) s^\alpha \, ds . \tag{B.1.27}
$$

2. *If* $t^{+\varepsilon+\alpha} k(t)$ *is integrable on* $(1, \infty)$ *for some* $\varepsilon > 0$, *then* $\int_1^\infty k(s) f(ts) \, ds < \infty$ *for* $t > 0$, *and*

$$
\lim_{t \to \infty} \int_1^\infty k(s) \frac{f(ts)}{f(t)} \, ds = \int_1^\infty k(s) s^\alpha \, ds . \tag{B.1.28}
$$

Proof. (1) Note that for $0 < \varepsilon < \alpha + 1$ the function $t^{\alpha - \varepsilon} k(t)$ is integrable on $(0, 1)$. Since there exist $c > 1$ and $\varepsilon > 0$ such that $f(tx)/f(t) \leq cx^{\alpha - \varepsilon}$ for $tx \geq t_0, 0 < x \leq 1$, by Proposition B.1.9(5), we can apply Lebesgue's dominated convergence theorem to obtain

$$\int_{t_0/t}^{1} k(s) \frac{f(ts)}{f(t)} \, ds \to \int_{0}^{1} k(s) s^{\alpha} \, ds \, , \quad t \to \infty \, .$$

Furthermore,

$$\left| \int_{0}^{t_0/t} k(s) \frac{f(ts)}{f(t)} \, ds \right| \leq \frac{1}{tf(t)} \int_{0}^{t_0} \left| k \left(\frac{s}{t} \right) f(s) \right| \, ds \to 0 \, , \quad t \to \infty,$$

since k is bounded and $tf(t) \to \infty$, as $t \to \infty$.

(2) The second statement is proved in a similar way. ∎

Remark B.1.13 It is easy to see that the conditions in Theorem B.1.12(1) can be replaced by f bounded on $(0, 1)$ and $t^{\alpha - \varepsilon} k(t)$ integrable on $(0, 1)$ for some $\varepsilon > 0$.

Remark B.1.14 De Bruijn (1959) noted that for any slowly varying function L there exists an asymptotically unique slowly varying function $L^\star$, called the *conjugate slowly varying function*, satisfying $L(x) L^\star(xL(x)) \to 1$, $L^\star(x) L(xL^\star(x)) \to 1$, $x \to \infty$. Note that one can obtain $L^\star$ as follows: Define $h(x) := xL(x)$. Then $L^\star(x) \sim \bar{h}(x)/x$, $x \to \infty$. In special cases one has $L^\star(x) \sim 1/L(x)$, $x \to \infty$.

Example: $L(x) \sim (\log x)^{\alpha} (\log \log x)^{\beta}$, $x \to \infty$, $\alpha > 0$, $\beta \in \mathbb{R}$, i.e., if $h(x) \sim x(\log x)^{\alpha} (\log \log x)^{\beta}$, $x \to \infty$, then $\bar{h}(x) \sim x(\log x)^{-\alpha} (\log \log x)^{-\beta}$, $x \to \infty$. If we replace x by x^{γ} and take $\beta = 0$, we obtain $f(x) \sim x^{\gamma} (\log x)^{\delta}$, $\gamma > 0$, $\delta \in \mathbb{R}$, implies $\bar{f}(x) \sim \gamma^{\delta/\gamma} x^{1/\gamma} (\log x)^{-\delta/\gamma}$, $x \to \infty$.

B.2 Extended Regular Variation (ERV); The class Π

By way of introduction for the class Π, which is a generalization of the class RV, we formulate the RV property somewhat differently. A measurable function $f : \mathbb{R}^+ \to \mathbb{R}$ is RV if there exists a positive function a such that for all $x > 0$ the limit

$$\lim_{t \to \infty} \frac{f(tx)}{a(t)}$$

exists and is positive.

An obvious generalization is the following: Suppose $f : \mathbb{R}^+ \to \mathbb{R}$ is measurable and there exists a real function $a > 0$ such that for all $x > 0$ the limit

$$\lim_{t \to \infty} \frac{f(tx) - f(t)}{a(t)} \tag{B.2.1}$$

exists and the limit function is not constant (this is to avoid trivialities). First note that (B.2.1) is equivalent to the existence of

$$\psi(x) := \lim_{t \to \infty} \frac{f(tx) - f(t)}{a(t)} \qquad \text{(B.2.2)}$$

for all $x > 0$ with ψ not constant.

Next we identify the class of possible limit functions ψ.

Theorem B.2.1 *Suppose $f : \mathbb{R} \to \mathbb{R}$ is measurable and a is positive. If (B.2.2) holds with ψ not constant, then*

$$\psi(x) = c\frac{x^\gamma - 1}{\gamma} , \qquad x > 0 , \qquad \text{(B.2.3)}$$

for some $\gamma \in \mathbb{R}$, $c \neq 0$ (for $\gamma = 0$ read $\psi(x) = c \log x$). Moreover, (B.2.1) holds with a function a that is measurable and is RV_γ.

Proof. Since ψ is not constant, there exists $x_0 > 0$ such that $\psi(x_0) \neq 0$. From (B.2.1) it follows that we can choose $a(t) = (f(x_0 t) - f(t))/\psi(x_0)$. Hence without loss of generality we may assume a to be measurable. For $y > 0$ arbitrary we have

$$\frac{a(ty)}{a(t)} = \frac{\frac{f(tx_0 y) - f(t)}{a(t)} - \frac{f(ty) - f(t)}{a(t)}}{\frac{f(tx_0 y) - f(ty)}{a(ty)}} \to \frac{\psi(x_0 y) - \psi(y)}{\psi(x_0)} , \qquad t \to \infty .$$

Hence $A(y) := \lim_{t \to \infty} a(ty)/a(t)$ exists (and is nonnegative) for all $y > 0$. Since

$$\frac{a(txy)}{a(t)} = \frac{a(txy)}{a(tx)} \frac{a(tx)}{a(t)}$$

we have

$$A(xy) = A(x)A(y) \quad \text{for all} \quad x, y > 0 . \qquad \text{(B.2.4)}$$

Since a is measurable, the function A is measurable. Moreover, the only measurable solutions of Cauchy's functional equation (B.2.4) are $A(y) = y^\gamma$ for some $\gamma \in \mathbb{R}$ (see the proof of Theorem B.1.3) and $A(y) = 0$ for $y > 0$.

However, if $A(y) = 0$ for $y > 0$, then since $A(y)\psi(x) = \psi(xy) - \psi(y)$ for all $x, y > 0$, we have that ψ is constant, contrary to our assumption. Hence $a \in RV_\gamma$ for some $\gamma \in \mathbb{R}$. As a consequence we have

$$y^\gamma \psi(x) = \psi(xy) - \psi(y) \quad \text{for all} \quad x, y > 0 . \qquad \text{(B.2.5)}$$

If $\gamma = 0$ we have Cauchy's functional equation again and $\psi(y) = c \log x$ for some $c \neq 0$ and all $x > 0$.

Next suppose $\gamma \neq 0$. Interchanging x and y in (B.2.5) and subtracting the resulting relations, we get

$$\psi(x)(1 - y^\gamma) = \psi(y)(1 - x^\gamma) \quad \text{for} \quad x, y > 0 .$$

Hence $\psi(x)/(1 - x^\gamma)$ is constant, i.e., $\psi(x) = c(1 - x^\gamma)/\gamma$ for $x > 0$, with $c \neq 0$. ∎

The following theorem states that for $\gamma \neq 0$ relation (B.2.1) defines classes of functions we have met before. Note that it is sufficient to consider (B.2.3) with $c > 0$ since replacing f by $-f$ in (B.2.1) changes the sign of c.

Theorem B.2.2 *Suppose the assumptions of Theorem B.2.1 are satisfied with $\gamma \neq 0$ and $c > 0$, i.e.,*

$$\lim_{t \to \infty} \frac{f(tx) - f(t)}{a(t)} = c \frac{x^\gamma - 1}{\gamma} .$$

1. *If $\gamma > 0$ then $\lim_{t\to\infty} f(t)/a(t) = c/\gamma$, and hence $f \in RV_\gamma$.*
2. *If $\gamma < 0$ then $f(\infty) := \lim_{x\to\infty} f(x)$ exists, $\lim_{t\to\infty}(f(\infty) - f(t))/a(t) = -c/\gamma$, and hence $f(\infty) - f(x) \in RV_\gamma$.*

Proof. The proofs of Theorem B.2.9 and Corollary B.2.11 below can easily be adapted to show that if $\gamma > 0$ $(\gamma < 0)$ there is a nondecreasing (nonincreasing) function g such that

$$f(t) - g(t) = o(a(t)) , \qquad t \to \infty . \tag{B.2.6}$$

Since we may assume $a \in RV_\gamma$ (Theorem B.2.1), it follows that we also have

$$\lim_{t \to \infty} \frac{g(tx) - g(t)}{a(t)} = c \frac{x^\gamma - 1}{\gamma}. \tag{B.2.7}$$

It will become apparent that it is sufficient to prove the theorem for g. Take $y > 1$ arbitrarily and define $t_1 = 1$ and $t_{n+1} = t_n y$ for $n = 1, 2, \ldots$. We have, by (B.2.7),

$$\lim_{n \to \infty} \frac{g(t_{n+2}) - g(t_{n+1})}{g(t_{n+1}) - g(t_n)} = y^\gamma. \tag{B.2.8}$$

Suppose $\gamma > 0$. Then (B.2.8) immediately implies $g(t_n) \to \infty, n \to \infty$. Further, for any $\varepsilon > 0$ there exists n_0 such that for any $n > n_0$,

$$g(t_{n+2}) - g(t_{n_0+1}) = \sum_{k=n_0}^{n} \{g(t_{k+2}) - g(t_{k+1})\}$$

$$< y^\gamma (1 + \varepsilon) \sum_{k=n_0}^{n} \{g(t_{k+1}) - g(t_k)\}$$

$$= y^\gamma (1 + \varepsilon) \{(g(t_{n+1}) - g(t_{n_0}))\}$$

and a similar lower inequality. It follows that

$$\lim_{n \to \infty} \frac{g(t_{n+1})}{g(t_n)} = y^\gamma \tag{B.2.9}$$

and hence

$$a(t_n) \sim \frac{g(t_{n+1}) - g(t_n)}{c \frac{y^\gamma - 1}{\gamma}} \sim \frac{\gamma}{c} g(t_n) . \tag{B.2.10}$$

Further, for $x > 1$,

$$\frac{g(t_n x)}{g(t_n)} - 1 = \frac{g(t_n x) - g(t_n)}{g(t_n)} \sim \frac{g(t_n x) - g(t_n)}{c\, \frac{a(t_n)}{\gamma}} \to x^\gamma - 1, \quad n \to \infty. \quad \text{(B.2.11)}$$

For any $s > 0$ choose $n(s) \in \mathbb{N}$ such that $t_{n(s)} \le s < t_{n(s)+1}$. Then by (B.2.9) and (B.2.11),

$$\frac{g(sx)}{g(s)} \le \frac{g(t_{n(s)+1}x)}{g(t_{n(s)+1})} \frac{g(t_{n(s)+1})}{g(t_{n(s)})} \to x^\gamma y^\gamma, \quad n \to \infty.$$

Similarly

$$\frac{g(sx)}{g(s)} \ge \frac{g(t_{n(s)}x)}{g(t_{n(s)})} \frac{g(t_{n(s)})}{g(t_{n(s)+1})} \to x^\gamma y^{-\gamma} (n \to \infty).$$

Since $y > 1$ is arbitrary, we have proved $g \in RV_\gamma$. Combining with (B.2.7) gives $a(t)/g(t) \to \gamma/c, t \to \infty$. With (B.2.6) this implies $f(t) \sim ca(t)/\gamma, t \to \infty$, hence $f \in RV_\gamma$.

Suppose next $\gamma < 0$. Then (B.2.8) immediately implies $\lim_{n\to\infty} g(t_n) < \infty$. Write $h(x) := \lim_{t\to\infty} g(t) - g(x)$. We have

$$\frac{h(t_n)}{a(t_n)} = \sum_{k=n}^\infty \frac{g(t_{k+1}) - g(t_k)}{a(t_k)} \frac{a(t_k)}{a(t_n)}.$$

Choose $\varepsilon > 0$ and $y > (1 + \varepsilon)^{-1/\gamma}$. Note that since $a \in RV_\gamma$ the above expression is bounded above for $n \ge n_0$ by

$$\sum_{k=n}^\infty c \frac{y^\gamma - 1}{\gamma} (1 + \varepsilon) \left(y^\gamma (1 + \varepsilon)\right)^{k-n} = \frac{y^\gamma - 1}{1 - y^\gamma (1 + \varepsilon)} \frac{1 + \varepsilon}{\gamma} c,$$

which tends to $-c/\gamma$ as $\varepsilon \downarrow 0$. A similar lower bound is easily obtained, and we conclude that

$$\lim_{n\to\infty} \frac{h(t_n)}{a(t_n)} = -\frac{c}{\gamma}.$$

Further, for $x > 1$,

$$\frac{h(t_{n+1})}{h(t_n)} - 1 = \frac{a(t_n)}{h(t_n)} \frac{h(t_{n+1}) - h(t_n)}{a(t_n)} \to y^\gamma - 1, \quad n \to \infty.$$

The rest of the proof follows closely the case $\gamma > 0$. ∎

Definition B.2.3 A measurable function $f : \mathbb{R}^+ \to \mathbb{R}$ is said to be of *extended regular variation* if there exists a function $a : \mathbb{R}^+ \to \mathbb{R}^+$ such that for some $\gamma \in \mathbb{R}$ and all $x > 0$,

$$\lim_{t\to\infty} \frac{f(tx) - f(t)}{a(t)} = \frac{x^\gamma - 1}{\gamma}.$$

Notation: $f \in ERV$ or $f \in ERV_\gamma$. The function a is called an *auxiliary function* for f.

Definition B.2.4 A measurable function $f : \mathbb{R}^+ \to \mathbb{R}$ is said to belong to the class Π if there exists a function $a : \mathbb{R}^+ \to \mathbb{R}^+$ such that for $x > 0$,

$$\lim_{t \to \infty} \frac{f(tx) - f(t)}{a(t)} = \log x \ . \tag{B.2.12}$$

Notation: $f \in \Pi$ or $f \in \Pi(a)$. The function a is called an auxiliary function for f.

Example B.2.5 The functions $f(t) = \log t + o(1)$, $f(t) = (\log t)^\alpha (\log \log t)^\beta + o(\log t)^{\alpha - 1}$, $\alpha > 0$, $\beta \in \mathbb{R}$, $f(t) = \exp((\log t)^\alpha) + o(\log t)^{\alpha - 1} \exp((\log t)^\alpha)$, for $0 < \alpha < 1$, $f(t) = t^{-1} \log \Gamma(t) + o(1)$, as $t \to \infty$, are in Π. The functions $f(t) = [\log t]$, $f(t) = 2 \log t + \sin \log t$ are in RV_0 but not in Π.

Remark B.2.6   1. Note that any positive function a_1 is an auxiliary function for f if and only if $a_1(t) \sim a(t)$, $t \to \infty$.
2. For the definition of Π it is sufficient to require (B.2.12) for all x in a set A satisfying the following requirements: $\lambda(A) > 0$ and there exists a sequence $x_n \in A$, $n = 1, 2, \ldots$, such that $x_n \to 1$, $n \to \infty$.
3. We can weaken the definition of Π as follows: there exist functions $a : \mathbb{R}^+ \to \mathbb{R}^+$ and $g : \mathbb{R}^+ \to \mathbb{R}$ such that for $x > 0$,

$$\lim_{t \to \infty} \frac{f(tx) - g(t)}{a(t)} = \log x \ .$$

Theorem B.2.7 If $f \in \Pi(a)$, then $\lim_{t \to \infty} a(tx)/a(t) = 1$ for all $x > 0$. Moreover, (B.2.12) holds with a function a that is measurable and hence is RV_0.

Proof. This is a special case of Theorem B.2.1.   ■

Theorem B.2.8 If $f \in \Pi(a)$ and $g : \mathbb{R}^+ \to \mathbb{R}$ is measurable and satisfies

$$\lim_{t \to \infty} \frac{f(t) - g(t)}{a(t)} = c \tag{B.2.13}$$

for some $c \in \mathbb{R}$, then (B.2.12) is satisfied with f replaced by g; hence $g \in \Pi(a)$.

This follows immediately from (B.1.23) and (B.1.24). Obviously for a fixed auxiliary function a the relation (B.1.24) between functions $f, g \in \Pi(a)$ is an equivalence relation. We shall see below (Proposition B.2.15(3)) that any equivalence class contains a smooth Π-function.

Theorem B.2.9 (Uniform convergence theorem) If $f \in \Pi$, then for $0 < a < b < \infty$ relation (B.2.12) holds uniformly for $x \in [a, b]$.

Proof. Define $F(t) := f(e^t)$ and $A(t) := a(e^t)$. It is sufficient to deduce a contradiction from the following assumption: there exist $\delta > 0$ and sequences $t_n \to \infty$, $x_n \to 0$, $n \to \infty$, such that for all n,

$$\left| \frac{F(x_n + t_n) - F(t_n)}{A(t_n)} \right| > \delta \ .$$

Consider the sets

$$J := [-\delta/5, +\delta/5]\,,$$

$$Y_{1,n} = \{y \ : \ \left| \frac{F(t_n + y) - F(t_n)}{A(t_n)} \right| > \frac{\delta}{2}\,, \ y \in J\}\,,$$

$$Y_{2,n} = \{y \ : \ \left| \frac{F(t_n + x_n) - F(t_n + y)}{A(t_n)} \right| > \frac{\delta}{2}\,, \ y \in J\}\,.$$

The above sets are measurable for each n and $Y_{1,n} \cup Y_{2,n} = J$; hence either $\lambda(Y_{1,n}) \geq 1/2\lambda(J)$ or $\lambda(Y_{2,n}) \geq 1/2\lambda(J)$ (or both), where λ denotes Lebesgue measure. Define

$$Z_{1,n} = \left\{z \ : \ \left| \frac{F(t_n + x_n) - F(t_n + x_n - z)}{A(t_n)} \right| > \frac{\delta}{2}\,, \ x_n - z \in J\right\}\,.$$

Then $\lambda(Z_{1,n}) = \lambda(Y_{2,n})$.

Since $a \in RV_0$ (Theorem B.2.1) we have the inequality $A(t_n) \geq A(t_n + x_n - z)/2$ for $z \in Z_{1,n}$ and $n \geq n_0$ by Proposition B.1.9(5). As a consequence, $Z_{1,n} \subset Z_{2,n}$ for $n \geq n_0$, where $Z_{2,n}$ is defined by

$$Z_{2,n} := \left\{z \ : \ \left| \frac{F(t_n + x_n) - F(t_n + x_n - z)}{A(t_n + x_n - z)} \right| > \frac{\delta}{4}\,, \ x_n - z \in J\right\} \subset \left[-\frac{\delta}{4}, \frac{\delta}{4}\right]$$

for n sufficiently large since $x_n \to 0$. Hence we find that $\lambda\left(\lim_{n\to\infty} \sup Z_{2,n}\right) \geq \lambda\left(\lim_{n\to\infty} \sup Z_{1,n}\right) \geq \lambda(J)/2$ or $\lambda\left(\lim_{n\to\infty} \sup Y_{1,n}\right) \geq \lambda(J)/2$. This implies the existence of a real number x_0 contained in infinitely many $Y_{1,n}$ or infinitely many $Z_{2,n}$, which contradicts the assumption $\lim_{t\to\infty}(F(t + x_0) - F(t))/A(t) = x_0$. ∎

Corollary B.2.10 *If* $f \in \Pi(a)$, *for any* $\varepsilon > 0$ *there exist* $t_0, c > 0$ *such that for* $t \geq t_0, x \geq 1$,

$$\left| \frac{f(tx) - f(t)}{a(t)} \right| \leq cx^\varepsilon. \tag{B.2.14}$$

Hence $f(t)$ *is bounded for* $t \geq t_0$.

Proof. By the uniform convergence theorem (Theorem B.2.9) we have

$$-2 \leq \frac{f(tu) - f(t)}{a(t)} \leq 2 \tag{B.2.15}$$

for $t \geq t_1$ and $1 \leq u \leq e$. For $x > 1$ define $n \in \mathbb{N}$ by $e^n \leq x < e^{n+1}$. Then

$$\frac{f(tx) - f(t)}{a(t)} = \sum_{k=0}^{n-1} \frac{f(e^{k+1}t) - f(e^k t)}{a(e^k t)}\frac{a(e^k t)}{a(t)} + \frac{f(tx) - f(e^n t)}{a(e^n t)}\frac{a(e^n t)}{a(t)}\,.$$

Using (B.2.15) and the inequality $a(tx)/a(t) \leq c_1 x^\varepsilon$ for some $c_1 > 0, t \geq t_2$ (Proposition B.1.9(5)), we find that for $t \geq t_0 := \max(t_1, t_2)$,

$$\left| \frac{f(tx) - f(t)}{a(t)} \right| \leq 2c_1 \sum_{k=0}^{n} e^{\varepsilon k} \leq ce^{n\varepsilon} \leq cx^\varepsilon\,.$$

For the last statement, take $t = t_0$ in (B.2.14). ∎

Corollary B.2.11 *If $f \in \Pi(a)$, there exists a nondecreasing function g such that $f(t) - g(t) = o(a(t))$, $t \to \infty$. In particular, $g \in \Pi(a)$ by Theorem B.2.8.*

Proof. By Corollary B.2.10 the function f is locally integrable on $[t_0, \infty)$. Note that by Theorem B.2.9,

$$\lim_{t \to \infty} \int_1^e \frac{f(tx) - f(t)}{a(t)} \frac{dx}{x} = \int_1^e \log x \, \frac{dx}{x} = \frac{1}{2} . \tag{B.2.16}$$

Now choose $t_1 \geq t_0$ such that $f(ex) - f(x) > 0$ for $x > t_1$. Then

$$\int_1^e \frac{f(tx)}{x} dx = \int_{t_1}^{te} \frac{f(x)}{x} dx - \int_{t_1}^t \frac{f(x)}{x} dx$$

$$= \int_{t_1}^{et_1} \frac{f(x)}{x} dx + \int_{t_1}^t \frac{f(ex) - f(x)}{x} dx$$

$$=: g_0(t) .$$

Note that g_0 is nondecreasing and by (B.2.16),

$$\lim_{t \to \infty} \frac{g_0(t) - f(t)}{a(t)} = \frac{1}{2} .$$

Now $g_0 \in \Pi(a)$ by Theorem B.2.8. Define $g(t) := g_0(te^{-1/2})$. Then $g \in \Pi(a)$ and $g(t) - f(t) = o(a(t))$, $t \to \infty$. ∎

The following theorem gives a characterization of the class Π.

Theorem B.2.12 *Suppose $f : \mathbb{R}^+ \to \mathbb{R}$ is measurable. For $t_0 \geq 0$ let $\varphi : (t_0, \infty) \to \mathbb{R}$ be defined by*

$$\varphi(t) := f(t) - t^{-1} \int_{t_0}^t f(s) ds. \tag{B.2.17}$$

The following statements are equivalent:

1.

$$f \in \Pi. \tag{B.2.18}$$

2. *The function $\varphi : (t_0, \infty) \to \mathbb{R}$ is well-defined for some $t_0 \geq 0$ and eventually positive, and*

$$\lim_{t \to \infty} \frac{f(tx) - f(t)}{\varphi(t)} = \log x , \quad x > 0 . \tag{B.2.19}$$

3. *The function $\varphi : (t_0, \infty) \to \mathbb{R}$ is well-defined for $t \geq t_0$ and slowly varying at infinity.*

4. *There exists $\rho \in RV_0$ such that*

$$f(t) = \rho(t) + \int_{t_0}^t \rho(s) \frac{ds}{s} . \tag{B.2.20}$$

5. *There exist $c_1, c_2 \in \mathbb{R}$, $a_1, a_2 \in RV_0$ with $a_1(t) \sim a_2(t)$, $t \to \infty$, such that*

$$f(t) = c_1 + c_2 a_1(t) + \int_1^t a_2(s) \, \frac{ds}{s} \ . \tag{B.2.21}$$

If f satisfies (B.2.20) (or (B.2.21)) then $f \in \Pi(\rho)$ (or $f \in \Pi(a_2)$ respectively). Hence $\rho(t) \sim a_2(t) \sim \varphi(t)$, $t \to \infty$.

Proof. We start by proving that (1) implies (2). Suppose $f \in \Pi(a)$. Take t_0 as in Corollary B.2.10. Then $\varphi(t)$ is well-defined for $t \geq t_0$. Note that for $t \geq t_0$,

$$\frac{\varphi(t)}{a(t)} = \frac{t_0 f(t)}{t a(t)} + \int_{t_0/t}^1 \frac{f(t) - f(tu)}{a(t)} \, du \ . \tag{B.2.22}$$

From Corollary B.2.10 it follows that $f(t) = o(t^\beta)$, $t \to \infty$, for any $\beta > 0$ (take $t = t_0$ in (B.2.14)). Since $ta(t) \in RV_1$ (Theorem B.2.7), we have $f(t) = o(ta(t))$, $t \to \infty$.

We can apply Lebesgue's theorem on dominated convergence to the second term on the right-hand side in (B.2.22) since by Corollary B.2.10 for $tu \geq t_0$, $0 < u < 1$,

$$\left| \frac{f\left(\frac{tu}{u}\right) - f(tu)}{a(tu)} \right| \leq cu^{-\varepsilon},$$

and by Proposition B.1.9(5) for $tu \geq t_1$, $0 < u \leq 1$,

$$0 < \frac{a(tu)}{a(t)} \leq c_1 u^{-\varepsilon} \ .$$

Hence $\lim_{t \to \infty} \varphi(t)/a(t) = -\int_0^1 \log u \, du = 1$, which proves that (1) implies (2). For proving that (2) implies (3) see Theorem B.2.7.

Next we prove that (3) implies (4). By Fubini's theorem we have

$$\int_{t_0}^t \frac{\varphi(s)}{s} \, ds = \int_{t_0}^t \frac{f(s)}{s} \, ds - \int_{t_0}^t \int_{t_0}^s \frac{f(u)}{s^2} \, du \, ds$$

$$= \frac{1}{t} \int_{t_0}^t f(u) \, du = f(t) - \varphi(t) \ .$$

Hence (B.2.20) with $\rho = \varphi$.

Finally, we prove that (4) implies (1). By the uniform convergence theorem (Theorem B.1.4) for functions in RV,

$$\frac{f(tx) - f(t)}{a_2(t)} = c_2 \left(\frac{a_1(tx)}{a_1(t)} - 1 \right) \frac{a_1(t)}{a_2(t)} + \int_1^x \frac{a_2(tu)}{a_2(t)} \frac{du}{u} \to \log x \ ,$$

as $t \to \infty$, for all $x > 0$. ∎

Corollary B.2.13 *If $f \in \Pi$, then $\lim_{t \to \infty} f(t) =: f(\infty) \leq \infty$ exists. If the limit is infinite, then $f \in RV_0$. If the limit is finite, then $f(\infty) - f(t) \in RV_0$. Moreover,*

$$a(t) = o(f(t)) \ , \qquad as \ t \to \infty, \tag{B.2.23}$$

and when $f(\infty) < \infty$,

$$a(t) = o(f(\infty) - f(t)) \ , \qquad as \ t \to \infty \ .$$

Proof. Consider the representation (B.2.20). Theorem B.1.5 implies that $\rho(t) = o(\int_1^t \rho(s)/s\, ds), t \to \infty$. Hence, if $\int_1^\infty \rho(s)/s\, ds < \infty, \rho(t) \to 0, t \to \infty$, and $\lim_{t \to \infty} f(t) = c + \int_1^\infty \rho(s)/s\, ds$. Then $f(\infty) - f(t) = \int_t^\infty \rho(s)/s\, ds \in RV_0$ (Proposition B.1.9(4)). If $\int_1^\infty \rho(s)/s\, ds = \infty$, then $f(t) \sim \int_1^t \rho(s)/s\, ds \in RV_0$ (Proposition B.1.9(4)).

When comparing (B.2.12) and (B.2.19) one sees that $\varphi(t) \sim a(t)$, as $t \to \infty$. Now Theorem B.1.5 implies $\varphi(t) = o(f(t))$. Relation (B.2.23) follows. The second relation follows in a similar way. ∎

Remark B.2.14   1. Note that from the proof of Corollary B.2.13 it follows, using (B.2.17), that $\varphi(t) \sim a(t), t \to \infty$. As a consequence of Corollary B.2.13, the limit relation (B.2.13) above is strictly stronger than $f(t) \sim g(t), t \to \infty$.

2. Theorem B.2.12 is also true (and the proof not much different) with φ replaced by the function

$$t \int_t^\infty f(u) \frac{du}{u^2} - f(t) \, .$$

3. The result of Corollary B.2.11 is obtained again from Theorem B.2.12 by taking $g(t) = \int_{t_0}^{et} \rho(s)/s\, ds$ with ρ as in (B.2.21).

4. Suppose f is locally integrable on $\mathbb{R}^+$ and $a \in RV_0$. Then

$$\frac{f(tx) - f(t)}{a(t)} \to 0 \, , \quad t \to \infty \, , \tag{B.2.24}$$

for $x > 0$, and

$$\frac{f(t) - t^{-1} \int_0^t f(s)\, ds}{a(t)} \to 0 \, , \quad t \to \infty \, , \tag{B.2.25}$$

are equivalent. The proof follows closely the proof of Theorem B.2.12.

5. From Theorem B.2.12(5) it is clear that for any $a \in RV_0$, there exists a function f such that $f \in \Pi(a)$.

6. Let $t_1 \geq 0$ be such that f is locally integrable on (t_1, ∞). Then Theorem B.2.12 holds for any $t_0 \geq t_1$.

We mention some properties of functions that belong to the class Π.

Proposition B.2.15   *1. If $f, g \in \Pi$ then $f + g \in \Pi$. If $f \in \Pi$, and $h \in RV_\alpha$, $\alpha > 0$, then $f \circ h \in \Pi$, where $h \circ f$ denotes the composition of the two functions. If $f \in \Pi$, $\lim_{t \to \infty} f(t) = \infty$, and h is differentiable with $h' \in RV_\alpha$, $\alpha > -1$, then $h \circ f \in \Pi$.*

2. If $f \in \Pi(a)$ is integrable on finite intervals of $\mathbb{R}^+$ and the function f_1 is defined by

$$f_1(t) := t^{-1} \int_0^t f(s)\, ds \, , \quad t > 0 \, , \tag{B.2.26}$$

then $f_1 \in \Pi(a)$. Conversely, if $f_1 \in \Pi(a)$ and f is nondecreasing, then $f \in \Pi(a)$.

3. *If $f \in \Pi(a)$, there exists a twice differentiable function $\overline{f}$ with $-\overline{f}'' \in RV_{-2}$ such that*

$$\lim_{t \to \infty} \frac{f(t) - \overline{f}(t)}{a(t)} = 0 .$$ (B.2.27)

In particular $\overline{f}$ is eventually concave. As a consequence of this, if $f \in \Pi$ is bounded on finite intervals of $\mathbb{R}^+$ and $\lim_{t \to \infty} f(t) = \infty$, then $\sup_{0 < x \le t} f(x) - f(t) = o(a(t))$, $t \to \infty$.

4. *Suppose $f \in \Pi(a)$. For arbitrary $\delta_1, \delta_2 > 0$ there exists $t_0 = t_0(\delta_1, \delta_2)$ such that for $x \ge 1, t \ge t_0$,*

$$(1 - \delta_2)\frac{1 - x^{\delta_1}}{\delta_1} - \delta_2 < \frac{f(tx) - f(t)}{a(t)} < (1 + \delta_2)\frac{x^{\delta_1} - 1}{\delta_1} + \delta_2 .$$ (B.2.28)

Note that conversely, if f satisfies the above property, then $f \in \Pi(a)$.

5. *Suppose*

$$f(t) = f(t_0) + \int_{t_0}^t g(s) \, ds , \quad t > t_0 ,$$ (B.2.29)

with $g \in RV_{-1}$. Then $f \in \Pi$. Conversely, if $f \in \Pi$ satisfies (B.2.29) with g nonincreasing, then $g \in RV_{-1}$. Moreover, in this case $tg(t)$ is an auxiliary function for f. This property supplements a corresponding statement for functions in RV_α, $a \ne 0$ (cf. Proposition B.1.9(11)).

Proof. (1) The statement $f + g \in \Pi$ is a consequence of representation (B.2.20) since the sum of two slowly varying functions is slowly varying (see Proposition B.1.9(2)).

If $f \in \Pi(a)$ and $h \in RV_\alpha$, then for $x > 0$ we have

$$\lim_{t \to \infty} \frac{f\left(\frac{h(tx)}{h(t)} h(t)\right) - f(h(t))}{\alpha a(h(t))} = \log x$$

by the uniform convergence theorem (Theorem B.2.9).

For the last statement we expand the function h:

$$\frac{h(f(tx)) - h(f(t))}{a(t)h'(f(t))} = \frac{f(tx) - f(t)}{a(t)} \frac{h'(f(t) + \theta\{f(tx) - f(t)\})}{h'(f(t))}$$

for some $0 < \theta = \theta(x, t) < 1$. Now the second factor on the right-hand side tends to 1 as $t \to \infty$ since $h' \in RV_\alpha$ and $f \in RV_0$ (see Corollary B.2.13) by the uniform convergence theorem (Theorem B.2.9).

(2) Define $\varphi(t) := f(t) - t^{-1} \int_0^t f(s) \, ds$ for $t > 0$. If $f \in \Pi(a)$, we have by Theorem B.2.12

$$\lim_{t \to \infty} \frac{f(t) - f_1(t)}{a(t)} = \lim_{t \to \infty} \frac{\varphi(t)}{a(t)} = 1.$$

As a consequence, $f_1 \in \Pi(a)$ (see Theorem B.2.8).

Conversely, suppose $f_1 \in \Pi(a)$. Then for $x > 0$ we have by definition $\int_0^t \varphi(s)/s \, ds = f_1(t)$ and hence

$$\frac{f_1(tx) - f_1(t)}{a(t)} = \int_1^x \frac{\varphi(ts)}{a(t)} \frac{ds}{s} \, .$$

Now fix $x > 1$. Since $f_1 \in \Pi(a)$ the above expression tends to $\log x$ as $t \to \infty$. Since f is nondecreasing, $t\varphi(t)$ is nondecreasing. This implies

$$(1 - x^{-1}) \frac{\varphi(t)}{a(t)} \le \int_1^x \frac{\varphi(ts)}{a(t)} \frac{ds}{s} \, ,$$

for $t > 0$, hence

$$\limsup_{t \to \infty} \frac{\varphi(t)}{a(t)} \le \frac{\log x}{1 - x^{-1}} \quad \text{for} \quad x > 1 \, .$$

Similarly we find that

$$\liminf_{t \to \infty} \frac{\varphi(t)}{a(t)} \ge \frac{-\log x}{x^{-1} - 1} \quad \text{for} \quad 0 < x < 1 \, .$$

Finally, let $x \to 1$ to obtain $\varphi(t) \sim a(t), t \to \infty$, which implies $\varphi \in RV_0$. The proof is finished by application of Theorem B.2.12.

(3) We may assume without loss of generality that f is integrable on finite intervals of $\mathbb{R}^+$. Define the functions f_i for $i = 1, 2, 3$ recursively by

$$f_i(t) := t^{-1} \int_0^t f_{i-1}(s) \, ds$$

for $t > 0$, where $f_0 = f$. Repeated application of Theorems B.2.8 and B.2.12 gives

$$f(t) - f_3(t) = \sum_{i=0}^2 \{f_i(t) - f_{i+1}(t)\} \sim 3a(t) \, , \quad t \to \infty \, .$$

Hence $f_3 \in \Pi(a)$ by Theorem B.2.8. Define $\overline{f}$ by $\overline{f}(t) := f_3(e^3 t)$. Then $f(t) - \overline{f}(t) = o(a(t)), t \to \infty$. Furthermore, $\overline{f}$ is twice differentiable and

$$t^2 f_3''(t) = (f_1(t) - f_2(t)) - 2(f_2(t) - f_3(t)) \sim -a(t) \, , \quad t \to \infty \, ,$$

by Theorem B.2.12.

(4) From Remark B.2.14(2) it follows that there exist functions a_0, b such that $a_0(t) \sim a(t), b(t) = o(a(t)), t \to \infty$, and

$$f(t) = \int_{t'}^t \frac{a_0(s)}{s} ds + b(t) \, , \quad \text{for} \quad t > t' \, . \tag{B.2.30}$$

Then for all $\varepsilon, \delta_1, \delta_3, \delta_4 > 0$ there exists $t = t_0(\varepsilon, \delta_1, \delta_3, \delta_4)$ such that for all $t \ge t_0, x \ge 1$ we have

$$f(tx) - f(t) = \int_1^x \frac{a_0(ts)}{s} ds + \frac{b(tx)}{a(tx)} a(tx) - b(t)$$

$$\le \left((1 + \delta_3) \int_1^x s^{\delta_1 - 1} \, ds + \varepsilon(1 + \delta_4) x^{\delta_1} + \varepsilon \right) a(t)$$

$$= \left\{ (1 + \delta_3 + \varepsilon(1 + \delta_4)\delta_1) \frac{x^{\delta_1} - 1}{\delta_1} + \varepsilon(2 + \delta_4) \right\} a(t)$$

using $a_0(t) \sim a(t), b(t) = o(a(t))$, and Proposition B.1.9(5). Hence f satisfies the stated upper inequality if we take ε, δ_3, and δ_4 such that $\max(\delta_3 + \varepsilon(1 + \delta_4)\delta_1, \varepsilon(2 + \delta_4)) = \delta_2$. The proof of the lower inequality is similar.

(5) We give the proof of the first statement. The proof of the other statement is similar. Let

$$\frac{f(tx) - f(t)}{tg(t)} = \int_1^x \frac{g(ts)}{g(t)} \, ds . \tag{B.2.31}$$

If $g \in RV_{-1}$, then the right-hand side in (B.2.31) tends to $\log x$, as $t \to \infty$ by the uniform convergence theorem for regularly varying functions (Theorem B.1.4). Next suppose $f \in \Pi(a)$. We have

$$\frac{f(tx) - f(t)}{a(t)} = \frac{tg(x)}{a(t)} \int_1^x \frac{g(ts)}{g(t)} \, ds ,$$

and the integral is at most $x - 1$ when $x > 1$. Hence for $x > 1$, since $f \in \Pi$, we get

$$\liminf_{t \to \infty} \frac{tg(t)}{a(t)} \geq \frac{\ln x}{x - 1} .$$

Similarly we find that $\limsup_{t \to \infty} tg(t)/a(t) \leq (\ln x)/(x - 1)$ for $0 < x < 1$. Let $x \to 1$ to obtain $tg(t) \sim a(t), t \to \infty$, and the last function is slowly varying by Theorem B.2.7. ∎

Remark B.2.16 A special case of the current section is obtained when the auxiliary function a satisfies $a(t) \to \rho > 0, t \to \infty$.

Note that the specialization of Theorem B.2.12 then gives the following statement: Suppose $g : \mathbb{R}^+ \to \mathbb{R}^+$ is measurable. Then $g \in RV_\rho$ if and only if $\log g$ is locally integrable on (t_0, ∞) for some $t_0 > 0$ and

$$\lim_{t \to \infty} \int_{t_0/t}^1 \log \left(\frac{g(ts)}{g(t)} \right) \, ds = \int_0^1 \log s^\rho \, ds = -\rho .$$

This can be seen by applying Theorem B.2.12 for $f(t) = \log g(t)$.

A uniform inequality in the spirit of Proposition B.1.10 is as follows.

Proposition B.2.17 If $f \in \Pi(a)$, there exists a positive function a_0 with $a_0(t) \sim a(t) \to \infty$ such that for all $\varepsilon, \delta > 0$ there is a $t_0 = t_0(\varepsilon, \delta)$ such that for $t, tx \geq t_0$,

$$\left| \frac{f(tx) - f(t)}{a_0(t)} - \log x \right| \leq \varepsilon \max(x^\delta, x^{-\delta}) .$$

Proof. We choose $a_0(t) := f(t) - t^{-1} \int_0^t f(s) \, ds$. Then by Theorem B.2.12, $a(t) \sim a_0(t), t \to \infty$, and

$$f(t) = a_0(t) + \int_0^t a_0(s) \, \frac{ds}{s} .$$

We have

$$\frac{f(tx) - f(t)}{a_0(t)} - \log x = \frac{a_0(tx)}{a_0(t)} - 1 + \int_1^x \left(\frac{a_0(tu)}{a_0(t)} - 1\right)\frac{du}{u}.$$

Hence, using the inequalities of Proposition B.1.10,

$$\left|\frac{f(tx) - f(t)}{a_0(t)} - \log x\right| \leq \varepsilon \max(x^\delta, x^{-\delta}) + \int_{\min(x,1)}^{\max(x,1)} \max(u^{\delta-1}, u^{-\delta-1})\, du$$

$$\leq \varepsilon\left(1 + \frac{1}{\delta}\right)\max(x^\delta, x^{-\delta}).$$ ∎

Combining Theorem B.2.2, Proposition B.1.10, and Proposition B.2.17 we obtain the following theorem:

Theorem B.2.18 (Drees (1998)) *Suppose for a measurable function f and a positive function a we have*

$$\lim_{t\to\infty}\frac{f(tx) - f(t)}{a(t)} = \frac{x^\gamma - 1}{\gamma}$$

for all $x > 0$, where γ is a real parameter, i.e., $f \in \mathrm{ERV}_\gamma$. Then for all $\varepsilon, \delta > 0$ there is a $t_0 = t_0(\varepsilon, \delta)$ such that for $t, tx \geq t_0$,

$$\left|\frac{f(tx) - f(t)}{a_0(t)} - \frac{x^\gamma - 1}{\gamma}\right| \leq \varepsilon x^\gamma \max(x^\delta, x^{-\delta}),$$

where

$$a_0(t) := \begin{cases} \gamma f(t), & \gamma > 0, \\ -\gamma(f(\infty) - f(t)), & \gamma < 0, \\ f(t) - t^{-1}\int_0^t f(s)\, ds, & \gamma = 0. \end{cases}$$

The following result is a generalization of part of Theorem B.2.12 (the kernel function k below is constant in Theorem B.2.12).

Theorem B.2.19 *Suppose $f \in \Pi(a)$ is integrable on finite intervals of $\mathbb{R}^+$.*

1. If the measurable function $k : \mathbb{R}^+ \to \mathbb{R}$ is bounded on $(0, 1)$, then

$$\int_0^1 k(s)\frac{f(ts) - f(t)}{a(t)}\, ds \to \int_0^1 k(s)\log s\, ds, \quad t \to \infty. \tag{B.2.32}$$

2. If $t^\varepsilon k(t)$ is integrable on $(1, \infty)$ for some $\varepsilon > 0$, then

$$\int_1^\infty k(s) f(ts)\, ds < \infty, \quad \text{for} \quad t > 0$$

and

$$\int_1^\infty k(s)\frac{f(ts) - f(t)}{a(t)}\, ds \to \int_1^\infty k(s)\log s\, ds, \quad t \to \infty.$$

Proof. (1) Note that for $0 < \varepsilon < 1$ the function $t^{-\varepsilon}k(t)$ is integrable on $(0, 1)$. We proceed as in the first part of the proof of Theorem B.2.12. Applying Corollary B.2.10 we have

$$\int_{t_0/t}^{1} k(s)\, \frac{f(ts) - f(t)}{a(t)}\, ds \to \int_{0}^{1} k(s) \log s\, ds$$

by Lebesgue's theorem on dominated convergence. Since k is bounded, $ta(t) \in RV_1$, and $f(t) = o(t^{1/2})$, $t \to \infty$, we have

$$\int_{0}^{t_0/t} k(s)\, \frac{f(ts) - f(t)}{a(t)}\, ds = \frac{\int_{0}^{t_0} k\left(\frac{s}{t}\right) f(s)\, ds - f(t) \int_{0}^{t_0} k\left(\frac{s}{t}\right)\, ds}{ta(t)} \to 0\, ,$$

as $t \to \infty$.

(2) The second statement is proved in a similar way. ∎

Remark B.2.20 Theorem B.2.19(1) also holds under the alternative conditions f bounded on $(0, 1)$ and $\int_{0}^{1} s^{-\varepsilon}k(s) < \infty$ for some $\varepsilon > 0$.

Theorem B.2.21 *Suppose that f is nondecreasing and ϕ is its left-continuous inverse function. Let γ be a real parameter. Equivalent are:*

1. *There exists a positive function a such that for $x > 0$,*

$$\lim_{t \to \infty} \frac{f(tx) - f(t)}{a(t)} = \frac{x^\gamma - 1}{\gamma}\, ,$$

2. *There exists a positive function g such that for all x for which $1 + \gamma x > 0$,*

$$\lim_{t \uparrow x^*} \frac{\phi(t + xg(t))}{\phi(t)} = (1 + \gamma x)^{1/\gamma}\, ,$$

where $x^* = \lim_{t \to \infty} f(t)$.

The functions a and g are connected as follows: $g(t) = a(\phi(t))$.

Proof. Assume (1). For $\varepsilon > 0$ and all t,

$$f(\phi(t) - \varepsilon) \le t \le f(\phi(t) + \varepsilon)\, .$$

It follows that

$$\frac{(1 - \varepsilon)^\gamma - 1}{\gamma} \leftarrow \frac{U((1 - \varepsilon)\phi(t)) - U(\phi(t))}{a(\phi(t))} \le \frac{t - U(\phi(t))}{a(\phi(t))}$$
$$\le \frac{U((1 + \varepsilon)\phi(t)) - U(\phi(t))}{a(\phi(t))} \to \frac{(1 + \varepsilon)^\gamma - 1}{\gamma}$$

as $t \uparrow x^*$ and consequently

$$\lim_{t \uparrow x^*} \frac{t - U(\phi(t))}{a(\phi(t))} = 0\, .$$

Hence by (B.2.21) for all $x > 0$,

$$\lim_{t \uparrow x^*} \frac{U(x\,\phi(t)) - t}{a(\phi(t))} = \frac{x^\gamma - 1}{\gamma},$$

and by Lemma 1.1.1,

$$\lim_{t \uparrow x^*} \frac{\phi(t + x\,a(\phi(t)))}{\phi(t)} = (1 + \gamma x)^{1/\gamma},$$

i.e., (2) holds. The converse is similar. ∎

B.3 Second-Order Extended Regular Variation (2ERV)

Recall that a measurable function f is of extended regular variation ($f \in$ ERV) if for some $\gamma \in \mathbb{R}$ and positive function a,

$$\lim_{t \to \infty} \frac{f(tx) - f(t)}{a(t)} = \frac{x^\gamma - 1}{\gamma} \tag{B.3.1}$$

for all $x > 0$. Since one often needs to control the speed of convergence in (B.3.1), it is useful to build a theory in which the convergence rate in (B.3.1) is the same for all $x > 0$, that is, there exists a positive function A with $\lim_{t \to \infty} A(t) = 0$ such that

$$H(x) := \lim_{t \to \infty} \frac{\frac{f(tx) - f(t)}{a(t)} - \frac{x^\gamma - 1}{\gamma}}{A(t)} \tag{B.3.2}$$

exists for all $x > 0$. First we want to exclude a more or less trivial case. If $H(x) = c(x^\gamma - 1)/\gamma$ for some $c \in \mathbb{R}$, the limit relation can be reformulated as

$$\lim_{t \to \infty} \frac{\frac{f(tx) - f(t)}{a(t)(1 + cA(t))} - \frac{x^\gamma - 1}{\gamma}}{A(t)} = 0,$$

which seems to be insufficiently informative. Hence we exclude this case.

Our first theorem identifies the form of the limit function in (B.3.2).

Theorem B.3.1 (de Haan and Stadtmüller (1996)) *Suppose that for some measurable function f and positive functions a and A the limit (B.3.2) exists for all $x > 0$ where the limit function is not a multiple of $(x^\gamma - 1)/\gamma$. Then there exist real constants c_1, c_2 and a parameter $\rho \leq 0$ such that for all $x > 0$,*

$$\lim_{t \to \infty} \frac{\frac{f(tx) - f(t)}{a(t)} - \frac{x^\gamma - 1}{\gamma}}{A(t)} = c_1 \int_1^x s^{\gamma - 1} \int_1^s u^{\rho - 1}\,du\,ds + c_2 \int_1^x s^{\gamma + \rho - 1}\,ds\ . \tag{B.3.3}$$

Moreover, for $x > 0$,

$$\lim_{t\to\infty} \frac{\frac{a(tx)}{a(t)} - x^\gamma}{A(t)} = c_1 x^\gamma \frac{x^\rho - 1}{\rho} \tag{B.3.4}$$

and

$$\lim_{t\to\infty} \frac{A(tx)}{A(t)} = x^\rho . \tag{B.3.5}$$

In view of the restrictions discussed before the theorem, the constant c_1 cannot be zero if $\rho = 0$.

Remark B.3.2 Relation (B.3.3) holds with a replaced by a_1 and A replaced by A_1 if and only if $A_1(t) \sim A(t)$ and $a_1(t)/a(t) - 1 = o(A(t))$, $t \to \infty$.

Remark B.3.3 Alternatively, one can write (B.3.2) as

$$\lim_{t\to\infty} \frac{f(tx) - f(t) - a(t)(x^\gamma - 1)/\gamma}{a_1(t)} = H(x)$$

with $a_1 := aA$. This will be used in the proof.

Definition B.3.4 A function f satisfying the conditions of Theorem B.3.1 is said to be of *second-order extended regular variation*. Notation: $f \in$ 2ERV or $f \in$ 2ERV$_{\gamma,\rho}$. The parameter ρ is called the *second-order parameter*.

Proof (of Theorem B.3.1). Consider for $x, y > 0$ the identity

$$x^{-\gamma} \left\{ \frac{f(txy) - f(t) - a(t)((xy)^\gamma - 1)/\gamma}{a_1(t)} - \frac{f(tx) - f(t) - a(t)(x^\gamma - 1)/\gamma}{a_1(t)} \right\}$$

$$= \frac{f(txy) - f(tx) - a(tx)(y^\gamma - 1)/\gamma}{a_1(tx)} \frac{(tx)^{-\gamma} a_1(tx)}{t^{-\gamma} a_1(t)}$$

$$+ \frac{y^\gamma - 1}{\gamma} \frac{(tx)^{-\gamma} a(tx) - t^{-\gamma} a(t)}{t^{-\gamma} a_1(t)} .$$

Letting $t \to \infty$ on both sides, we obtain

$$x^{-\gamma} (H(xy) - H(x))$$

$$= \lim_{t\to\infty} \left\{ H(y)(1 + o(1)) \frac{(tx)^{-\gamma} a_1(tx)}{t^{-\gamma} a_1(t)} + \frac{y^\gamma - 1}{\gamma} \frac{(tx)^{-\gamma} a(tx) - t^{-\gamma} a(t)}{t^{-\gamma} a_1(t)} \right\} . \tag{B.3.6}$$

By assumption, there exist $y_1, y_2 \in \mathbb{R}$ such that $(H(y_1), (y_1^\gamma - 1)/\gamma)$ and $(H(y_2), (y_2^\gamma - 1)/\gamma)$ are linearly independent; hence with $\lambda = (y_1^\gamma - 1)/(y_2^\gamma - 1)$ we obtain that $H(y_1) - \lambda H(y_2) \neq 0$. Now we subtract λ times (B.3.6) at argument $y = y_2$ from (B.3.6) with argument $y = y_1$ and we obtain

$$\lim_{t\to\infty} \left\{ (H(y_1) - \lambda H(y_2))(1 + o(1)) \frac{(tx)^{-\gamma} a_1(tx)}{t^{-\gamma} a_1(t)} \right\}$$

$$= x^{-\gamma} \left\{ (H(xy_1) - H(x)) - \frac{y_1^\gamma - 1}{y_2^\gamma - 1} (H(xy_2) - H(x)) \right\} , \quad x > 0 .$$

From this, we conclude that $\lim_{t\to\infty}(tx)^{-\gamma}a_1(tx)/(t^{-\gamma}a_1(t))$ exists for all $x > 0$. Since the limit should be finite for all $x > 0$, it must be positive for all $x > 0$. Hence $t^{-\gamma}a_1(t)$ is regularly varying with index ρ, say. The existence of this limit, together with relation (B.3.6), implies that $\lim_{t\to\infty}((tx)^{-\gamma}a(tx) - t^{-\gamma}a(t))/(t^{-\gamma}a_1(t))$ also exists for all $x > 0$. Hence we obtain (B.3.4) for some $c_1 \in \mathbb{R}$ by Theorem B.2.1. From Theorem B.2.2 we must have $\rho \le 0$.

As a result, we obtain the following functional equation for H:

$$H(xy) = H(y)x^{\rho+\gamma} + H(x) + c_1 x^\gamma \frac{y^\gamma - 1}{\gamma} \frac{x^\rho - 1}{\rho} \quad \text{for} \quad x, y > 0 . \quad \text{(B.3.7)}$$

A simple calculation verifies that

$$H_1(x) = c_1 \int_1^x s^{\gamma-1} \int_1^s u^{\rho-1} \, du \, ds$$

is a solution of (B.3.7). Obviously, the function $G(x) = H(x) - H_1(x)$ satisfies the homogeneous equation

$$G(xy) = G(x) + G(y)x^{\rho+\gamma} , \quad \text{for } x, y > 0 .$$

If $\rho + \gamma = 0$, this is Cauchy's equation, having the unique solution $G(x) = c_2 \log x$ in the class of measurable functions. If $\rho + \gamma \ne 0$, by symmetry we obtain

$$G(xy) = G(y) + G(x)y^{\rho+\gamma}$$

and hence $G(x)(1 - y^{\rho+\gamma}) = G(y)(1 - x^{\rho+\gamma})$ for $x, y > 0$, which implies that with some $c_2 \in \mathbb{R}$, $G(x) = c_2(x^{\rho+\gamma} - 1)/(\rho + \gamma)$. Hence the general solution is

$$c_1 \int_1^x s^{\gamma-1} \int_1^s u^{\rho-1} \, du \, ds + c_2 \int_1^x s^{\rho+\gamma-1} \, ds . \quad \blacksquare$$

Remark B.3.5 1. The function A, describing the rate of convergence in (B.3.2), is regularly varying with index ρ. So if $\rho < 0$ we have an algebraic speed of convergence in (B.3.2). In case $\rho = 0$ it is much slower, for example, logarithmic.
2. From (B.3.3) we see that $H(x)$ can be written as

$$H(x) = \begin{cases} \frac{1}{\rho}\left(\frac{x^{\gamma+\rho}-1}{\gamma+\rho} - \frac{x^\gamma-1}{\gamma}\right) , & \rho \ne 0, \\ \frac{1}{\gamma}\left(x^\gamma \log x - \frac{x^\gamma-1}{\gamma}\right) , & \rho = 0, \gamma \ne 0, \\ \frac{1}{2}(\log x)^2 , & \rho = \gamma = 0, \end{cases} \quad \text{(B.3.8)}$$

if we replace a and A in (B.3.3) with $\tilde{a}$ and $\tilde{A}$, respectively,

$$\tilde{A}(t) := (c_1 + c_2\rho) A(t),$$
$$\tilde{a}(t) := a(t) (1 + c_2 A(t)) .$$

This corresponds to $c_1 = 1$ and $c_2 = 0$ in (B.3.3). This representation has been chosen in such a way that $H(1) = 0$ and $H'(x) = x^{\gamma-1}(x^\rho - 1)/\rho$.

Next we prove that in some cases we can relate 2ERV functions to classes of functions we have met before.

Theorem B.3.6 (de Haan and Stadtmüller (1996)) *Suppose that the function f satisfies the conditions of Theorem B.3.1. Then:*

1. *in case $\rho = \gamma = 0$,*

$$f(t) = h(t) + \int_0^t \frac{h(s)}{s} \, ds$$

   with $h \in \Pi$;
2. *in case $\rho = 0$, $\gamma \neq 0$,*

$$\pm t^{-\gamma} \tilde{f}(t) \in \Pi, \quad where \quad \tilde{f}(t) := \begin{cases} f(t), & \gamma > 0, \\ f(\infty) - f(t), & \gamma < 0; \end{cases}$$

3. *in case $\rho < 0$, there exists some nonzero constant c such that*

$$\pm \left(f(t) - c \, \frac{t^\gamma - 1}{\gamma} \right) \quad is\ in \quad \mathrm{ERV}_{\gamma + \rho} \, .$$

Conversely, any of the properties (1), (2), (3) implies that f is in 2ERV.

Remark B.3.7 More precisely: In case $\rho = 0$ and $\gamma > 0$,

$$\lim_{t \to \infty} \frac{(tx)^{-\gamma} f(tx) - t^{-\gamma} f(t)}{c_1 t^{-\gamma} f(t) A(t)} = \log x \tag{B.3.9}$$

or, equivalently,

$$\lim_{t \to \infty} \frac{\frac{f(tx)}{f(t)} - x^\gamma}{c_1 A(t)} = x^\gamma \log x \, . \tag{B.3.10}$$

In case $\rho = 0$ and $\gamma < 0$, (B.3.9) and (B.3.10) hold with f replaced by $f(\infty) - f(t)$.

In case $\rho < 0$, $\lim_{t \to \infty} t^{-\gamma} a(t) = c$ exists in $(0, \infty)$, and

$$\lim_{t \to \infty} \frac{\left\{ f(tx) - c \, \frac{(tx)^\gamma - 1}{\gamma} \right\} - \left\{ f(t) - c \, \frac{t^\gamma - 1}{\gamma} \right\}}{a(t) A(t) / \left(\frac{c_1}{\rho} + c_2 \right)} = \frac{x^{\gamma + \rho} - 1}{\gamma + \rho} \, .$$

Remark B.3.8 1. It follows from the representations of Theorem B.3.6 that (B.3.3) holds locally uniformly in $(0, \infty)$ by the properties of the function classes to which the f's belong in the different cases. The representations also lead to Potter bounds for relation (B.3.3).
2. Note that in most cases the existence of a second-order relation makes the first-order relation simpler; for example, in case $\rho < 0$ and $\rho + \gamma = 0$ one has $f(t) \sim c_3 t^\gamma$, as $t \to \infty$.

Proof (of Theorem B.3.6). (1) For $\gamma = \rho = 0$ relation (B.3.3) implies

$$\lim_{t \to \infty} \frac{f(txy) - f(ty) - f(tx) + f(t)}{a_1(t)} = (\log x)(\log y)$$

for $x, y > 0$ with $a_1 := aA$. Hence $g_x \in \Pi(a_1 \log x)$ for $x > 0$, where

$$g_x(t) := f(tx) - f(t) .$$

Now Theorem B.2.12 implies that

$$\lim_{t \to \infty} \frac{h_x(t)}{a_1(t) \log x} = 1, \tag{B.3.11}$$

where

$$h_x(t) := g_x(t) - \frac{1}{t} \int_0^t g_x(s) \, ds = h(tx) - h(t)$$

with

$$h(t) := f(t) - \frac{1}{t} \int_0^t f(s) \, ds .$$

Hence (B.3.11) translates into

$$\lim_{t \to \infty} \frac{h(tx) - h(t)}{a_1(t)} = \log x$$

for $x > 0$, i.e., $h \in \Pi(a_1)$. Finally, by Fubini's theorem,

$$f(t) = h(t) + \int_0^t \frac{h(s)}{s} ds . \tag{B.3.12}$$

Conversely, if (B.3.12) holds and $h \in \Pi(a_1)$ then for $x > 0$,

$$\lim_{t \to \infty} \frac{f(tx) - f(t) - (h(t) + a_1(t)) \log x}{a_1(t)}$$

$$= \lim_{t \to \infty} \frac{h(tx) - h(t)}{a_1(t)} + \int_1^x \left(\frac{h(st) - h(t)}{a_1(t)} - 1 \right) \frac{ds}{s}$$

$$= \log x + \int_1^x (\log s - 1) \frac{ds}{s}$$

$$= \frac{1}{2} (\log x)^2 .$$

(2) We can assume (if necessary change the function a somewhat) that $c_2 = 0$ in (B.3.3). Write (B.3.4) as

$$\lim_{t \to \infty} \frac{\left(\frac{a(tx)}{\gamma} - \frac{a(t)}{\gamma} \right) / a(t) - \frac{x^\gamma - 1}{\gamma}}{A(t)} = \frac{c_1}{\gamma} x^\gamma \log x$$

and subtract from (B.3.3), i.e.,

$$\lim_{t\to\infty} \frac{\frac{f(tx)-f(t)}{a(t)} - \frac{x^\gamma-1}{\gamma}}{A(t)} = \frac{c_1}{\gamma}\left(x^\gamma \log x - \frac{x^\gamma - 1}{\gamma}\right).$$

We get

$$\lim_{t\to\infty} \frac{\{f(tx) - \gamma^{-1}a(tx)\} - \{f(t) - \gamma^{-1}a(t)\}}{a_1(t)} = \frac{-c_1}{\gamma}\frac{x^\gamma - 1}{\gamma} \qquad (B.3.13)$$

with $a_1 := aA$ as before. Consider $\gamma > 0$ first. By Theorem B.2.2,

$$\lim_{t\to\infty} \frac{f(t) - \gamma^{-1}a(t)}{a_1(t)} = -\frac{c_1}{\gamma^2}. \qquad (B.3.14)$$

Then we have for $x > 0$,

$$\lim_{t\to\infty} \frac{f(tx) - \gamma^{-1}a(tx)}{a_1(t)} = -\frac{c_1}{\gamma^2}x^\gamma. \qquad (B.3.15)$$

Subtract (B.3.14) from $x^{-\gamma}$ times (B.3.15) to obtain

$$\lim_{t\to\infty} \frac{(tx)^{-\gamma}f(tx) - t^{-\gamma}f(t)}{t^{-\gamma}a_1(t)} - \frac{1}{\gamma}\frac{(tx)^{-\gamma}a(tx) - t^{-\gamma}a(t)}{t^{-\gamma}a_1(t)} = 0.$$

The result follows in view of (B.3.4). The proof for $\gamma < 0$ is similar. Conversely, if, for example, $t^{-\gamma}f(t) \in \Pi(\tilde{a})$, then

$$\lim_{t\to\infty} \frac{f(tx) - f(t) - \gamma f(t)(x^\gamma - 1)/\gamma}{t^\gamma \tilde{a}(t)} = x^\gamma \log x.$$

If $f(\infty) := \lim_{t\to\infty} f(t)$ is finite and for example $-t^{-\gamma}(f(\infty) - f(t)) \in \Pi(\tilde{a})$, then

$$\lim_{t\to\infty} \frac{f(tx) - f(t) + \gamma\,(f(\infty) - f(t))\,(x^\gamma - 1)/\gamma}{t^\gamma \tilde{a}(t)} = x^\gamma \log x.$$

Hence $f \in 2ERV$.

(3) In view of Theorem B.2.2, relation (B.3.4) implies for $\rho < 0$ and $c_1 \neq 0$ in (B.3.3) that

$$c := \lim_{t\to\infty} t^{-\gamma}a(t)$$

exists in $(0, \infty)$ and

$$\lim_{t\to\infty} \frac{c - t^{-\gamma}a(t)}{t^{-\gamma}a_1(t)} = \frac{-c_1}{\rho},$$

i.e.,

$$a(t) = ct^\gamma + \frac{c_1}{\rho}a_1(t)(1 + o(1)).$$

Hence for $x > 0$,

$$\lim_{t \to \infty} \frac{f(tx) - f(t) - \left\{ ct^\gamma + c_1 \rho^{-1} a_1(t)(1 + o(t)) \right\} (x^\gamma - 1)/\gamma}{a_1(t)}$$
$$= \frac{c_1}{\rho} \left(\frac{x^{\gamma + \rho} - 1}{\gamma + \rho} - \frac{x^\gamma - 1}{\gamma} \right) + c_2 \frac{x^{\gamma + \rho} - 1}{\gamma + \rho} ,$$

i.e.,

$$\lim_{t \to \infty} \frac{\left\{ f(tx) - c \frac{(tx)^\gamma - 1}{\gamma} \right\} - \left\{ f(t) - c \frac{t^\gamma - 1}{\gamma} \right\}}{a_1(t)} = \left(\frac{c_1}{\rho} + c_2 \right) \frac{x^{\gamma + \rho} - 1}{\gamma + \rho} .$$

$$(B.3.16)$$

Note that in order not to have a trivial limit we need to have $c_1/\rho + c_2 \neq 0$. Conversely, if a function f satisfies (B.3.16), then

$$\lim_{t \to \infty} \frac{f(tx) - f(t) - c \, t^\gamma \frac{x^\gamma - 1}{\gamma}}{(c_1/\rho + c_2) a_1(t)} = \frac{x^{\gamma + \rho} - 1}{\gamma + \rho}$$

for $x > 0$, i.e., $f \in 2ERV$.

For $\rho < 0$ and $c_1 = 0$, we have by Lemma B.3.9 below that

$$a(t) = ct^\gamma + o(a_1(t)) , \quad t \to \infty ,$$

with $c \neq 0$ since $a \in RV_\gamma$ and $a_1 \in RV_{\gamma + \rho}$. Hence (B.3.16) holds, with $c_1 = 0$. ∎

Lemma B.3.9 (Ash, Erdős and Rubel (1974)) *Let r be a regularly varying function with index less than zero. If for some measurable function q,*

$$\lim_{t \to \infty} \frac{q(tx) - q(t)}{r(t)} = 0$$

for all $x > 0$, then

$$C := \lim_{t \to \infty} q(t)$$

exists (finite) and

$$C - q(t) = o(r(t)) , \quad t \to \infty .$$

Proof. Take $\Lambda > 1$. For $n = 1, 2, \ldots$ and $y \geq x$,

$$\frac{|q(y) - q(x)|}{r(x)} \leq \frac{r(y)}{r(x)} \sum_{k=1}^{n} \frac{|q(\Lambda^k y) - q(\Lambda^{k-1} y)|}{r(\Lambda^{k-1} y)} \frac{r(\Lambda^{k-1} y)}{r(y)}$$
$$+ \frac{|q(\Lambda^n y) - q(\Lambda^n x)|}{r(\Lambda^n x)} \frac{r(\Lambda^n x)}{r(x)} + \sum_{k=1}^{n} \frac{|q(\Lambda^k x) - q(\Lambda^{k-1} x)|}{r(\Lambda^{k-1} x)} \frac{r(\Lambda^{k-1} x)}{r(x)} .$$

Now

$$\limsup_{x \to \infty} \frac{r(\Lambda x)}{r(x)} =: \delta^2 < 1 , \qquad \lim_{x \to \infty} \sup_{y \geq x} \frac{r(y)}{r(x)} = 1 .$$

Hence with $\varepsilon > 0$ for $x \geq x_0$ and $y \geq x$,

$$\frac{r(\Lambda x)}{r(x)} \leq \delta , \qquad \frac{r(y)}{r(x)} \leq 2 , \qquad \frac{|q(\Lambda x) - q(x)|}{r(x)} \leq \varepsilon,$$

and for $y \geq x \geq x_0$,

$$\frac{|q(y) - q(x)|}{r(x)} \leq 2\varepsilon \sum_{k=1}^{n} \delta^{k-1} + \delta^n + \varepsilon \sum_{k=1}^{n} \delta^{k-1} \to \frac{3\varepsilon}{1 - \delta} \tag{B.3.17}$$

as $n \to \infty$. Since $\lim_{x \to \infty} r(x) = 0$, for $\varepsilon > 0$ we can find x_1 such that

$$|q(y) - q(x)| < \varepsilon$$

for $y \geq x \geq x_1$. Hence $C := \lim_{x \to \infty} q(x)$ exists by Cauchy's criterion. Taking $y \to \infty$ in (B.3.17) we conclude that

$$C - q(x) = o(r(x)) , \qquad x \to \infty . \qquad \blacksquare$$

We finish the main theorems by establishing uniform inequalities for the class 2ERV.

Theorem B.3.10 (Drees (1998); cf. Cheng and Jiang (2001)) *Let f be a measurable function. Suppose for some positive function a, some positive or negative function A with $\lim_{t \to \infty} A(t) = 0$, and parameters $\gamma \in \mathbb{R}$, $\rho \leq 0$, that*

$$\lim_{t \to \infty} \frac{\frac{f(tx) - f(t)}{a(t)} - \frac{x^\gamma - 1}{\gamma}}{A(t)} = \int_1^x s^{\gamma - 1} \int_1^s u^{\rho - 1} \, du \, ds \tag{B.3.18}$$

for all $x > 0$, i.e., $f \in 2\text{ERV}$. Then for all $\varepsilon, \delta > 0$ there exists $t_0 = t_0(\varepsilon, \delta)$ such that for all $t, tx \geq t_0$,

$$\left| \frac{\frac{f(tx) - f(t)}{a_0(t)} - \frac{x^\gamma - 1}{\gamma}}{A_0(t)} - \Psi_{\gamma, \rho}(x) \right| \leq \varepsilon \max \left(x^{\gamma + \rho + \delta}, x^{\gamma + \rho - \delta} \right), \tag{B.3.19}$$

where

$$\Psi_{\gamma, \rho}(x) := \begin{cases} \frac{x^{\gamma + \rho} - 1}{\gamma + \rho} , & \rho < 0, \\ \frac{1}{\gamma} x^\gamma \log x , & \gamma \neq \rho = 0 , \\ \frac{1}{2} (\log x)^2 , & \gamma = \rho = 0 , \end{cases} \tag{B.3.20}$$

$$a_0(t) := \begin{cases} ct^\gamma , & \rho < 0, \\ -\gamma(f(\infty) - f(t)) , & \gamma < \rho = 0, \\ \gamma f(t) , & \gamma > \rho = 0, \\ \hat{f}(t) + \hat{\hat{f}}(t) , & \gamma = \rho = 0 , \end{cases} \tag{B.3.21}$$

with $c := \lim_{t\to\infty} t^{-\gamma} a(t)$,

$$A_0(t) := \begin{cases} -(\gamma + \rho)\frac{\overline{f}(\infty)-\overline{f}(t)}{a_0(t)} & \gamma + \rho < 0, \ \rho < 0, \\[2mm] (\gamma + \rho)\frac{\overline{f}(t)}{a_0(t)}, & \gamma + \rho > 0, \ \rho < 0, \\[2mm] \frac{\hat{\overline{f}}(t)}{a_0(t)}, & \gamma + \rho = 0, \ \rho < 0, \\[2mm] \frac{\hat{\overline{f}}(t)}{\overline{f}(t)}, & \gamma \neq \rho = 0, \\[2mm] \frac{\hat{\overline{f}}(t)}{a_0(t)}, & \gamma = \rho = 0, \end{cases} \tag{B.3.22}$$

where for an integrable function h,

$$\hat{h}(t) := h(t) - \frac{1}{t}\int_0^t h(s)\, ds \tag{B.3.23}$$

and

$$\overline{f}(t) := \begin{cases} f(t) - c\,\frac{t^\gamma - 1}{\gamma}, & \rho < 0, \\[2mm] t^{-\gamma}(f(\infty) - f(t)), & \gamma < \rho = 0, \\[2mm] t^{-\gamma} f(t), & \gamma > \rho = 0, \\[2mm] \hat{f}(t), & \gamma = \rho = 0. \end{cases} \tag{B.3.24}$$

Moreover,

$$\left| \frac{\frac{a_0(tx)}{a_0(t)} - x^\gamma}{A_0(t)} - x^\gamma\,\frac{x^\rho - 1}{\rho} \right| \leq \varepsilon \max(x^{\gamma+\rho+\delta}, x^{\gamma+\rho-\delta}) . \tag{B.3.25}$$

Proof. From Theorem B.3.6 it is clear that for all cases except when $\gamma = \rho = 0$ the limit relation (B.3.1) can be reformulated as a relation of extended regular variation (first order) for a simple transform of f. Hence those cases are covered by Theorem B.2.18. It remains to consider the case $\gamma = \rho = 0$. We use Theorem B.3.6:

$$f(t) = h(t) + \int_0^t h(s)\, ds$$

with $h \in \Pi$. Now Theorem B.2.18 implies

$$\left| \frac{h(tx) - h(t)}{k(t)} - \log x \right| \leq \varepsilon \max(x^\delta, x^{-\delta})$$

for some positive function k and therefore

$$\left| \frac{\frac{f(tx)-f(t)}{a_0(t)} - \log x}{A_0(t)} - \frac{\log^2 x}{2} \right|$$

$$\leq \left| \frac{h(tx) - h(t)}{k(t)} - \log x \right| + \int_1^x \left| \frac{h(tu) - h(t)}{k(t)} - \log u \right| \frac{du}{u}$$

$$\leq \varepsilon \left(1 + \frac{1}{\delta}\right) \max(x^\delta, x^{-\delta}) . \qquad \blacksquare$$

Corollary B.3.11 *It follows that if* $\rho < 0$ *or* $\rho = 0, \gamma < 0$, *then*

$$\lim_{\substack{t\to\infty \\ x=x(t)\to\infty}} \frac{\frac{f(tx)-f(t)}{a(t)}\frac{\gamma}{x^\gamma-1}-1}{A(t)} = -\frac{1}{\rho+\gamma_-}$$

with $\gamma_- := \min(\gamma, 0)$.

Remark B.3.12 Analogous of Theorems B.3.1, B.3.6, and B.3.10 for *third*-order extended regular variation can be found in the Appendix of Fraga Alves, de Haan and Lin (2003).

Next we show that a function f satisfying the second-order condition is always close to a smooth function of the same kind.

Theorem B.3.13 *Suppose*

$$\lim_{t\to\infty} \frac{f(tx)-f(t)-a(t)(x^\gamma-1)/\gamma}{a_1(t)} = H(x) \qquad (B.3.26)$$

with H not a multiple of $(x^\gamma-1)/\gamma$. Then there exists a twice differentiable function f_1 with

$$\lim_{t\to\infty}\left(f(t)-f_1(t)\right)/a_1(t) = 0 , \qquad (B.3.27)$$

$$\lim_{t\to\infty}\left(a(t)-t\,f_1'(t)\right)/a_1(t) = 0 , \qquad (B.3.28)$$

$$\lim_{t\to\infty} a(t)A_1(t)/a_1(t) = 1 ; \qquad (B.3.29)$$

with $h = f$ or $h = f_1$ we have

$$\lim_{t\to\infty} \frac{\frac{h(tx)-h(t)}{tf_1'(t)}-\frac{x^\gamma-1}{\gamma}}{A_1(t)} = H(x)$$

and such that with

$$A_1(t) := \frac{t\,f_1''(t)}{f_1'(t)} - \gamma + 1 , \qquad (B.3.30)$$

we have

$$\operatorname{sign}\left(A_1(t)\right) \text{ is constant eventually} , \qquad (B.3.31)$$

$$\lim_{t\to\infty} A_1(t) = 0 , \qquad (B.3.32)$$

$$|A_1| \in RV_\rho . \qquad (B.3.33)$$

Proof. For the case $\gamma = \rho = 0$ the proof is given in Lemma B.3.14 below. For other values of γ and ρ separate proofs apply. As an example we give the proof for $\rho = 0$, $\gamma > 0$.

Assume that the function a_1 is positive (for negative a_1 a similar proof applies). Then (B.3.26) implies by Theorem B.3.6,

$$\lim_{t\to\infty} \frac{(tx)^{-\gamma} f(tx) - t^{-\gamma} f(t)}{t^{-\gamma} a_1(t)/\gamma} = \log x \qquad (B.3.34)$$

for all $x > 0$; hence (B.3.26) holds with $a(t) = \gamma f(t) + \gamma^{-1} a_1(t)$.

Now (B.3.34) says that the function $t^{-\gamma} f(t)$ is in the class Π; hence by Proposition B.2.15(3) there is a function g_1 with

$$\lim_{t\to\infty} \frac{t^{-\gamma} f(t) - g_1(t)}{t^{-\gamma} a_1(t)} = \lim_{t\to\infty} \frac{f(t) - t^{\gamma} g_1(t)}{a_1(t)} = 0 \qquad (B.3.35)$$

and such that

$$\lim_{t\to\infty} \frac{t g_1''(t)}{g_1'(t)} = -1 . \qquad (B.3.36)$$

Combining (B.3.34), (B.3.35), and

$$\lim_{t\to\infty} \frac{g_1(tx) - g_1(t)}{t g_1'(t)} = \log x ,$$

we obtain

$$\lim_{t\to\infty} \frac{t g_1'(t)}{t^{-\gamma} a_1(t)/\gamma} = 1 ,$$

i.e.,

$$\lim_{t\to\infty} \frac{\gamma t^{\gamma+1} g_1'(t)}{a_1(t)} = 1 . \qquad (B.3.37)$$

We take $f_1(t) := t^{\gamma} g_1(t)$. Then (B.3.27) holds by (B.3.35). Further,

$$\frac{a(t) - t f_1'(t)}{a_1(t)} = \frac{\gamma f(t) + \gamma^{-1} a_1(t) - t f_1'(t)}{a_1(t)}$$

$$= \gamma \frac{f(t) - t^{\gamma} g_1(t)}{a_1(t)} + \gamma^{-1} + \frac{\gamma t^{\gamma} g_1(t) - t (t^{\gamma} g_1(t))'}{a_1(t)}$$

$$= \gamma \frac{f(t) - t^{\gamma} g_1(t)}{a_1(t)} + \gamma^{-1} - \frac{t^{\gamma+1} g_1'(t)}{a_1(t)} \to 0,$$

$t \to \infty$, by (B.3.35) and (B.3.37). Hence (B.3.28) holds. Finally, by (B.3.28), (B.3.36), and (B.3.37),

$$a(t) A_1(t) \sim t f_1'(t) A_1(t)$$

$$= t^2 f_1''(t) - (\gamma - 1) t f_1'(t)$$

$$= (\gamma + 1) t^{\gamma+1} g_1'(t) + t^{\gamma+2} g_1''(t)$$

$$= t^{\gamma+1} g_1'(t) \left\{ \frac{t g_1''(t)}{g_1'(t)} + \gamma + 1 \right\}$$

$$\sim \gamma t^{\gamma+1} g_1'(t) \sim a_1(t) , \quad t \to \infty .$$

Hence (B.3.31), (B.3.32), (B.3.33), and (B.3.29) hold.   ∎

Lemma B.3.14 (A.A. Balkema) *Let ϕ be measurable and for all x,*

$$\phi(t+x) = a_0(t) + a_1(t)x + a_2(t)\frac{x^2}{2} + o(a_2(t)), \quad t \to \infty. \tag{B.3.38}$$

Since this relation is essentially 2ERV, we know that it holds locally uniformly. Let γ be a C^2 function with compact support that satisfies

$$\int x^k \gamma(x) \, dx = 0 \quad for \quad k = 1, 2,$$

$$\int x^k \gamma(x) \, dx = 1 \quad for \quad k = 0,$$

and define $\psi = \phi \star \gamma$ as

$$\psi(t) := \int \phi(t+s)\gamma(s) \, ds$$

and for $m = 1, 2, \ldots,$

$$\psi^{(m)} = \psi^{(m-1)} \star \psi$$

with $\psi^{(0)} = \psi$.
Then $\psi^{(m)}$ satisfies, for all x,

$$\lim_{t \to \infty} \frac{\psi^{(m)}(t+x)}{\psi^{(m)}(t)} = 1. \tag{B.3.39}$$

For $k = 0, 1, 2$,

$$a_k(t) - \psi^{(k)}(t) = o(a_2(t)),$$

and in particular,

$$\phi(t) - \psi(t) = o(a_2(t)), \quad as \quad t \to \infty.$$

Proof. If p is a polynomial of degree $k \leq 2$, then so is $\tilde{p} := p \star \gamma$. Indeed, $\tilde{p}^{(k)} = p^{(k)} \star \gamma$ is constant. Since $\psi^{(k)} = \phi \star \gamma^{(k)}$ for $k \leq 2$, we have

$$\psi^{(2)}(t+x) = \int \phi(t+x-s)\gamma^{(2)}(s) \, ds$$

$$= \int \left\{ a_0(t) + (x-s)a_1(t) + \frac{(x-s)^2}{2}a_2(t) \right\} \gamma^{(2)}(s) \, ds$$

$$= \int o(a_2(t))\gamma^{(2)}(s) \, ds$$

$$= a_2(t) \int \frac{s^2}{2}\gamma^{(2)}(s) \, ds + \int o(a_2(t))\gamma^{(2)}(s) \, ds$$

(since $p \star \gamma^{(2)} = p^{(2)} \star \gamma$ vanishes for any polynomial of degree < 2). The first term equals $a_2(t)$ and is independent of x. The second term is $o(a_2(t))$ since γ has bounded support by the local uniformity in (B.3.38). This proves (B.3.39).

Now write as above, with $x = 0, k = 0, 1$,

$$\psi^{(k)}(t) = \int \left\{ a_o(t) - s a_1(t) + \frac{s^2}{2} a_2(t) \right\} \gamma^{(k)}(s) \, ds + \int o(a_2(t)) \gamma^{(k)}(s) \, ds \; .$$

The second integral is $o(a_2(t))$ and the first integral reduces to

$$a_k(t) \int \frac{(-s)^k}{k!} \gamma^{(k)}(s) \, ds = a_k(t)$$

since $\int (s^j/j!) \gamma^{(k)}(s) \, ds$ vanishes for $j < k$ (note that $p \star \gamma^{(k)}(s) = 0$ for all s if $p < k$) and

$$\int \frac{(-s)^j}{j!} \gamma^{(k)}(s) \, ds = \int \frac{(-s)^{j-k}}{(j-k)!} \gamma(s) \, ds$$

for $j \geq k$. ∎

Remark B.3.15 A simpler case of second-order behavior is related to regular variation rather than extended regular variation. Suppose $f \in RV_\gamma$ for some $\gamma \in \mathbb{R}$ and there is some positive or negative function A such that

$$\lim_{t \to \infty} \frac{\frac{f(tx)}{f(t)} - x^\gamma}{A(t)} =: H(x) \tag{B.3.40}$$

exists for $x > 0$ with H not constant. If (B.3.40) holds, we say that the function f is of *second-order regular variation*. Then

$$\lim_{t \to \infty} \frac{(tx)^{-\gamma} f(tx) - t^{-\gamma} f(t)}{t^{-\gamma} f(t) A(t)} = H(x);$$

hence the function $\pm t^{-\gamma} f(t)$ is extended regularly varying, $H(x) = (x^\rho - 1)/\rho$ for some $\rho \leq 0$, and the theory of Section B.2 applies. Using the inequalities of Theorem B.2.18 one then gets, if $\rho < 0$,

$$\lim_{\substack{t \to \infty \\ x = x(t) \to \infty}} \frac{\frac{f(tx)}{x^\gamma f(t)} - 1}{A(t)} = -\frac{1}{\rho} \; .$$

In view of applications in Chapters 3, 4, and 5 we now show how the second-order condition for f translates into a second-order condition for $\log f$.

Lemma B.3.16 *Let f be a measurable function with $\lim_{t \to \infty} f(t) = f(\infty) \in (0, \infty]$. Assume there exist functions, a positive and A with $\lim_{t \to \infty} A(t) = 0$ and not changing sign eventually, such that*

$$\lim_{t\to\infty} \frac{\frac{f(tx)-f(t)}{a(t)} - \frac{x^\gamma-1}{\gamma}}{A(t)} = H_{\gamma,\rho}(x), \tag{B.3.41}$$

where

$$H_{\gamma,\rho}(x) = \frac{1}{\rho}\left(\frac{x^{\gamma+\rho}-1}{\gamma+\rho} - \frac{x^\gamma-1}{\gamma}\right) \qquad (\rho \le 0).$$

Suppose $\gamma \ne \rho$.
  Then

$$\lim_{t\to\infty} \frac{\frac{a(t)}{f(t)} - \gamma_+}{A(t)} = \begin{cases} 0, & \gamma < \rho \le 0, \\ \pm\infty, & \rho < \gamma \le 0 \text{ or } (0 < \gamma < -\rho \text{ and } l \ne 0) \text{ or } \gamma = -\rho, \\ \frac{\gamma}{\gamma+\rho}, & (0 < \gamma < -\rho \text{ and } l = 0) \text{ or } \gamma > -\rho \ge 0, \end{cases} \tag{B.3.42}$$

*where for $\gamma > 0$ we define $l := \lim_{t\to\infty} f(t) - a(t)/\gamma$. Furthermore, in case $\gamma > 0$
assume $\rho < 0$, and we have*

$$\lim_{t\to\infty} \frac{\frac{\log f(tx)-\log f(t)}{a(t)/f(t)} - \frac{x^{\gamma_-}-1}{\gamma_-}}{Q(t)} = H_{\gamma_-,\rho'}(x), \tag{B.3.43}$$

where

$$Q(t) = \begin{cases} A(t), & \gamma < \rho \le 0, \\ \gamma_+ - \frac{a(t)}{f(t)}, & \rho < \gamma \le 0 \text{ or } (0 < \gamma < -\rho \text{ and } l \ne 0) \text{ or } \gamma = -\rho, \\ \frac{\rho}{\gamma+\rho}A(t), & (0 < \gamma < -\rho \text{ and } l = 0) \text{ or } \gamma > -\rho > 0, \end{cases}$$

$|Q(t)| \in RV_{\rho'}$, *and*

$$\rho' = \begin{cases} \rho, & \gamma < \rho \le 0, \\ \gamma, & \rho < \gamma \le 0, \\ -\gamma, & (0 < \gamma < -\rho \text{ and } l \ne 0), \\ \rho, & (0 < \gamma < -\rho \text{ and } l = 0) \text{ or } \gamma \ge -\rho > 0. \end{cases}$$

*If $\gamma > 0$ and $\rho = 0$, the limit in (B.3.43) equals zero for any $Q(t)$ satisfying
$A(t) = O(Q(t))$ or equivalently $a(t)/f(t) - \gamma = O(Q(t))$.*

Proof. We start with the proof of (B.3.42) and separately analyze the cases $\gamma < 0$,
$\gamma = 0, 0 < \gamma < -\rho, \gamma = -\rho$, and $\gamma > -\rho$.
  Start with $\gamma < 0$. Then $a(t)/A(t) \in RV_{\gamma-\rho}$ and by assumption $f(\infty) > 0$.
Hence

$$\lim_{t\to\infty} \frac{\frac{a(t)}{f(t)} - \gamma_+}{A(t)} = \lim_{t\to\infty} \frac{a(t)}{f(t)A(t)} = \begin{cases} 0, & \gamma - \rho < 0, \\ \pm\infty, & \gamma - \rho > 0. \end{cases}$$

Next consider $\gamma = 0$. Then $\rho < 0$ since we assume $\gamma \ne \rho$. Then $f(t) \in RV_0$ and
from (B.3.4) and Theorem B.1.6, there exists $\lim_{t\to\infty} a(t)$ and it is positive. Hence,
$\lim_{t\to\infty} a(t)/(f(t)A(t)) = \pm\infty$.
  Next we consider the various possibilities when γ is positive. Note that from
(B.3.4),

$$\lim_{t\to\infty} \frac{(tx)^{-\gamma}a(tx) - t^{-\gamma}a(t)}{t^{-\gamma}a(t)A(t)} = \frac{x^\rho - 1}{\rho}, \tag{B.3.44}$$

and combining this with (B.3.41), we have

$$\lim_{t\to\infty} \frac{\left(f(tx) - \frac{a(tx)}{\gamma}\right) - \left(f(t) - \frac{a(t)}{\gamma}\right)}{-\frac{1}{\gamma}a(t)A(t)} = \frac{x^{\gamma+\rho} - 1}{\gamma + \rho}. \tag{B.3.45}$$

Then if $\gamma + \rho > 0$,

$$\lim_{t\to\infty} \frac{f(t) - \frac{a(t)}{\gamma}}{-\frac{1}{\gamma}a(t)A(t)} = \frac{1}{\gamma + \rho},$$

that is,

$$\lim_{t\to\infty} \frac{\frac{a(t)}{f(t)} - \gamma_+}{A(t)} = \frac{\gamma}{\gamma + \rho}.$$

Next if $\gamma + \rho < 0$, there exists $l := \lim_{t\to\infty} f(t) - a(t)/\gamma$ and

$$\lim_{t\to\infty} \frac{\left(f(t) - \frac{a(t)}{\gamma}\right) - l}{\frac{1}{\gamma}a(t)A(t)} = -\frac{1}{\gamma + \rho}.$$

Hence

$$\lim_{t\to\infty} \frac{\frac{a(t)}{f(t)} - \gamma_+}{A(t)} = \lim_{t\to\infty} \frac{\gamma a(t)}{f(t)} \left(-\frac{\left(f(t) - \frac{a(t)}{\gamma}\right) - l}{a(t)A(t)} - \frac{l}{a(t)A(t)} \right)$$

$$= \begin{cases} \pm\infty, & 0 < \gamma < -\rho \text{ and } l \neq 0, \\ \frac{\gamma}{\gamma+\rho}, & 0 < \gamma < -\rho \text{ and } l = 0. \end{cases}$$

Finally, consider $\gamma = -\rho > 0$. From (B.3.45) it follows that $-a(t)A(t)/\gamma = o\left(f(t) - a(t)/\gamma\right)$ and so

$$\lim_{t\to\infty} \frac{\frac{a(t)}{f(t)} - \gamma_+}{A(t)} = \lim_{t\to\infty} \frac{\gamma a(t)}{f(t)} \frac{f(t) - \frac{a(t)}{\gamma}}{-a(t)A(t)} = \pm\infty.$$

We have now proved (B.3.42).

For the proof of (B.3.43), first note that $\lim_{t\to\infty} a(t)/f(t) = \gamma_+$ (cf. Lemma 1.2.9). Then for $\gamma \leq 0$, from (B.3.41),

$$\log\left(\frac{f(tx)}{f(t)}\right) = \log\left(1 + \frac{a(t)}{f(t)}\left\{\frac{x^\gamma - 1}{\gamma} + A(t)H_{\gamma,\rho}(x) + o(A(t))\right\}\right)$$

$$= \frac{a(t)}{f(t)}\left\{\frac{x^\gamma - 1}{\gamma} + A(t)H_{\gamma,\rho}(x) + o(A(t))\right\}$$

$$-\frac{1}{2}\left(\frac{a(t)}{f(t)}\right)^2\left\{\frac{x^\gamma-1}{\gamma}+A(t)H_{\gamma,\rho}(x)+o(A(t))\right\}^2$$
$$+o\left(\left(\frac{a(t)}{f(t)}\right)^2\right);$$

hence

$$\lim_{t\to\infty}\frac{\log f(tx)-\log f(t)}{\frac{a(t)}{f(t)}}-\frac{x^\gamma-1}{\gamma}$$

$$=A(t)H_{\gamma,\rho}(x)+o(A(t))-\frac{1}{2}\frac{a(t)}{f(t)}\left(\frac{x^\gamma-1}{\gamma}\right)^2+o\left(\frac{a(t)}{f(t)}\right).$$

The result follows for this case.

Now consider $\gamma>0$. Again from (B.3.41),

$$x^{-\gamma}\frac{f(tx)}{f(t)}=x^{-\gamma}+\frac{a(t)}{f(t)}\frac{1-x^{-\gamma}}{\gamma}+x^{-\gamma}\frac{a(t)}{f(t)}\left\{A(t)H_{\gamma,\rho}(x)+o(A(t))\right\}$$

$$=1+\left(x^{-\gamma}-1\right)\left(1-\frac{a(t)}{\gamma f(t)}\right)+x^{-\gamma}\frac{a(t)}{f(t)}\left\{A(t)H_{\gamma,\rho}(x)+o(A(t))\right\};$$

hence

$$\lim_{t\to\infty}\frac{\log f(tx)-\log f(t)}{\frac{a(t)}{f(t)}}-\log x$$

$$=-\left(\log x+\frac{x^{-\gamma}-1}{\gamma}\right)\frac{f(t)}{a(t)}\left(\frac{a(t)}{f(t)}-\gamma\right)$$

$$+x^{-\gamma}A(t)H_{\gamma,\rho}(x)+o(A(t))+o\left(\frac{a(t)}{f(t)}-\gamma\right).$$

Combining this with (B.3.42), the result follows for $\gamma>0$. ∎

Remark B.3.17 Relation (B.3.45) is true for $\gamma\neq0$ and $\rho\leq0$.

Remark B.3.18 From Lemma B.3.16 it follows that

$$q_{\gamma,\rho}=\lim_{t\to\infty}\frac{a(t)/f(t)-\gamma_+}{Q(t)}=\begin{cases}0, & \gamma<\rho\leq0,\\ \gamma/\rho, & (\lim_{t\to\infty}f(t)-a(t)/\gamma=0\\ & \text{and }0<\gamma<-\rho)\text{ or }\gamma\geq-\rho>0,\\ -1, & (\lim_{t\to\infty}f(t)-a(t)/\gamma\neq0\\ & \text{and }0<\gamma<-\rho)\text{ or }\rho<\gamma\leq0.\end{cases}$$
$$(B.3.46)$$

Next we consider the special case of f nondecreasing and give equivalent conditions in terms of $\phi:=f^{\leftarrow}$ (the inverse function of f). This is relevant for extreme-value statistics where (B.3.3) is a condition in terms of the quantile function (so the inverse of a probability distribution) and one wants to have conditions in terms of the distribution function itself.

Theorem B.3.19 *Suppose that f is nondecreasing and ϕ is its left-continuous inverse function. Then (B.3.3) is equivalent to*

$$\lim_{t\uparrow f(\infty)} \frac{\frac{\phi(t+xa(\phi(t)))}{\phi(t)} - (1+\gamma x)^{1/\gamma}}{A(\phi(t))} = -(1+\gamma x)^{-1+1/\gamma} H((1+\gamma x)^{1/\gamma}) \quad \text{(B.3.47)}$$

locally uniformly for $x \in (-1/\max(0,\gamma), 1/\max(-\gamma, 0))$, where H is the limit function in (B.3.3)

Remark B.3.20   1. The result is also true for the right-continuous inverse of f .
  2. In case $\gamma = 0$ we define $(1+\gamma x)^{1/\gamma} = e^x$.
  3. For specific parameters we can give more specific statements, such as, for example:
  (a) if $\alpha = 0$, $\gamma > 0$, then $\pm t^{-1/\gamma}\phi(t) \in \Pi$;
  (b) if $\alpha = 0$, $\gamma < 0$, then $\pm t^{1/\gamma}\phi(f(\infty) - t^{-1}) \in \Pi$;
  (c) if $\alpha < 0$, $\gamma = 0$, then $\lim_{t\to\infty}(e^{-c(t+x)}\phi(t+x)) - c/(e^{-ct}\phi(t) - c) = e^{\alpha x}$, $x \in \mathbb{R}$.

Proof. Since by Remark B.3.8(1) relation (B.3.3) holds locally uniformly we can replace x by $x(t) = 1 + \varepsilon A(t)$ in (B.3.3) and get

$$\lim_{t\to\infty} \frac{f(t+t\varepsilon A(t)) - f(t) - a(t)((1+\varepsilon A(t))^\gamma - 1)/\gamma}{a(t)A(t)} = 0 , \quad \text{(B.3.48)}$$

hence

$$\lim_{t\to\infty} \frac{f(t+t\varepsilon A(t)) - f(t)}{a(t)A(t)} = \varepsilon . \qquad\blacksquare$$

Applying this for $\varepsilon > 0$ and $\varepsilon < 0$ and using $f((\phi(t))^-) \le t \le f((\phi(t))^+)$ we obtain $\lim_{t\uparrow f(\infty)}\{f(\phi(t)) - t\}/\{a(\phi(t))A(\phi(t))\} = 0$. This and (B.3.3) imply that

$$\lim_{t\uparrow f(\infty)} \frac{\frac{f(\phi(t)x)-t}{a(\phi(t))} - \frac{x^\gamma-1}{\gamma}}{A(\phi(t))} = H(x) .$$

Now $F_t(x) := (f(\phi(t)x) - t)/a(\phi(t))$, $x > 0$, $t < f(\infty)$, is a family (with respect to t) of nondecreasing functions; furthermore, $(x^\gamma - 1)/\gamma$ has a positive continuous derivative, the function $H(\cdot)$ is continuous, and the function A satisfies $A(\phi(t)) \to 0$, $t \uparrow f(\infty)$. Therefore we can apply an obvious generalization of Vervaat's lemma (see Lemma A.0.2) to deduce (B.3.47). The converse implication is similar.

B.4 ERV with an Extra Parameter

Definition B.4.1 Let $f_s(t)$ be a measurable function for $0 \le s \le 1$ and $t > 0$. We say that the function f is *jointly regularly varying* if there exists a continuous function γ defined on $[0, 1]$ such that for $x > 0$,

$$\lim_{t\to\infty} \sup_{0\le s\le 1} \left| \frac{f_s(tx)}{f_s(t)} - x^{\gamma(s)} \right| = 0 \ . \tag{B.4.1}$$

We define analogously *joint extended regular variation*:

$$\lim_{t\to\infty} \sup_{0\le s\le 1} \left| \frac{f_s(tx) - f_s(t)}{a_s(t)} - \frac{x^{\gamma(s)} - 1}{\gamma(s)} \right| = 0, \tag{B.4.2}$$

where $a_s(t)$ is positive. The definition of *joint (extended) regular variation of second-order* is analogous. The function γ is called the *index function*.

This concept of joint (extended) regular variation is used in Chapter 9 and we develop some properties that are needed in that chapter.

Theorem B.4.2 *Suppose f is jointly regularly varying. For any positive $\varepsilon, \delta > 0$ there exists $t_0 = t_0(\varepsilon, \delta)$ such that for t, $tx \ge t_0$, $0 \le s \le 1$,*

$$\left| \frac{f_s(tx)}{f_s(t)} - x^{\gamma(s)} \right| \le \varepsilon x^{\gamma(s)} \max\left(x^\delta, x^{-\delta}\right) \tag{B.4.3}$$

and

$$(1 - \varepsilon)x^{\gamma(s)} \min\left(x^\delta, x^{-\delta}\right) \le \frac{f_s(tx)}{f_s(t)} \le (1 + \varepsilon)x^{\gamma(s)} \max\left(x^\delta, x^{-\delta}\right) \ . \tag{B.4.4}$$

Proof. Clearly it is sufficient to prove the statements for $\gamma = 0$ for all s. The first step is to prove that (B.4.1) holds locally uniformly for $x \in (0, \infty)$. It is sufficient to deduce a contradiction from the following assumption: suppose there exist $\delta > 0$ and sequences $s_n \to s_0$, $t_n \to \infty$, $x_n \to 0$, as $n \to \infty$, such that

$$\left| F_{s_n}(t_n + x_n) - F_{s_n}(t_n) \right| > \delta$$

for $n = 1, 2, \ldots$, where $F_s(x) := \log f_s(e^x)$. The rest of the proof of the local uniformity is exactly like the proof of Theorem B.2.9. Now we know that for any $\varepsilon \in (0, 1)$, there exists a t_0 such that if $t \ge t_0$, $x \in [1, e]$, $s \in [0, 1]$,

$$|\log f_s(tx) - \log f_s(t)| < \varepsilon \ .$$

Take any $x \ge 1$. We write $x = e^n y$, $y \in [1, e)$ for some nonnegative integer n. Then

$$|\log f_s(tx) - \log f_s(t)|$$

$$\le \sum_{i=1}^{n} \left| \log f_s(te^i) - \log f_s(te^{i-1}) \right| + \left| \log f_s(te^n y) - \log f_s(te^n) \right|$$

$$\le (n + 1)\varepsilon_1 \le \varepsilon_1 \log x + \varepsilon_1 \ .$$

For the last inequality we use $\log x > n$. This proves (B.4.4) for $x \ge 1$. By interchanging the role of t and tx in case $x < 1$, we get the full statement (B.4.4). The proof that (B.4.4) implies (B.4.3) is analogous to the proof of Proposition B.1.10. ∎

Next we shall prove uniform inequalities for joint extended regular variation.

Theorem B.4.3 *Suppose f is jointly extended regularly varying, i.e., (B.4.2) holds. For any $\varepsilon, \delta > 0$ there exists $t_0 = t_0(\varepsilon, \delta)$ such that for $t, tx \geq t_0$, $0 \leq s \leq 1$,*

$$\left| \frac{f_s(tx) - f_s(t)}{a_s(t)} - \frac{x^{\gamma(s)} - 1}{\gamma(s)} \right| \leq \varepsilon \left(1 + x^{\gamma(s)} \max \left(x^\delta, x^{-\delta} \right) \right) . \tag{B.4.5}$$

Moreover, the function $a_s(t)$ is jointly regularly varying with index function γ.

Proof. The proof of the last statement is analogous to that of Theorem B.2.1 and it is left to the reader.

The defining relation (B.4.2) holds locally uniformly for $x \in (0, \infty)$. This can be proved in the same way as for the corresponding result in the previous theorem, now following the proof of Theorem B.2.9.

By Theorem B.4.2 for any $\varepsilon, \delta > 0$ we can find t_0 such that for $t, tx \geq t_0$ we have

$$\left| \frac{a_s(tx)}{a_s(t)} - x^{\gamma(s)} \right| \leq \varepsilon x^{\gamma(s)} \max \left(x^\delta, x^{-\delta} \right) , \tag{B.4.6}$$

$$(1 - \varepsilon) x^{\gamma(s)} \min \left(x^\delta, x^{-\delta} \right) \leq \frac{a_s(tx)}{a_s(t)} \leq (1 + \varepsilon) x^{\gamma(s)} \max \left(x^\delta, x^{-\delta} \right), \tag{B.4.7}$$

and

$$\left| \frac{f_s(ty) - f_s(t)}{a_s(t)} - \frac{y^{\gamma(s)} - 1}{\gamma(s)} \right| \leq \varepsilon \tag{B.4.8}$$

for $y \in [1, e]$. Take any $x > 1$. We write $x = e^n y$, where $y \in [1, e]$ with some nonnegative integer n. Note that

$$\frac{f_s(te^n y) - f_s(t)}{a_s(t)} - \frac{(e^n y)^{\gamma(s)} - 1}{\gamma(s)}$$

$$= \left(\frac{f_s(te^n y) - f_s(te^n)}{a_s(te^n)} - \frac{y^{\gamma(s)} - 1}{\gamma(s)} \right) \frac{a_s(te^n)}{a_s(t)}$$

$$+ \sum_{i=0}^{n-1} \left(\frac{f_s(te^{i+1}) - f_s(te^i)}{a_s(te^i)} - \frac{e^{\gamma(s)} - 1}{\gamma(s)} \right) \frac{a_s(te^i)}{a_s(t)}$$

$$+ \left(\frac{a_s(te^n)}{a_s(t)} - e^{n\gamma(s)} \right) \frac{y^{\gamma(s)} - 1}{\gamma(s)} + \sum_{i=0}^{n-1} \left(\frac{a_s(te^i)}{a_s(t)} - e^{i\gamma(s)} \right) \frac{e^{\gamma(s)} - 1}{\gamma(s)} .$$

Applying (B.4.6), (B.4.7), (B.4.8), and $(y^{\gamma(s)} - 1)/\gamma(s) \leq (e^{\gamma(s)} - 1)/\gamma(s)$, we get with

$$C := \sup_{0 \leq s \leq 1} \left(1 + \varepsilon + \frac{e^{\gamma(s)} - 1}{\gamma(s)} \right) \frac{\gamma(s) + \varepsilon}{e^{\gamma(s) + \varepsilon} - 1}$$

$$= \left(1 + \varepsilon + \frac{e^{\sup \gamma(s)} - 1}{\sup \gamma(s)} \right) \frac{\varepsilon + \inf \gamma(s)}{e^{\varepsilon + \inf \gamma(s)} - 1}$$

that

$$\left| \frac{f_s(te^n y) - f_s(t)}{a_s(t)} - \frac{(e^n y)^{\gamma(s)} - 1}{\gamma(s)} \right|$$

$$\leq \varepsilon \sum_{i=0}^{n} (1 + \varepsilon) e^{i(\gamma(s)+\varepsilon)} + \sum_{i=0}^{n} \varepsilon \, e^{i(\gamma(s)+\varepsilon)} \frac{e^{\gamma(s)} - 1}{\gamma(s)}$$

$$= \varepsilon \left(1 + \varepsilon + \frac{e^{\gamma(s)} - 1}{\gamma(s)} \right) \frac{e^{(n+1)(\gamma(s)+\varepsilon)} - 1}{e^{\gamma(s)+\varepsilon} - 1}$$

$$\leq \varepsilon \, C \frac{e^{(n+1)(\gamma(s)+\varepsilon)} - 1}{\gamma(s) + \varepsilon}.$$

Now on the set $\{s \in [0, 1] : \gamma(s) + \varepsilon \leq -\sqrt{\varepsilon}\}$ we have

$$\frac{e^{(n+1)(\gamma(s)+\varepsilon)} - 1}{\gamma(s) + \varepsilon} \leq \frac{1}{-\gamma(s) - \varepsilon} \leq \frac{1}{\sqrt{\varepsilon}},$$

and on the set $\{s \in [0, 1] : \gamma(s) + \varepsilon \geq -\sqrt{\varepsilon}\}$ we have

$$\frac{e^{(n+1)(\gamma(s)+\varepsilon)} - 1}{\gamma(s) + \varepsilon} \leq \frac{e^{(n+1)(\gamma(s)+\varepsilon+2\sqrt{\varepsilon})}}{\sqrt{\varepsilon}}$$

(consider $\gamma(s) + \varepsilon < 0$ and $\gamma(s) + \varepsilon > 0$ separately and use that $(1 - e^{-x})/x \leq 1$ for $x > 0$ and that $(e^x - 1)/x$ is increasing for $x > 0$).

Hence (with $y = xe^{-n}$ as before)

$$\left| \frac{f_s(te^n y) - f_s(t)}{a_s(t)} - \frac{(e^n y)^{\gamma(s)} - 1}{\gamma(s)} \right| \leq \varepsilon \, C \left(\frac{1}{\sqrt{\varepsilon}} + \frac{1}{\sqrt{\varepsilon}} e^{(n+1)(\gamma(s)+\varepsilon+2\sqrt{\varepsilon})} \right)$$

$$\leq C \left(\sqrt{\varepsilon} + \sqrt{\varepsilon} e^{\gamma(s)+\varepsilon+2\sqrt{\varepsilon}} x^{\gamma(s)+\varepsilon+2\sqrt{\varepsilon}} \right)$$

since $\log x > n$. This proves (B.4.5). ∎

Theorem B.4.4 *Suppose f is jointly extended regularly varying, i.e., (B.4.2) holds. Moreover, suppose f is positive. Then*

$$\lim_{t \to \infty} \sup_{0 \leq s \leq 1} \left| \frac{a_s(t)}{f_s(t)} - \gamma_+(s) \right| = 0, \tag{B.4.9}$$

where $\gamma_+(s) := \max(0, \gamma(s))$. For any $\varepsilon, \delta > 0$ there exists $t_0 = t_0(\varepsilon, \delta)$ such that if $t, tx \geq t_0$, $0 \leq s \leq 1$,

$$\left| \frac{\log f_s(tx) - \log f_s(t)}{a_s(t)/f_s(t)} - \frac{x^{\gamma_-(s)} - 1}{\gamma_-(s)} \right| \leq \varepsilon \left(1 + x^{\gamma_-(s)} \max(x^\delta, x^{-\delta}) \right), \tag{B.4.10}$$

where $\gamma_-(s) := \min(0, \gamma(s))$.

Proof. For (B.4.9) we need to prove that if $t_n \to \infty$, $s_n \to s_0$,

$$\frac{f_{s_n}(t_n)}{a_{s_n}(t_n)} \to \begin{cases} 1/\gamma(s_0), & \gamma(s_0) > 0, \\ \infty, & \gamma(s_0) \leq 0. \end{cases}$$

Suppose $\gamma(s_0) > 0$. By Theorem B.4.3 for any $\varepsilon \in (0, \gamma(s_0))$ there is a t_0 such that for $t_n \geq t_0$,

$$\left| \frac{f_{s_n}(t_n) - f_{s_n}(t_0)}{a_{s_n}(t_n)} - \frac{1 - \left(\frac{t_0}{t_n}\right)^{\gamma(s_n)}}{\gamma(s_n)} \right| \leq \varepsilon \left(1 + \left(\frac{t_0}{t_n}\right)^{\gamma(s_n)-\varepsilon} \right).$$

It follows that

$$\lim_{n \to \infty} \frac{f_{s_n}(t_n) - f_{s_n}(t_0)}{a_{s_n}(t_n)} = \frac{1}{\gamma(s_0)}. \tag{B.4.11}$$

Now Theorem B.4.3 implies (take $t = t_0$ and $x \to \infty$ in the statement of the theorem) that

$$\lim_{t \to \infty} a_s(t) = \infty \tag{B.4.12}$$

uniformly for those s with $\gamma(s) > \gamma(t_0)/2$. Combining (B.4.11) and (B.4.12) gives

$$\lim_{n \to \infty} \frac{f_{s_n}(t_n)}{a_{s_n}(t_n)} = \frac{1}{\gamma(s_0)}.$$

Next suppose $\gamma(s_0) \leq 0$. By Theorem B.4.3 the function $\psi_s(t)$ defined by

$$\psi_s(t) := \int_1^\infty (f_s(tx) - f_s(t)) \frac{dx}{x^2} = t \int_t^\infty f_s(x) \frac{dx}{x^2} - f_s(t)$$

is well defined for those s for which $\gamma(s) < 1$. By partial integration

$$\int_{t_0}^t \psi_s(u) \frac{du}{u} = \int_{t_0}^t \int_u^\infty f_s(x) \frac{dx}{x^2} du - \int_{t_0}^t f_s(u) \frac{du}{u}$$

$$= t \int_t^\infty f_s(x) \frac{dx}{x^2} - t_0 \int_{t_0}^\infty f_s(x) \frac{dx}{x^2}$$

$$= \psi_s(t) + f_s(t) - t_0 \int_{t_0}^\infty f_s(x) \frac{dx}{x^2},$$

i.e.,

$$f_s(t) = \int_{t_0}^t \psi_s(u) \frac{du}{u} - \psi_s(t) + t_0 \int_{t_0}^\infty f_s(x) \frac{dx}{x^2}. \tag{B.4.13}$$

Next we shall prove

$$\lim_{t \to \infty} \sup_{s \in E_c} \left| \frac{\psi_s(t)}{a_s(t)} - \frac{1}{1 - \gamma(s)} \right| = 0 \tag{B.4.14}$$

for $0 < c < 1$, where $E_c := \{s \in [0, 1] : \gamma(s) \le c\}$. This implies that ψ is jointly regularly varying and also that it is sufficient to prove (B.4.9) where a is replaced with ψ.

Now,

$$\frac{\psi_s(t)}{a_s(t)} = \int_1^\infty \frac{f_s(tx) - f_s(t)}{a_s(t)} \frac{dx}{x^2}.$$

For any $\varepsilon \in (0, c)$, by Theorem B.4.3 there exists t_0 such that for $t, tx \ge t_0$,

$$\left| \int_1^\infty \left(\frac{f_s(tx) - f_s(t)}{a_s(t)} - \frac{x^{\gamma(s)} - 1}{\gamma(s)} \right) \frac{dx}{x^2} \right| \le \varepsilon \int_1^\infty \left(1 + x^{\gamma(s)+\varepsilon} \right) \frac{dx}{x^2}.$$

Hence

$$\lim_{t \to \infty} \frac{\psi_s(t)}{a_s(t)} = \int_1^\infty \frac{x^{\gamma(s)} - 1}{\gamma(s)} \frac{dx}{x^2} = \frac{1}{1 - \gamma(s)}$$

uniformly for $s \in E_c$. Hence it is sufficient to prove

$$\lim_{t_n \to \infty} \frac{f_{s_n}(t_n)}{\psi_{s_n}(t_n)} = \infty$$

for $\gamma(s_0) \le 0$, where ψ is jointly regularly varying with index function γ. Now by (B.4.13),

$$\liminf_{n \to \infty} \frac{f_{s_n}(t_n)}{\psi_{s_n}(t_n)} \ge \liminf_{n \to \infty} \int_{t_0/t_n}^1 \frac{\psi_{s_n}(t_n u)}{\psi_{s_n}(t_n)} \frac{du}{u} - 1$$

$$\ge \liminf_{n \to \infty} \int_{t_0/t_n}^1 (1 - 2\varepsilon) u^{\gamma(s_n)-1+2\varepsilon} \, du - 1$$

$$\ge \liminf_{n \to \infty} \int_{t_0/t_n}^1 (1 - 2\varepsilon) u^{3\varepsilon-1} \, du - 1$$

$$= \liminf_{n \to \infty} (1 - 2\varepsilon) \frac{1 - (t_0/t_n)^{3\varepsilon}}{3\varepsilon} - 1$$

$$= \frac{1 - 2\varepsilon}{3\varepsilon} - 1,$$

which tends to infinity as $\varepsilon \downarrow 0$. The proof of (B.4.9) is complete.

For (B.4.10) by Theorem B.4.3 we need to prove only that for any $s_n \to s_0$, $x_n \to x_0 > 0$, $t_n \to \infty$,

$$\lim_{n \to \infty} \frac{\log f_{s_n}(t_n x_n) - \log f_{s_n}(t_n)}{a_{s_n}(t_n)/f_{s_n}(t_n)} = \frac{x_0^{\gamma_-(s_0)} - 1}{\gamma_-(s_0)}.$$

We write

$$\lim_{n \to \infty} \frac{\log f_{s_n}(t_n x_n) - \log f_{s_n}(t_n)}{a_{s_n}(t_n)/f_{s_n}(t_n)} = \lim_{n \to \infty} \frac{\log \left(\frac{f_{s_n}(t_n x) - f_{s_n}(t_n)}{a_{s_n}(t_n)} \frac{a_{s_n}(t_n)}{f_{s_n}(t_n)} + 1 \right)}{a_{s_n}(t_n)/f_{s_n}(t_n)}.$$

For $\gamma(s_0) > 0$, $\lim_{n \to \infty} a_{s_n}(t_n)/f_{s_n}(t_n) = \gamma(s_0)$; hence this converges to

$$\frac{1}{\gamma(s_0)} \log \left(\frac{x_0^{\gamma(s_0)} - 1}{\gamma(s_0)} \gamma(s_0) + 1 \right) = \log x_0 \, .$$

For $\gamma(s_0) \leq 0$, $\lim_{n \to \infty} a_{s_n}(t_n)/f_{s_n}(t_n) = 0$; hence the limit equals

$$\lim_{n \to \infty} \frac{f_{s_n}(t_n x_n) - f_{s_n}(t_n)}{a_{s_n}(t_n)} = \frac{x_0^{\gamma(s_0)} - 1}{\gamma(s_0)} \, . \qquad \blacksquare$$

References

1. K. Aarssen and L. de Haan: On the maximal life span of humans. *Mathematical Population Studies* **4**, 259–281 (1994).
2. M. Ancona-Navarrete and J. Tawn: A comparison of methods for estimating the extremal index. *Extremes* **3**, 5–38 (2000).
3. B.C. Arnold, N. Balakrishnan, and H.N. Nagaraja: *A First Course in Order Statistics.* Wiley, New York (1992).
4. J.M. Ash, P. Erdős and L.A. Rubel: Very slowly varying functions. *Aeq. Math.* **10**, 1–9 (1974).
5. A.A. Balkema and L. de Haan: A.s. continuity of stable moving average processes with index < 1. *Ann. Appl. Probab.* **16**, 333–343 (1988).
6. A.A. Balkema, L. de Haan, and R.L. Karandikar: Asymptotic distributions of the maximum of n independent stochastic processes. *J. Appl. Prob.* **30**, 66–81 (1993).
7. O. Barndorff-Nielsen: On the limit behaviour of extreme order statistics. *Ann. Math. Statist.* **34**, 992–1002 (1963).
8. J. Beirlant and J.L. Teugels: Asymptotics of Hill's estimator. *Theory Probab. Appl.* **31**, 463–469 (1986).
9. J. Beirlant, P. Vynckier, and J. L. Teugels: Tail index estimation, Pareto quantile plots and regression diagnostics. *J. Amer. Statist. Association* **91**, 1659–1667 (1996).
10. J. Beirlant, J.L. Teugels, and P. Vynckier: *Practical Analysis of Extreme Values.* Leuven University Press, Leuven, Belgium (1996).
11. P. Billingsley: *Convergence of Probability Measures.* Wiley, New York (1968).
12. P. Billingsley: *Weak Convergence of Measures: Applications in Probability.* SIAM, Philadelphia (1971).
13. P. Billingsley: *Probability and Measure.* Wiley, New York (1979).
14. L. Breiman: *Probability.* Addison-Wesley (1968); Republished by SIAM, Philadelphia (1992).
15. B. Brown and S. Resnick: Extreme values of independent stochastic processes. *J. Appl. Probab.* **14**, 732–739 (1977).
16. N. G. de Bruijn: Pairs of slowly oscillating functions occurring in asymptotic problems concerning the Laplace transform. *Nw. Arch. Wisk.* **7**, 20–26 (1959).
17. S. Cheng and C. Jiang: The Edgeworth expansion for distributions of extreme values. *Science in China* **44**, 427–437 (2001).
18. K.L. Chung: *A Course in Probability Theory.* 2nd Edition, Academic Press, New York–London (1974).

19. M. Csörgő and L. Horváth: *Weighted Approximations in Probability and Statistics.* John Wiley & Sons, Chichester, England (1993).

20. D. J. Daley and D. Vere-Jones: *An Introduction to the Theory of Point Processes.* Springer, Berlin (1988).

21. J. Danielsson, L. de Haan, L. Peng, and C. G. de Vries: Using a bootstrap method to choose the sample fraction in tail index estimation. *J. Multivariate Analysis* **76**, 226–248 (2001).

22. A.L.M. Dekkers and L. de Haan: Optimal choice of sample fraction in extreme-value estimation. *J. Multivariate Analysis* **47**, 173–195 (1993).

23. A.L.M. Dekkers, J.H.J. Einmahl, and L. de Haan: A moment estimator for the index of an extreme-value distribution. *Ann. Statist.* **17**, 1833–1855 (1989).

24. D. Dietrich, L. de Haan, and J. Hüsler: Testing extreme value conditions. *Extremes* **5**, 71–85 (2002).

25. G. Draisma, H. Drees, A. Ferreira, and L. de Haan: Bivariate tail estimation: dependence in asymptotic independence. *Bernoulli* **10**, 251–280 (2004).

26. G. Draisma, L. de Haan, L. Peng, and T.T. Pereira: A bootstrap based method to achieve optimality in estimating the extreme value index. *Extremes* **2**, 367–404 (1999).

27. H. Drees: On smooth statistical tail functionals. *Scand. J. Statist.* **25**, 187–210 (1998).

28. H. Drees: Weighted approximations of tail processes for β-mixing random variables. *Ann. Appl. Probab.* **10**, 1274–1301 (2000).

29. H. Drees: Tail empirical processes under mixing conditions. In: H.G. Dehling, T. Mikosch, and M. Sorensen (eds.) *Empirical Process Techniques for Dependent Data.* Birkhäuser, Boston, 325–342 (2002).

30. H. Drees: Extreme quantile estimation for dependent data with applications to finance. *Bernoulli* **9**, 617–657 (2003).

31. H. Drees, A. Ferreira, and L. de Haan: On maximum likelihood estimation of the extreme value index. *Ann. Appl. Probab.* **14**, 1179–1201 (2003).

32. H. Drees, L. de Haan, and D. Li: On large deviations for extremes. *Stat. Prob. Letters* **64**, 51–62 (2003).

33. H. Drees, L. de Haan, and D. Li: Approximations to the tail empirical distribution function with application to testing extreme value conditions. To appear in *J. Statist. Plann. Inference* (2006).

34. H. Drees and E. Kaufmann: Selecting the optimal sample fraction in univariate extreme value estimation. *Stoch. Proc. Appl.* **75**, 149–172 (1998).

35. W.F. Eddy and J.D. Gale: The convex hull of a spherically symmetric sample. *Adv. Appl. Prob.* **13**, 751–763 (1981).

36. J.H.J. Einmahl: Multivariate empirical processes. PhD thesis, CWI Tract **32**, Amsterdam (1987).

37. J.H.J. Einmahl: The empirical distribution function as a tail estimator. *Statistica Neerlandica* **44**, 79–82 (1990).

38. J.H.J. Einmahl: A Bahadur-Kiefer theorem beyond the largest observation. *J. Multivariate Anal.* **55**, 29–38 (1995).

39. J.H.J. Einmahl: Poisson and Gaussian approximation of weighted local empirical processes. *Stoch. Proc. Appl.* **70**, 31–58 (1997).

40. J.H.J. Einmahl, L. de Haan, and V. Piterbarg: Non-parametric estimation of the spectral measure of an extreme value distribution. *Ann. Statist.* **29**, 1401–1423 (2001).

41. J.H.J. Einmahl and T. Lin: Asymptotic normality of extreme value estimators on $C[0, 1]$. *Ann. Statist.* **34**, 469–492 (2006).

42. P. Embrechts, C. Klüppelberg, and T. Mikosch: *Modelling Extremal Events for Insurance and Finance.* Springer-Verlag, Berlin Heidelberg (1997).

43. P. Embrechts, L. de Haan, and X. Huang: Modelling Multivariate Extremes. In: P. Embrechts (ed.) *Extremes and Integrated Risk Measures*. Risk Waters Group, 59–67 (2000).

44. M. Falk: Some best estimators for distributions with finite endpoint. *Statistics* **27**, 115–125 (1995).

45. M. Falk, J. Hüsler, and R.-D. Reiss: *Laws of Small Numbers: Extremes and Rare Events*. Birkhäuser, Basel (1994).

46. W. Feller: *An Introduction to Probability Theory and Its Applications*. Vol.1, 3rd edition, John Wiley & Sons, New York (1968).

47. A. Ferreira and C. de Vries: Optimal confidence intervals for the tail index and high quantiles. Discussion paper, Tinbergen Institute, the Netherlands (2004).

48. R.A. Fisher and L.H.C. Tippett: Limiting forms of the frequency distribution of the largest or smallest member of a sample. *Proc. Cambridge Philos. Soc.* **24**, 180–190 (1928).

49. M.I. Fraga Alves, M.I. Gomes, and L. de Haan: A new class of semi-parametric estimators of the second order parameter. *Portugalia Mathematica* **60**, 193–213 (2003).

50. M.I. Fraga Alves, L. de Haan and Tao Lin: Estimation of the parameter controlling the speed of convergence in extreme value theory. *Math. Methods Statist.* **12**, 155–176 (2003).

51. M. Fréchet: Sur la loi de probabilité de l'écart maximum. *Ann. Soc. Math. Polon.* **6**, 93–116 (1927).

52. J. Geffroy: Contributions à la theorie des valeurs extrêmes. *Publ. Inst. Statist. Univ. Paris* 7 **8**, 37–185 (1958).

53. J.L. Geluk and L. de Haan: Regular variation, Extensions and Tauberian Theorems. *CWI Tract* **40**, Amsterdam (1987).

54. E. Giné, M. G. Hahn, and P. Vatan: Max-infinitely divisible and max-stable sample continuous processes. *Probab. Th. Rel. Fields* **87**, 139–165 (1990).

55. B.V. Gnedenko: Sur la distribution limite du terme maximum d'une série aléatoire. *Ann. Math.* **44**, 423–453 (1943).

56. L. de Haan: A spectral representation for max-stable processes. *Ann. Prob.* **12**, 1194–1204 (1984).

57. L. de Haan and A. Hordijk: The rate of growth of sample maxima. *Ann. Math. Stat.* **43**, 1185–1196 (1972).

58. L. de Haan and T.T. Pereira: Spatial extremes: the stationary case. *Ann. Statist.* To appear (2006).

59. L. de Haan and J. Pickands: Stationary min-stable stochastic processes. *Probab. Th. Rel. Fields* **72**, 477–492 (1986).

60. L. de Haan and S.I. Resnick: Estimating the limit distribution of multivariate extremes. *Commun. Statist. - Stochastic Models* **9**, 275–309 (1993).

61. L. de Haan and S.I. Resnick: Second order regular variation and rates of convergence in extreme value theory. *Ann. Prob.* **24**, 119–124 (1996).

62. L. de Haan, S.I. Resnick, H. Rootzén, and C. de Vries: Extremal behaviour of solutions to a stochastic difference equation with applications to ARCH-processes. *Stoch. Proc. Appl.* **32**, 213–224 (1989).

63. L. de Haan and A.K. Sinha: Estimating the probability of a rare event. *Ann. Statist.* **27**, 732–759 (1999).

64. L. de Haan and U. Stadtmüller: Generalized regular variation of second order. *J. Australian Math. Soc.* (Series A) **61**, 381–395 (1996).

65. W.J. Hall and J.A. Wellner: The rate of convergence in law of the maximum of an exponential sample. *Statist. Neerlandica* **33**, 151–154 (1979).

66. P.R. Halmos: *Measure Theory*. Springer (1950).

67. E. Hewitt and K. Stromberg: *Real and Abstract Analysis*. Springer (1969).

68. B.M. Hill: A simple general approach to inference about the tail of a distribution. *Ann. Statist.* **3**, 1163–1174 (1975).

69. J.R.M. Hosking and J.R. Wallis: Parameter and quantile estimation for the generalized Pareto distribution. *Technometrics* **29**, 339–349 (1987).

70. J. Hüsler and D. Li: On testing extreme value conditions. Accepted for publication in *Extremes* (2005).

71. J. Hüsler and R.-D. Reiss: Maxima of normal random vectors: between independence and complete dependence. *Stat. Prob. Letters* **7**, 283–286 (1989).

72. P. Jagers: Aspects of random measures and point processes. In: P. Ney and S. Port (eds.) *Advances in Probability and Related Topics*. Marcel Dekker, New York (1974).

73. D.W. Jansen and C.G. de Vries: On the frequency of large stock returns: Putting booms and busts into perspective. *Review of Economics and Statistics* **73**, 18–24 (1991).

74. A.F. Jenkinson: The frequency distribution of annual maximum (or minimum) values of meteorological elements. *Quart. J. Roy. Meteorol. Soc.* **81**, 158–171 (1955).

75. H. Joe: *Multivariate Models and Dependence Concepts*. Chapman & Hall, London (1997).

76. O. Kallenberg: *Random Measures*. 3rd edition, Akademic-Verlag, Berlin (1983).

77. M.J. Klass: The Robbins-Siegmund criterion for partial maxima. *Ann. Prob.* **13**, 1369–1370 (1985).

78. M.R. Leadbetter, G. Lindgren, and H. Rootzén: *Extremes and Related Properties of Random Sequences and Processes*. Springer, Berlin (1983).

79. A. Ledford and J.A. Tawn: Statistics for near independence in multivariate extreme values. *Biometrika* **83**, 169–187 (1996).

80. A. Ledford and J.A. Tawn: Modelling dependence within joint tail regions. *J. Royal Statist. Soc. Ser. B* **59**, 475–499 (1997).

81. A. Ledford and J.A. Tawn: Concomitant tail behaviour for extremes. *Adv. Appl. Prob.* **30**, 197–215 (1998).

82. M.J. Martins: Heavy tails estimation — variants to the Hill estimator. PhD thesis (in Portuguese), University of Lisbon, Portugal (2000).

83. D.M. Mason: Laws of large numbers for sums of extreme values. *Ann. Prob.* **10**, 754–764 (1982).

84. D. G. Mejzler: On the problem of the limit distribution for the maximal term of a variational series. *L'vov Politechn. Inst. Naucn. Zp. (Fiz.-Mat.)* (in Russian) **38**, 90–109 (1956).

85. R. von Mises: La distribution de la plus grande de n valeurs. *Rev. Math. Union Interbalcanique* **1**, 141–160 (1936) Reproduced in: *Selected Papers of Richard von Mises*, Amer. Math. Soc., Vol. 2, 271–294 (1964).

86. R.B. Nelsen: *An Introduction to Copulas*. Springer-Verlag, New York (1998).

87. J. Pickands III: Maxima of stationary Gaussian processes. *Z. Wahrsch. verw. Gebiete* **7**, 190–233 (1967).

88. J. Pickands III: Sample sequences of maxima. *Ann. Math. Stat.* **38**, 1570–1574 (1967).

89. J. Pickands III: Statistical inference using extreme order statistics. *Ann. Statist.* **3**, 119–131 (1975).

90. J. Pickands III: Multivariate Extreme Value Distributions. Proceedings: 43rd Session of the International Statistical Institute. Book 2, Buenos Aires, Argentina, 859–878 (1981).

91. H.S.A. Potter: The mean value of a Dirichlet series II. *Proc. London Math. Soc.* **47**, 1–19 (1942).

92. J.W. Pratt: On interchanging limits and integrals. *Ann. Math. Statist.* **31**, 74–77 (1960).

93. A. Rényi: On the theory of order statistics. *Acta Mathematica Scient. Hungar.* tomus IV, 191–227 (1953).

94. S.I. Resnick: *Extreme Values, Regular Variation and Point Processes.* Springer-Verlag, New York (1987).

95. S.I. Resnick and R. Roy: Random USc functions, max-stable processes and continuous choice. *Ann. Appl. Probab.* **1**, 267–292 (1991).

96. H. Rootzén: Attainable rates of convergence for maxima. *Statist. Prob. Letters* **2**, 219–221 (1984).

97. H. Rootzén: The tail empirical process for stationary sequences. Report 1995:9, Mathematical Statistics, Chalmers University of Technology (1995).

98. H. L. Royden: *Real Analysis.* 2nd edition, Macmillan, New York (1968).

99. G. Shorack and J. Wellner: *Empirical Processes with Applications to Statistics.* Wiley, New York (1986).

100. M. Sibuya: Bivariate extreme statistics. *Ann. Inst. Statist. Math. Tokyo,* **11**, 195–210 (1960).

101. B. Smid and A. J. Stam: Convergence in distribution of quotients of order statistics. *Stoch. Proc. Appl.* **3**, 287–292 (1975).

102. N.V. Smirnov: Limit distributions for the terms of a variational series. In Russian: *Trudy Mat. Inst. Steklov.* **25** (1949). Translation: *Transl. Amer. Math. Soc.* **11**, 82–143 (1952).

103. R.L. Smith: Uniform rates of convergence in extreme value theory. *Adv. in Appl. Probab.* **14**, 543–565 (1982).

104. R.L. Smith: Estimating tails of probability distributions. *Ann. Statist.* **15**, 1174–1207 (1987).

105. W. Vervaat: Functional limit theorems for processes with positive drift and their inverses. *Z. Wahrsch. verw. Gebiete* **23**, 245–253 (1971).

Index